SOCIETA' ITALIANA DI FISICA

RENDICONTI
DELLA
SCUOLA INTERNAZIONALE DI FISICA «ENRICO FERMI»

LXXX CORSO

a cura di A. R. OSBORNE
Direttore del Corso
e di P. MALANOTTE RIZZOLI
Segretaria del Corso

VARENNA SUL LAGO DI COMO
VILLA MONASTERO
7 - 19 LUGLIO 1980

Argomenti di fisica degli oceani

1982

SOCIETÀ ITALIANA DI FISICA
BOLOGNA - ITALY

ITALIAN PHYSICAL SOCIETY

PROCEEDINGS
OF THE
INTERNATIONAL SCHOOL OF PHYSICS « ENRICO FERMI »

Course LXXX

edited by A. R. OSBORNE
Director of the Course
and by P. MALANOTTE RIZZOLI
Scientific Secretary

VARENNA ON LAKE COMO
VILLA MONASTERO
7th-19th JULY 1980

Topics in Ocean Physics

1982

NORTH-HOLLAND PUBLISHING COMPANY, AMSTERDAM · NEW YORK · OXFORD

Publishers:

NORTH-HOLLAND PUBLISHING COMPANY
AMSTERDAM - NEW YORK - OXFORD

Sole distributors for the U. S. A. and Canada:

ELSEVIER SCIENCE PUBLISHING COMPANY INC.
52 VANDERBILT AVENUE, NEW YORK, N. Y. 10017

Technical Editor
P. PAPALI

Library of Congress Cataloging in Publication Data

International School of Physics « Enrico Fermi » (1980: Varenna, Italy)
Topics in ocean physics.

(Proceedings of the International School of Physics « Enrico Fermi »; course 80).
At head of title: Italian Physical Society.
Added t.p. title: Argomenti di fisica degli oceani.
1. Ocean waves—Congresses. 2. Ocean currents—Congresses. 3. Hydrodynamics—Congresses.
I. Osborne, A. R. II. Rizzoli, P. Malanotte. III. Società italiana di fisica. IV. Title: Argomenti di fisica degli oceani. V. Series.
GC200.158 1980 551.47 82-3432

ISBN 0-444-86160-2 AACR2

Printed in Italy

INDICE

Introduction.

The International School of Physics « Enrico Fermi » presented for the first time in the Summer of 1980 a course in the field of physical oceanography: « Topics in Ocean Physics ». The School, held under the auspices of the Italian Physical Society (S.I.F.), took place in the stimulating environment of Villa Monastero in the small town of Varenna on Lake Como in northern Italy. The two-week course was presented to a selected group of young scientists many of whom had formal training in physics from Italian universities. A major goal of the School was to introduce to the participating students a selected set of interesting and exciting, high-level topics in ocean physics and thereby attempt to woo them from other physical disciplines into the field of physical oceanography! A second objective was to provide some basis for the selection of a suitable topic for those students who had recently decided to begin studying and working in the field of ocean physics, which due to many recent economic and environmental reasons will hopefully become a vital area of research for Mediterranean countries.

Thus, as the organization of the school began to unfold, the need for exposing the students to several stimulating topics, rather than a narrow subfield, became clear. The School was to provide a forum for the introduction of exciting ideas rather than a format for tutorials on a single subject. A broad panorama of the field was, however, not possible or desirable and we, therefore, focused on several restricted topics whose specific content evolved according to the availability of lecturers:

1) the dynamics of mesoscale and large-scale flows,

2) nonlinear wave mechanics,

3) mixed-layer physics,

4) models of ocean surface waves.

The lecturers were requested to discuss recent and timely results in order to give the students a perspective of the field with emphasis on recent developments. The somewhat broad spectrum of selected topics allowed for considerable cross-fertilization between subfields not only among the students but also among the lecturers as well.

Part I contains papers on the dynamics of large-scale and mesoscale motions. Allen ROBINSON discusses the results of POLYMODE and other related studies which have formed the basis of our present knowledge and ideas about the dynamics of ocean currents and large-scale circulation. The central points of discussion include the general circulation of the mid-latitude oceans, the intense flow of the major ocean currents like the Gulf Stream and Kuroshio, and mesoscale eddies. Rick SALMON presents a theoretical introduction to the field of geostrophic turbulence which bridges both geophysical fluid dynamics and statistical turbulence theory. The quasi-geostrophic approximation and the methods of statistical mechanics are the basis for the physics of quasi-random motions of rotating stratified turbulence over topography. A. D. KIRWAN *et al.* detail results of observations of mesoscale and large-scale circulation of the oceans made by satellite-tracked drogued buoys. The study includes analysis of the large-scale circulation in the Eastern Pacific, the Kuroshio and mesoscale eddies of the east Australian coast. Paola MALANOTTE RIZZOLI develops a general approach to the nonlinear modeling of solitary mesoscale oceanic eddies. She illustrates how the method can be used to discover the variety of circumstances in which planetary solitary waves may be found and clarifies the relationship among various types of solutions. After discussing their stabilities, she also provides the scientific framework in which nonlinear coherent structures can be of bearing and importance.

Part II contains contributions regarding recent advances in the field of nonlinear wave mechanics. Henry YUEN surveys the field of instabilities of nonlinear waves in deep water. Envelope solitons, the Zakharov equation and higher-order stochastic models are among the recent topics discussed. Harvey SEGUR gives an introduction to the highly technical mathematical methods which have successfully resulted in exact solutions to certain classes of nonlinear evolution equations. He discusses the role of the scattering transform in the solutions to these equations and the concept of the soliton as a natural consequence of the theory. The scattering transform is discussed in its role as a generalized extension of the Fourier transform to nonlinear problems. Joe HAMMACK gives an exposition on weakly nonlinear wave systems which have length scales small compared to the Kelvin-Rossby radius of deformation and time scales small compared to the inertial period. He discusses in this context soliton solutions to the Korteweg-de Vries and nonlinear Schrödinger equations, and shows laboratory data demonstrating the existence of these two types of solitons. He also publishes, for the first time, results on his recent investigations of edge waves. Finally OSBORNE and BURCH discuss internal-soliton observations made west of Thailand, results which appeared previously in *Science*.

In Part III Peter NIILER presents a review of recent observations of the physical processes of mixed-layer formation in the upper ocean during storms. He provides evidence that strong mechanical agitation of the water column, which results in the mixed layer, occurs both by vertical overturning due to vertical shear of horizontal currents as well as the action of surface wave breaking.

Part IV is devoted entirely to linear wave models of ocean surface waves. Leon BORGMAN discusses probability laws for random ocean surface waves and gives methods for their computer simulation. With others he elaborates on methods for measuring the ocean surface wave directional spectrum from a tilt-and-roll buoy. Methods for the determination of long-time extremal statistics of random waves are also surveyed. Luigi CAVALERI presents an exposition on the characteristics of wind waves and their generation and dissipation mechanisms. Mathematical models for wave generation are also discussed. A. R. OSBORNE describes the computer simulation of linear random waves and compares these results to known theories and to data.

It is our hope that this volume projects some of the feeling for the stimulating intellectual and cultural atmosphere that existed for two brief weeks at Villa Monastero in the Summer of 1980.

We would like to thank Prof. C. CASTAGNOLI, President of the Italian Physical Society, whose support made this School possible. Dr. G. WOLZAK, Secretary of the School, was of invaluable service in its organization. We are very thankful to P. PAPALI for his expert handling of the editing and the publication of this volume.

A. R. OSBORNE and P. MALANOTTE RIZZOLI

[illegible]

breaking.

[illegible]

It is our hope that these Proceedings will [illegible]

[illegible]

A. R. OSBORNE and P. MALANOTTE-RIZZOLI

Part I

THE DYNAMICS OF MESOSCALE AND LARGE-SCALE FLOWS

Dynamics of Ocean Currents and Circulation: Results of POLYMODE and Related Investigations.

A. R. Robinson

Harvard University - Cambridge, Mass.

Within the world oceans there occur a great variety of physical phenomena and associated dynamical problems. These range over many decades of time and space scales. Some of the kinematically identifiable phenomena can be successfully and simply isolated for dynamical study while others are intrinsically parts of a coupled or multicomponent system. The coupling is often nonlinear, as is to be expected for a large-scale fluid flow. Much of what is known about the oceanic dynamics today has been learned quite recently as the result of vigorous scientific activity involving new ideas and instruments. Novel sampling techniques now are not only importantly quantifying aspects of interesting physics, but are still revealing qualitatively new phenomena and suggesting new interactions and coupling mechanisms.

The general approach to the formulation of an oceanic dynamical study is generally heuristic and involves idealizations. The phenomenon of interest may be isolated in a variety of ways, *e.g.* by the identification of a general physical process, by restricting attention to a particular geographical region or depth of the ocean or range of time and space scales. This approach leads almost always to a situation in which the physical problem for the phenomenon under explicit description requires closure by assumptions about (or parameterizations of) unresolved scales or interconnections to other regions. These remarks apply, of course, to the design of experiments and interpretation of measurements as well as to the construction of models and theories.

Here I will discuss the general circulation, *i.e.* the long-time average planetary-scale flow, of the mid-latitude oceans. This involves the movement of the water in the major subtropical gyres over the large region of the open ocean from the sea surface to the bottom and the intense flow of the major ocean currents such as the Gulf Stream and Kuroshio. Moreover, I will focus attention on recent advances in ocean current dynamics resulting from studies of the spatial and temporal variability of ocean currents which occurs on scales of tens to hundreds of kilometers and weeks to months. Motions of these scales are called mesoscale (or synoptic scale) eddies and are the oceanic analogue to atmospheric weather systems. Eddies contain the dominant kinetic energy over much of the ocean and eddy-mean field interactions are believed to be of central importance to the maintenance and character of the general circulation.

The classical picture of the dynamics of the general circulation was based on coarsely sampled and grossly averaged observations and did not attempt to take what little was known or suspected about the variability into account. Because of the geometry and rotation of the Earth and the nature of the atmospheric and solar forcing of the oceans, the polar mid-latitude and equatorial oceans were known to have special dynamics. Major achievements were the rationalization of the existence of the main subtropical gyres, western boundary currents (Gulf Stream) and the main thermocline [1]. The deep flow in the open ocean had not been directly measured and was believed to be steady and less than about a centimeter per second. Prototype deep direct measurements by neutrally buoyant subsurface floats in 1959 ([2], sect. **1**) suggested the existence of eddies, but more than a decade passed before instruments, technique and scientists were ready to undertake the investigations of the last decade which are essentially the basis of our present knowledge and ideas.

Concomitant with the classical picture the original numerical models of the general circulation were of coarse resolution [3] and invoked large, usually constant eddy viscosities which are now believed to be a physically unacceptable parameterization of some important aspects of mean mesoscale effects on the general circulation. Present models are of fine enough resolution to explicitly represent eddy scale motions. Such models relate effectively to the contemporary observational data base. Since physically important statistics of the eddies and the circulation appear to require data records of many years' duration, a symbiosis between models and observations (including some model simulations of long time duration) is of particular importance now in order to advance our understanding of the general circulation.

Although considerable progress has been made in understanding currents and circulation in the past several years, much more remains to be done and this scientific subject is still in an early stage of development. But some information and ideas do now exist which should endure and problems are presently being attacked by techniques and methodology of increasing sophistication. Simply stated, the present research objectives are i) the description of the kinematics and dynamics of the physical fields and ii) the construction of a valid theoretical/numerical circulation model. The elements of the *description* include identification of synoptic features, scales and statistics, local dynamical balances, general physical processes, energy (etc.) sources, transports and sinks, and interaction mechanisms. The *models* summarize the present state of knowledge and facilitate theoretical/experimental interaction. Moreover, they are the basis for the application of our physical knowledge of the ocean to other scientific areas and to practical problems which involve physical transport mechanisms, dispersion and boundary exchanges. Examples are nutrient and planktonic transports in biology, the transports of chemicals of both natural and anthropomorphic origin, the movement of purposely or otherwise deposited waste materials (including radioactive material). Of especial importance are oceanic heat fluxes and exchanges. On the large scale these fluxes contribute to climatic balances and changes of the coupled atmosphere-ocean system; on smaller scales they affect microclimate and influence the possibility of exploitation of the ocean for energy production.

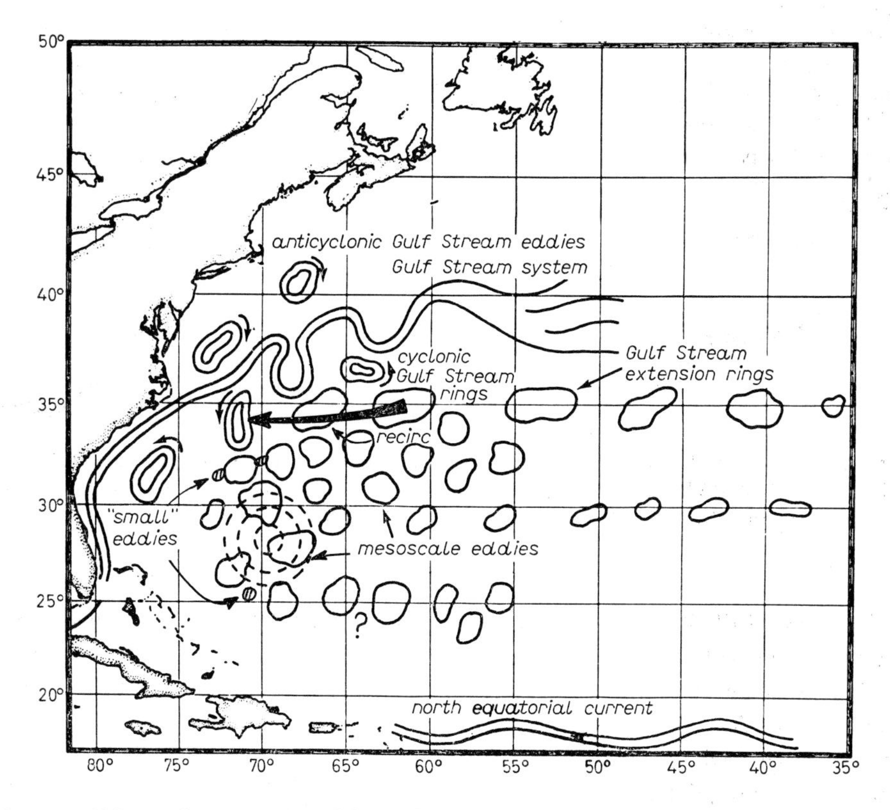

Fig. 1. – Schematic map of mid-latitude circulation and synoptic types of variable ocean currents (after [4], fig. 1).

A schematic representation of circulation features and synoptic types of eddy currents is presented in fig. 1 on a map of the Northwestern Atlantic Ocean, which has been the scene of the most intensive recent scientific explorations. The location of major field programs in the 1970's dedicated to eddy research throughout the North Atlantic is shown in fig. 2. These include the USSR mooring array POLYGON (1970), the US-UK MODE-I (Mid Ocean Dynamics Experiment, 1971-1974), the Canadian (Bedford Institute of Oceanog-

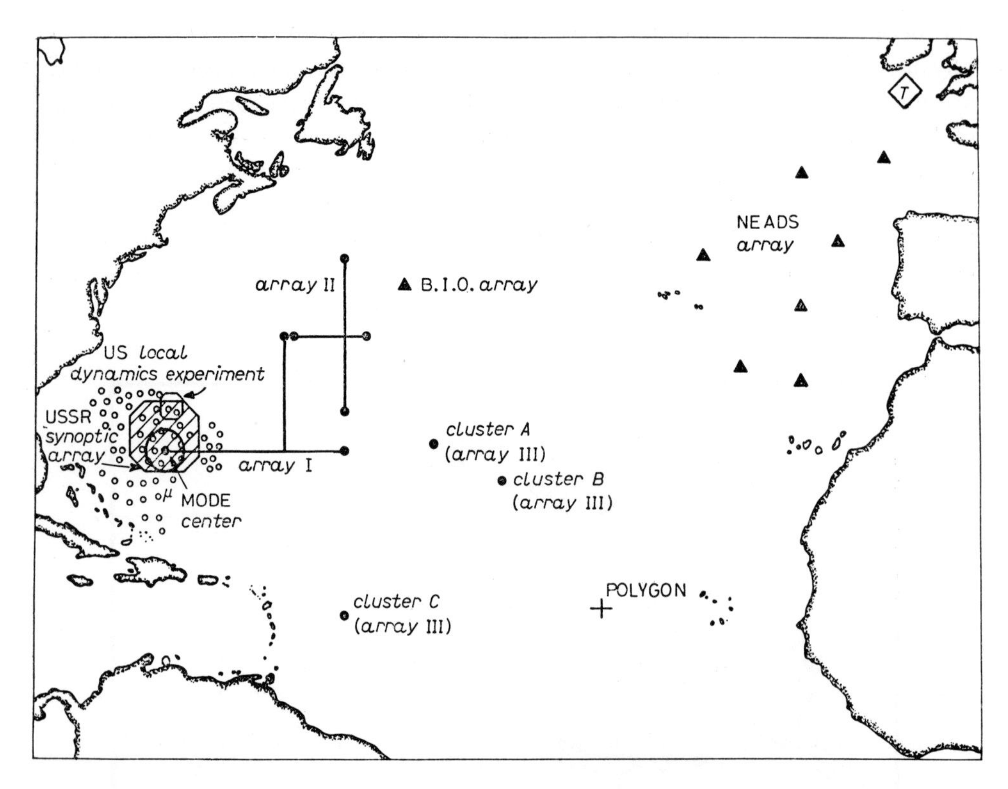

Fig. 2. – Location of major recent experiments on eddy currents in the North Atlantic: ○○○ region of SOFAR float work, μ Meddy, T Tourbillon experiment (courtesy of R. Heinmiller, US POLYMODE Executive Office, MIT).

raphy) near Gulf Stream mooring array, the US SOFAR float observational region during and after MODE-I, the NEADS (North East Atlantic Dynamics Study) mooring sites maintained from 1976-1979 co-operatively by English, French and West German scientists, and the synoptic eddy study « Tourbillon » (France, UK, 1979). POLYGON is summarized by [5], MODE-I is substantially reviewed by the MODE Group [2] and preliminary reports on the BIO array, the NEADS experiment and Tourbillon are given, respectively, by [6-8].

The other experiments on the map of fig. 2 are components of a very large effort, the joint US-USSR POLYMODE Program briefly described by [9]. Details of the POLYMODE experimental program are given in table I, and the instrumentation employed is listed in table II. Instrumental intercomparisons have been carried out throughout the program and, now that the field experimentation has ended, continuing efforts are under way to ensure the scientific unification of the composite data set. In addition to the experimental effort, a large and varied number of theoretical and modeling studies have and

TABLE I. – POLYMODE *experimental program.*

Statistical/geographical	Local dynamics	Synoptic
Objectives	*Objectives*	*Objectives*
Determination of energy levels and space and time scales of the eddy field throughout the western North Atlantic. Possible determination of higher-order statistics (momentum and heat fluxes, Reynolds stresses) and of mean flows.	Determination of the three-dimensional structure and time evolution of the variability field in a region larger than an eddy diameter: anisotropies, propagation characteristics, phase relationships, vertical coupling, energy partition, etc. Investigation of smallest-scale quasi-geostrophic fields. Direct determination of the local dynamical balance in the eddy field.	Mapping of the eddy field over a region encompassing one or more eddies for several eddy cycles. To explore and determine the dominant and/or controlling dynamical balances governing eddy motions and interactions.
Components	*Components*	*Components*
Moored arrays of current meters and temperature recorders. Long XBT lines. Extended SOFAR float array.	Density surveys. Two-level SOFAR float array. Near-surface cyclesonde moorings. Moored arrays of current meters and temperature recorders. Electromagnetic profiler surveys. Shipboard meteorological measurements.	Moored array of current meters and temperature recorders. Density surveys. XBT mapping. SOFAR float array.
Time scale	*Time scale*	*Time scale*
Several years ~1975-1978.	Several months, starting from May 1978.	One year, July 1977-September 1978.

are being carried out under the POLYMODE program. These range from relevant fundamental studies in geostrophic turbulence to eddy-resolved general circulation models of the North Atlantic and involve analytical and numerical research efforts. The POLYMODE program will be formally concluded by a final scientific symposium scheduled to be held in the USA in the summer of 1981.

The next set of fig. 3-5 illustrate features of the type of mid-ocean eddies found in MODE-I and POLYMODE. The Eulerian time series of fig. 3*a*) illustrate a richness of scales, intermittency and the occurrence of « events », such as the well-defined main MODE-I eddy itself. The Lagrangian trajectories (fig. 3*b*)) are of the SOFAR floats neutrally bouyant at 1500 m depth observed

TABLE II. - POLYMODE *instrumentation.*

Class	Type	Comments
Current meters	VACM	U.S.—Subsurface moorings
	Alekseev	U.S.S.R.—Surface moorings
	TSIIT	
	TSIT	
Temperature recorders	Wunsch/Dahlen TPR temperature recorder	U.S.—Subsurface moorings U.S.S.R.—Surface moorings
XBT	T-5 and T-7	U.S.—Used on U.S. and U.S.S.R. research vessels
Shipboard profilers	W.H.O.I./Brown CTD	U.S.
	Plessy STD	
	AIST CTD	U.S.S.R.
	ISTOK CTD	
Free profilers	Sanford electromagnetic absolute velocity profiler	U.S.
Moored profiler	cyclesonde	U.S.
SOFAR floats	floats	U.S.
	autonomous listening stations	U.S.—Developed for POLYMODE
Shipboard meteorology	standard measurements	U.S.S.R.

during the entire period from September 1972 to June 1976. This is the well-known so-called « spaghetti diagram » which upon study reveals a number of mechanisms of particle behavior during dispersion. The vertical shear profiles of eddy current (fig. 4) are indicative of energy primarily in the 1st baroclinic mode (2nd vertical eigenmode of a linearized quasi-geostrophic model based on the local observed density profile), which, together with the barotropic mode (1st eigenmode), usually accounts for over 90 % of the energy. The maps of fig. 4 show the comparison of the mid-thermocline eddy field with that at 1500 m, which, as the zero crossing of the first baroclinic mode, is regarded as the « barotropic mode window ». The barotropic scales are faster and shorter than the baroclinic one. The small circles on the POLYMODE map represent XBT stations. The MODE observational circle is displayed for comparison. The POLYMODE data region is large enough to contain more than one eddy and repeated synoptic observations were carried out for over a year.

Features of the synoptic description of the more energetic eddy currents

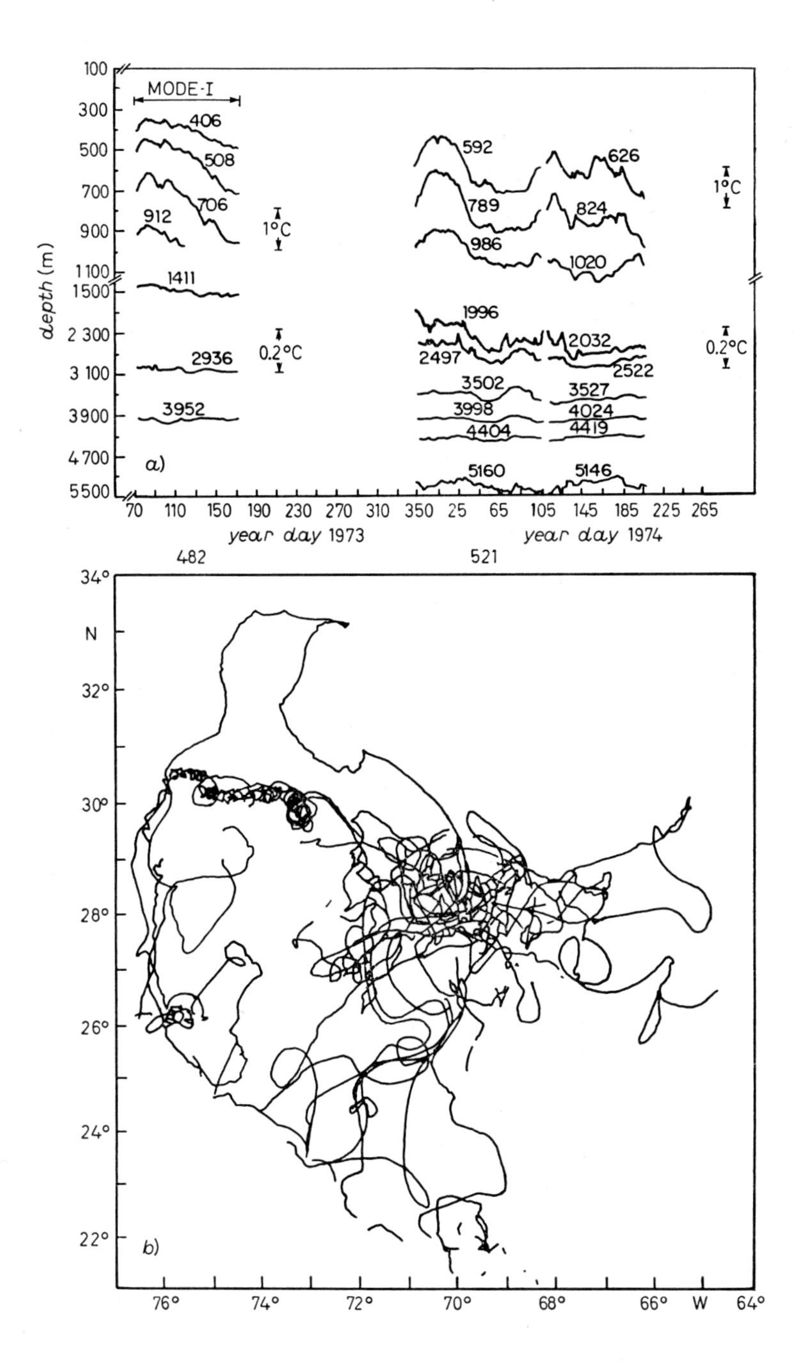

Fig. 3. – Measurements recording mid-ocean eddy currents: *a*) time series from temperature recorders moored at various depths, *b*) composite trajectories of SOFAR floats. (From [2], *a*) fig. 6*b*), *b*) fig. 7.)

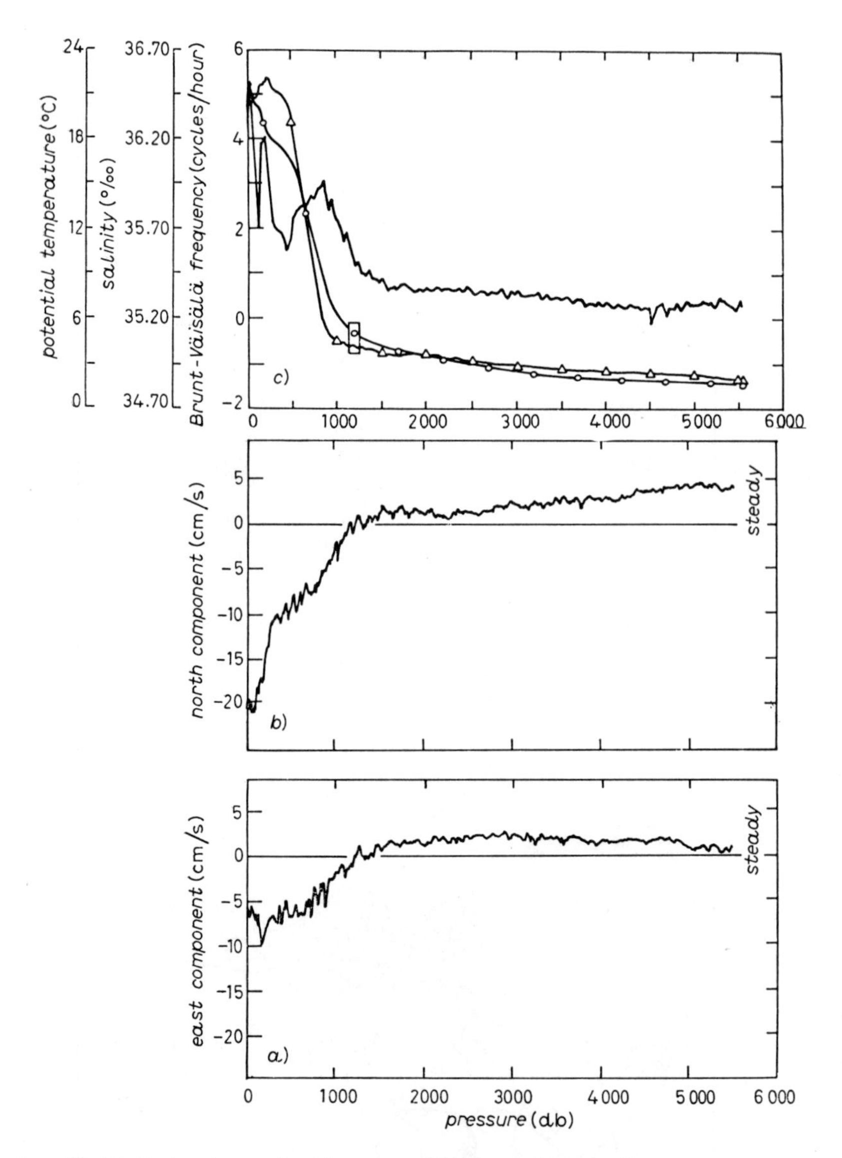

Fig. 4. – Vertical structure of mid-ocean eddies: *a*), *b*) velocity components; *c*) temperature, salinity and buoyancy frequency. (Extracts from [2], fig. 12 and 13.)

of the Gulf Stream and ring system are now beginning to emerge as a new data set is assembled. Figure 6 shows a remarkable « weather map » for the region constructed from data composite over 4 months, which would correspond roughly to a week of « atmospheric » time. The data, which include observations from 4 oceanographic cruises combined with other hydro and XBT measurements available from the NODC, show nine cold-core cyclonic rings

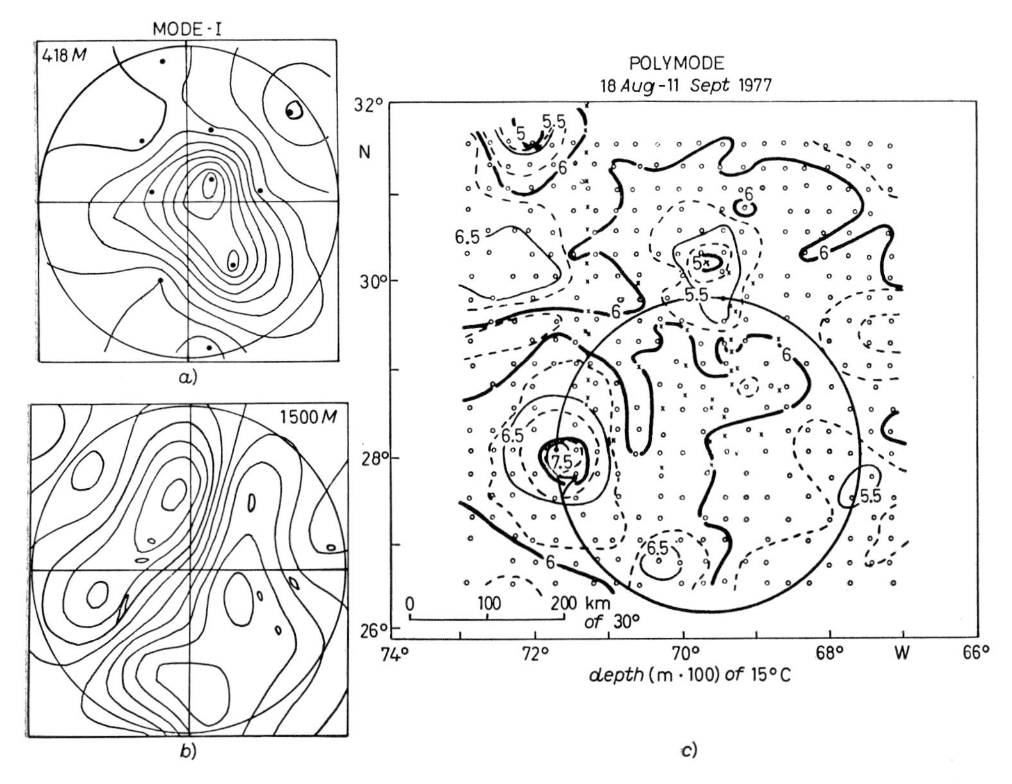

Fig. 5. – Sample synoptic maps from MODE-I and POLYMODE (from [10], fig. 1).

to the south and 3 warm-core anticyclonic rings north of the meandering current.

Generally, the new data now suggest new ideas about ring kinematics and dynamics. Only a few years ago cold-core rings were thought to decay slowly in the Sargasso Sea while gradually drifting to the southwest. Now the picture involves vigorous multiple ring-stream (and ring-ring) interactions. After formation, rings move generally westward (upstream) and make contact with the current again. The resultant interaction can involve advection, partial or total coalescence, permanent absorption (ring extinction) or subsequent downstream rebirth of the ring. The ring-stream interaction appears to involve extended external amounts of upper-layer water. This fact together with the multiple ring-stream interactions is modifying both the conceptual role and quantitative estimates of rings in the large-scale general-circulation heat transport. An interesting example of ring-stream and ring-ring interaction is shown in fig. 7. On the 18 April 1977 a cold-core ring has just reattached to the Gulf Stream. After advection and temporary absorption it is isolated again after about a month. During the cold-core ring reattachment process a warm-core ring was spawned and shed to the north of the Stream.

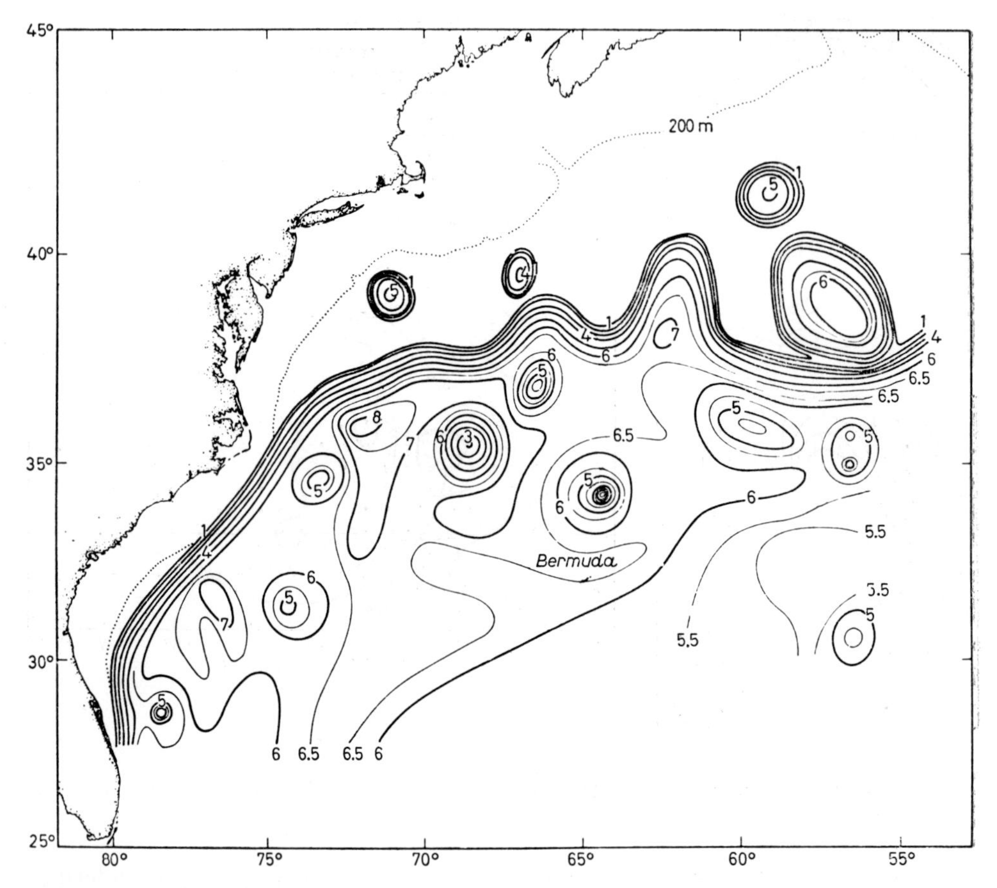

Fig. 6. – Almost synoptic maps of Gulf Stream current and ring system from composite measurements over four months (from [11], fig. 1a)).

This type of data, together with the statistics of some measurements available under the current and in the « near field » region of the Gulf Stream and the results of models (which will be mentioned below), implicates the Gulf Stream system as a region of strong eddy-near field interaction as a source region for eddy energy which is exported to the open mid-oceanic region. Strong currents and relatively sparse data make this region still difficult for research. This contributed to the choice of mid-ocean and near-field environmental locations for the POLYMODE program components. However, exploratory measurements and the development of new instruments and techniques (long-range SOFAR floats, new current meters and mooring technology, etc.) should make it feasible to carry out major experimentation here by the mid 1980's. Of especial importance are satellite measurements including infra-red [12] sea surface height measurements by radar altimetry. In fact, near-surface currents obtained from satellite-tracked drogues already form an important part

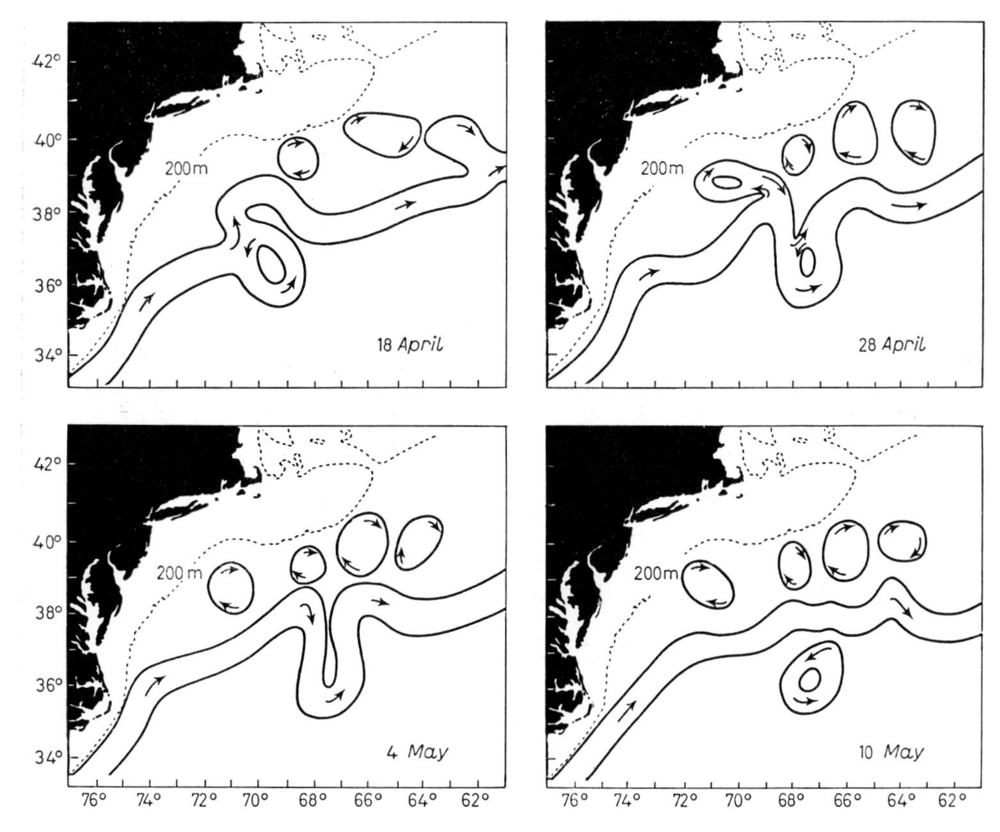

Fig. 7. – Example of strong Gulf Stream current and ring interaction involving reattachment, advection and rebirth (from [13], fig. 2*a*)).

of the new data base for the ring-stream system. A sample trajectory is shown in fig. 8. The drogue of that figure remained in the cold-core ring of fig. 7 for 8 months and completed 86 loops.

The synoptic experiment of POLYMODE provides a unique descriptive basis for the analysis of eddy-eddy interaction mechanisms and regional mid-ocean eddy dynamics. Within the large XBT [14] and hydrographic survey region of fig. 5, Soviet scientists maintained for over a year's duration a mapping array of 19 surface moorings with a separation of 72 km with current meters at 100, 400, 700, 1400 m depths. More than fifteen eddies and several elongated crest and trough features were identified. Nine maps spanning five months of objectively analyzed streamfunction for the low-pass-filtered (2 days) velocity at 700 m in the main thermocline are shown in fig. 9. Numbers below the maps are day and month of 1978; the first map is centered on January 14, etc. Each hatch mark is 25 km and the maps are 300 km on a side. Preliminary analysis reveals [15] i) periods when eddies were close packed

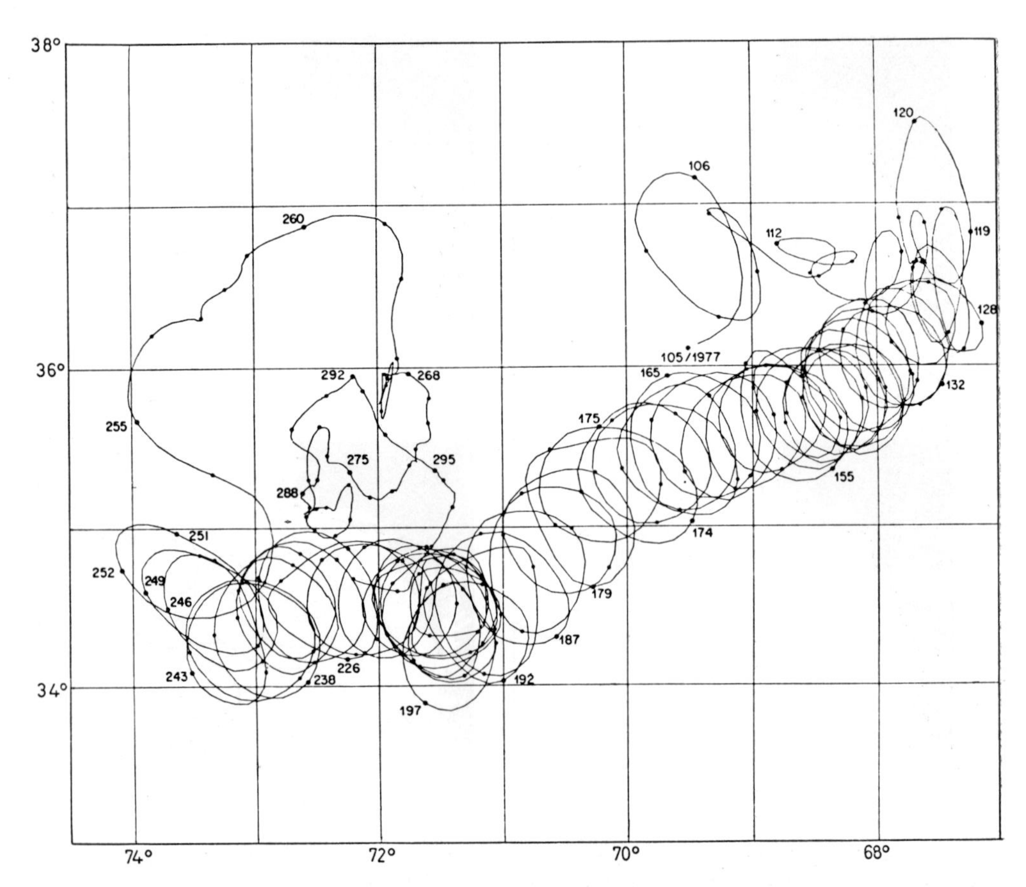

Fig. 8. – Trajectory of satellite-tracked buoy trapped in Gulf Stream ring (from [13], fig. 3*a*)).

in a near checkerboard pattern, ii) relatively quiescent periods with only a solitary eddy in the region, iii) direct energy exchange events between eddies of the same sign, iv) merger (splitting of 2 (1) eddies of the same sign to form 1 (2) eddy). Analysis, interpretation and modeling of this rich data set will, of course, require considerable time. The surface mooring systems are known to yield quantitatively accurate directions and, therefore, phase diagrams. Although the recorded speeds [2, 16] of the meso (synoptic) scale eddy currents are too fast due to a rectification of high-frequency noise, a preliminary intercomparison of the USSR current meter speeds with US SOFAR float measurements indicate that a uniform calibration factor of about 1.75:1 can be successfully employed almost uniformly [17].

The POLYMODE Local Dynamics Experiment (LDE) involved very intensive measurements in a smaller region overlapping the northern part of the synoptic region. It employed innovative arrays of moored, free floating and

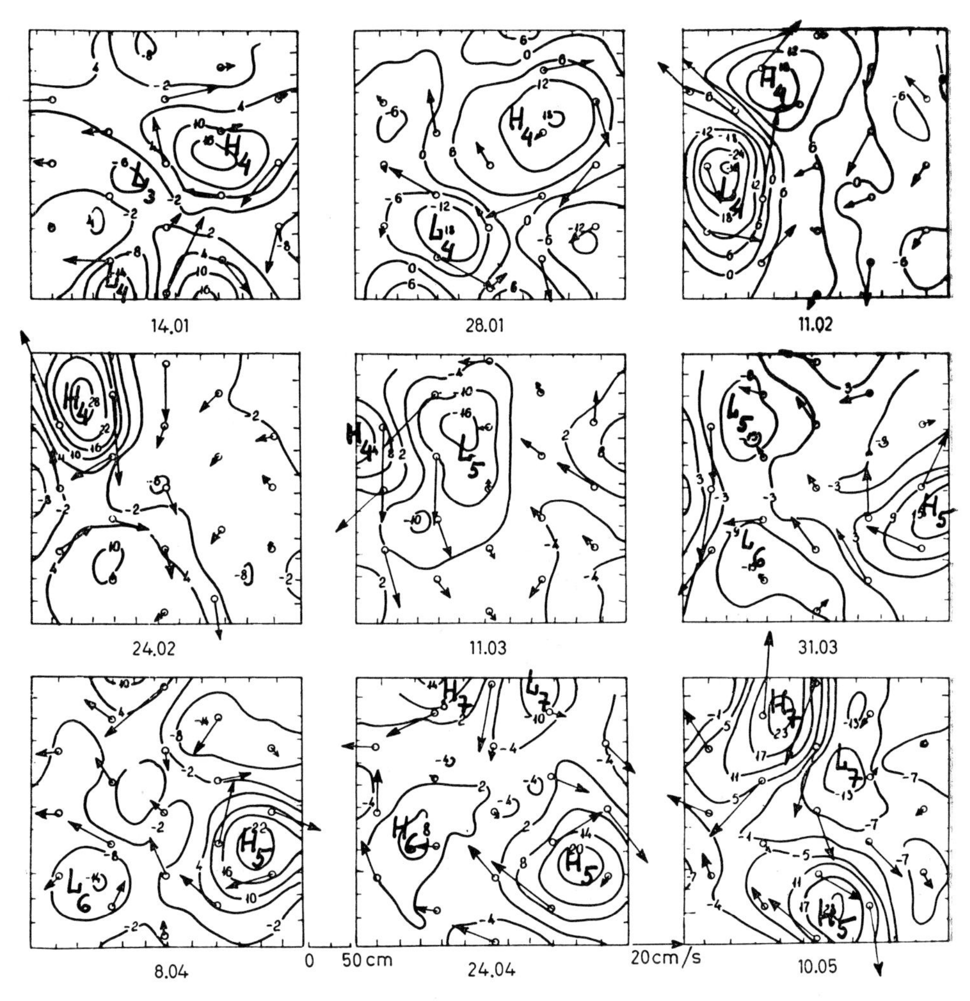

Fig. 9. – Maps from the POLYMODE synoptic experiment of evolving eddy currents at 700 m depth over a five-month period (from [15], fig. 4).

shipborne measurements carefully designed by objective techniques including model simulations [18] in order to measure as accurately higher derivatives of fields for the purpose of determining local balances such as contributions to potential-vorticity conservation, etc. Analysis is under way; results should be of importance to the efficient design of future experiments elsewhere as well as of local interest. During the LDE SOFAR floats were used at 700 m in the main thermocline as well as at 1500 m in the sound channel as was only the case in MODE-I, which was made possible by instrumental developments in the interim. Figure 10 shows trajectories of two floats launched near each other (1 degree is approximately 100 km). The smoother trajectory is characteristic of the behavior of the floats in the large cluster of which these

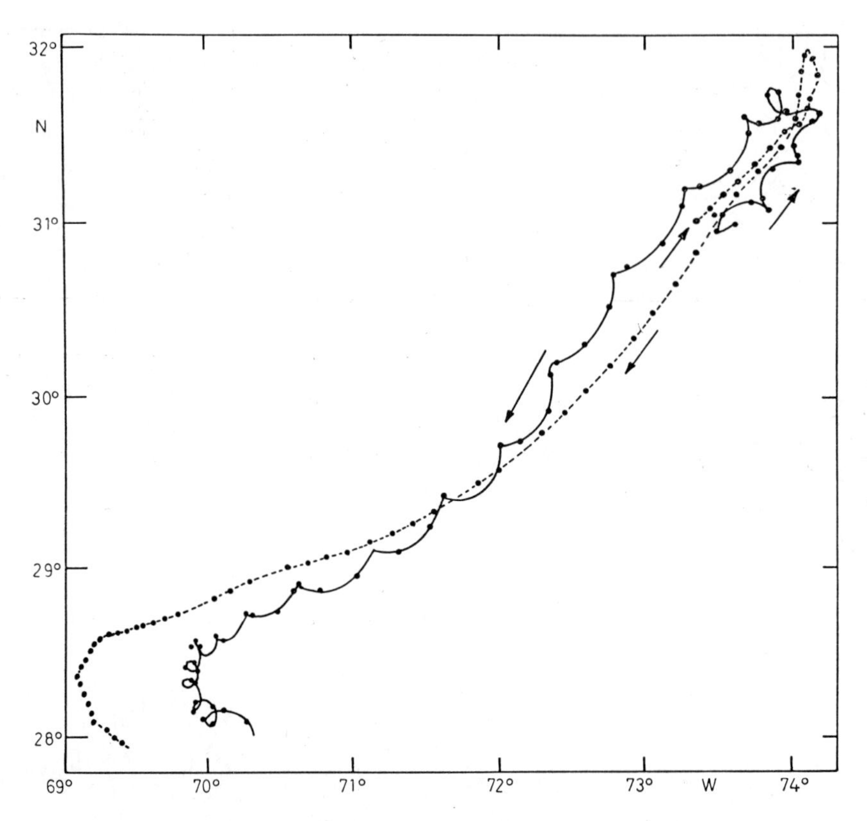

Fig. 10. – A pair of neighbor SOFAR float tracks, one of which is trapped in a small eddy discovered in the POLYMODE local dynamics experiment: ——— float 52, – – – float 53 (from [9], fig. 2).

two were a part. The clusters at both levels first moved coherently with a direction reversal from northeastward to southwestward in a manner believed to be consistent with Rossby wave dispersion [19]. It is hoped that the observations of the later breaking-up of the clusters during dispersion will provide valuable information on the nature of the smaller scales of quasi-geostrophic turbulence in mid-ocean.

The difference of behavior of the two floats shown in fig. 10 is striking and illustrates a major finding of the LDE scientists, *viz.* the discovery of *small* mesoscale features. The irregular trajectory is now known to have been caused by the trapping of the float in a small eddy or rotating lens. Several such features were found during the LDE [9, 20]. These features which are now known to exist in the upper or mid thermocline or deep water have radii on the order of several tens of kilometers and extend vertically for a few hundred to almost a thousand meters. Their most remarkable aspect is their very long lifetimes which are indicated by the distinctly anomalous properties of the water in these features compared to that of the surrounding sea. On the basis of a no-mixing hypothesis, lifetimes of more than one to several *years* are indicated; such features correspondingly must have traveled and maintained their integ-

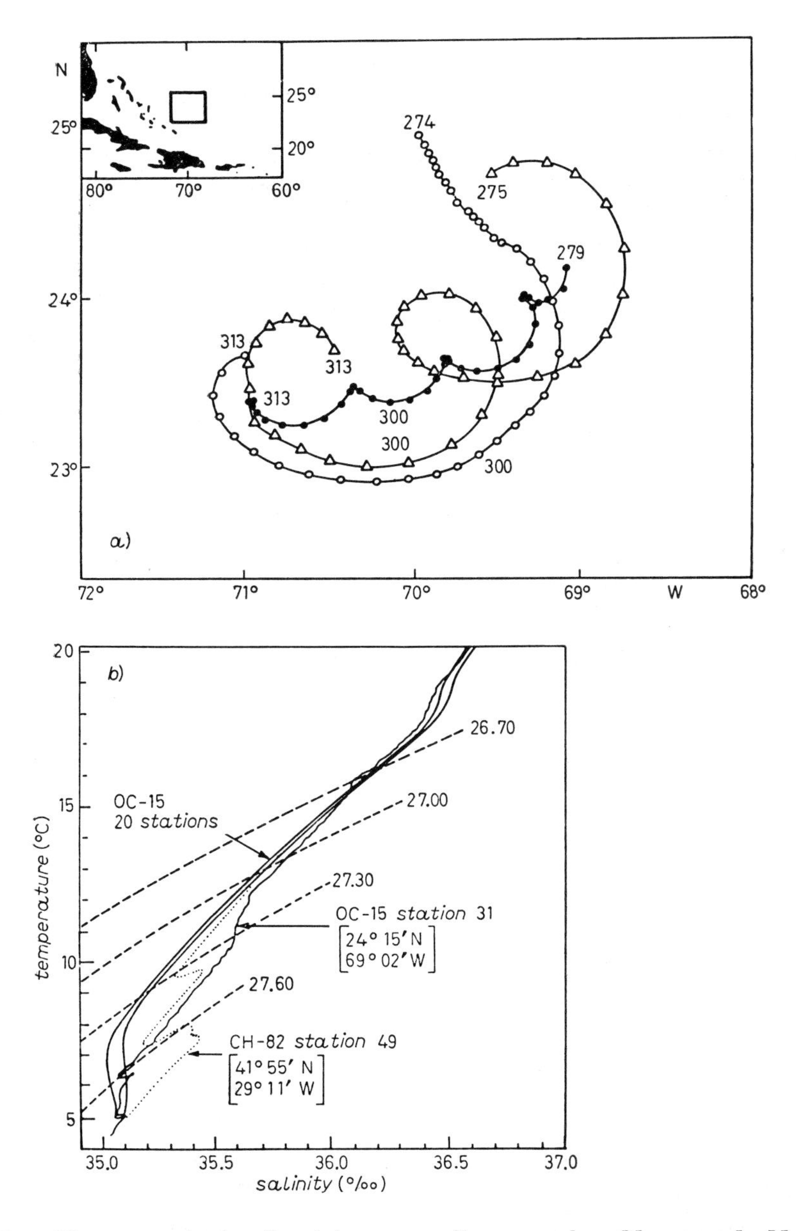

Fig. 11. – Measurements in the intense, small mesoscale eddy named « Meddy »: *a*) SOFAR float tracks: ○ float 11, depth 700 m; ● float 12, depth 850 m; △ float 14, depth 700 m; *b*) anomalous *T-S* compared with surrounding water and similar Mediterranean water elsewhere. (From [21].)

rity over distances of thousands of kilometers. The first of such small meso-scale features was in fact observed prior to the LDE experiment by SOFAR float and hydrographic measurements as shown in fig. 11, in a location to the south of MODE-I near the Bahamas and indicated as the « Meddy » in fig. 2. The nickname « Meddy » was adopted because the anomalous lens contains water which was probably of Mediterranean Sea origin 6000 km distant. The 20 station T-S envelope is for external water and station 31 is through the lens. It is not yet known how frequently or where these features occur, but their existence indicates a dramatic scale dependence of mixing and stirring processes in the general circulation the implication of which must be understood.

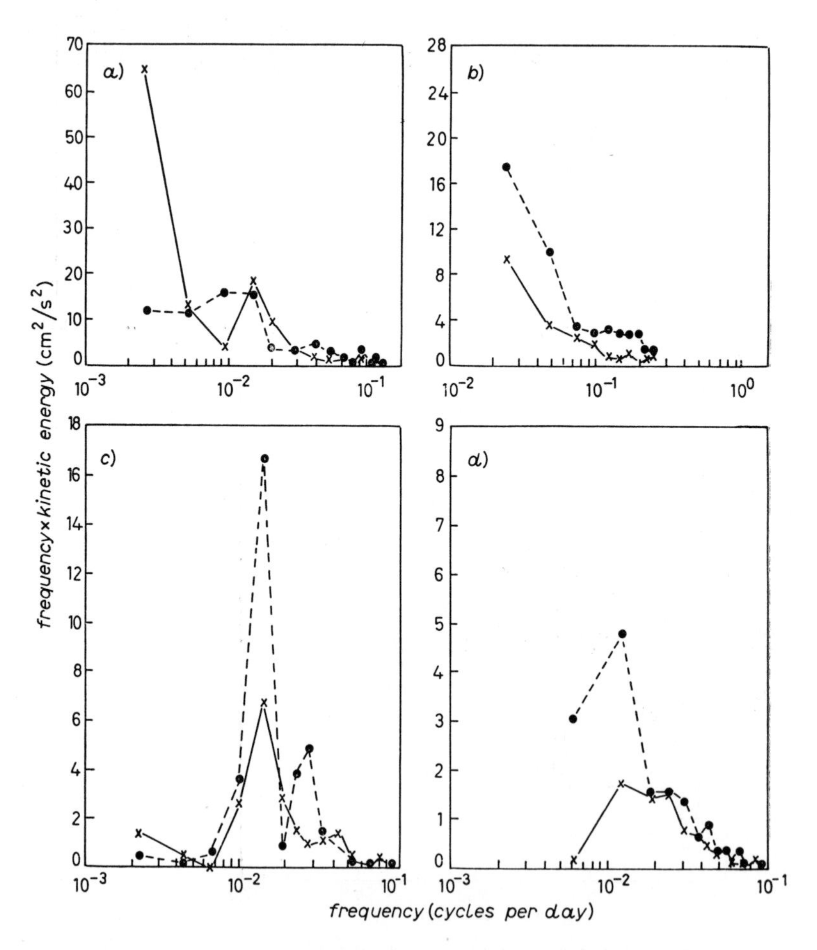

Fig. 12. – Frequency spectra of kinetic energy in mid-ocean eddies in the thermocline and the deep water: *a*) mode center, 500 m, ——— U, – – – V; *b*) mode east, 500 m, ——— U, – – – V; *c*) mode center, 4000 m, ——— U, – – – V; *d*) mode east, 4000 m, ——— U, – – – V (from [2], fig. 10*a*), *c*)).

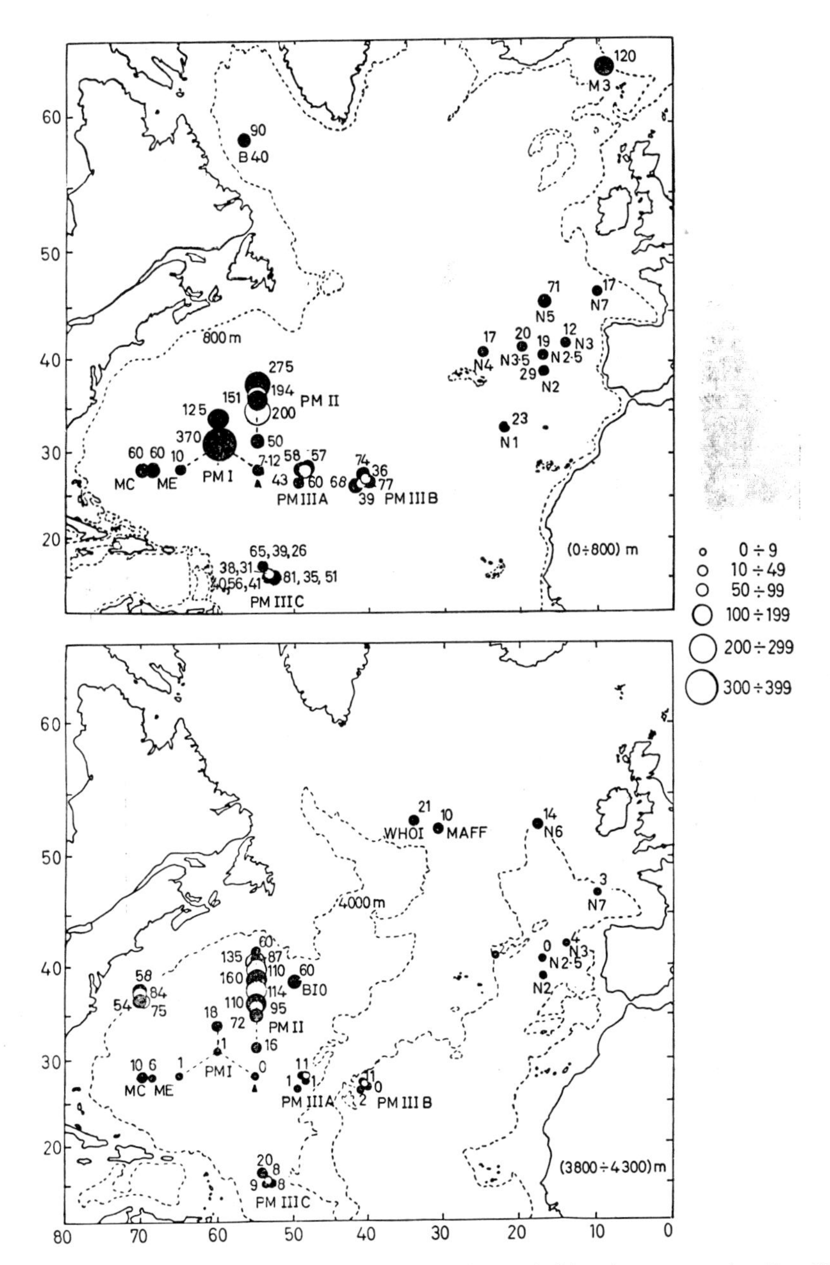

Fig. 13. – Maps showing geophysical distribution of kinetic energy in the North Atlantic thermocline and deep water (courtesy of R. R. DICKSON, MAFF, Lowestoft, UK).

The kinetic-energy spectra and maps presented in fig. 12 and 13 illustrate findings from the POLYMODE statistical geographic experiment, NEADS, and related measurements. The spectra are from two long-term moorings spaced 100 km apart left in place at the end of the MODE-I experiment; all records indicate

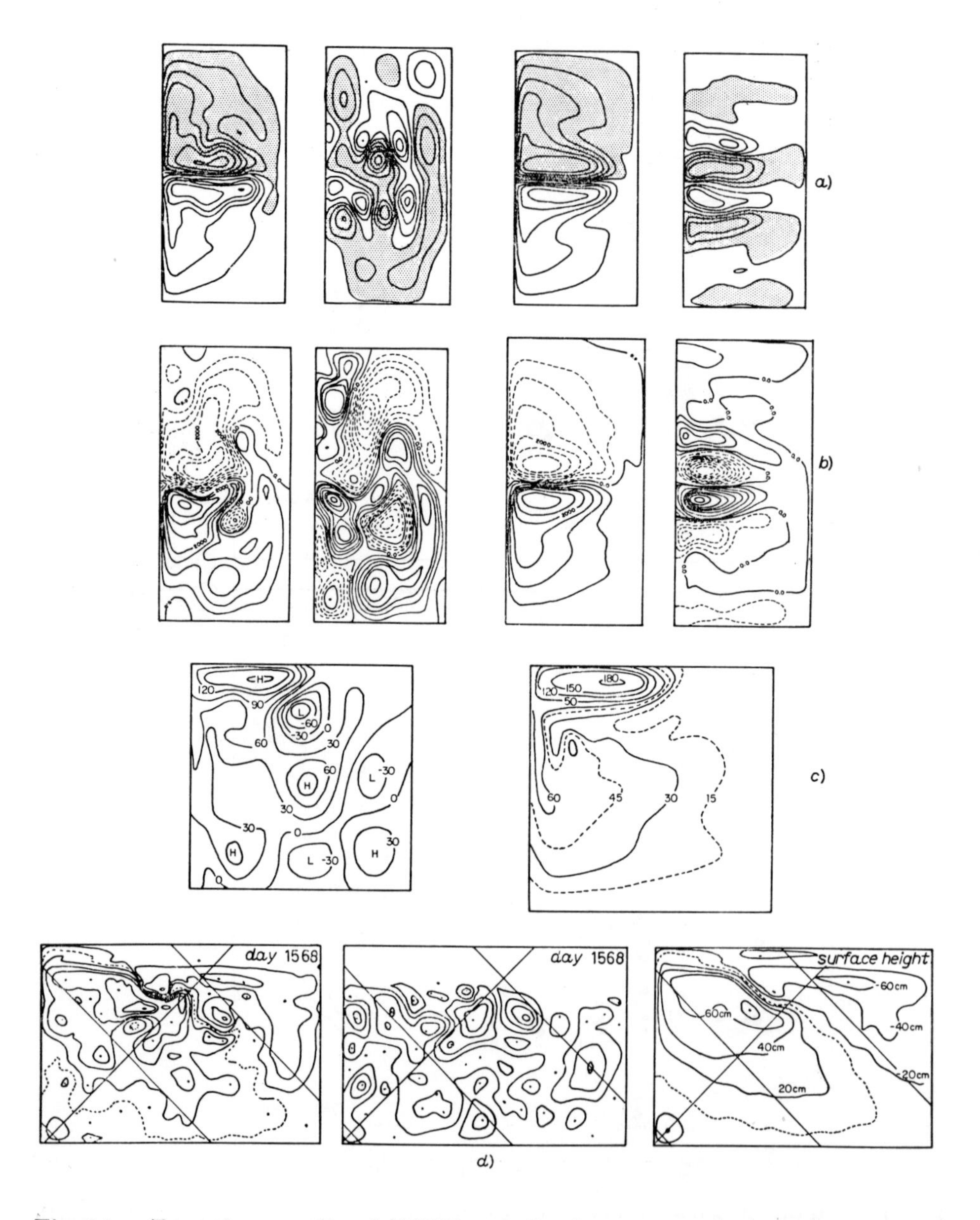

Fig. 14. – Exemplary results of EGCM: *a*), *b*) show a sample instantaneous and the mean fields for each of the two layers of a primitive equation and quasi-geostrophic model, respectively; *c*) an instantaneous and the mean transport streamfunction fields of a multilayer primitive-equation model; *d*) surface and deep instantaneous dynamic topographies and the mean surface height from a 2-gyre model with a continental shelf. (From [2], fig. 18.)

an eddy energy-containing band from about 50 to 150 days, and the upper thermocline records are « red ». By the mid 1970's it had become apparent that, while eddies were present almost everywhere they had been looked for in the North Atlantic, their characteristics varied considerably from place to place. To determine the distribution of these properties and the influences of boundaries, topography, strong currents, possible surface forcing, etc. were the main reasons for undertaking the statistical geographical experiment. A synthetical mapping of eddy kinetic-energy distribution from direct measurements is under way and preliminary results are shown in fig. 13. Although different filters were used in the data on these maps, almost all exclude two-day and higher frequencies. Higher energies exist generally in the western basin relative to the eastern and the energies increase significantly as the Gulf Stream is approached. Records are seen to exist under the Stream in fig. 13*b*), *e.g.* at 55° W longitude (see also [2], subsect. **3**.4) but not in the current's thermocline because of mooring difficulties. When these maps are overlain on maps of bottom topography and compared depth by depth, they are consistent with the tendency for relatively flat bottoms and strong energy levels to inhibit baroclinicity (the vertical variation with depth of the horizontal velocity) as indicated by modeling results ([2], subsect. **4**.8).

A comprehensive overview of the theoretical status of eddy dynamics will not be attempted, but some results will be demonstrated from the new generation of ocean current and circulation models which have high horizontal resolution (on the order of 10 to 40 km) and explicitly resolve and spontaneously generate eddies. The eddy-resolving general circulation models (EGCM's) relate directly to the totality of statistically summarized field data and their geophysical distribution. EGCM's have been recently overviewed by the MODE Group ([2], subsect. **4**.9) and by ROBINSON *et al.* [4]; fig. 14-16 are from those reviews. EGCM experiments have been carried out for the most part in idealized geometry consisting of one gyre (model subtropical) or two gyres (model subtropical and subpolar) which are created by the smooth steady latitudinal wind stress patterns (*e.g.* half or full sinusoidal) used to force the motion. The model equations are « quasi-geostrophic » or « primitive », the models are two layer or several layers in the vertical. Some multilayer primitive-equation models have been forced by a surface heat flux smoothly and simply varying with latitude in addition to winds. The flow is usually initiated from a state of rest with a prescribed stratification as a function of depth which is horizontally uniform. Such flows have been studied analytically and numerically with coarse resolution prior to the advent of EGCM's and the resulting circulations are smooth and steady [3]. In EGCM's, however, during the spin-up the large-scale flow goes unsteady, eddies are spontaneously generated with oceanically interesting model time and space scales. The models are then run long enough to approach (hopefully) a statistical steady state. After the steady state has been reached, the experiment must be run for several model years in order to obtain a simulated data record for scientific analysis. In addition to the gross quantities mentioned above, a single experimental run requires the specification of subgrid scale parameters associated with dissipation, etc., and of various computational parameters. Results were early found to be parameter dependent and relatively sensitive to subgrid scale

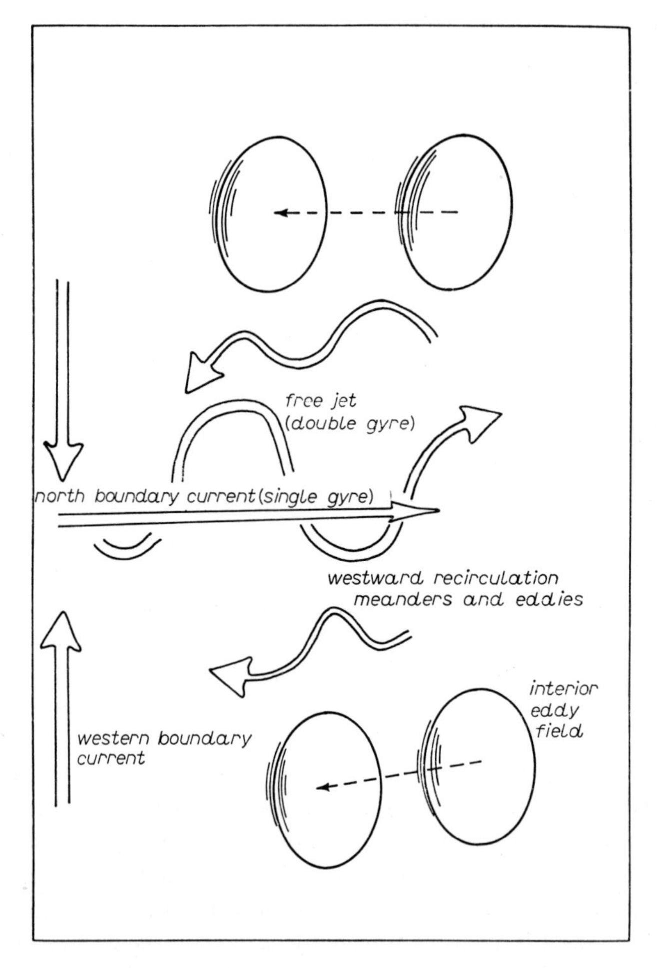

Fig. 15. – Schematic summary representation of EGCM results (from [4], fig. 10).

specifications. Some sample « snapshots » and long-time averaged fields are shown in fig. 14 and the structure of the instantaneous flow is schematized in fig. 15.

The large amount of data required to be analyzed together with the diversity of results from different experiments presents a challenging task for interpretation and synthesis. Techniques for the inference of the fundamental physics

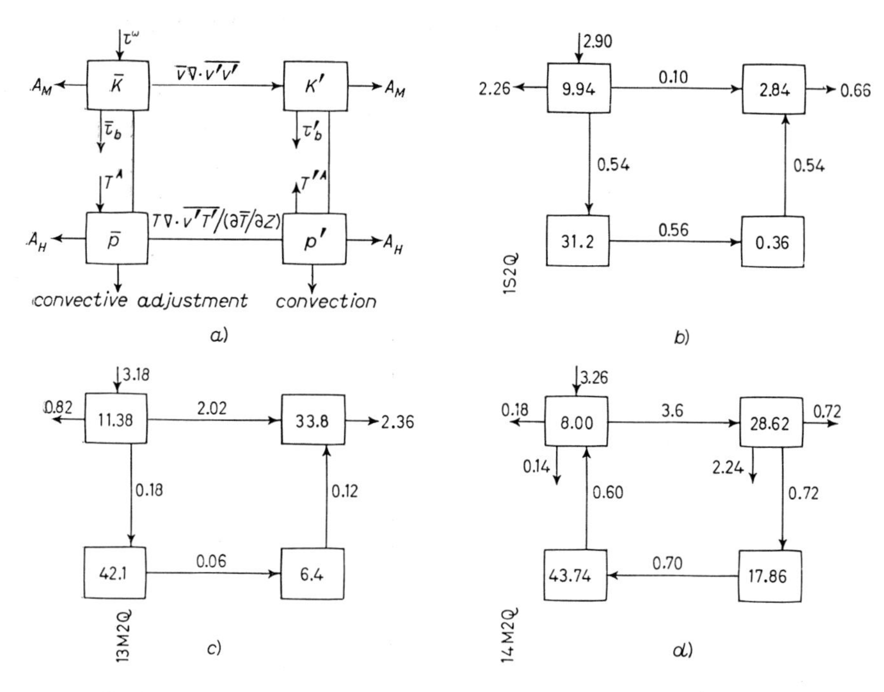

Fig. 16. – Basin-integrated energy transfer diagram from some EGCM experiments (from [4], fig. 12).

of the circulation from massive amounts of data have been and are being worked out by modelers and scientists, a difficult task but antitheoretical to the one which usually plagues field oceanographers, who must deal with frustratingly sparse data sets. One important method employed is the study of energy budgets and conversion. Although dynamical processes such as energetics are known to be quite inhomogeneous in the model flows, overall integrated basin averages are commonly calculated as a simple starting point and are illustrated in fig. 16. The basic interpretation of such diagrams is in terms of the mean (barred) and eddy (primed) kinetic (K) and potential (P) energies and the transfers between them. Arrows show direction of transfer and numbers above the arrows quantify the transfer rates. Eddy kinetic energy K' can be produced either by a transfer from $\bar{K}$ directly or by a transfer from $\bar{P}$ indirectly through P'. The former is a barotropic instability mechanism. Either or both production mechanisms may be seen to occur in various experiments. Moreover, due to vertical and geographical inhomogeneities, the basin averages may not be characteristic of any actual flow region. More detailed and revealing energy analyses have been carried out by several investigators for partial regions and depth ranges which have been reviewed by HARRISON [22]. Some interesting generalizations can be formulated: i) eddy energy is generated in the intense current regions of the flow (Gulf Stream separation and/or return flow) by a finite-amplitude process related to barotropic, baroclinic, or mixed instability, ii) eddy energy is exported to the open-ocean regions, iii) a simulated Gulf Stream recirculation is generated by the models and iv) permanent deep currents can be produced by indirect eddy-dependent mechanisms, *i.e.* not driven locally directly by eddy flux divergences [23, 24]. A comparison of some EGCM results with observations is shown in fig. 17*a*);

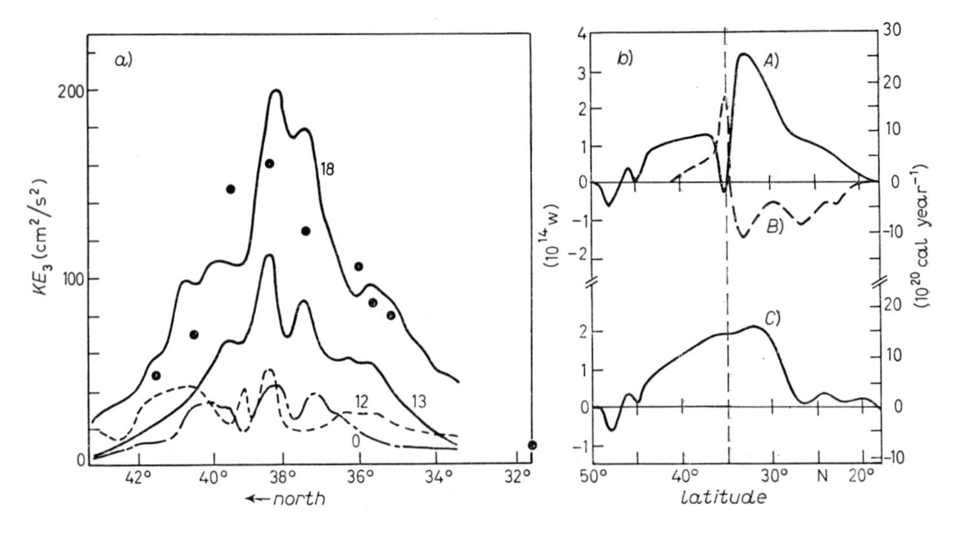

Fig. 17. – Some statistics of EGCM experiments: *a*) deep kinetic energy *vs.* latitude for several experiments compared to WHOI moored current meter measurements; *b*) zonally averaged northward heat transport in experiment *d*) of fig. 14: *A*) mean, *B*) eddy, *C*) total. (*a*) courtesy of W. HOLLAND and W. SCHMITZ; W. SCHMITZ: private communication; *b*) from A. J. SEMTNER as quoted by [25], fig. 39.)

the different experiments are for different domain sizes and bottom frictions. The lower curves correspond to some experiments of the type shown in fig. 14; the more recent experiments show better agreement as a result of iteration of such parameter choices after the verification attempt with available current meter data. Figure 17*b*) shows a calculation of heat flux in an EGCM, *viz.* 14*d*). This kind of result is of great interest for climate dynamics studies, since real-ocean heat transports are not well known or easily measured. At this point in the development of models, a sensitivity of such a secondary statistics of interest to model parameters must unfortunately still be expected. Although results today are impressive, much remains to be done including geometric, topography and forcing studies and a considerable effort is required to begin to move forward in the realm of parameterization of eddy effects on the larger scales.

Finally I mention the second kind of eddy-resolving numerical models, those which have generally the highest resolution but which are limited only to a region of the flow of interest or to the exploration of an idealized process. They are reviewed by the MODE Group ([2], subsect. 4·8) and very well introduced by the lectures of Salmon in this series. The initial models used periodic boundary conditions and were of great interest in the interpretation of MODE-I results and the design of the POLYMODE LDE. Figures 18 and 19 illustrate the simulated vorticity balance studies mentioned above.

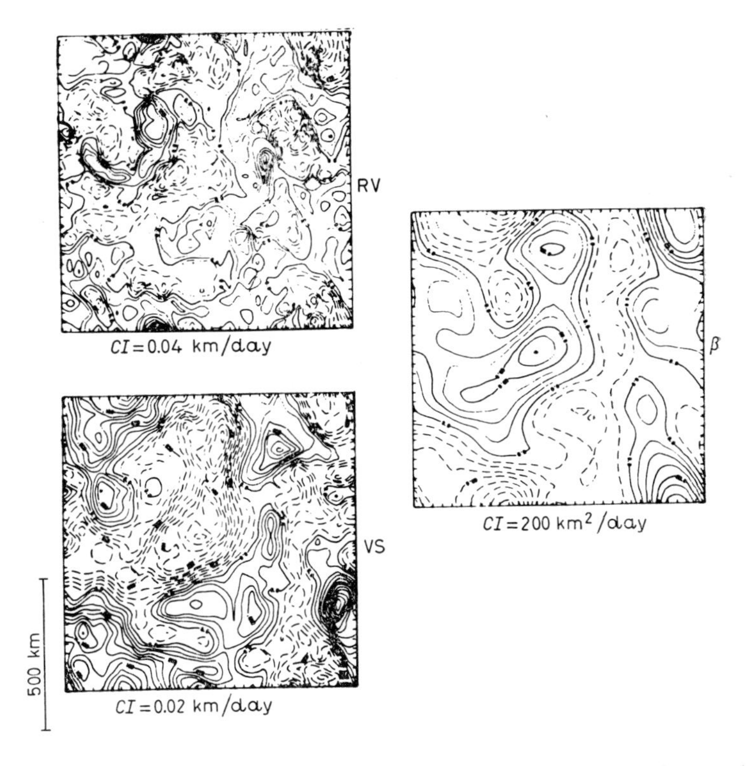

Fig. 18. – Typical fields at 450 m from a spin-down process experiment used for a balance-of-terms study. β: streamfunction; RV: relative vorticity; VS: vortex stretching. (From [26], fig. 5.)

Figure 20 shows a recent result from a regional dynamical/forecast model which used « open-ocean » boundary conditions which correspond to a general parameterization of the interconnection conditions across the fluid-fluid boundaries. Curve *a*) is the result of a simulated forecast experiment under realistic conditions; the r.m.s. domain error is controlled to be about 10 %, although new boundary data are supplied only from a coarsely sampled strip and not at every time step as is formally required.

Having reviewed the results of recent scientific investigation and mentioned some specific ongoing studies of interest, I would like to conclude with a general summary of presently and foreseeably important research tasks and goals.

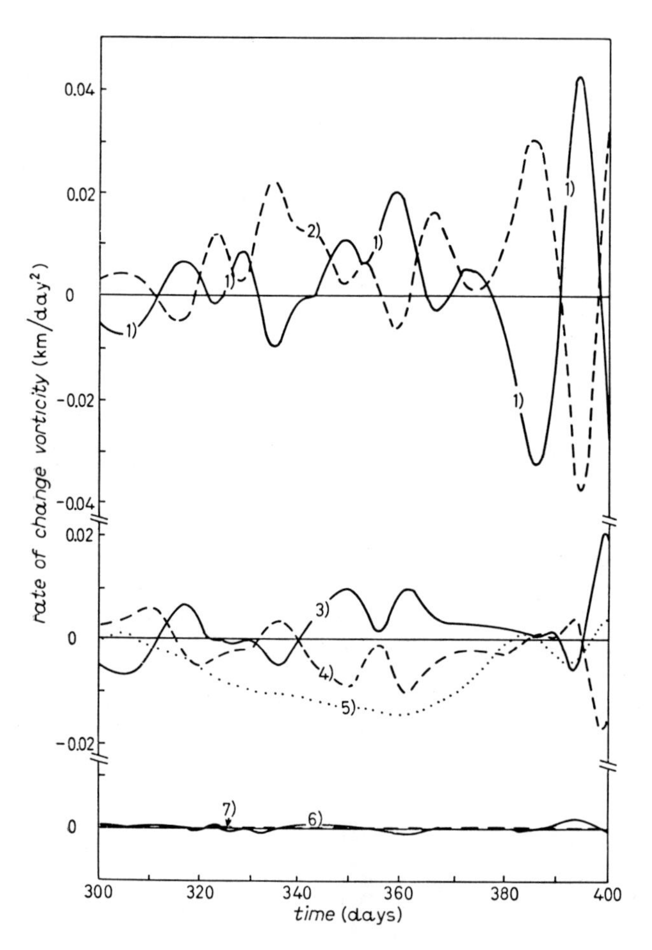

Fig. 19. – Simulated time series from a point in the experiment illustrated in fig. 18 for the various terms in the vorticity balance (layer 2, point A): 1) time change of relative vorticity RV, 2) advection of RV, 3) time change of vortex stretching VS, 4) advection of VS, 5) the planetary β-effect, 6), 7) negligible higher-order frictional terms (from [26], fig. 17).

First of all, large amounts of new data collected in POLYMODE and related programs require analysis, interpretation and *synthesis*. This process should lead to an *improvement of models*, progress in model verification and the development of entirely new aspects, such as models of small mesoscale features and simulations of the statistics of soliton transports. *Efficient detailed exploration* of currents and eddies in *new oceanic regions* of interest is scientifically important and should be feasible in terms of understanding gained as the result of recent major efforts. A continued effort in the area of *instrument development* and scientific trial is needed. *Applications of the advances in physics and dynamics* to geochemical and tracer problems, biological process studies, acoustics,

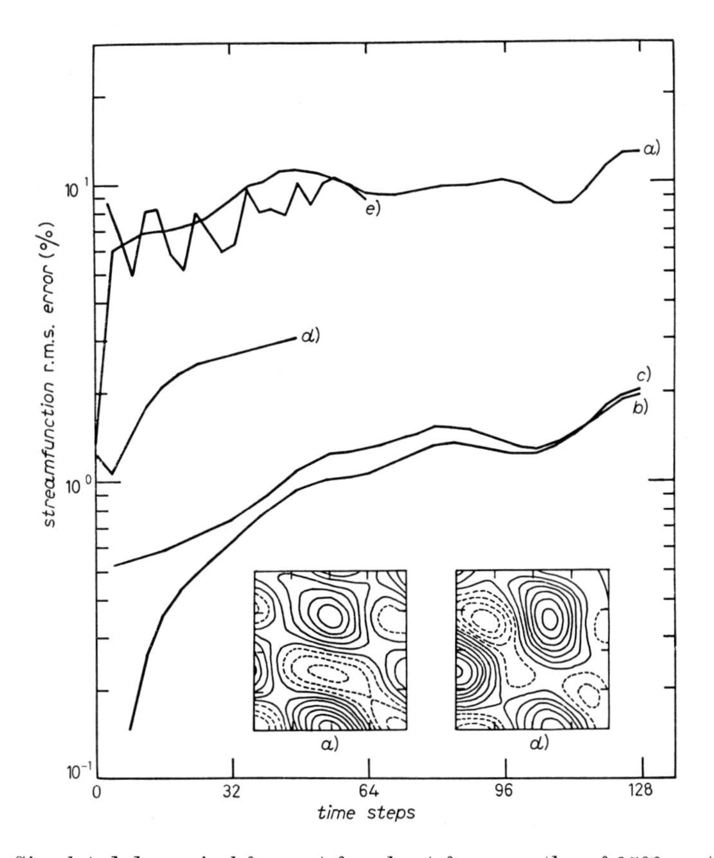

Fig. 20. – Simulated dynamical forecast for about four months of 1500 m streamfunction with a barotropic open-ocean model. Inserts show the eddy field in the 500 km domain at the start and end of the forecast. See text. (From [10], fig. 10 and 20.) An interesting comparison of periodic process model results with observations has been carried out by SCHMITZ and OWENS (1979).

climate dynamics and other applications should be vigorously pursued with dynamicists themselves necessarily playing an active role, especially at early stages of such efforts.

By the mid-1980's an *intensive and extensive study of the Gulf Stream* current, ring, extension and recirculation *system* should be initiated. It should serve to determine the kinematics and dynamics of energy production, transport and dissipation and the physics of interactions between scales in that region which has been identified as of great importance to the dynamics of the circulation and to eddies.

Major long-range objectives which should be achievable to a large extent by the end of the 1980's are: i) the implementation of a *descriptive predictive system* for meso (synoptic) scale currents over large regions of the ocean, which should consist of a numerical-dynamical model, a statistical model and a real-time observational network including subsurface and remotely sensed components and ii) the construction and operation of a verified *large-scale circulation model* with effective and efficient parameterization of meso (synoptic) scale processes for long-time-scale applications including coupling to an atmospheric model.

* * *

These lectures were based in part upon a presentation in December 1979 at the IAPSO/IUGG XVII General Assembly, Canberra (POLYMODE and Advances in Ocean Current Dynamics, Abstract, *Proces-Verbaux* No. 15, Canberra, IAPSO-IUGG, p. 97) for which several POLYMODE scientists made preliminary results available. For these results and their discussions I am grateful. The preparation of this review was sponsored by the National Science Foundation (OCE-77-28289) and the Office for Naval Research (N00014-75-C-0225). It is a pleasure to thank the Italian Physical Society, the summer school Chairman, Dr. A. OSBORNE, and Secretary, Dr. P. MALANOTTE RIZZOLI, for the opportunity of presenting this lecture in an attractive and stimulating environment.

REFERENCES

[1] H. STOMMEL: *The Gulf Stream: A Physical and Dynamical Description*, 2nd edition (Berkeley, Cal., 1965).
[2] MODE GROUP: *Deep-Sea Res.*, **25**, 859 (1978).
[3] NATIONAL ACADEMY OF SCIENCES: *Numerical models of Ocean circulation*, in *Proceedings of the Durban Symposium* (Sponsored by the Ocean Science Board), hard-cover book available from the NAS Publishing Office, Wash., D.C., USA (1975).
[4] A. R. ROBINSON, D. E. HARRISON and D. B. HAIDVOGEL: *Dyn. Atmos. Oceans*, **3**, 143 (1979).
[5] L. M. BREKHOUSKI, K. N. FEDEROV, L. M. FOMIN and N. N. KOSHLYAKOV: *Deep-Sea Res.*, **18**, 1189 (1971).
[6] R. HENDRY and B. RENINGER: Dauson 78-035 Cruise Report, *POLYMODE News*, No. 59, unpublished manuscript (1978).
[7] J. GOULD, R. DICKSON, J. MULLER and C. MAILLARD: *Measurements of mesoscale variability in the northeast Atlantic*, Abstract, *Proces-Verbaux* No. 15, Canberra, IAPSO-IUGG, p. 101 (1979).
[8] C. DEVERDIERE: *The Tourbillon experiment, POLYMODE News*, No. 73, unpublished manuscript (1980).
[9] B. K. HARTLINE: *Science*, **205**, 571 (1979).
[10] A. R. ROBINSON and D. B. HAIDVOGEL: *J. Phys. Oceanogr.*, **12**, 1909 (1980).
[11] P. L. RICHARDSON, R. E. CHENEY and L. V. WORTHINGTON: *J. Geophys. Res.*, **83**, 6136 (1978).
[12] G. A. MAUL and S. R. BAIG: *Satellite measurements of Gulf Stream meanders. POLYMODE News*, No. 7, unpublished manuscript (1976).
[13] P. L. RICHARDSON: *J. Phys. Oceanogr.*, **10**, 90 (1979).

[14] G. METCALF and E. BARANOV *et al.*: *POLYMODE Synoptic XBT Mapping, POLYMODE News*, No. 64, unpublished manuscript (1979).
[15] Y. GRACHEV, M. KOSHLYAKOV, T. TIKHOMIROVA and V. YENIKEYEV: *Synoptic eddy field in the POLYMODE area, September 1977-May 1978, POLYMODE News*, No. 69, unpublished manuscript (1979).
[16] SCOR WORKING GROUP 21: *An intercomparison of some current meters*. - II: *Report on an experiment carried out from the research vessel Akademik Kurchatov, March-April 1970, by the Working Group on Continuous Current Velocity Measurements*, UNESCO, Technical Papers in Marine Sciences, No. 17 (1974).
[17] S. RISER: private communication (1980).
[18] J. MCWILLIAMS and R. HEINMILLER: *The POLYMODE local dynamics experiment: objectives, location and plan*, Office of IDOE, NSF and ONR, unpublished manuscript (1978).
[19] J. PRICE: private communication (1980).
[20] C. EBBESMEYER, J. MCWILLIAMS, T. ROSSBY and B. TAFT: private communication (1980).
[21] S. MCDOWELL and T. ROSSBY: *Science*, **202**, 1085 (1978).
[22] D. E. HARRISON: *Rev. Geophys. Space Phys.*, **17**, 974 (1979).
[23] W. R. HOLLAND: *Dyn. Atmos. Oceans*, **3**, 11 (1979).
[24] P. B. RHINES and W. HOLLAND: *Dyn. Atmos. Oceans*, **3**, 289 (1979).
[25] Y. MINTZ: *On the simulation of the oceanic general circulation*, in *Report of JOC Study Conference on Climate Models*, edited by W. L. GATES, Vol. **2** (Geneva, 1979), p. 607.
[26] W. B. OWENS: *J. Phys. Oceanogr.*, **9**, 337 (1979).

Geostrophic Turbulence.

R. SALMON

Scripps Institution of Oceanography (A-025) - La Jolla, Calif., 92093

1. – Introduction.

Highly nonlinear, quasi-random motion in rapidly rotating, stably stratified fluid is called « geostrophic turbulence ». The subject has relevance to large-scale flow in the Earth's oceans and atmosphere. The theory of geostrophic turbulence bridges the distinct fields of geophysical fluid dynamics and statistical turbulence theory. The present lectures offer a self-contained but brisk introduction to both of these fields.

The quasi-geostrophic equations (sect. **2**) form the basis for the theory. These equations approximate the general equations of fluid motion in the limit of small Rossby number. The Rossby number may be defined as the ratio of the Earth's rotation period (a day) to a characteristic time scale of the flow. The operative equation expresses the conservation of a scalar quantity called potential vorticity following fluid particles.

The quasi-geostrophic potential-vorticity equation generalizes the vorticity equation governing two-dimensional Navier-Stokes turbulence (sect. **3**). The latter subject has aroused intense theoretical interest, in part for its geophysical relevance, but more perhaps for the opportunity it affords to test deductive theories of homogeneous turbulence against relatively high-resolution numerical experiments. Numerical simulations of three-dimensional turbulence are severely limited in spatial resolution by the computing capacity of even the most powerful modern machines.

The vast differences between two- and three-dimensional Navier-Stokes turbulence set the tone for our subject. The differences arise because vortex stretching, which is the primary mechanism of energy transfer from large to small scales of motion in three dimensions, is absent in two-dimensional flow. In two dimensions the simultaneous conservation of energy and vorticity actually implies a transfer of energy from the small to the large scales of motion. Simple arguments based upon energy and potential-vorticity conservation expose analogous distinctive properties of the quasi-geostrophic equations (sect. **4**).

The complexity of turbulence invites statistical analysis. Classical statistical mechanics predicts the ideal states of « absolute equilibrium » towards which nonlinearities acting alone would drive the flow (sect. **5**). These states are strictly nonrealizable, but they indicate the qualitative role of nonlinear interactions in realistic, nonequilibrium flows. The quasi-geostrophic equilibrium states are unusual and instructive. For example, equilibrium nonequatorial flow is nearly depth-invariant at the largest scales of motion. Nonequilibrium (closure) theory is more complicated (sect. **6**), but has proved useful in conjunction with direct computer simulations of the equations of motion. Beta-plane turbulence (sect. **7**) is a case in point.

This is not a comprehensive review paper. Rather, topics were selected and ordered to illustrate the important ideas in logical sequence. The material in sect. **5** on « the equatorial funneling effect » is new.

2. – The quasi-geostropic equations.

We employ the quasi-geostrophic equations for a system comprised of two immiscible fluid layers with slightly different uniform densities (fig. 1). The upper surface at $z = n(x, y, t)$ is free and the lower rigid boundary lies at $z = -D + d(x, y)$, where $D = D_1 + D_2$, $D_i = \overline{h}_i$, $h_i(x, y, t)$ is the vertical thickness of the i-th fluid layer, and the overbar denotes horizontal average over the flow domain. The horizontal boundaries are rigid vertical walls or absent altogether.

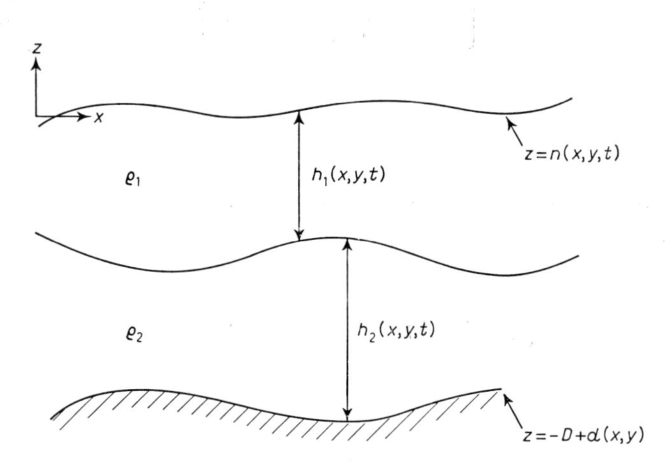

Fig. 1. – The geometry of the two-layer fluid.

Let $(L, U, L/U)$ be the scale for the (horizontal variability, horizontal velocity, time). If either

$$(D/L)^2 \ll 1 \qquad \text{or} \qquad R_0(D/L)^2 \ll 1 \,, \tag{2.1}$$

where $R_0 \equiv U/f_0 L$ is the « Rossby number », then the motion is hydrostatic, the horizontal velocity is depth-invariant within each layer, and the governing equations are the « shallow water » equations:

$$\frac{\partial \boldsymbol{u}_i}{\partial t} + \boldsymbol{u}_i \cdot \nabla \boldsymbol{u}_i + f\hat{k} \times \boldsymbol{u}_i = \begin{cases} -g\nabla n\,, & i = 1\,, \\ -g\nabla n + g'\nabla h_1\,, & i = 2\,, \end{cases} \tag{2.2}$$

and

$$\nabla \cdot \boldsymbol{u}_i + \frac{\partial w_i}{\partial z} = 0\,. \tag{2.3}$$

Here $\boldsymbol{u}_i = (u_i, v_i)$ is the horizontal and w_i the vertical velocity of the i-th layer, the (x, y, z) axis points (east, north, up), g is gravity, g' reduced gravity, and the Coriolis parameter

$$f = f_0 + \beta y\,, \tag{2.4}$$

where

$$f_0 = 2\Omega \sin \varphi_0\,, \qquad \beta = 2\Omega \cos \varphi_0 / r_{\mathrm{E}}\,,$$

$\Omega = 2\pi$ day^{-1}, φ_0 is a reference latitude, and r_{E} is the radius of Earth. We assume $\bar{n} = 0$ and $|n| \ll h_i$. Then the vertical boundary conditions are

$$\begin{cases} w_1 = 0 & \text{at } z = 0\,, \\ w_i = -\left(\dfrac{\partial h_1}{\partial t} + \boldsymbol{u}_i \cdot \nabla h_1\right) & \text{at } z = -h_1 \end{cases} \tag{2.5}$$

and

$$w_2 = \boldsymbol{u}_2 \cdot \nabla d \qquad \text{at } z = -D + d\,.$$

Taking the vertical component of the curl ($\hat{k} \cdot \nabla \times$) of (2.2) and using (2.3) yields the vorticity equations

$$\left(\frac{\partial}{\partial t} + \boldsymbol{u}_i \cdot \nabla\right)(\varrho_i + f) = (\varrho_i + f)\frac{\partial w_i}{\partial z}\,, \tag{2.6}$$

where

$$\varrho_i = \frac{\partial v_i}{\partial x} - \frac{\partial u_i}{\partial y} \tag{2.7}$$

is the relative vorticity. Integrate (2.6) through each layer and use (2.5). The result is

$$\left(\frac{\partial}{\partial t} + \boldsymbol{u}_i \cdot \nabla\right) q_i = 0\,, \tag{2.8}$$

where

$$q_i = (\varrho_i + f)/h_i \tag{2.9}$$

is the potential vorticity. In general, (2.8) is a single equation in the three dependent variables u_i, v_i and h_i. However, under the additional scaling assumptions of a rapid, slowly varying rotation rate,

$$R_0 \ll 1 \quad \text{and} \quad \beta L/f_0 \ll 1\,, \tag{2.10}$$

(2.8) reduces to an equation in a single unknown, and potential-vorticity conservation determines the full dynamics. The most convincing derivation is an expansion in powers of the Rossby number (see [1], p. 386). We shall be less formal. If (2.10) hold, then the horizontal momentum balance is geostrophic, that is, between the pressure gradient and Coriolis terms in (2.2). Thus

$$\boldsymbol{u}_i \approx \hat{k} \times \nabla \psi_i\,, \tag{2.11}$$

where

$$\psi_1 = gn/f_0 \tag{2.12}$$

and

$$\psi_2 = gn/f_0 - g' h_1/f_0\,. \tag{2.13}$$

Then

$$\varrho_i \approx \nabla^2 \psi_i\,, \tag{2.14}$$

$$h_1 \approx D_1 + f_0(\psi_1 - \psi_2)/g' \tag{2.15}$$

and

$$h_2 \approx D_2 + f_0(\psi_2 - \psi_1)/g' - d\,. \tag{2.16}$$

Substitution of (2.12)-(2.16) closes (2.8). But

$$|\varrho_i|/f_0 = O(R_0) \tag{2.17}$$

and

$$|\beta y|/f_0 = O(\beta L/f_0)\,, \tag{2.18}$$

which are both small parameters. If, in addition, we assume

$$|D_i - h_i|/D_i \ll 1\,, \tag{2.19}$$

then

$$q_i \approx \frac{f_0}{D_i}\left[1 + \frac{\varrho_i}{f_0} + \frac{\beta y}{f_0} + \frac{D_i - h_i}{D_i}\right] \tag{2.20}$$

and (2.8) reduces to the form used throughout these lectures:

$$\frac{\partial q_i}{\partial t} + J(\psi_i, q_i) = 0\,, \qquad J(A, B) \equiv \frac{\partial A}{\partial x}\frac{\partial B}{\partial y} - \frac{\partial B}{\partial x}\frac{\partial A}{\partial y}\,, \tag{2.21}$$

where

$$\begin{cases} q_1 = \nabla^2\psi_1 + F_1(\psi_2 - \psi_1) + f\,, \\ q_2 = \nabla^2\psi_2 + F_2(\psi_1 - \psi_2) + f + b(x, y)\,, \\ F_i \equiv f_0^2/g' D_i \qquad \text{and} \qquad b \equiv f_0\, d/D_2\,. \end{cases} \tag{2.22}$$

The assumption of two immiscible fluid layers separated by a sharp change in density is appropriate for much of the ocean. However, (2.21), (2.22) are closely analogous to the corresponding equation for a continuously stratified fluid, *viz.*

$$\begin{cases} \dfrac{\partial q}{\partial t} + J(\psi, q) = 0\,, \\ q = \nabla^2\psi + \dfrac{\partial}{\partial z}\left(\dfrac{f^2}{\mathcal{N}^2}\dfrac{\partial \psi}{\partial z}\right) + f\,, \end{cases} \tag{2.23}$$

where $\mathcal{N}(z)$ is the buoyancy (or Väisälä) frequency. In particular (2.21), (2.22) are vertical-finite-difference analogs of (2.23). (See [1], p. 396.) Thus (2.21), (2.22) govern atmospheric motion as well. The rigid-lid condition (2.5*a*) becomes a crude model of the tropopause. More surprising is the fact that two vertical degrees of freedom appear adequate to resolve low-Rossby-number motions in both ocean and atmosphere. An explanation will be offered in sect. **4**. The parameters F_i in (2.22) may be specified in terms of the depth ratio

$$\Delta \equiv D_1/D_2 = F_2/F_1 \tag{2.24}$$

and the internal deformation radius

$$k_R^{-1} \equiv (F_1 + F_2)^{-\frac{1}{2}} \tag{2.25}$$

(units of length). Typical mid-latitude values are $\Delta = 1/7$, $k_R^{-1} = 50$ km for the ocean and $\Delta = 1$, $k_R^{-1} = 500$ km for the atmosphere. In these lectures we assume $\Delta = 1$ (for algebraic simplicity) corresponding to a linear mean density gradient. Then

$$F_1 = F_2 = k_R^2/2 . \tag{2.26}$$

In simply connected geometry, the appropriate horizontal boundary conditions are $\psi_i = 0$. If islands are present, more general boundary conditions are required, but these are of no interest here.

In addition to potential vorticity, (2.21) conserves the total energy (proportional to)

$$\overline{\nabla\psi_1 \cdot \nabla\psi_1} + \overline{\nabla\psi_2 \cdot \nabla\psi_2} + k_R^2\overline{(\psi_1 - \psi_2)^2}/2 . \tag{2.27}$$

The terms in (2.27) represent the kinetic energy in the top layer, the kinetic energy in the bottom layer and the « available potential energy ». The latter vanishes when the interface is flat (cf. eqs. (2.15), (2.16)), which is the state of minimum potential energy. Currents with horizontal length scale L and order U shear across the interface have a ratio of available potential to kinetic energy of order $L^2k_R^2$, which can be very large. Such large-scale baroclinic currents exist and can be stable because of the rotation. Conversion from potential to kinetic energy is much less efficient in rotating than in nonrotating flow and takes the form of a « baroclinic instability » that prefers the length scale k_R^{-1}. Much of the special flavor of geophysical fluid dynamics derives from these facts. Linear stability theory offers a quantitative description in the small-amplitude regime, but more general arguments based on integral conservation properties alone lead qualitatively to the same results. The conservation arguments are valuable because they extend to realistic, nonlinear flow.

3. – Two-dimensional turbulence.

The vorticity equation for a single layer of uniformly rotating (f = const) Newtonian fluid over a flat bottom ($d = 0$) is

$$\frac{\partial\varrho}{\partial t} + J(\psi, \varrho) = v\nabla^2\varrho , \tag{3.1}$$

where

$$\varrho = \nabla^2\psi . \tag{3.2}$$

Two-dimensional turbulence governed by (3.1) is the prototype for geostrophic turbulence. Its distinctive property is a tendency for smaller eddies to feed their energy into the larger scales of motion [2, 3]. To isolate the role of nonlinear interactions specialize (3.1) to inviscid flow ($v = 0$). The inviscid equations conserve energy

$$\overline{\nabla\psi\cdot\nabla\psi}=\int_0^\infty \mathscr{E}(k,t)\,\mathrm{d}k\,, \tag{3.3}$$

and « enstrophy »

$$\overline{\varrho^2}=\int_0^\infty k^2\,\mathscr{E}(k,t)\,\mathrm{d}k\,, \tag{3.4}$$

where $\mathscr{E}(k,t)$ is the energy in wave number k (summed over all directions) at time t. Suppose $\mathscr{E}(k,0)$ is concentrated at k_1 (fig. 2A)). If the energy subsequently spreads out to other wave numbers, then simultaneous conservation of the zeroth and second moments of $\mathscr{E}(k)$ implies that more energy moves

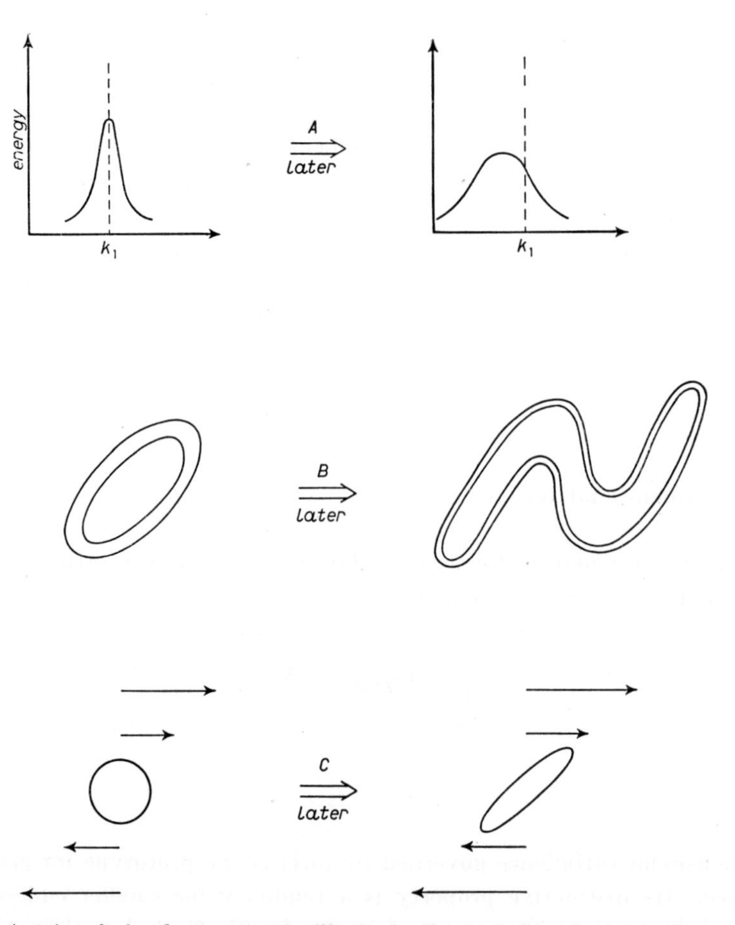

Fig. 2. – A trio of simple arguments illustrate the transfer of energy from small to large scales in two-dimensional flow.

toward wave numbers less than k than toward higher wave numbers. Precisely,

$$\frac{\mathrm{d}}{\mathrm{d}t}k_{\mathrm{E}} = -\frac{1}{2k_1}\frac{(\mathrm{d}/\mathrm{d}t)\int_0^\infty (k-k_1)^2 \mathscr{E}(k,t)\,\mathrm{d}k}{\int_0^\infty \mathscr{E}(k)\,\mathrm{d}k}, \tag{3.5}$$

where

$$k_{\mathrm{E}} \equiv \int_0^\infty k\mathscr{E}(k,t)\,\mathrm{d}k \bigg/ \int_0^\infty \mathscr{E}(k)\,\mathrm{d}k \tag{3.6}$$

is a wave number that characterizes the energy-containing scales of motion. The right side of (3.5) is negative if the dispersion of $\mathscr{E}(k)$ is increasing. By similar reasoning enstrophy moves chiefly to higher wave numbers. In three-dimensional turbulence, enstrophy is not conserved, and energy moves to high wave numbers.

The Fourier-space argument gives no hint at the mechanics of these transfers, but consider the following: If $v = 0$, then isolines of vorticity are also material lines, that is, always connect the same fluid particles. Consider two such neighboring lines of constant vorticity (fig. 2B)). If, as expected, these material lines lengthen on the average in time, then mass conservation requires their average separation to decrease, because the area between the lines is constant in two-dimensional flow. Thus the mean square vorticity gradient,

$$\overline{\nabla\varrho\cdot\nabla\varrho} = \int_0^\infty k^4 \mathscr{E}(k,t)\,\mathrm{d}k\,, \tag{3.7}$$

increases, moving enstrophy into higher wave numbers.

Again, consider an initially isotropic small-scale eddy in a large-scale uniform shear (fig. 2C)). At a later time the eddy has been strained into the shape at the right, the Reynolds momentum flux $\overline{u_1' u_2'}$ is up the mean momentum gradient, and energy moves from the eddy into the mean flow. Crudely, if k is the wave number associated with the eddy and E is its energy, then, by vorticity conservation on particles, k^2E remains constant. But k increases as the eddy is stretched, and, therefore, E must decrease. The energy which is lost by the eddy must, by conservation of total energy, show up as an increase in the energy of the straining field.

Now consider the flow which develops when friction is restored and an external source injects energy continuously into the fluid at an intermediate wave number k_1. By all of the above arguments, we expect energy to move leftward (*i.e.* toward lower wave numbers) and enstrophy rightward from k_1.

Let ε_1 be the injection rate of energy (per unit mass per unit time) and n_1 the corresponding enstrophy injection rate. Let $\varepsilon(k)$ and $n(k)$ be the rates of nonlinear energy and enstrophy transfer past k (positive toward higher k). The viscosity in (3.1) is scale-selective. Let k_D be the highest wave number below which this viscosity is negligible. Then, if we assume quasi-steady statistics in the « inertial range » on $k_1 < k < k_D$, the nonlinear transfer of energy and enstrophy must be independent of k, *i.e.*

$$\varepsilon(k) = \varepsilon(k_D) \quad \text{and} \quad n(k) = n(k_D)\,. \tag{3.8}$$

Also

$$k_D^2\, \varepsilon(k_D) < n(k_D)\,, \tag{3.9}$$

by definition of k_D. Thus, if $v \to 0$ ($k_D \to \infty$), then $\varepsilon(k_D) \to 0$, since $n(k_D)$ cannot exceed n_1.

Inertial-range theory [4, 5] hypothesizes that, as $k_D/k_1 \to \infty$, the inertial range behaves like a « cascade » in which direct enstrophy transfer occurs only between eddies of comparable size. (Such transfer is also said to be « local » in wave number.) Towards the middle of the inertial range, which is many cascade steps removed from both forcing and viscosity, the energy spectrum depends plausibly only on k and $n = n(k)$. It then follows from dimensional requirements that

$$\mathscr{E}(k) = C n^{2/3} k^{-3}, \qquad k_1 \ll k \ll k_D\,, \tag{3.10}$$

where C is a universal dimensionless constant. Similar arguments suggest the existence of an energy inertial range on $k \ll k_1$ in which $n(k)$ is zero, $\varepsilon = \varepsilon(k)$ is (a negative) constant,

$$\mathscr{E}(k) = C' |\varepsilon|^{2/3} k^{-5/3}, \qquad k \ll k_1\,, \tag{3.11}$$

and C' is another universal constant. Numerical experiments generally support these ranges [6], but the theoretical question of existence of « true » inertial ranges is both extremely thorny and probably irrelevant to the Earth's geophysical fluids. We regard (3.10) and (3.11) as conceptually useful idealizations.

There is a close correspondence between the k^{-1} enstrophy spectrum in the enstrophy-cascading inertial range and Batchelor's [7] k^{-1} spectrum for the variance of a passively advected scalar quantity. The passive scalar concentration obeys the same equation (3.1) as ϱ, but without (3.2). Batchelor's argument assumes only that the scales of motion contributing to the r.m.s.

strain rate σ are large compared to the length scales of scalar variability under consideration, and that the straining is persistent in time. If

$$\psi = \sigma xy \tag{3.12}$$

locally in the vicinity of $x = y = 0$, and the scalar field is a local sinusoid with wave number k, then, after a sufficient time,

$$\frac{dk}{dt} = \sigma k \tag{3.13}$$

following particles. Let $\Lambda(k)$ be the spectrum of scalar variance. The variance in a wave number band with initial width Δk centered on wave number k must, by conservation of variance, obey

$$\Lambda(k)\,\Delta k = \Lambda(\gamma k)\cdot(\gamma\,\Delta k)\,, \tag{3.14}$$

where $\gamma = \exp[\sigma t]$. The right side of (3.14) is the spectrum times the band width after time t. If the statistics are steady, then (3.14) holds for arbitrary γ. Thus

$$\Lambda(k) \propto k^{-1}. \tag{3.15}$$

Note that this argument does not require a localness-in-wave-number hypothesis (rather the opposite), so that Batchelor's theory might help explain why a k^{-3} energy spectrum is often observed when the conditions for a true inertial range seem lacking.

Halikas studied two-dimensional turbulence in a specially designed plexiglass tank on the rotating table at the Scripps Institution of Oceanography (plate 1). The tank radius and depth are comparable, so that the flow is two-dimensional only because the Rossby number is small. In a series of experiments, a vertical cylinder moving slowly relative to the tank released a column of dye in its wake. The dye streak (viewed from above) further demonstrates the dramatically different behavior of two- and three-dimensional turbulence. In nonrotating flow (plate 2), the dye sheet rapidly increases its surface area, molecular diffusion is very efficient, and the dye concentration quickly becomes uniform over a wide area. In rotating flow (plate 3), the fluid motion is columnar, the dye sheet increases its surface area much more slowly, and the dye remains concentrated in a small volume of fluid. Because of the vastly different dilution rates, Halikas' flow visualization technique actually works better in rotating than in nonrotating flow. This experiment is relevant to diffusion from a point source in the geophysical fluids.

Plate 1. – Cylindrical tank on the rotating table. The vertical dye-filled column is moved by an arm attached to the axis of the tank. The tank diameter is 125 cm.

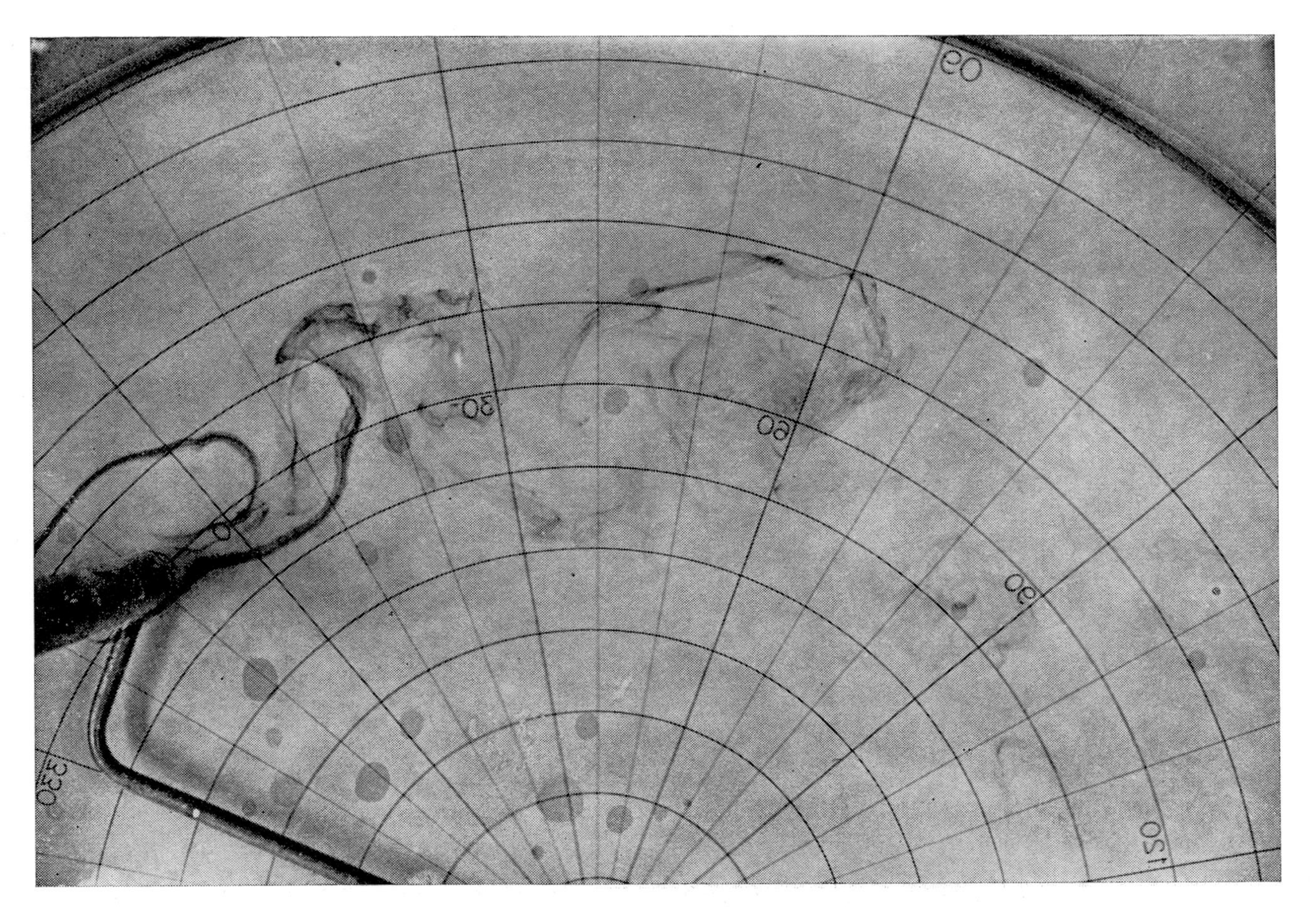

Plate 2. – Nonrotating flow. Towing period 78 s.

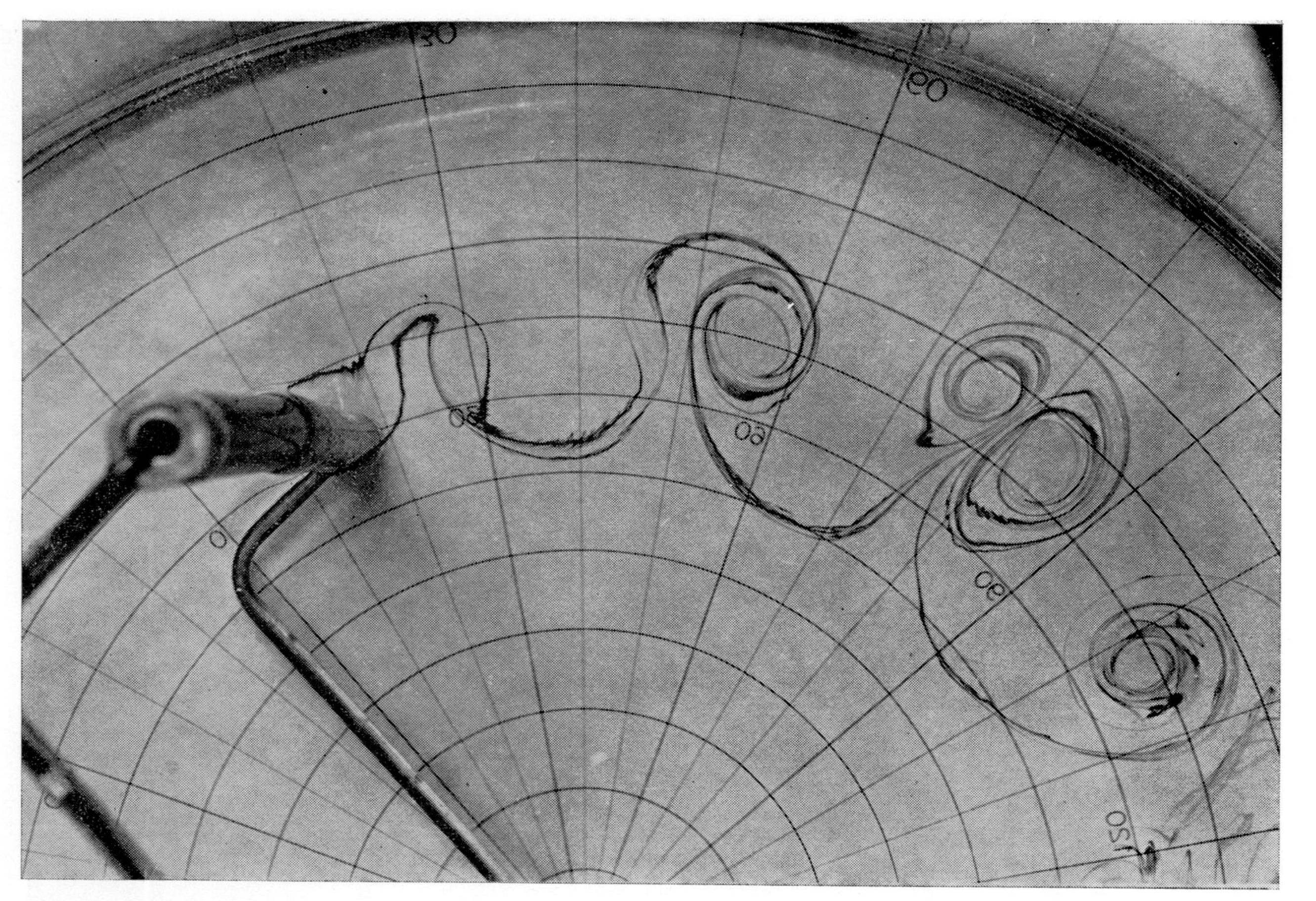

Plate 3. – Rotating flow. Towing period 78 s. Rotation period 3.55 s.

4. – Two-layer rotating turbulence.

The quasi-geostrophic equations (2.21), (2.22) for two-layer flow include the effects of density stratification, bottom topography and spatially variable rotation rate. The latter, particularly, has interesting consequences in the presence of horizontal boundaries. The remainder of these lectures will entertain each of these features in isolation. The present section treats horizontally unbounded stratified flow over a flat bottom [8-10]. The governing equations are (2.21), (2.22) with f constant, $D_1 = D_2$, and $b = 0$.

The discussion is simplest if we exchange ψ_1, ψ_2 for the vertical-mode variables

$$\psi = (\psi_1 + \psi_2)/2 \tag{4.1}$$

and

$$\tau = (\psi_1 - \psi_2)/2\,, \tag{4.2}$$

where ψ will be called the barotropic streamfunction and τ the baroclinic. Note that the upper thermocline displacement is proportional to

$$D_1 - h_1 = -2f_0\,\tau/g' \tag{4.3}$$

and that τ is temperaturelike in the sense that $\tau > 0$ implies a lower-than-average vertically averaged density. The (inviscid) dynamics become

$$\frac{\partial}{\partial t}\nabla^2\psi + J(\psi, \nabla^2\psi) + J(\tau, \nabla^2\tau) = 0 \tag{4.4}$$

and

$$\frac{\partial}{\partial t}\nabla^2\tau + J(\psi, \nabla^2\tau) + J(\tau, \nabla^2\psi) = k_{\mathrm{R}}^2\left[\frac{\partial\tau}{\partial t} + J(\psi, \tau)\right]. \tag{4.5}$$

Let the streamfunctions be expanded in Fourier series:

$$(\psi, \tau) = \sum_{\boldsymbol{k}} (\psi_{\boldsymbol{k}}, \tau_{\boldsymbol{k}}) \exp[i\boldsymbol{k}\cdot\boldsymbol{x}]\,, \tag{4.6}$$

and define

$$U(\boldsymbol{k}) = k^2|\psi_{\boldsymbol{k}}|^2 \qquad \text{and} \qquad E(\boldsymbol{k}) = (k^2 + k_{\mathrm{R}}^2)|\tau_{\boldsymbol{k}}|^2. \tag{4.7}$$

We call $U(\boldsymbol{k})$ the barotropic energy in wave number $\boldsymbol{k}$ and $E(\boldsymbol{k})$ the total baroclinic energy. The latter consists of baroclinic kinetic energy and available potential energy in the ratio k^2/k_{R}^2.

The quadratic integral invariants of the motion are the total energy and the potential enstrophy $\overline{q_i^2}$ of each layer. It is convenient to replace the latter two by their sum and difference, which are, of course, also conserved. Then conservation of the energy and sum enstrophy are equivalent to

$$\frac{\mathrm{d}}{\mathrm{d}t}\sum_{\boldsymbol{k}}[U(\boldsymbol{k})+E(\boldsymbol{k})]=0 \tag{4.8}$$

and

$$\frac{\mathrm{d}}{\mathrm{d}t}\sum_{\boldsymbol{k}}[k^2U(\boldsymbol{k})+(k^2+k_{\mathrm{R}}^2)E(\boldsymbol{k})]=0\,. \tag{4.9}$$

Conservation of the difference potential enstrophy,

$$\sum_{\boldsymbol{k}}k^2(k^2+k_{\mathrm{R}}^2)\,\psi_{\boldsymbol{k}}\,\tau_{-\boldsymbol{k}}\,, \tag{4.10}$$

puts a restriction on energy transfer between layers. However, our arguments require only (4.8) and (4.9). The analogy with two-dimensional turbulence is already apparent. The energies enter (4.8), (4.9) precisely as in two-dimensional turbulence, except that the baroclinic energy $E(\boldsymbol{k})$ has an effective square wave number of $k^2+k_{\mathrm{R}}^2$. By analogy with two-dimensional turbulence, we expect the total energy to move toward lower effective wave numbers. Thus (4.8) and (4.9) imply that nonlinear interactions barotropize the flow.

The quadratic integral invariants of the motion are significant because they are conserved by individual wave number triads. Let $\boldsymbol{k}$, $\boldsymbol{p}$, $\boldsymbol{q}$ be any three horizontal wave numbers that sum vectorially to zero:

$$\boldsymbol{k}+\boldsymbol{p}+\boldsymbol{q}=0\,. \tag{4.11}$$

The dynamics permits two types of triad in the two-layer fluid. One type consists of three barotropic components and the other consists of one barotropic and two baroclinic components. Energy transfer in the two types of triad is constrained by the detailed forms of (4.8) and (4.9), namely

$$\begin{cases}\dot U(\boldsymbol{k})+\dot U(\boldsymbol{p})+\dot U(\boldsymbol{q})=0\,,\\ k^2\dot U(\boldsymbol{k})+p^2\dot U(\boldsymbol{p})+q^2\dot U(\boldsymbol{q})=0\,,\end{cases} \tag{4.12}$$

and

$$\begin{cases}\dot U(\boldsymbol{k})+\dot E(\boldsymbol{p})+\dot E(\boldsymbol{q})=0\,,\\ k^2\dot U(\boldsymbol{k})+(p^2+k_{\mathrm{R}}^2)\dot E(\boldsymbol{p})+(q^2+k_{\mathrm{R}}^2)\dot E(\boldsymbol{q})=0\,,\end{cases} \tag{4.13}$$

where the tendencies in (4.12), (4.13) are those resulting from interactions with other members in the triad. (Thus $\dot{U}(\boldsymbol{k})$ is different in (4.12) and (4.13).)

The constraints (4.12) on the UUU triad are the same as in two-dimensional turbulence. In (4.12) suppose $k \leqslant p \leqslant q$. The constraints (4.12) prevent energy transfer between different scales of motion in both extremely local ($k \approx p$) and extremely nonlocal ($k \ll p$) triads. In three-dimensional turbulence only (4.12a) holds and any shape triad can transfer energy between scales.

For UEE triads the deformation radius is a critical length scale. When $k, p, q \gg k_R$ the constraints (4.13) are the same as (4.12), so that energy transfer in the UEE triads is the same as in two-dimensional turbulence. This agrees with the physical picture of two-layer flow on scales smaller than the deformation radius as being essentially uncoupled single layers which see the interface as a rigid boundary. However, the situation when $k, p, q \ll k_R$ is similar to three-dimensional turbulence. In this case energy transfer is between the two baroclinic components only and may be either local ($p = O(q)$) or nonlocal ($p \ll q$). Because of k_R in (4.13b), the enstrophy constraint poses no inhibition against rightward transfer of baroclinic energy on wave numbers less than k_R. If $k = O(k_R)$ in (4.13), then energy can be transferred to the barotropic component $U(\boldsymbol{k})$ as well.

An extension of the above ideas yields the direction of wave number energy transfer in forced-equilibrium flow. Consider a hypothetical two-layer fluid with a minimum wave number $k_0 \ll k_R$ near which stirring and/or heating forces act. « Dissipation » is also present, but, for simplicity, it is limited to only two regions of the spectrum: near k_0, where it models the loss of large-scale energy to (Ekman-type) frictional boundary layers, and near $k_D \gg k_R$, where it parameterizes the transition from two- to three-dimensional flow on scales too small to feel the Earth's rotation. In the general inertial region $k_0 < k < k_D$ neither forcing nor dissipation is significant. The fluid is assumed to reach a statistically steady state in which the transfer of total energy and sum enstrophy past k is the same for every k on $k_0 < k < k_D$. On scales smaller than the deformation radius, the dynamics are those of uncoupled single layers in which the general invariants reduce to ordinary kinetic energy and enstrophy. Since the only energy source is at k_0, we expect k^{-3} enstrophy inertial ranges on $k_R < k < k_D$. If k_D/k_R is large, then the energy reaching k_D is negligible. This means that the large scales must adjust to a state in which the net energy production at k_0 is zero, that is

$$\dot{U}(\boldsymbol{k}_0)_{nc} + \dot{E}(\boldsymbol{k}_0)_{nc} = 0 \,, \tag{4.14}$$

where the subscripts nc denote tendencies due to forcing and dissipation. Because of (4.14) the potential-enstrophy production is proportional to

$$k_R^2 \dot{E}(\boldsymbol{k}_0)_{nc} \,. \tag{4.15}$$

The potential-enstrophy production equals the transfer rate of potential enstrophy across $k_0 < k < k_D$, and it must be positive since there is only an enstrophy sink at k_D. Thus (4.15) is positive, $\dot{U}(\boldsymbol{k}_0)_{nc}$ must be negative by (4.14), and internal interactions must transfer large-scale energy from baroclinic to barotropic mode. However, by the reasoning given above, large-scale baroclinic components can transfer energy only between themselves. The transfer from baroclinic to barotropic mode must, therefore, occur on $k_0 < k < k_R$ as a rightward transfer of baroclinic energy in *UEE* triads and an equal and opposite leftward transfer of barotropic energy in *UUU* triads. The conversion of baroclinic to barotropic energy can occur near k_R. Figure 3 summarizes the energy flow in wave number space. Potential-enstrophy transfer, which may be deduced by similar arguments, is indicated by dashed arrows. Figure 3 differs from the wave number energy flow diagrams conventionally drawn in meteorology in that both the upper horizontal and vertical arrows represent a conversion of potential to kinetic energy.

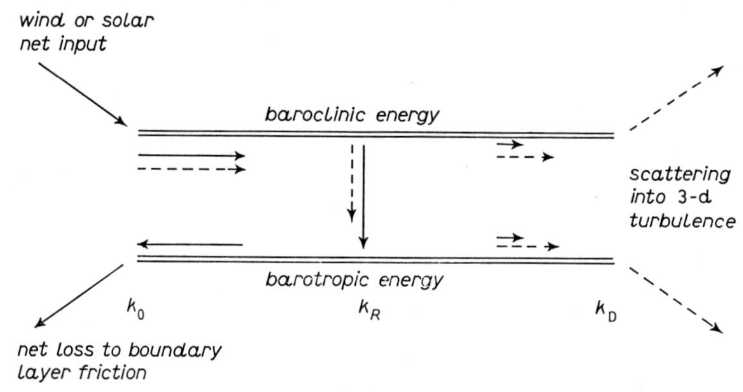

Fig. 3. – The wave number energy flow diagram for a two-layer system. The solid arrows represent energy flow and the dashed arrows sum potential-enstrophy flow.

The generalization of these ideas to multilayer or continuously stratified flow is straightforward. The multilayer system conserves

$$\int_0^\infty [E_0(k) + E_1(k) + E_2(k) + \dots]\, \mathrm{d}k \tag{4.16}$$

and

$$\int_0^\infty [k^2 E_0(k) + (k^2 + k_1^2)\, E_1(k) + (k^2 + k_2^2)\, E_2(k) + \dots]\, \mathrm{d}k\,, \tag{4.17}$$

where $E_n(k)$ is the energy in horizontal wave number k and vertical mode n, k_n^{-1} is the n-th internal deformation radius, and $k_1 = k_R$. In a uniformly stratified ocean,

$$k_n = n\pi f/\mathcal{N}D\,. \tag{4.18}$$

We expect net energy transfer into modes (k, n) with lower total wave numbers $k^2 + k_n^2$. But k_n increases with n. Observations do indeed suggest that the barotropic $(n = 0)$ and first baroclinic $(n = 1)$ modes hold most of the energy in nonequatorial geostrophic flow [11].

Now consider what happens as k_n varies with latitude through its dependence on f. As the equator is approached, the k_n vanish, removing the inhibition against high vertical-mode numbers. Since the total wave number $\sqrt{k^2 + k_n^2}$ of each mode (k, n) is smaller than its value in mid-latitudes, the total energy at the equator will be greater than in mid-latitude. Thus the same logic that predicts negative eddy viscosity in two dimensions suggests that a uniformly excited ocean would transfer energy equatorward and into high vertical wave numbers. Recent observations of the equatorial ocean reveal that the equator does contain a surprising amount of low-frequency energy in high vertical wave numbers [12]. These large-amplitude equatorial motions have previously been explained as a forced response to local winds [13]. The above arguments suggest, however, that energy at higher latitudes could be the source for equatorial motion. However, the quasi-geostrophic approximation itself breaks down at the equator, so that these ideas require some generalization. We shall return to this «equatorial funneling effect» in the following section.

5. – Entropy and absolute equilibrium.

Classical statistical mechanics can apply to the macroscopic motions of ideal fluids and other continuous fields [2, 14, 15]. The macroscopic analog of thermal equilibrium is physically nonrealizable, but it anticipates the role played by nonlinear interactions in realistic nonequilibrium flow. We commence with a swift review of the basics, adopting the information theory viewpoint [16]. This approach emphasizes the guessing nature of the whole subject.

Consider a general mechanical system whose precise state is specified by the value of N real numbers $[y_1, y_2, \dots, y_N]$ and whose dynamics is governed by N first-order ordinary differential equations of the form

$$\dot{y}_i = G_i(y_1, y_2, \dots, y_N)\,. \tag{5.1}$$

This form encompasses our fluid equations if $N = \infty$. For example, let the y_i be defined by

$$\psi = \sum_i (y_i/k_i)\,\varphi_i(\boldsymbol{x})\,, \tag{5.2}$$

where $\varphi_i(\boldsymbol{x})$ is a normalized eigenfunction satisfying

$$\nabla^2\varphi_i + k_i^2\varphi_i = 0 , \qquad \overline{\varphi_i^2} = 1 , \tag{5.3}$$

and $\varphi_i = 0$ on the boundary of the fluid, and k_i^2 is the associated eigenvalue. Then (3.1) transforms to

$$\dot{y}_i = \sum_{j,l} A_{ijl}\, y_j y_l - v_i y_i , \tag{5.4}$$

where

$$A_{ijl} = (k_j/k_i k_l)\,\overline{\varphi_i J(\varphi_l, \varphi_j)} \tag{5.5}$$

and

$$v_i = v k_i^2 . \tag{5.6}$$

The N-dimensional space spanned by the y_i is called phase space, and each point in phase space represents a possible state of the system as a whole. Every realization of (5.4) traces out a trajectory in phase space.

Let the joint-probability distribution of the y_i in an ensemble of realizations of (5.4) be

$$P(y_1, y_2, \ldots, y_N, t) . \tag{5.7}$$

Since the moving phase points that represent individual realizations of (5.4) can neither be created nor destroyed,

$$\frac{\partial P}{\partial t} + \sum_i \frac{\partial}{\partial y_i}(\dot{y}_i P) = 0 , \tag{5.8}$$

where $\dot{y}_i$ is given by (5.4). Equation (5.8) is analagous to the continuity equation in fluid mechanics with P (the density of phase points) replacing the ordinary fluid density, $\dot{y}_i$ (the i-direction velocity of phase points) replacing the fluid velocity, and the summation over N phase-space dimensions replacing the summation over three physical-space dimensions. If (5.4) is such that

$$\sum_i \partial\dot{y}_i/\partial y_i = 0 , \tag{5.9}$$

then (5.8) reduces to

$$\frac{\partial P}{\partial t} + \sum_i \dot{y}_i \frac{\partial P}{\partial y_i} = 0 , \tag{5.10}$$

which is the analog of the continuity equation for an incompressible fluid. Hamiltonian systems satisfy (5.9) automatically provided that the y_i are chosen to be generalized co-ordinates and momenta. For the fluid systems considered here, (5.9) holds because A_{ijl} vanishes whenever two of its indices are equal. (The fluid equations can, in fact, be derived from Hamilton's principle (see, for example [17], p. 3). However, a close connection between the corresponding canonical equations (which involve functional derivatives) and (5.9) has not been demonstrated. In practice, it seems easiest to verify (5.9) directly as needed.)

Equation (5.10) implies that the volume occupied by a collection of phase particles always containing the same particles remains constant in time. This contraint, while important, is not confining. Consider, for example, a two-dimensional phase space in which P is initially constant within a compact region and zero outside (fig. 4). In a wide class of physical systems, which are

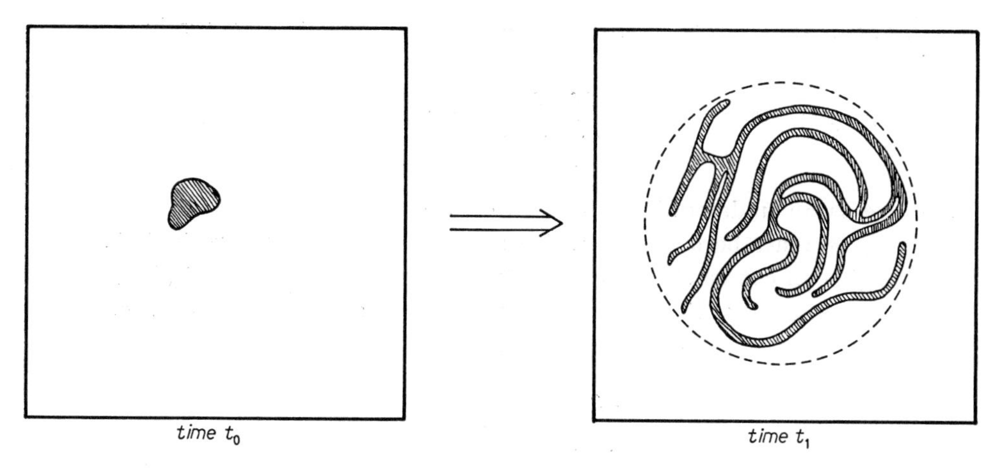

Fig. 4. – Mixing in a two-dimensional phase space.

said to « mix », the initially compact region « spreads out » by developing filamentous arms which gradually « fill up » the phase space. Now P is typically sought for computing the statical average of phase functions $F(y_1, y_2, ...,$ $..., y_N, t)$, *viz.*

$$\langle F\rangle = \int\int ... \int FP \prod_i \mathrm{d}y_i \,. \tag{5.11}$$

Suppose that P does indeed evolve from time t_0 to t_1, as shown in fig. 4. In practice, it is usually impossible to calculate $P(t_1)$ accurately from (5.10) for use in (5.11). However, it is obvious that, for any F that depends smoothly on its arguments, $\langle F\rangle$ at t_1 can be calculated to good accuracy by replacing $P(t_1)$ with a density function that is constant over the circular region of fig. 4. We, therefore, distinguish between $P(t)$, the probability density obtained from (5.10) by solution of (5.4) for a entire ensemble of systems, and $\hat{P}(t)$, the « phenomenological » or practical density, which can be considered a smoothed version of $P(t)$. Statistical mechanics seeks $\hat{P}(t)$ without first calculating $P(t)$. The mixing property of the dynamics motivates the concept of $\hat{P}(t)$, but it also imposes the consistency requirement that $\hat{P}(t)$ ought to be progressively more « spread out » at successively later times. This is a qualitative statement of the second law.

Entropy is a functional of $\hat{P}$ which measures the spread of $\hat{P}$, that is, by how much $\hat{P}$ differs from a delta-function in phase space. If the y_i can assume only discrete values y_{il}, then the entropy takes the form

$$S = -\sum_i \sum_l \hat{P}_{il} \ln \hat{P}_{il} \tag{5.12}$$

([18], p. 9). If the y_i assume continuous values, then (5.12) generalizes to

$$S = -\int\int ... \int \hat{P} \ln \hat{P} \cdot \mathcal{M} \prod_i \mathrm{d}y_i \,, \tag{5.13}$$

where $\mathcal{M}$ is an undetermined measure of the phase space. Because of (5.9), however, $\mathcal{M}$ must be a constant. To see this, specialize to the inviscid case and imagine an extremely clever observer for whom integration of (5.4), (5.10) presents no difficulty. For such an observer, $P = \hat{P}$ at all times, and the entropy should not increase. However, (5.10) implies that

$$\int\int ... \int P \ln P \cdot \mathcal{M} \prod_i \mathrm{d}y_i \tag{5.14}$$

can remain constant only if $\mathcal{M}$ is a constant.

Practical knowledge about a mixing system based upon partial (*i.e.* statistical) specification of its initial state always decreases in time, and the entropy, therefore, increases. However, information about quantities which are constants of the motion is never degraded in time. Absolute equilibrium is defined to be the state in which only certain integral constants of the motion are known. (Only linear and quadratic invariants are easily handled. However, these seem to be the most important.) Unfortunately, if $v \neq 0$, then (5.4) has no constants of the motion. Viscous ensembles cannot, therefore, approach absolute equilibrium. However, if the « mixing time » is short compared to the viscous time for at least a subset of phase co-ordinates, then qualitative features of absolute equilibrium can appear in realistic, dissipative flows.

We begin by calculating absolute equilibrium for two-dimensional turbulence. The constants of the motion are the energy

$$E = \sum_i E_i \equiv \sum_i y_i^2 \tag{5.15}$$

and the enstrophy

$$\Omega = \sum_i \Omega_i \equiv \sum_i k_i^2 y_i^2 . \tag{5.16}$$

To discover $\hat{P}$, maximize S subject to the requirements

$$\langle 1 \rangle = 1 , \qquad \langle E \rangle = E_0 , \qquad \langle \Omega \rangle = \Omega_0 , \tag{5.17}$$

where E_0 and Ω_0 are known initial values. The result is

$$\hat{P} = \exp[\lambda - \alpha E - \gamma \Omega] , \tag{5.18}$$

where λ, α, γ are constants determined from (5.17). Directly from (5.18)

$$\langle y_i^2 \rangle = \tfrac{1}{2}(\alpha + \gamma k_i^2)^{-1}, \tag{5.19}$$

which shows that the quantity $\alpha E_i + \gamma \Omega_i$ is equipartitioned among the modes. Equation (5.19) implies an energy spectrum proportional to

$$k/(\alpha + \gamma k^2) , \tag{5.20}$$

which corresponds to infinite total energy and enstrophy, because the integral of (5.20) diverges at large wave number. In three dimensions energy alone is equipartitioned, the equilibrium spectrum goes like k^2, and the divergence is even more catastrophic. These equilibrium states are, therefore, realizable only if the system is artificially truncated to finite N, as if, for example, all modes with wave numbers larger than some arbitrarily chosen cut-off k_c were excluded from the dynamics. The truncated system still satisfies (5.9). A thought experiment in which k_c is raised by finite increments, with the system allowed to re-equilibrate at each new value of k_c, then suggests that nonlinear interactions in three-dimensional turbulence would, acting by themselves, pass all of the energy out to infinite wave number. This conclusion seems rather tame, but similarly spirited interpretations of the formally divergent equilibrium states corresponding to more complicated dynamics yield useful and sometimes surprising predictions [19].

Consider, for example, the flow of a single layer of rotating fluid over topography. The dynamics are

$$\frac{\partial q}{\partial t} + J(\psi, q) = 0 , \tag{5.21}$$

where

$$q = \nabla^2 \psi + h(x, y) \tag{5.22}$$

and

$$h = f(y) + b(x, y) . \tag{5.23}$$

The notation is the same as in sect. **2**. In terms of the coefficients y_i defined by (5.2), the quadratic invariants are the energy,

$$E = \sum_i y_i^2 , \tag{5.24}$$

and the potential enstrophy (less a constant),

$$\Omega = \sum_i k_i^2 y_i^2 - 2 \sum_i k_i h_i . \tag{5.25}$$

Again the equilibrium state is multivariate Gaussian, but now there is a non-zero mean flow,

$$\langle y_i \rangle = \gamma k_i h_i / (\alpha + \gamma k_i^2) , \tag{5.26}$$

which is locked to the topography. The energy in mode i,

$$\langle y_i^2 \rangle = \tfrac{1}{2}(\alpha + \gamma k_i^2)^{-1} + \langle y_i \rangle^2 , \tag{5.27}$$

is enhanced by the energy in the average contour current. The transform of (5.26),

$$\left(\frac{\alpha}{\gamma} - \nabla^2\right) \langle \psi \rangle = h , \tag{5.28}$$

is useful if $h(x, y)$ has coherent form. Note that $\langle \psi \rangle$ is an exact steady solution to (5.21), (5.22).

If the nonlinear terms in (5.21) are suppressed, then the enstrophy invariant reduces to $\overline{h\nabla^2\psi}$, the mean flow vanishes if the latter is initially zero, and absolute equilibrium has energy equipartition. The topography scatters energy into higher wave numbers, but the linear dynamics prevents vortex stretching on slopes from accumulating vorticity of preferred sign on topographic peaks and troughs.

Numerical simulation of (5.21) with random topography and constant f confirm all of these qualitative predictions [20]. Figure 5 shows the enstrophy spectrum after 2.5 turn-over times in three independent experiments with identical, sharply peaked initial spectra. The three experiments correspond to *a*) no topography (long dashes), *b*) topography and nonlinearity of equal strength (solid line) and *c*) no nonlinearity (short dashes). The topography spectrum is hatched. The induced mean flow in case *b*) shows up as a spectral

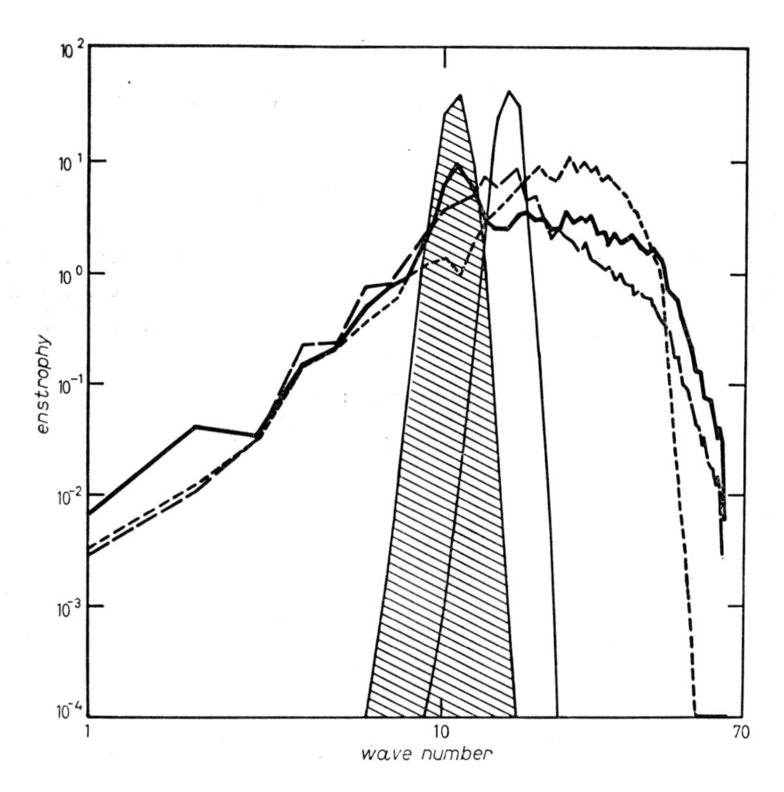

Fig. 5. – Enstrophy spectra of two-dimensional turbulence over topography after 2.5 turn-overs beginning from a narrow spectral peak. The three experiments correspond to no topography (long dashes, $R_0/\delta = \infty$), topography and nonlinearity of equal strength (solid, $R_0/\delta \approx 1$) and no nonlinearity (short dashes, $R_0/\delta \simeq 0$). The topography spectrum is hatched. Courtesy of G. HOLLOWAY.

bump at the topography wave numbers and as a visually apparent correlation between the topography and instantaneous streamfunction (fig. 6). Topographic scattering increases the enstrophy at high wave numbers in both cases *b*) and *c*), but more so in *c*), where the enstrophy transfer to high wave numbers is not linked to energy transfer to low wave numbers by an enstrophy invariant.

An interesting special case of (5.21), (5.22) is beta-plane flow in a rectangular flat-bottomed ocean, $0 < x < L$, $0 < y < L$. The mean-flow equation is

$$\left(\frac{\alpha}{\gamma} - \nabla^2\right)\langle\psi\rangle = \beta(y - y_0) \tag{5.29}$$

with $\langle\psi\rangle = 0$ on boundaries. The constants α, γ are determined from

$$E_0 = \langle E\rangle = \tfrac{1}{2}\sum_i (\alpha + \gamma k_i^2)^{-1} + \overline{\nabla\langle\psi\rangle\cdot\nabla\langle\psi\rangle} \tag{5.30}$$

and

$$\Omega_0 = \langle \Omega \rangle = \tfrac{1}{2} \sum_i k_i^2 (\alpha + \gamma k_i^2)^{-1} + \overline{\langle \nabla^2 \psi \rangle}^2 + 2\overline{f \langle \nabla^2 \psi \rangle} , \tag{5.31}$$

where E_0 and Ω_0 are the initial energy and enstrophy. (The constant y_0 can be considered a Lagrange multiplier corresponding to a third possible integral invariant, the average potential vorticity $\bar{q}$.) Let the initial state be a random field concentrated at wave number $k_* = O(L^{-1})$ and uncorrelated with latitude. Then $\Omega_0 = k_*^2 E_0$ and the initial Rossby number is small if $E_0 \ll \beta^2 L^4$.

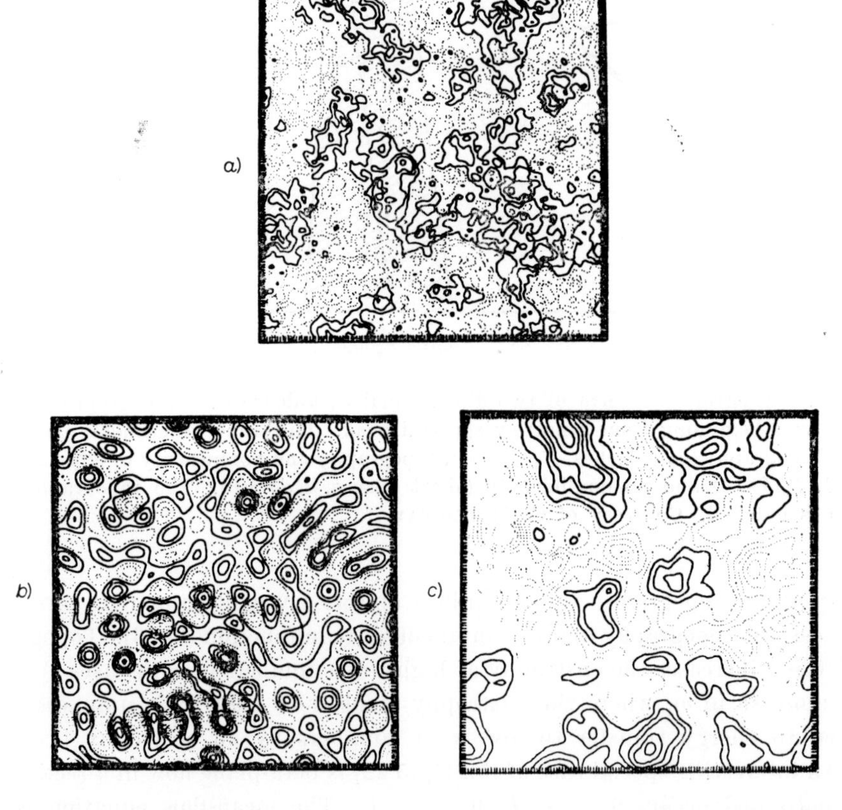

Fig. 6. – The topography (a)), initial streamfunction (b)) and the streamfunction after four turn-overs (c)). Courtesy of G. HOLLOWAY.

The equilibrium mean flow exceeds the initial bounds on energy or enstrophy unless

$$0 < \frac{\gamma}{\alpha} \equiv l^2 \ll L^2 . \tag{5.32}$$

Thus consistent solutions to (5.29) have a westward interior flow with speed βl^2 and boundary layers of thickness l and characteristic speed βLl. The total mean kinetic energy is shared equally by the boundary layers and the interior flow, but the boundary layers constitute the sole contribution of the mean flow to the enstrophy. This mean flow is the same as that derived by FOFONOFF [21] in another context. The transient energy is spread uniformly throughout the basin and the transient spectrum has a (broad) peak at wave number l^{-1}. Thus the eddies have the same scale as the boundary layer thickness and cannot grow larger without breaking the constraints. Note that the scale l is unrelated to the deformation radius. This small-scale energy originates from Rossby-wave reflection at the western boundary [22].

Multilayer equilibria show how energy gets spread among the vertical modes. For example, absolute equilibrium for the two-layer system with equal layer depths and the same initial energy spectrum in each layer has barotropic energy

$$U(\boldsymbol{k}) = 1/(\alpha + \gamma k^2) \tag{5.33}$$

and baroclinic energy

$$E(\boldsymbol{k}) = 1/[\alpha + \gamma(k^2 + k_{\mathrm{R}}^2)] \,. \tag{5.34}$$

Thus

$$E(k) = U(\sqrt{k^2 + k_{\mathrm{R}}^2}) \,. \tag{5.35}$$

In all cases of interest, the equilibrium $U(k)$ increases with decreasing k. Thus the equilibrium flow is nearly barotropic on scales larger than the deformation radius. Paradoxically, then, depth-invariant flow constitues the « most disordered » state for large-scale ocean currents. Strong barotropic flows are in fact observed. RHINES [23] describes a six-month mooring record in which a transient 20 cm/s current remained nearly depth-invariant through 4000 m depth over the entire observing period. The site was in the energetic Gulf Stream extension region, over a smooth abyssal plain. The generalization of (5.33), (5.34) to a N-layer system is

$$E_n(\boldsymbol{k}) = 1/[\alpha + \gamma(k^2 + k_n^2)] \,, \tag{5.36}$$

where k_n is given by (4.18). The higher vertical modes contain decreasing amounts of energy in equilibrium.

We next consider absolute equilibria for flows with variable Väisälä frequency and Coriolis parameter. These equilibria suggest how far the previously postulated « equatorial funneling effect » would proceed. Because $\mathcal{N}(z)$ and $f(y)$ are spatially inhomogeneous, it is now more natural to formulate the problem in physical space than in transform space. First, consider continuous stratification. The dynamics are

$$\frac{\partial q}{\partial t} + J(\psi, q) = 0 , \tag{5.37}$$

where

$$q = \mathscr{L}[\psi] + f(y) \tag{5.38}$$

and

$$\mathscr{L}[\psi] = \nabla^2 \psi + \frac{\partial}{\partial z}\left[\frac{f^2}{\mathscr{N}^2}\frac{\partial \psi}{\partial z}\right]. \tag{5.39}$$

The boundary conditions are

$$\partial\psi/\partial z = 0 \qquad \text{at } z = 0, -D , \tag{5.40}$$

corresponding to zero vertical velocity. It is temporarily convenient to expand ψ and q in the eigenmodes $\varphi_i(x, y, z)$ of the operator $\mathscr{L}$. These obey

$$\mathscr{L}[\varphi_i] + \lambda_i^2 \varphi_i = 0 \tag{5.41}$$

and

$$\frac{\partial \varphi_i}{\partial z} = 0 \qquad \text{at } z = 0, -D , \tag{5.42}$$

where i is the index for the eigenmodes and λ_i^2 is the corresponding eigenvalue. Then, if

$$\psi = \sum_i a_i \varphi_i \qquad \text{and} \qquad f = \sum_i f_i \varphi_i , \tag{5.43}$$

it follows from (5.39) that

$$q = \sum_i (-\lambda_i^2 a_i + f_i)\varphi_i . \tag{5.44}$$

Since $\mathscr{L}$ is self-adjoint, the φ_i can have the usual properties of orthonormality

$$\iiint \mathrm{d}\boldsymbol{x}\, \varphi_i(\boldsymbol{x})\, \varphi_j(\boldsymbol{x}) = \delta_{ij} \tag{5.45}$$

and completeness

$$\sum_i \varphi_i(\boldsymbol{x})\, \varphi_i(\boldsymbol{x}_0) = \delta(\boldsymbol{x} - \boldsymbol{x}_0) . \tag{5.46}$$

Motion in the phase space spanned by the a_i is nondivergent. The quadratic integral invariants are the energy

$$E = \sum_i \lambda_i^2 a_i^2 \tag{5.47}$$

and the enstrophy at level z

$$\Omega(z) = \sum_i \sum_j (\lambda_i^2 \lambda_j^2 a_i a_j - 2\lambda_i^2 a_i f_j) \iint \mathrm{d}x\, \mathrm{d}y\, \varphi_i(\boldsymbol{x})\, \varphi_j(\boldsymbol{x}) + \text{constants}\,. \tag{5.48}$$

The equilibrium probability distribution is

$$\hat{P} \propto \exp\left[-\alpha E - \int \gamma(z)\, \Omega(z)\, \mathrm{d}z\right], \tag{5.49}$$

where α and $\gamma(z)$ are Lagrange multipliers chosen such that $\langle E\rangle$ and $\langle\Omega(z)\rangle$ match their initial values. Since (5.49) is multivariate Gaussian, the equilibrium flow statistics are completely specified by the first and second moments $\langle a_i\rangle$ and $\langle a_i a_j\rangle$, or by the equivalent physical-space moments $\langle\psi(\boldsymbol{x})\rangle$ and $\langle\psi(\boldsymbol{x})\,\psi(\boldsymbol{x}_0)\rangle$. It is possible to get differential equations governing the latter directly from manipulations on (5.49). To derive these equations, note that (5.49) may be written in a standard form

$$P \propto \exp\left[-\tfrac{1}{2} \sum_i \sum_j A_{ij} a_i' a_j'\right], \tag{5.50}$$

where

$$A_{ij} = 2\alpha\lambda_i^2 \delta_{ij} + 2\lambda_i^2 \lambda_j^2 B_{ij}\,, \tag{5.51}$$

$$B_{ij} = \iiint \gamma(z)\, \varphi_i(\boldsymbol{x})\, \varphi_j(\boldsymbol{x})\, \mathrm{d}\boldsymbol{x} \tag{5.52}$$

and

$$a_i' \equiv a_i - \langle a_i\rangle\,. \tag{5.53}$$

Then

$$\langle a_i\rangle = 2 \sum_j \sum_l A_{ij}^{-1} \lambda_j^2 B_{jl} f_l \tag{5.54}$$

and

$$\langle a_i' a_j'\rangle = A_{ij}^{-1}, \tag{5.55}$$

where A^{-1} is the inverse of the matrix A. Now (5.55) implies that

$$\sum_i \sum_j \left[\sum_l A_{il} \langle a_l' a_j'\rangle\, \varphi_i(\boldsymbol{x})\, \varphi_j(\boldsymbol{x}_0)\right] = \sum_i \varphi_i(\boldsymbol{x})\, \varphi_i(\boldsymbol{x}_0) = \delta(\boldsymbol{x} - \boldsymbol{x}_0) \tag{5.56}$$

by (5.46). Substitution of (5.51) into the left-hand side of (5.56) and straightforward manipulations using (5.41)-(5.52) reduce (5.56) to

$$2\,\mathscr{L}[\gamma\,\mathscr{L} - \alpha]\, R(\boldsymbol{x}, \boldsymbol{x}_0) = \delta(\boldsymbol{x} - \boldsymbol{x}_0)\,, \tag{5.57}$$

where

$$R(\boldsymbol{x}, \boldsymbol{x}_0) \equiv \langle \psi'(\boldsymbol{x})\, \psi'(\boldsymbol{x}_0) \rangle \tag{5.58}$$

and $\mathscr{L}$ operates on $\boldsymbol{x}$. Similarly, the transform of (5.54) is

$$(\gamma \mathscr{L} - \alpha) \langle \psi \rangle + \gamma f = 0 \,, \tag{5.59}$$

which is analogous to (5.28). Equations (5.57), (5.59) completely determine the equilibrium state. For example, the equilibrium transient kinetic energy at $\boldsymbol{x}$ is

$$[\nabla \cdot \nabla_0 R(\boldsymbol{x}, \boldsymbol{x}_0)]_{\boldsymbol{x}_0 = \boldsymbol{x}} \,, \tag{5.60}$$

where

$$\nabla = \left(\frac{\partial}{\partial x}, \frac{\partial}{\partial y}\right) \quad \text{and} \quad \nabla_0 = \left(\frac{\partial}{\partial x_0}, \frac{\partial}{\partial y_0}\right). \tag{5.61}$$

The development leading up to (5.57), (5.59) can also be carried out directly in physical space, without reference to the eigenmodes of $\mathscr{L}$. Then, instead of the infinite-dimensional integrations, one has to deal with functional derivatives. Here the eigenmodes are purely a device for avoiding functional methods. An alternate device (which is less efficient in this case) would be to write the averages as functional integrals over physical space, replace derivatives by finite differences, perform the integrations required to get the moments of $\psi(\boldsymbol{x})$ and then pass back from differences to differentials. The same equations result.

Qualitative properties of the equilibrium state are obvious from (5.57). The operator $\mathscr{L}$ has the form of a diffusion operator with vertical-diffusion coefficient $f^2/\mathscr{N}^2$. Thus $R(\boldsymbol{x}, \boldsymbol{x}_0)$ corresponds to the equilibrium « temperature » distribution with a delta-function source at $\boldsymbol{x}_0$. The vertical boundaries are nonconducting. (The analogy is inexact because the diffusion operator acts twice. Also $\gamma(z)$ has general z-dependence.) We anticipate maximum transient energy in regions of strong stratification and small Coriolis parameter.

To investigate the funneling effect in its simplest form, consider N-layer quasi-geostrophic flow in an equatorial ($f = \beta y$) channel between walls at $y = \pm \pi L/2$. Let the layer depths and density jumps all be equal. Nondimensionalize the horizontal distance by L, time by L/U (where U is a characteristic velocity), vertical distance by the fluid depth D, and the streamfunction by UL. The nondimensional dynamics are

$$\frac{\partial q_i}{\partial t} + J(\psi_i, q_i) + \beta^* \frac{\partial \psi_i}{\partial x} = 0 \,, \tag{5.62}$$

where

$$q_i = \mathscr{L}_i[\psi_j]\,, \tag{5.63}$$

$$\mathscr{L}_i[\psi_j] = \nabla^2\psi_i + F_0 y^2 \sum_j T_{ij}\psi_j\,, \tag{5.64}$$

$$F_0 = N\beta^2 L^4/g'D\,, \qquad \beta^* = \beta L^2/U\,,$$

and

$$[T_{ij}] = \begin{bmatrix} -1 & +1 & & & & & \\ 1 & -2 & 1 & & & 0 & \\ & 1 & -2 & 1 & & & \\ & & \ddots & \ddots & \ddots & & \\ & 0 & & & 1 & -2 & 1 \\ & & & & & +1 & -1 \end{bmatrix}.$$

Let the boundary conditions be

$$\psi(x, y) = \psi(x + 2\pi, y) \tag{5.65}$$

and

$$\psi\left(x, \pm\frac{\pi}{2} + y\right) = -\psi\left(x, \pm\frac{\pi}{2} - y\right). \tag{5.66}$$

The analog of (5.57) is

$$2\mathscr{L}_i[\gamma_j\,\mathscr{L}_j[R_{mn}(\boldsymbol{x}, \boldsymbol{x}_0)] - \alpha R_{jn}(\boldsymbol{x}, \boldsymbol{x}_0)] = \delta(\boldsymbol{x} - \boldsymbol{x}_0)\,\delta_{in}\,. \tag{5.67}$$

Suppose γ_j is independent of j (corresponding to an approximately uniform vertical initial energy distribution). It is then convenient to introduce the vertical-mode variables $\hat{\psi}_s$ defined by

$$\psi_i = \sum_{s=0}^{N-1} \hat{\psi}_s(x, y)\,\varphi_s(i)\,, \tag{5.68}$$

where

$$\varphi_s(i) = \sqrt{\frac{2-\delta_{s0}}{N}}\cos\left[\left(i - \frac{1}{2}\right)\frac{s\pi}{N}\right], \qquad i = 1, \ldots, N\,, \; s = 0, \ldots, N-1\,. \tag{5.69}$$

We find that

$$\langle\hat{\psi}_s(x, y)\,\hat{\psi}_{s_0}(x_0, y_0)\rangle = \hat{R}_s(x, y; x_0, y_0)\,\delta_{ss_0} \tag{5.70}$$

obeys

$$\hat{\mathscr{L}}_s(\hat{\mathscr{L}}_s - \alpha/\gamma)\,\hat{R}_s = \frac{1}{2\gamma}\,\delta(x - x_0)\,\delta(y - y_0)\,, \tag{5.71}$$

where

$$\hat{\mathscr{L}}_s \equiv \nabla^2 - \varepsilon_s^{-2} y^2 \tag{5.72}$$

and

$$\varepsilon_s = \left[2\sqrt{F_0} \sin \frac{s\pi}{2N} \right]^{-1} \tag{5.73}$$

is the (nondimensional) s-th deformation radius. In this geometry $\sum_i \overline{q_i^2}$ and $\sum_i \overline{q_i y}$ are separately conserved. If the latter is zero initially, then there is no equilibrium mean flow, and the enstrophy invariant reduces to

$$\Omega = \sum_i \overline{\mathscr{L}_i[\psi_j]^2} = \sum_s \overline{\hat{\mathscr{L}}_s(\hat{\psi}_s)^2} \equiv \sum_s \hat{\Omega}_s \,. \tag{5.74}$$

Fig. 7. – The vertical-mode streamfunctions $\hat{\psi}_s$ in the equatorial channel ($0<x<2\pi$, $-\pi/2<y<\pi/2$) for the barotropic mode $s=0$ (bottom), $s=1$, $s=3$ and $s=5$ (top). The equator lies along the axis of the channel. *a*) weeks $=0$, *b*) weeks $=3.346$, *c*) weeks $=23.421$.

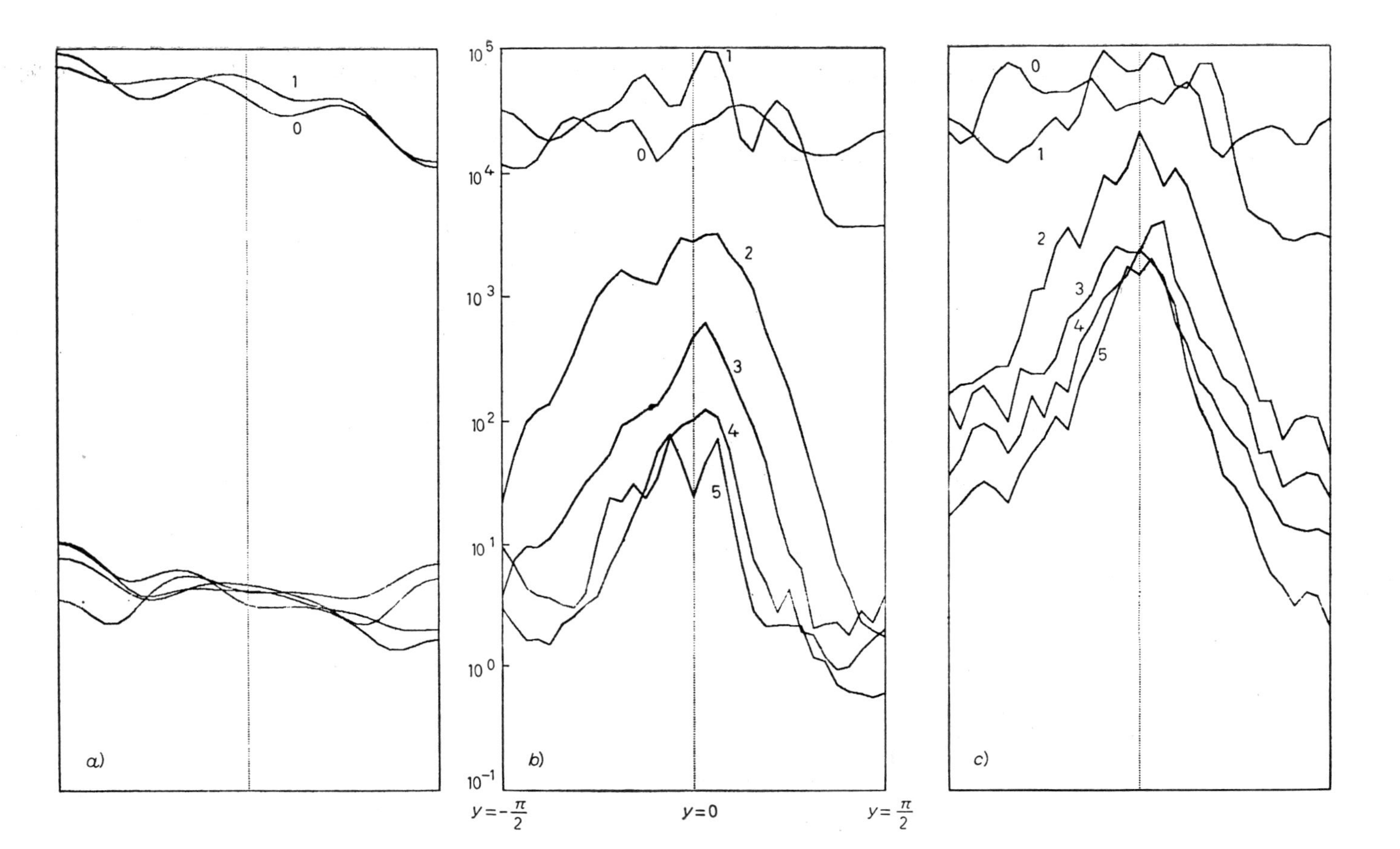

Fig. 8. - The kinetic energy averaged over x in the vertical modes $s = 0, 1, 2, 3, 4, 5$. The equator lies at $y = 0$. *a*) weeks = 0, *b*) weeks = 10, *c*) weeks = 30.

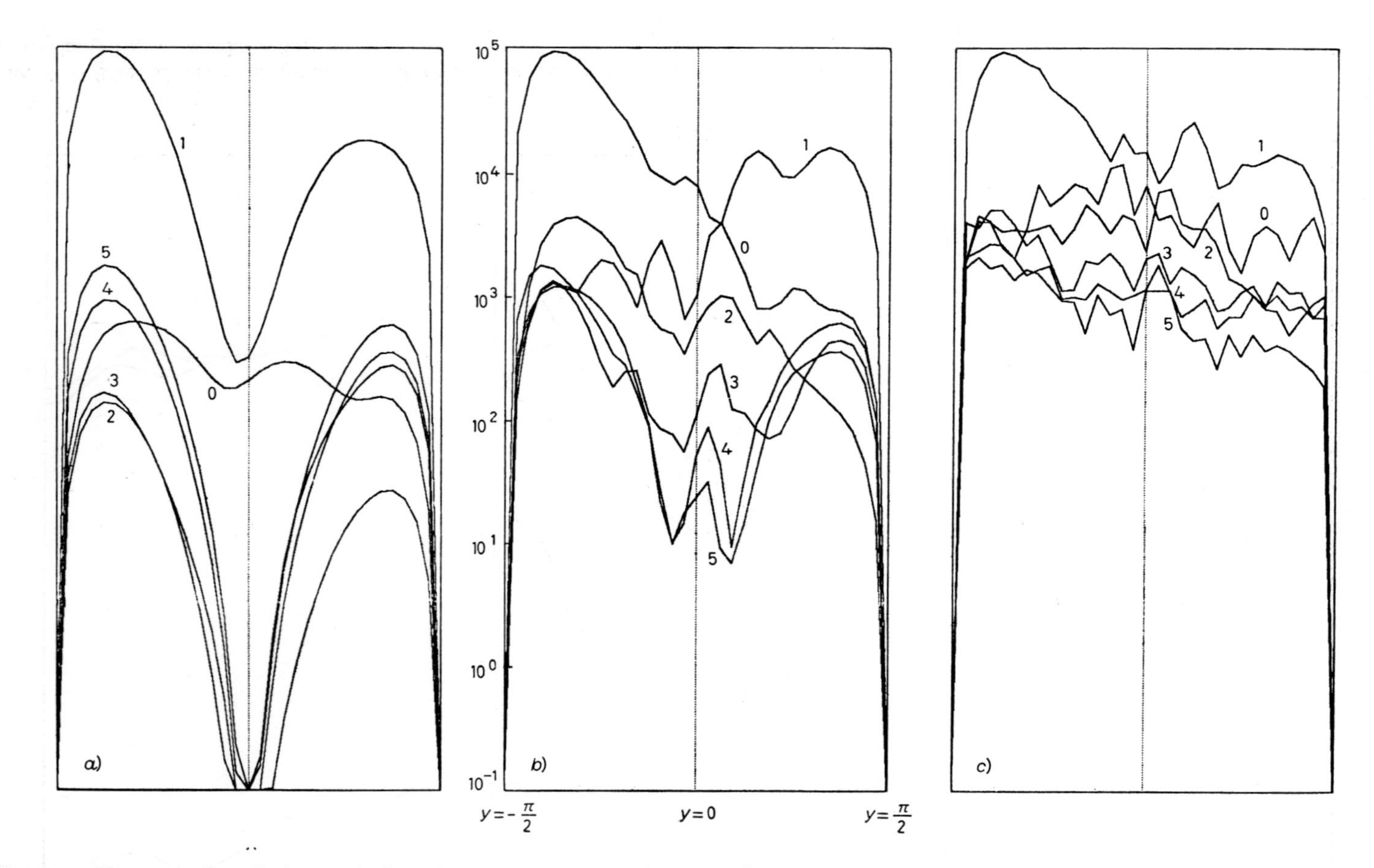

Fig. 9. – The quantity Ω_s in vertical modes $s = 0, 1, 2, 3, 4, 5$, *a*) weeks = 0, *b*) weeks = 10, *c*) weeks = 30.

For the barotropic mode ($s = 0$), the equilibrium energy is independent of latitude and maximum at wave number α/γ. For the baroclinic modes ($s \neq 0$), $\varepsilon_s^2 \alpha/\gamma$ is small and

$$(\nabla^2 - \varepsilon_s^{-2} y^2)^2 \hat{R}_s = \frac{1}{2\gamma} \delta(\boldsymbol{x} - \boldsymbol{x}_0) \tag{5.75}$$

has solutions of the form

$$\hat{R}_s = \frac{\varepsilon_s}{2\gamma} \Phi\left(\frac{\boldsymbol{x}}{\varepsilon_s^{\frac{1}{2}}}, \frac{\boldsymbol{x}_0}{\varepsilon_s^{\frac{1}{2}}}\right), \tag{5.76}$$

where

$$(\nabla^2 - y^2)^2 \Phi(\boldsymbol{x}, \boldsymbol{x}_0) = \delta(\boldsymbol{x} - \boldsymbol{x}_0) \tag{5.77}$$

(independent of ε_s). Thus the equatorial peak in the kinetic energy in vertical mode s has latitudinal width $\varepsilon_s^{\frac{1}{2}}$, or dimensionally

$$\left(\frac{g'H}{\beta^2}\right)^{\frac{1}{4}} \sin^{-\frac{1}{2}} \frac{s\pi}{2N}, \tag{5.78}$$

where H is the layer thickness. This region narrows with increasing s, but (5.60) and (5.76) imply that the equilibrium kinetic energy at $y = 0$ is independent of s in this case. Thus equatorial confinement is sharpest for the higher vertical modes. For $|y| \gg \varepsilon_s^{\frac{1}{2}}$ (5.75) is solved approximately by treating y^2 as a constant on the left-hand side. To the same order of approximation $\hat{\Omega}_s$ is independent of latitude.

Figures 7-9 show the evolution of a six-layer inviscid quasi-geostrophic ocean in the equatorial beta-plane channel. The initial streamfunction is random and the kinetic energy is invariant with depth and equally divided between the barotropic and first baroclinic modes. The channel is 6000 by 3000 km and the energy density is $1.2 \cdot 10^8$ erg·cm. After only 23 weeks, high vertical-mode kinetic energy is strongly trapped at the equator and $\hat{\Omega}_s$ is nearly independent of latitude.

6. – Closure.

Real turbulence is often far from the state of absolute equilibrium considered in the previous section. Unfortunately, no fully satisfactory nonequilibrium theory exists. Straightforward averaging of the equations of motion yields an unclosed hierarchy of statistical moment equations in which the evolution equation for the n-th moment always contains the $(n+1)$-th moment. This is the « closure problem ». All known methods for closing the moment equations involve additional physical assumptions not deducible from the equations of motion.

This section discusses relatively simple second-moment single-time closures which are representative of the closures used in application-oriented studies. These closures (and others of the « direct interaction » group) can be « derived » by a variety of perturbationlike procedures, none of which pretends to mathematical rigor. The derivations lead to moment equations which always resemble Boltzmann's collision equation for a gas of hard spheres. This structure is virtually guaranteed, if the moment equations are to satisfy certain consistency properties. However, the best justification for these closures is the usually good agreement between their predictions and direct computer simulations of the equations of motion [24].

A traditional goal of closure theory is to extend the spatial resolution (or Reynolds number) beyond that attainable in direct numerical simulations of the equations of motion. However, quasi-geostrophic dynamics obtain over less than three wave number octaves in the ocean and two in the atmosphere. This resolution is within the reach of modern computers. Still, closure is a useful tool for interpreting the statistics obtained by averaging the direct simulations. For example, closure theory can compute the effects of one statistic on another in a way that permits analytical simplifications after the dominant terms are identified. Unfortunately, the complexity of the theory can increase prohibitively if statistical symmetries (such as spatial homogeneity) are relaxed.

Let the equations of motion again be

$$\dot{y}_i = \sum_{j,l} A_{ijl} y_j y_l - v_i y_i \,, \tag{6.1}$$

where

$$A_{ijl} = A_{ilj} \,, \tag{6.2}$$

with no loss in generality. The coupling coefficients also satisfy the Liouville property

$$A_{ijj} = 0 = A_{iij} \,. \tag{6.3}$$

For simplicity we assume

$$\langle y_i \rangle = 0 \qquad \text{and} \qquad \langle y_i(t)\, y_j(t) \rangle = Y_i(t)\, \delta_{ij} \,, \tag{6.4}$$

which are equivalent to horizontal statistical homogeneity.

The eddy-damped Markovian (EDM) model [25, 26] closes the moment hierarchy by discarding fourth cumulants. A « cumulant » is the difference between a moment and the value it would have if all the variables were Gaussian. Direct averaging of (6.1) yields

$$\left(\frac{\mathrm{d}}{\mathrm{d}t} + 2v_i\right) Y_i = 2 \sum_{j,l} A_{ijl} \langle y_i y_j y_l \rangle \tag{6.5}$$

and

$$(6.6)\qquad \left(\frac{\mathrm{d}}{\mathrm{d}t}+v_i+v_j+v_l\right)\langle y_i y_j y_l\rangle =$$

$$=\sum_{m,n}\{A_{imn}\langle y_m y_n y_j y_l\rangle + A_{jmn}\langle y_m y_n y_i y_l\rangle + A_{lmn}\langle y_m y_n y_j y_i\rangle\}\,.$$

If A, B, C, D are Gaussian variables, then

$$(6.7)\qquad \langle ABCD\rangle = \langle AB\rangle\langle CD\rangle + \langle AC\rangle\langle BD\rangle + \langle AD\rangle\langle BC\rangle\,.$$

Apply this factorization to the right-hand side of (6.6), solve (6.6) for $\langle y_i y_j y_l\rangle$, and substitute the result into (6.5). Then (6.5) becomes

$$(6.8)\qquad \left(\frac{\mathrm{d}}{\mathrm{d}t}+2v_i\right)Y_i=\sum_{j,l}\int_0^t \mathrm{d}s\,\exp\left[-(t-s)(v_i+v_j+v_l)\right]\cdot$$

$$\cdot\{4(A_{ijl})^2\,Y_j(s)\,Y_l(s)+8A_{ijl}A_{jil}\,Y_l(s)\,Y_i(s)\}\,.$$

Unfortunately, numerical solutions of (6.8) lead to unphysical large negative energy ($Y_i<0$) when the Reynolds number is large. To see why, identify subscript i with wave number $\boldsymbol{k}$, and imagine that the initial spectrum has a sharp peak (fig. 10). Soon after $t=0$, the right-hand side of (6.8) is large and negative for wave numbers inside the peak and large and positive for those outside. At even later times, the magnitude of the term in curly brackets in (6.8) decreases, but the fluid still remembers the initial large tendencies, because the time integration in (6.8) runs all the way back to $t=0$. The limit of large Reynolds number corresponds to v_i, v_j, $v_l\to 0$ in (6.8), and in this limit the values of the term in curly brackets in the distant past ($s\ll t$) are weighted equally with those of the near past ($s\approx t$). Consequently the large initial tendencies are never forgotten and the plunging spectral peak shoots right through the zero level. The cause of this « sling shot » effect is easily seen to be the inefficiency of memory cut-off in the model as the $v_i\to 0$.

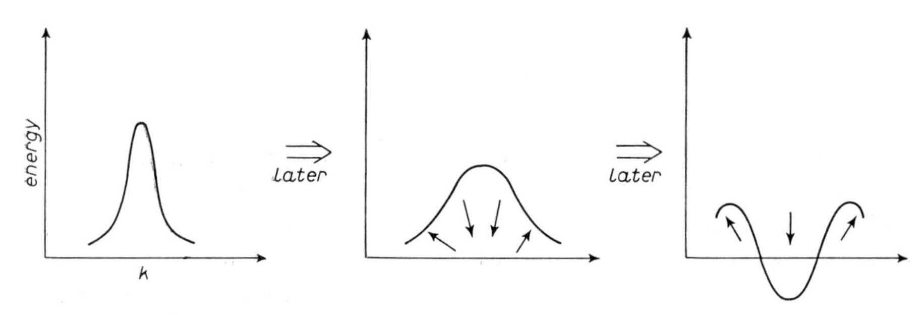

Fig. 10. – Failure of quasi-normal closure (6.8).

In real turbulence, past states are forgotten not because of the viscosity but as a result of nonlinear scrambling of the fluid by itself. At large Reynolds numbers the memory time should become independent of the viscosity, in contradiction with (6.8). These remarks suggest that (6.8) can be repaired by replacing v_i, v_j, v_l, on the right-hand side only, with augmented viscosities μ_i, μ_j, μ_l which measure the scrambling effect of the flow on itself. The μ_i differ from the familiar « eddy viscosity » of fluid mechanics in that they enter (6.6) but not (6.5) and, therefore, do not directly augment the dissipation of energy. Since the μ_i ought to be increasing functions of the Reynolds number, at large Reynolds number it might become accurate to replace the term in curly brackets of (6.8) (with the μ_i inserted) with its value at $s = t$. The latter (Markovization) step eliminates nonsimultaneous covariances. The result is

$$\left(\frac{\mathrm{d}}{\mathrm{d}t} + 2v_i + 2n_i\right)Y_i = \sum_{j,l} \theta_{ijl}\{4(A_{ijl})^2 Y_j Y_l\} , \tag{6.9}$$

where

$$n_i = -4 \sum_{j,l} \theta_{ijl} A_{ijl} A_{jil} Y_l , \tag{6.10}$$

$$\frac{\mathrm{d}}{\mathrm{d}t}\theta_{ijl} = 1 - (\mu_i + \mu_j + \mu_l)\theta_{ijl} \tag{6.11}$$

and

$$\theta_{ijl}(0) = 0 , \tag{6.12}$$

corresponding to precisely Gaussian initial conditions.

Equations (6.9)-(6.12) comprise the EDM. It is easy to show that the EDM does not allow $Y_i < 0$. Note that

$$\theta_{ijl} \to \frac{1}{\mu_i + \mu_j + \mu_l} \qquad \text{as } t \to \infty . \tag{6.13}$$

The parameters μ_i can be prescribed heuristically. For example,

$$\mu_k^2 = \int_0^k k^2 \mathscr{E}(k, t)\, \mathrm{d}k \tag{6.14}$$

equates the memory time to the time required for larger-scale flow to strain an eddy of size k^{-1}.

The EDM has several desirable consistency properties. First, it automatically conserves any quadratic invariant which is conserved by the exact dynamics. Second, it predicts only realizable statistics. (Negative energies can never appear.) Third, EDM is consistent with the second law. This third property requires elaboration. From the point of view of EDM, the $Y_i(0)$ are given initially and the closure predicts $Y_i(t)$ at all future times. However, the initial conditions correspond to many possible exact specifications of the y_i which together fill a finite volume in phase space like that of fig. 4. Since the averages predicted by EDM are phenomenological (no pretense is made of actually solving the equations of motion), EDM must be consistent with the principle of continuous increase in entropy. To calculate the entropy at time t corresponding to a knowledge of all the $Y_i(t)$, maximize S subject to the normalization requirement and

$$\langle y_i^2 \rangle = Y_i\,, \qquad \text{all } i\,. \tag{6.15}$$

The result is

$$\hat{P} = \exp\left[\lambda - \sum_i \alpha_i y_i^2\right], \tag{6.16}$$

where

$$\lambda = \tfrac{1}{2}\sum_i \ln(\alpha_i/\pi) \tag{6.17}$$

and

$$\alpha_i = 1/2\,Y_i\,. \tag{6.18}$$

The entropy is thus

$$S = -\frac{1}{2}\sum_i\left[\ln\left(\frac{\alpha_i}{\pi}\right) + \frac{1}{2}\right], \tag{6.19}$$

and we must, therefore, require

$$\frac{\mathrm{d}}{\mathrm{d}t}\sum_i \ln Y_i \geqslant 0\,, \tag{6.20}$$

where the tendencies are those resulting from the nonlinear terms in the closure. Note that absolute equilibrium may be deduced directly by maximizing the sum in (6.20) subject to total-energy and enstrophy conservation.

To see that EDM satisfies these consistency properties, note that (6.9), (6.10) can also be obtained by the following formal procedure. Let $v_i = 0$ and expand the exact dynamics in a Taylor series about $t = 0$:

$$y_i(t) = y_i(0) + t\sum_{j,l} A_{ijl}\, y_j(0)\, y_l(0) + t^2 \sum_{j,l}\sum_{m,n} A_{ijl} A_{lmn}\, y_j(0)\, y_m(0)\, y_n(0) + \dots . \tag{6.21}$$

Let the $y_i(0)$ be Gaussian. Then squaring and averaging (6.21) gives

$$Y_i(t) = Y_i(0) + 2t^2 \sum_{j,l} (A_{ijl})^2\, Y_j(0)\, Y_l(0) + \\ + 4t^2 \sum_{j,l} A_{ijl} A_{jil}\, Y_l(0)\, Y_i(0) + O(t^4)\,. \tag{6.22}$$

Truncate (6.22) after $O(t^2)$, differentiate with respect to time, and then restore all arguments to t. After these alterations (6.22) reads

$$\dot{Y}_i(t) = \sum_{j,l} t\{4(A_{ijl})^2\, Y_j(t)\, Y_l(t) + 8A_{ijl}A_{jil}\, Y_l(t)\, Y_i(t)\}\,, \tag{6.23}$$

which becomes identical with (6.9), (6.10) if the factor t is replaced by the triad decorrelation time θ_{ijl}. The conservation and realizability properties hold for (6.23), because the same properties hold term by term for the Taylor series expansion of the exact dynamics. The generalization to (6.9), (6.10) is easy. (Sum the contributions of individual triads.) The inequality (6.20) can be verified directly, but also as follows. The difference between the entropy at time t and time zero is

$$S(t) - S(0) = -\int [\hat{P}(t)\ln \hat{P}(t) - \hat{P}(0)\ln \hat{P}(0)] \tag{6.24}$$

(by definition),

$$= -\int [\hat{P}(t)\ln \hat{P}(t) - P(0)\ln P(0)] \tag{6.25}$$

(because the initial distribution is Gaussian by assumption),

$$= -\int [\hat{P}(t)\ln \hat{P}(t) - P(t)\ln P(t)] \tag{6.26}$$

(by (5.9)). Thus

$$S(t) - S(0) \geqslant 0\,, \tag{6.27}$$

because $\hat{P}(t)$ is, by definition, the distribution that maximizes $S(t)$. As $t \to 0$, (6.23) becomes exact. Therefore, (6.23) must be consistent with $\mathrm{d}S(t)/\mathrm{d}t > 0$.

It is useful to realize that the EDM closure is exact for the model equation

$$\left(\frac{\mathrm{d}}{\mathrm{d}t} + 2n_i\right) y_i = W(t) \sum_{j,l} (\theta_{ijl})^{\frac{1}{2}} y_j^{\mathrm{G}} y_l^{\mathrm{G}}\,, \tag{6.28}$$

where n_i is given by (6.10), $W(t)$ is a white-noise process with

$$\langle W(t)\, W(t')\rangle = 2\delta(t - t')\,, \tag{6.29}$$

and y_i^{G} is a Gaussian field with the same covariance as y_i. Conversely, if (6.28) is proposed *a priori* as a stochastic model for (6.1), then the only choice for the nonrandom damping coefficient n_i that satisfies the consistency properties is that given by (6.10).

The derivation of (6.23) by expansion in time is very similar to Hasselmann's original derivation of his weak wave equations. The difference is, loosely speaking, that the leftover factor of t in (6.23) is «used up» by the requirement that the interactions be resonant. In fact, waves incorporate easily into EDM [27]. Suppose, for example, that (6.1) is generalized to

$$\dot{y}_i + \sqrt{-1}\,\omega_i = \sum_{j,l} A_{ijl}\, y_j y_l\,, \qquad \omega_i = \text{const}\,. \tag{6.30}$$

If the steps leading to (6.9)-(6.12) are repeated, (6.9), (6.10) are unchanged, but (6.13) becomes

$$\theta_{ijl} \to \frac{\mu_i + \mu_j + \mu_l}{(\mu_i + \mu_j + \mu_l)^2 + (\omega_i + \omega_j + \omega_l)^2} \qquad \text{as } t \to \infty\,. \tag{6.31}$$

If weak turbulence corresponds to the limit $\mu \to 0$, then

$$\theta_{ijl} \to \pi\delta(\omega_i + \omega_j + \omega_l) \tag{6.32}$$

and EDM reduces to the weak wave equations (assuming the possibility of tertiary resonant interactions). However, it may be unrealistic to eliminate off-resonant interactions completely. For example, particular dispersion relations might allow no pathway to absolute equilibrium.

The above closures can be viewed as abridgments of Kraichnan's [28] direct interaction approximation (DIA). The independent variables in the DIA are the nonsimultaneous covariances $\langle y_i(t)\, y_i(t')\rangle$ and an averaged Green's function which measures the response of each mode to an infinitesimal excitation in the presence of all the other modes. Loosely speaking, this response function replaces the decorrelation time factors θ_{ijl} above. The DIA is entirely self-contained; there are no unspecified parameters or undetermined constants. Unfortunately, however, the DIA demands integrations back over the lag times and these greatly increase computing requirements. Additionally, the decorrelation time implied by the DIA response functions confuses the memory time scale for true flow distortion with the advective time scale for sweeping of small structures past a fixed point by larger eddies in the flow. This defect is sometimes described by the statement that «the DIA is not invariant with respect to random Galilean transformations». Because of Galilean noninvariance, the DIA erroneously predicts a $k^{-3/2}$ inertial-range power law instead of the $k^{-5/3}$ Kolmogorov range.

The Markovian closures require many fewer computations and storages than DIA and the freedom to specify the decorrelation times μ_i^{-1} offers a way to ensure Galilean invariance. The test field model [29] separates the effects of random displacements from true distortion by relating the decorrelation time to the time required for an initially incompressible flow to develop a compressible component if pressure forces (and the constraint to conserve mass) are suddenly switched off. For a sufficiently steep spectrum the test field prescription reduces to (6.14). Unfortunately, the test field model has no clear generalization to more complicated dynamics. However, many results are rather insensitive to the precise form of θ_{ijl}.

Closure applications relevant to geostrophic turbulence include studies of the two-dimensional inertial ranges [30], eddy viscosity [31], predictability [32, 33], the decay of anisotropy [34], two-dimensional turbulence on the beta-plane [20, 27] and over random topography [35, 36], decaying stratified quasi-geostrophic turbulence [37] and forced two-layer turbulence [10, 38]. The next section considers beta-plane turbulence in some detail.

7. – Beta-plane turbulence.

Consider two-dimensional turbulence on the beta-plane, governed by

$$\frac{\partial \varrho}{\partial t} + J(\psi, \varrho) + \beta \frac{\partial \psi}{\partial x} = v\nabla^2 \varrho , \tag{7.1}$$

where

$$\varrho = \nabla^2 \psi . \tag{7.2}$$

These equations differ from (3.1) only in the presence of the linear beta term, which gives rise to Rossby waves. The Rossby-wave properties of dispersion and anisotropy make beta-plane turbulence an ideal testing ground for general ideas about waves in turbulence. Let the flow be infinitely periodic, that is

$$\psi(x, y) = \psi(x + L_p, y) = \psi(x, y + L_p) \tag{7.3}$$

for all x and y and some L_p. Then the integral invariants (energy and enstrophy) and absolute-equilibrium states are the same as for two-dimensional turbulence, and are independent of beta. The nonequilibrium flow is, however, drastically affected by the Rossby waves. At the scales of motion at which the wave steepness (the ratio of particle to phase speeds) falls below unity, wave dispersion inhibits nonlinear interactions and the flow favors zonal (east-west) currents. The supposedly isotropic final state is thus approached via phase-space pathways which exhibit considerable anisotropy.

The Fourier transform of (7.1) is

$$\left(\frac{\mathrm{d}}{\mathrm{d}t} + i\omega_{\boldsymbol{k}}\right)\psi_{\boldsymbol{k}} + \sum_{\boldsymbol{p}+\boldsymbol{q}=\boldsymbol{k}} (p_x q_y - q_x p_y)\, q^2/k^2\, \psi_{\boldsymbol{p}}\, \psi_{\boldsymbol{q}} = 0 , \tag{7.4}$$

where

$$\omega_{\boldsymbol{k}} = -\beta k_x / k^2. \tag{7.5}$$

In the small-amplitude limit, solutions to (7.4) are superposed Rossby waves with the dispersion relation (7.5). These waves all have westward phase velocities and wave periods which decrease as the wavelength increases (for fixed direction of propagation). For fixed frequency, nondimensionalize $\boldsymbol{k}$ by β/ω. Then (7.5) is a circle on the (k_x, k_y)-plane with center at $(-\frac{1}{2}, 0)$ and unit diameter. The group velocity is

$$\left(\frac{\partial\omega}{\partial k_x}, \frac{\partial\omega}{\partial k_y}\right) = \frac{\omega^2}{\beta}\left(\frac{2k^2-1}{k^2}, \pm\frac{\sqrt{1-k^2}}{k}\right). \tag{7.6}$$

Thus long waves ($k < 2^{-1/2}$) carry energy westward at a speed that approaches $\omega^2/\beta k^2$ as $k \to 0$. Short waves move energy eastward at a much slower speed. The region affected by an initially concentrated disturbance, therefore, lies principally to the west. This property explains why intense boundary currents like the Gulf Stream are found at western boundaries.

For motions of length scale L and characteristic velocity U, the size ratio of the beta term to the nonlinear terms in (7.1) is $\beta L^2/U$. RHINES [9, 22] proposed that the wave number $k_\beta \equiv \sqrt{\beta/U}$ typically separates a « wave regime » at $k < k_\beta$ from a « turbulence regime » at $k > k_\beta$. In the ocean, $k_\beta^{-1} \approx 70$ km. Suppose that the energy is initially concentrated at wave number $k_1 \gg k_\beta$. By the reasoning of sect. **3**, which holds whether beta is present or not, the energy moves steadily into lower wave numbers. When the energy reaches k_β, however, Rossby waves begin to propagate freely, as the wave period becomes shorter than an eddy turn-over time. The transition is abrupt because the Rossby-wave dispersion relation (7.5) and the « dispersion relation » for the turbulence, $\omega = Uk$, have opposite slopes. The analog of (6.8) for (7.4) is

$$\begin{aligned}\frac{\mathrm{d}}{\mathrm{d}t}U_{\boldsymbol{k}} = \sum_{\boldsymbol{p}+\boldsymbol{q}=\boldsymbol{k}} \frac{|\boldsymbol{k}\times\boldsymbol{p}|^2}{k^2p^2q^2}\int_0^t &\mathrm{d}s \exp[-(\mu_{\boldsymbol{q}}+\mu_{\boldsymbol{p}}+\mu_{\boldsymbol{k}})(t-s)]\cdot\\ &\cdot\cos[(\omega_{\boldsymbol{q}}+\omega_{\boldsymbol{p}}+\omega_{-\boldsymbol{k}})(t-s)]\cdot\\ &\cdot\{(p^2-q^2)^2\,U_{\boldsymbol{p}}(s)\,U_{\boldsymbol{q}}(s)-2(q^2-p^2)(k^2-p^2)\,U_{\boldsymbol{k}}(s)\,U_{\boldsymbol{p}}(s)\},\end{aligned} \tag{7.7}$$

where $U_{\boldsymbol{k}}$ is the energy in wave number $\boldsymbol{k}$ and $\mu_{\boldsymbol{k}}^{-1}$ is the time for the turbulence to deform an eddy of size k^{-1}. In the wave regime, $\omega \gg \mu$ and the oscillating factor in (7.7) greatly reduces energy transfer unless

$$\omega_{\boldsymbol{p}} + \omega_{\boldsymbol{q}} = \omega_{\boldsymbol{k}}, \tag{7.8}$$

which is the condition for resonance. We thus expect energy transfer to lower wave numbers to proceed much more slowly in the wave regime. Nondispersive waves would satisfy (7.8) automatically.

Markovization of (7.7) gives the analog of (6.9), namely

$$\frac{d}{dt} U_k = \sum_{p+q=k} \frac{|k \times p|}{k^2 p^2 q^2} \theta_{kpq}\{\text{same as (7.7)}\}, \tag{7.9}$$

where

$$\theta_{kpq} = \frac{\mu_k + \mu_p + \mu_q}{(\mu_k + \mu_p + \mu_q)^2 + (\omega_p + \omega_q + \omega_{-k})^2}. \tag{7.10}$$

Again,

$$\theta_{kpq} \to \pi\delta(\omega_p + \omega_q + \omega_{-k}) \tag{7.11}$$

in the limit of weak turbulence ($\mu/\omega \to 0$). The resonance condition (7.8) can be viewed as a selection rule on frequencies analogous to

$$p + q = k \tag{7.12}$$

for wave numbers. The two together are a formidable constraint on triads that can transfer energy [39]. However, perfect resonance is too stringent a requirement. Instead, (7.10) shows that, if (7.8) is satisfied to within $O(\mu)$, then the interaction is as good as resonant. In the turbulence regime ($\mu \gg \omega$) the efficiency of energy transfer decreases with increasing μ, but in the wave regime ($\mu \ll \omega$) the efficiency can increase with μ, because the turbulence allows energy transfer in slightly off-resonant triads.

The appearance of anisotropy at low wave numbers is also expected. Consider a triad of wave vectors satisfying (7.12) with $p \geqslant q \geqslant k$. We continue to be interested in the case in which average energy transfer is into the lowest wave number k. Let $\varepsilon = k/p$ be small. Then

$$q = -p(1 + O(\varepsilon)) \tag{7.13}$$

and

$$\omega_p + \omega_q + \omega_{-k} = -\frac{\beta}{k}\left(\frac{k_x}{k} + O(\varepsilon^2)\right). \tag{7.14}$$

The ω_p, ω_q terms are of minor importance for two reasons. First, they cancel in sign. Second, the wave frequency is inversely proportional to wave number. If k lies within the wave regime, then this reasoning suggest dominant transfer into wave vectors k with $k_x/k_y = O(\varepsilon^2)$. Weak wave theory predicts that only the highest-frequency component is unstable to infinitesimal perturbations in an isolated triad [40]. RHINES [22] thus notes that zonal anisotropy can also be anticipated as the necessary consequence of energy transfer to low wave numbers *and* low frequencies.

Numerical experimental [9, 20, 22, 27] confirm these qualitative predictions. Figure 11 shows the enstrophy spectra after six turn-over times in three independent experiments with zero ($k_\beta = 0$), moderate ($k_\beta \approx 2$) and strong beta ($k_\beta \approx 4$). The common initial condition is a narrow spectral peak at $k = 10$. Zonal anisotropy is apparent in both the streamfunction and vorticity

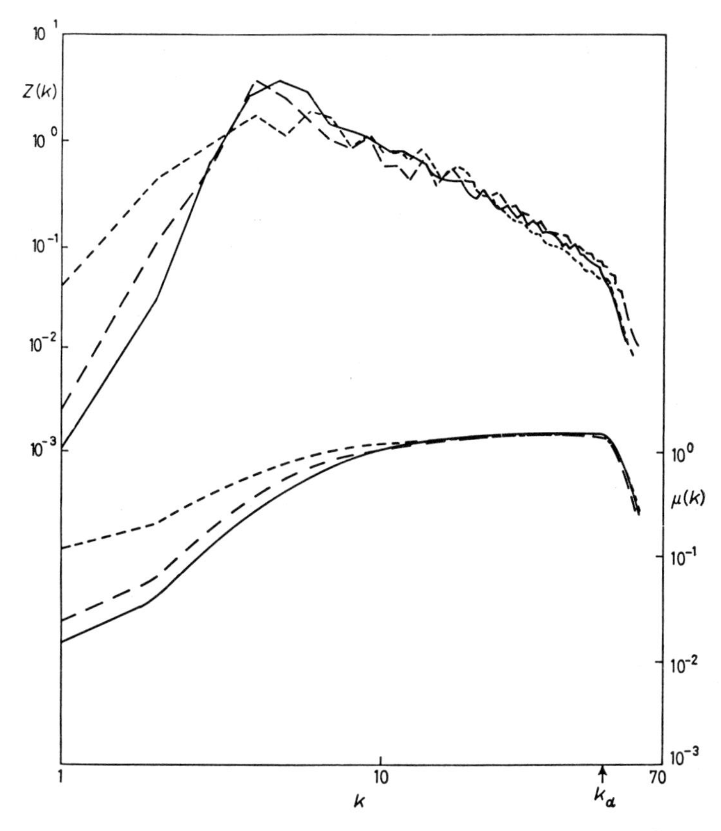

Fig. 11. – Enstrophy spectra after six turn-overs of beta-plane turbulence with zero (short dashes), moderate ($= 12.5$) (long dashes) and strong ($= 25$) (solid) beta. Courtesy of G. HOLLOWAY.

fields in the beta cases (fig. 12). HOLLOWAY [27] examines these experiments using the closure (7.9) with μ_k given by the test field model. He approximates the modal enstrophy spectrum $\langle \varrho_k \varrho_{-k} \rangle$ with the truncated angular expansion

$$2\pi k \langle \varrho_k \varrho_{-k} \rangle = Z(k)\big(1 - R(k) \cos 2\varphi_k\big) , \tag{7.15}$$

where φ_k is the angle between $\boldsymbol{k}$ and the k_x-axis. Positive $R(k)$ corresponds to zonal flow. Figure 13 shows $R(k)$ for the experiment with moderate beta along with closure estimates of $R(k)$ for three values of the adjustable para-

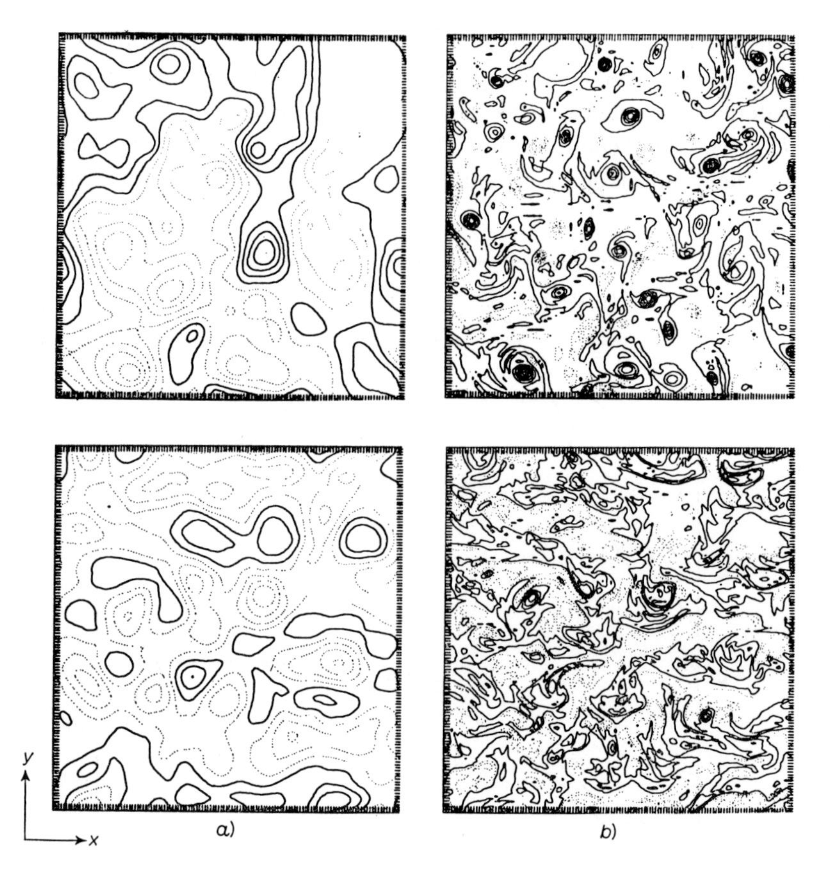

Fig. 12. – Streamfunction (a)) and vorticity (b)) fields after six turn-overs in the experiments with zero (top) and strong ($=25$) beta (bottom). Courtesy of G. HOLLOWAY.

meter in the test field model. As expected, zonal anisotropy is largest for $k < k_\beta$, but $R(k)$ has a secondary maximum at large k. Closure theory traces the high-wave-number anisotropy to the indirect effect of straining of the small-scale eddies by the large-scale zonal motion. The importance of this phenomenon is anticipated by Herring's [34] study of the decay of anisotropy in two-dimensional turbulence.

An isolated energetic patch within a broad quiescent region of fluid offers maximum contrast to the initially narrow spectral peak. Numerical experiments without beta confirm that both the patch diameter and dominant eddy size increase until the patch holds too few eddies to maintain turbulence [9]. Noting that the westward group velocity of large Rossby waves is proportional to the square of the wavelength, RHINES [22] conjectures that a turbulent patch on the beta-plane would, after an initial period of possibly slow growth, radiate energy quickly over a wide area. Turbulence would cease as particle speeds decrease within the growing patch. Approximately the maxim holds:

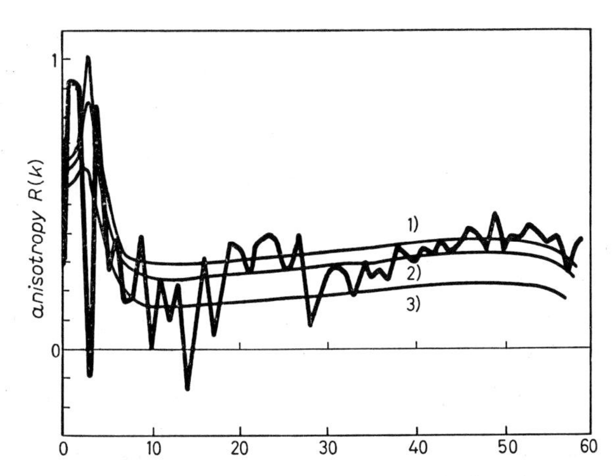

Fig. 13. – The simulated (heavy curve) and theoretical anisotropy spectrum $R(k)$ for beta-plane turbulence. The theoretical curves corresponds to three values of the adjustable parameter in the test field model: 1) $\lambda = 0.6$, 2) $\lambda = 0.7$, 3) $\lambda = 1.0$. Courtesy of G. HOLLOWAY.

Waves move energy rapidly through physical space, but inhibit energy transfer in wave number space. Turbulence the opposite.

The end state of beta-plane turbulence is controversial. (The question may be well be irrelevant to the relatively highly damped geophysical fluids, but not perhaps to the circulation on Jupiter [41].) The absolute-equilibrium state is a « best guess » by the criteria in sect. **5**, but it could be misleading if « mixing » in phase space ceases after the fluid enters the wave regime. RHINES [22] suggests steady zonal currents with alternating flow directions and length scale k_β^{-1} that would satisfy the linear-stability criterion

$$\beta - u_{yy} \neq 0 .$$

Rossby wave solitons are another, intriguing possibility.

8. – Comments.

Two themes pervade these lectures. The first is that weak, integral statements of fundamental conservation principles can anticipate distinctive flow characteristics. The idea succeeds because potential-enstrophy conservation is a strong constraint on quasi-geostrophic flow, and because higher-order invariants (ultimately, potential-vorticity conservation on particles) have apparently spotty projections on the phase-space manifolds corresponding to fixed values of the energy and enstrophy.

The second theme is the common tendency for natural systems to seek a state of maximum disorder. This principle crept almost unnoticed into the discussion of sect. **3** in the suppositions that a narrow spectral peak would broaden, and that material lines and wave vectors lengthen on the average. It attained the status of a mathematical inequality (the second law) in sect. **5**, and in sect. **6** combined with the weak-conservation principle to form the basis for closure.

The statistical-moment hierarchy was closed at the considerable price of replacing the ensemble of realizations of the exact dynamics by an ensemble of realizations of a stochastic model equation which conserves squared vorticity only in the spatial and ensemble average, but respects the idea of continuous mixing in phase space. The trade-off is between resemblance to the true dynamics and the possibility of getting simple closed equations for the statistics. Ironically, these closures seem better suited to address the relatively broad issues that arise in the rich quasi-geostrophic dynamics than more esoteric questions

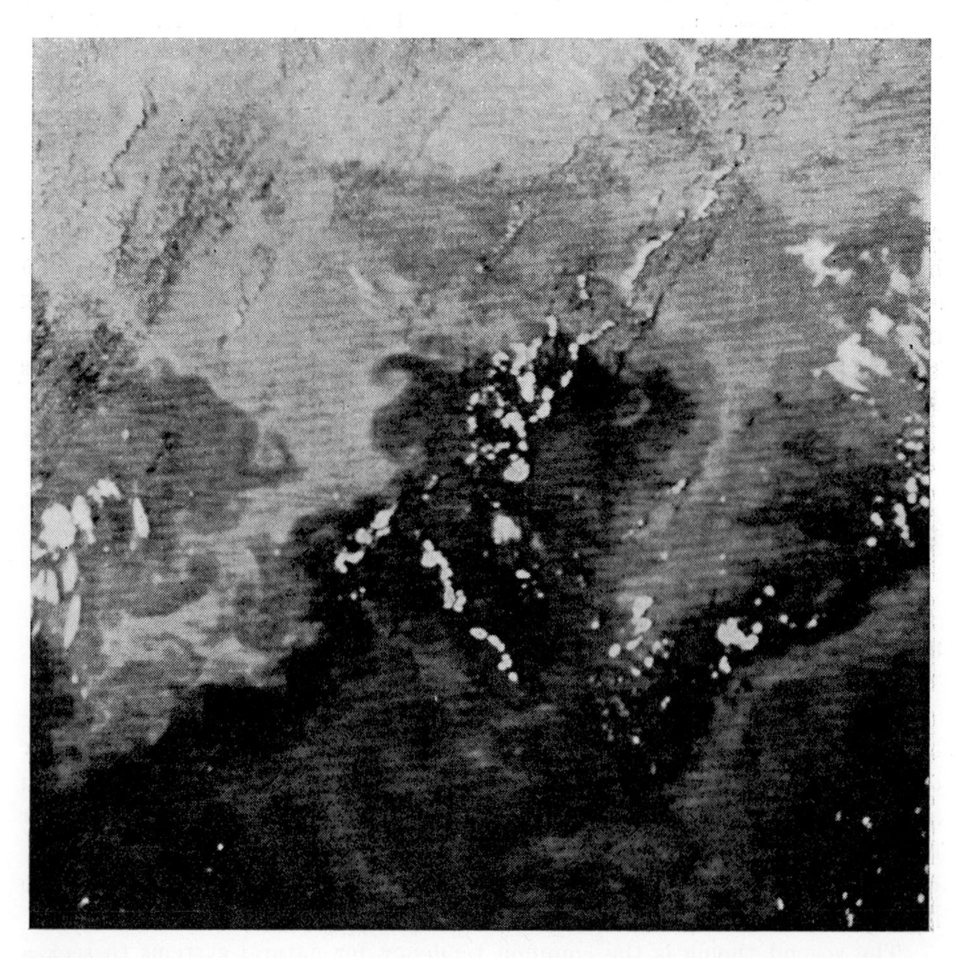

Plate 4.

about pure Navier-Stokes turbulence. However, applications to still more complicated, particularly inhomogeneous situations will require even further simplifications in the theory. The model viewpoint suggests a course: Retain the idea of a stochastic model, but relax even further the imposed requirements.

The broad goal of the theory is to deduce the major features of the observed circulation qualitatively but exclusively from first physical principles. Progress toward this end has been excruciatingly slow, consistently lagging the observations themselves. Modern observations show that the ocean is a perplexing mixture of order and apparent chaos. An infra-red photo of sea surface temperature offers a striking illustration (plate 4). The photo covers an area 560 km square centered on 29° N, 144° W in the Pacific. The maximum temperature contrast is about 1.5 °C. The photo was obtained by M. VAN WOERT of Scripps. It shows the subtropical front, a long semi-permanent boundary between cool (light-colored) northern water and warm (dark) water to the south. (The small white spots are clouds.) The explanation for the front may well lie within the realm of the linear quasi-laminar theories of basinscale wind-driven circulation, but the photo shows how the frontal boundary is broken up by turbulence. The large whorls have deformation scale.

* * *

I am funded by the National Science Foundation, IDOE (OCE 78-25670) as a part of Polymode.

REFERENCES

[1] J. PEDLOSKY: *Geophysical Fluid Dynamics* (New York, N. Y., 1979).
[2] L. ONSAGER: *Suppl. Nuovo Cimento*, **6**, 279 (1949).
[3] R. FJORTOFT: *Tellus*, **5**, 225 (1953).
[4] R. H. KRAICHNAN: *Phys. Fluids*, **10**, 1417 (1967).
[5] G. K. BATCHELOR: *Phys. Fluids*, **12**, 233 (1969).
[6] D. K. LILLY: *Phys. Fluids Suppl.*, **2**, 240 (1969).
[7] G. K. BATCHELOR: *J. Fluid Mech.*, **5**, 113 (1959).
[8] J. G. CHARNEY: *J. Atmos. Sci.*, **28**, 1087 (1971).
[9] P. B. RHINES: *The Sea*, Chap. VI (New York, N. Y., 1977), p. 189.
[10] R. SALMON: *Geophys. Astrophys. Fluid Dyn.*, **10**, 25 (1978).
[11] MODE GROUP: *Deep-Sea Res.*, **25**, 859 (1978).
[12] J. R. LUYTEN and J. C. SWALLOW: *Deep-Sea Res.*, **23**, 1005 (1976).
[13] C. WUNSCH: *J. Phys. Oceanogr.*, **7**, 497 (1977).
[14] E. HOPF: *J. Rat. Mech.*, **1**, 87 (1952).
[15] T. D. LEE: *Q. Appl. Math.*, **10**, 69 (1952).
[16] E. T. JAYNES: *Phys. Rev.*, **106**, 620 (1957).
[17] R. L. SELIGER and G. B. WHITHAM: *Proc. R. Soc. London Ser. A*, **305**, 1 (1968).
[18] A. I. KHINCHIN: *Mathematical Foundations of Information Theory* (New York, N. Y., 1957).

[19] R. SALMON, G. HOLLOWAY and M. C. HENDERSHOTT: *J. Fluid Mech.*, **75**, 691 (1976).
[20] G. HOLLOWAY: Ph. D. Dissertation, University of California San Diego (1976).
[21] N. P. FOFONOFF: *J. Mar. Res.*, **13**, 254 (1954).
[22] P. B. RHINES: *J. Fluid Mech.*, **69**, 417 (1975).
[23] P. B. RHINES: *Annu. Rev. Fluid Mech.*, **11**, 401 (1979).
[24] J. R. HERRING, S. A. ORSZAG, R. H. KRAICHNAN and D. C. FOX: *J. Fluid Mech.*, **66**, 417 (1974).
[25] S. A. ORSZAG: *J. Fluid Mech.*, **41**, 363 (1970).
[26] S. A. ORSZAG: *Fluid Dynamics*, edited by R. BALIAN and J. S. PEUBE (New York, N. Y., 1977).
[27] G. HOLLOWAY: *J. Fluid Mech.*, **82**, 747 (1977).
[28] R. H. KRAICHNAN: *J. Fluid Mech.*, **5**, 497 (1959).
[29] R. H. KRAICHNAN: *J. Fluid Mech.*, **47**, 513 (1971).
[30] R. H. KRAICHNAN: *J. Fluid Mech.*, **47**, 525 (1971).
[31] R. H. KRAICHNAN: *J. Atmos. Sci.*, **33**, 1521 (1976).
[32] C. E. LEITH: *J. Atmos. Sci.*, **28**, 145 (1971).
[33] C. E. LEITH and R. H. KRAICHNAN: *J. Atmos. Sci.*, **29**, 1041 (1972).
[34] J. R. HERRING: *J. Atmos. Sci.*, **32**, 2254 (1975).
[35] J. R. HERRING: *J. Atmos. Sci.*, **34**, 1731 (1977).
[36] G. HOLLOWAY: *J. Phys. Oceanogr.*, **8**, 414 (1978).
[37] J. R. HERRING: *J. Atmos. Sci.*, **37**, 969 (1980).
[38] R. SALMON: *Geophys. Astrophys. Fluid Dyn.*, **15**, 167 (1980).
[39] M. S. LONGUET-HIGGINS and A. E. GILL: *Proc. R. Soc. London Ser. A*, **299**, 120 (1967).
[40] K. HASSELMANN: *J. Fluid Mech.*, **30**, 737 (1967).
[41] G. P. WILLIAMS: *J. Atmos. Sci.*, **35**, 1399 (1978).

Observations of Large and Mesoscale Motions in the Near-Surface Layer.

A. D. KIRWAN (*)

Science Applications Incorporated - 4348 *Carter Creek Parkway, Bryan, Tex.* 77801

G. R. CRESSWELL

CSIRO Division of Fisheries and Oceanography
P. O. Box 21, *Cronulla, N.S.W.*, 2230 *Australia*

1. – Introduction.

1·1. *Purpose.* – The purpose here is to acquaint the student with some recent results in observations of the meso- and large-scale circulation of the surface layers of the oceans. These results are crucially dependent on remote-sensing technology, especially satellite position-fixing capability.

We first introduce some concepts which are peculiar to remote position-fixing technology. Then we discuss the results of some special studies on the meso- and large-scale circulation of the oceans. For our purposes, the meso-scale and large scale are characterized respectively by the Rossby radius of deformation and the basin size.

The observations discussed here conceptually are Lagrangian. However, no time is spent on important questions concerning Eulerian-Lagrangian transformations. The theoretical aspects of this problem have been known for over 200 years. The observations available now are neither accurate nor plentiful enough to shed new light on this.

In the subsequent discussion we will refer to a « drifter ». This is a measurement apparatus which is composed of three parts. The first is a surface buoy which responds in part to surface currents and waves. Its sole function is to provide a housing for a radio transmitter. The second component of a drifter is a drogue or drag body. The function of this body is to attach the measurement apparatus to a specific parcel of the ocean. In practice this is either a parachute or a window shade drogue. For reasons to be brought out later,

(*) Present address: Department of Marine Science, University of South Florida, 140, 17th Avenue South, St. Petersburg, Fla. 33701.

we prefer the former. The third element of a drifter is the cable which connects the drogue with the buoy. If the drogue depth is shallow, the cable drag is negligible when compared to the buoy response. But for deep drogued drifters the drag on the drogue line must be accounted for. This effect, however, is not important in the results discussed here.

1'2. *Background.*

1'2.1. Historical perspective. By virtually any standard, the study of ocean currents is one of the most esoteric branches of science. Open-ocean currents are very expensive to observe; the hypotheses being tested today deal with complicated physical processes that are unintelligible to anyone not active in the field, and there is usually no immediate practical use for the results.

This was not always the case. Even before the times of the Roman Empire the study of ocean currents was an important aspect of science with significant practical implications. Up until the introduction of steam-powered vessels most maritime nations had extensive programs for acquiring and synthesizing reports of ocean currents from the merchant marine.

The extent of this knowledge was quite extraordinary. Current systems were well described along shipping routes by the Carthageans in the Mediterranean and the Chinese in Asia long before the birth of Christ. As late as the eighteenth centrury, FRANKLIN prepared detailed and accurate maps of the Gulf Stream from merchant ship reports of currents, surface temperatures and seaweed. Curiously his work was disregarded by the British because he was a mere colonist. Even after the introduction of steam ships MAURY continued to study this current in the nineteenth century.

The early investigators were well aware not only of the large-scale steady-state component of the circulation, but of the mesoscale variability as well. They knew that strong currents tended to flow in narrow ribbons, that major current systems such as the Gulf Stream and Kuroshio had meanders, and that eddies spun off and often rejoined these currents. They were also very sensitive to the ocean response to the wind. Quantifying these responses is the goal of most modern research programs.

1'2.2. Early observational techniques. In addition to ship drift and sea surface temperatures, knowledge of ocean currents has come from trajectories of such things as ship wreck debris, seaweed, coconuts [1], drift bottles and drift cards. These have never been satisfactory as they only give two widely spaced points with no details on the variability. The cost of each data point, however, is small.

In the early 1960's, FUGLISTER and WORTHINGTON tracked buoys in the Gulf Stream by ship radar. The results were quite interesting, but the technique was awkward and expensive even by oceanographic standards. Frequent ship positioning was required (but not always available) and a dedicated ship was required. In those days a blue-water research vessel cost about $ 4000/day. It could track the buoys about 50 days with about 10 fixes a day. This comes out to $ 400/data point.

1'2.3. Remote sensing. From this it is clear that remote-sensing technology is a logical technique to apply to the problem of observing trajectories in the upper layers of the ocean. More data points are obtained than with drift cards and the prorated cost of data points declines dramatically.

The earliest attempt at remote sensing of ocean trajectories by satellite was performed during the IRLS experiment (1970). This was followed by the French EOLE experiment (1971-1974). Both of these used a range positioning system in which the drifter transmitted, with a suitable time delay, only after it had heard the satellite interrogation transmission. This technique was both expensive and awkward. It required a fair amount of intelligence on the drifter with only a clock and tape recorder on the satellite. In those days the intelligence was both expensive and unreliable.

The obvious solution was to concentrate the intelligence and reliability on the satellite and have the drifter transmit continuously. When the drifter is in view of the satellite, the latter will detect a Doppler shift of the frequency of the drifter transmission. This has been the approach followed on NIMBUS G (launched 1975) and the NOAA-ARGOS system (launched 1978).

The transmitter operates continuously at a prescribed frequency. The signal is coded, so that an identification and a number of data channels are included in the transmission. Examples of the information sent back on the data channels are sea temperature, air temperature, barometric pressure, battery voltage, drogue indicator.

A particular drifter is in view of the satellite anywhere from a few seconds to 15 to 20 min. The data cycle of the drifter transmitter is around one second every minute, so the satellite may receive one to 20 « hits » from the drifter. As the satellite is traveling quite fast with respect to the drifter, it will detect a Doppler shift in the transmission frequency. A shift can be caused by any transmitter located on the surface of a cone whose vertex is at the satellite at the time of the hit. The intersection of this cone with the Earth's surface gives the drifter position. In practice the dual position solution is readily resolved.

2. – Effect of wind and surface currents.

2'1. *Theoretical considerations.* – In this section we examine the bias introduced on drifter motions by surface winds and currents acting on a surface buoy which houses the antenna used to transmit to the satellite. The analysis follows closely that of [2].

Several assumptions are necessary to make the problem tractable. The least tenable of these may be to neglect any effect of surface waves. Wave data are not available regularly from remote areas. Thus, although not realistic, this assumption is at least consistent with the available environmental data. Normally surface meteorological data are available only at the synoptic times. Hence it is logical to assume that the wind and surface current are steady during synoptic periods.

Two drag bodies are identified: the surface buoy and the drogue. The former is acted upon by wind drag on its dry portion and surface current drag on its wetted portion; the latter by the current drag at its particular depth. The equations of motion for the drifter then reduce to the balance of these three drag forces:

$$\sum_{j=1}^{3} \boldsymbol{F}_j = 0 . \tag{1}$$

The general form for each of these drags is given by

$$\boldsymbol{F}_i = \varrho_i C_{Di} A_i |\boldsymbol{V} - \boldsymbol{V}_i| (\boldsymbol{V} - \boldsymbol{V}_i) . \tag{2}$$

Here ϱ_i, C_D, A and $\boldsymbol{V}$ refer respectively to the density, drag coefficient, the area and velocity seen by the fluid streaming past the appropriate drag element, and the velocity of the drifter. The subscript i refers to the i-th drag element, while $\boldsymbol{V}_i$ refers to the fluid velocity at the i-th drag element.

Substituting (2) into (1) yields a nonlinear algebraic equation for $\boldsymbol{V}$ in which the $\boldsymbol{V}_i$'s can be regarded as parameters of solution. Of course, ϱ_i, C_{Di} and A_i are prescribed *a priori*. KIRWAN *et al.* [2] have obtained numerical solutions to (1) and (2). Figure 1, taken from their paper, summarizes the results over a wide range of parameters for the NOAA Data Buoy Office (NDBO) buoy.

The calculations were performed by solving eqs. (1) and (2) for every orientation of the wind and surface current. As the direction of the wind and surface current are varied independently, double-valued solutions occur for the lower wind speeds.

The figure shows that substantial differences between the drifter velocity $\boldsymbol{V}$ and the fluid velocity at the drogue $\boldsymbol{V}_z$ can exist for higher wind speeds. For example, percentage errors in excess of 25% may occur under moderate wind and current shear conditions with the 9.5 m parachute.

The NDBO buoy is usually deployed with a window shade drogue. Calculations show that this drogue increases the error an additional 10% over that obtained with the 9.5 m parachute. This is a consequence of a smaller surface area.

The window shade has an even more serious shortcoming. Because of its excessive length, there is no well-defined reference level for the velocity.

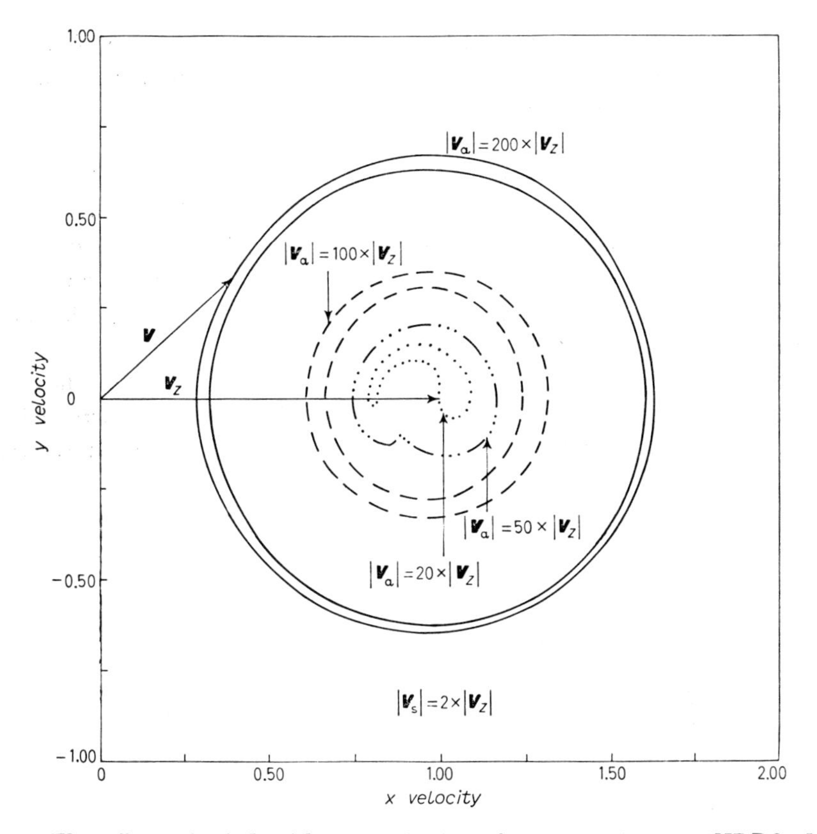

Fig. 1. – The effect of wind with a constant surface current on a NDBO drifter drogued by a 9.2 m parachute. The locus for the drifter velocity for a given wind speed occurs within the indicated region.

The structure of the velocity profile determines the total drag on this element. Because of its excessive length, it may well penetrate throught the mixed layer and hence respond to more than one dynamical regime. This greatly restricts its utility as a scientific tool.

An important special case occurs when the drifter is deployed without a drogue or it falls off. Then (1) and (2) can be inverted analytically. The result is

$$V = (V_s + KV_a)/(1 + K) \,. \tag{3}$$

The subscripts a and s refer to air and surface respectively and

$$K = \varrho_a C_{Da} A_a / \varrho_s C_{Ds} A_s \,.$$

Typically, the wet and the dry drag coefficient of a buoy are the same and the density of air is about one thousandth of that of water. Equation (3) then shows that the crucial factor in minimizing the wind is the ratio of the dry to wet area. For the NDBO buoy this ratio is almost 2, while for the ones used in the studies reported later the ratio is 0.6 or less.

2'2. *Experimental results.* – The first study published on full-scale observations under open-ocean conditions of wind drag on drifters was that of [2]. That study reported on an analysis of trajectories of five buoys deployed in the Florida Loop Current. Table I gives the hydrodynamic characteristics of the drifters and fig. 2 shows the trajectories. Figures 3 through 5 show the speeds and currents for each of the three legs.

These figures show that the corrected cross-stream speed records are very similar for all of the drifters. This is true even though the uncorrected cross-stream speeds of drifter 4 during legs 1 and 2 are in some cases a factor of

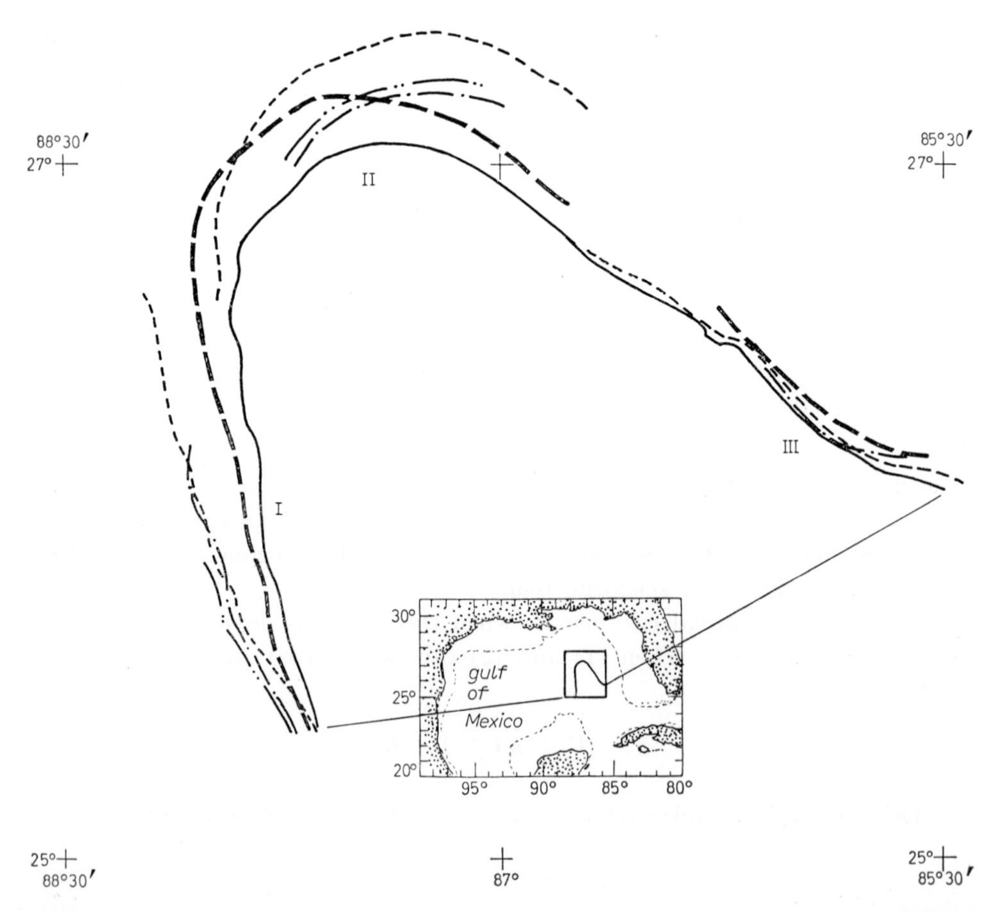

Fig. 2. – Trajectories for Gulf of Mexico windage test. The three different legs are indicated by Roman numerals: ——— buoy 1, — — — buoy 2, —·—·— buoy 3, - - - buoy 4, —··—··— buoy 5.

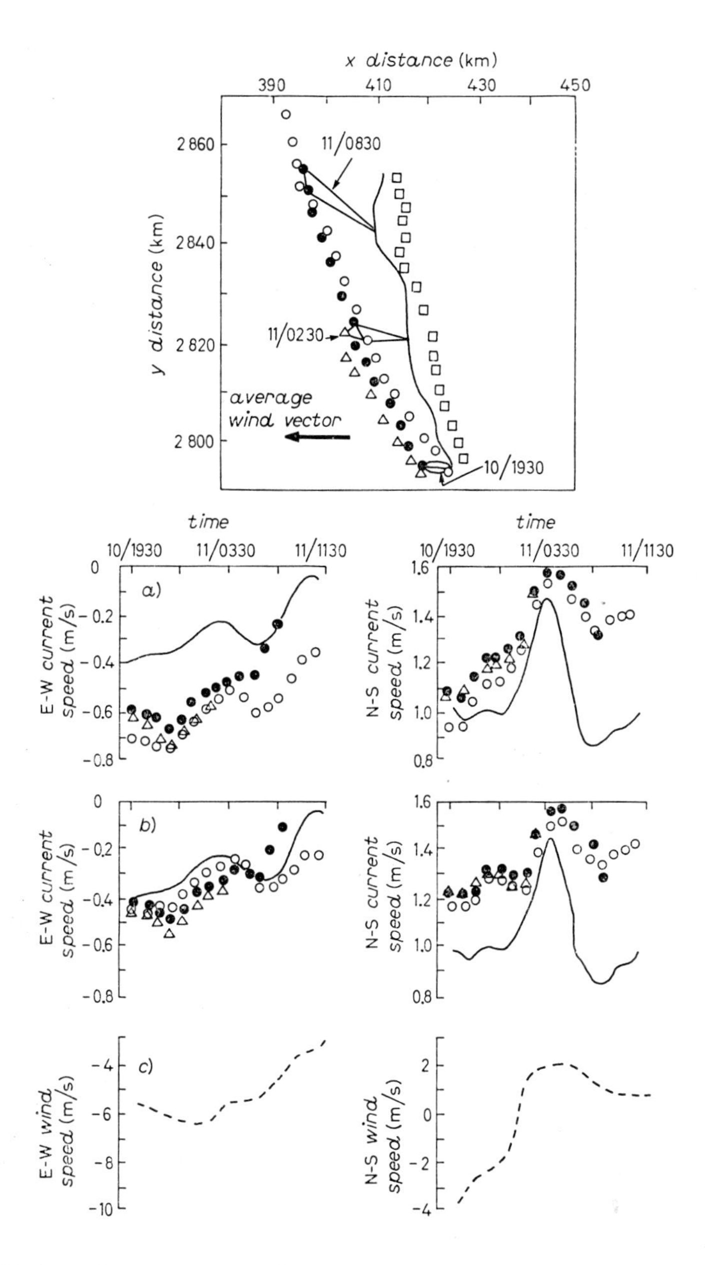

Fig. 3. – Trajectories, currents and wind records for leg 1. The top panel shows trajectories of the five drifters. The bottom panels show the measured current components (a)) as well as the corrected current components (b), c)). Also shown are the wind component. ▫ buoy 2, ▵ buoy 3, ○ buoy 4, • buoy 5.

TABLE I. – *Hydrodynamic characteristics of drifters used in loop current study.*

Drifter number	A_a (m²)	A_8 (m²)
1	cylinder ([a]) 9.755×10^{-1}	parachute 8.93×10
2	same as 1	same as 1
3	cylinder 1.366×10	cylinder 1.343×10
4	cylinder 3.581×10	cylinder 1.375×10
		rectangular pole $4.297\cdot10^{-10}$
5	cylinder 3.581×10	cylinder 1.375×10, vane 4.262×10

(*a*) C_D for cylinder is 1.2.

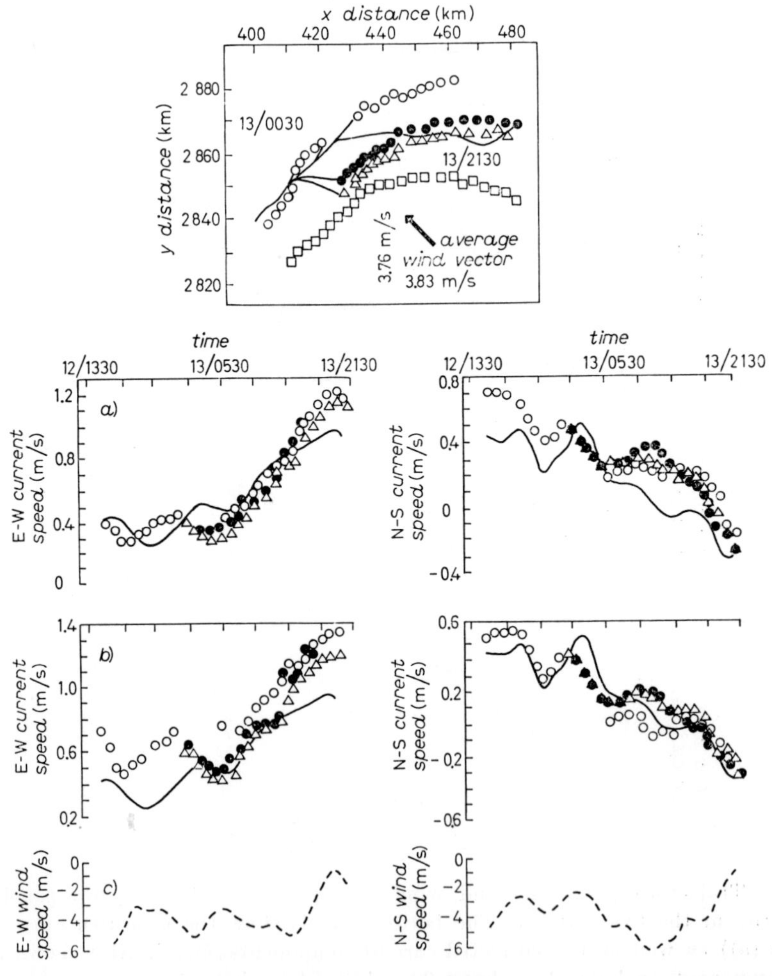

Fig. 4. – Trajectories, currents and wind records for leg 2. The top panel shows trajectories of the five drifters. The bottom panels show the measured current components (*a*)) as well as the corrected current components (*b*), *c*)). Also shown are the wind components. □ buoy 2, △ buoy 3, ○ buoy 4, • buoy 5.

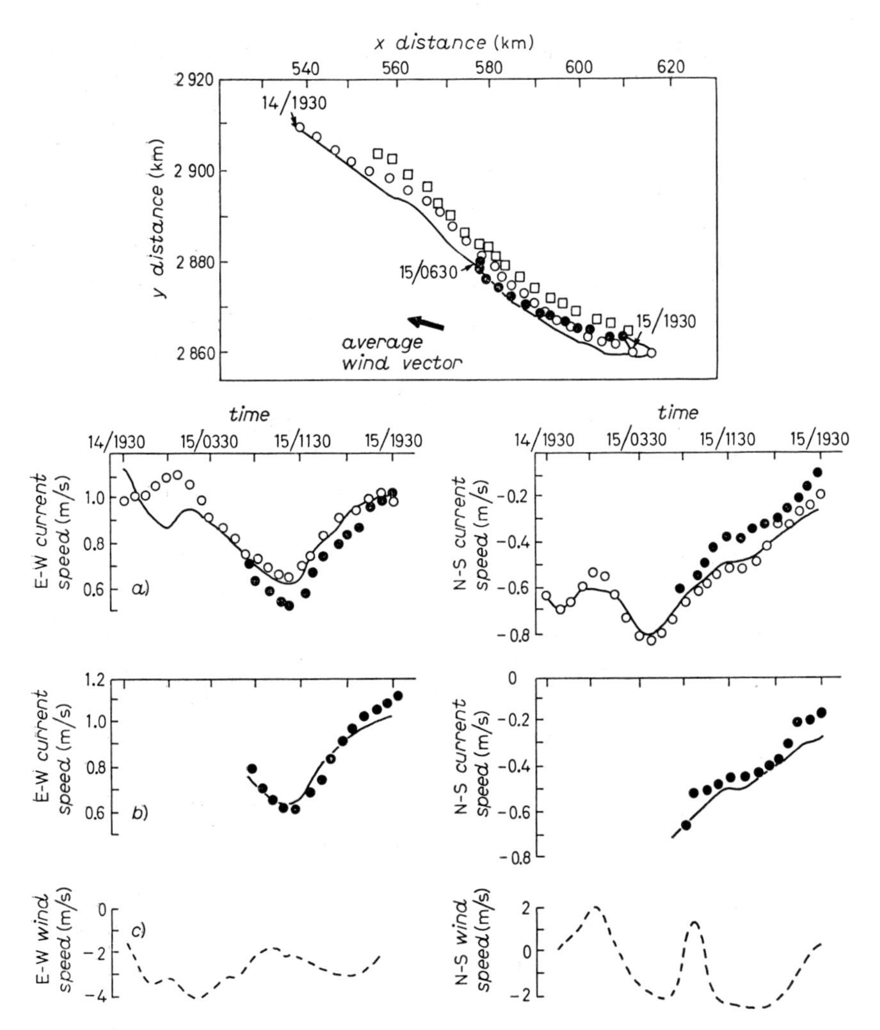

Fig. 5. – Trajectories, currents and wind records for leg 3. The top panel shows trajectories of the five drifters. The bottom panels show the measured current components (a)) as well as the corrected current components (b), c)). Also shown are the wind components. □ buoy 1, ○ buoy 4, • buoy 5.

2 greater than the velocities of drifter 1. After correction the speeds are within 15% over most of the trajectories. Drifter 4 was drogued during leg 3, and fig. 5 indicates that the effect of wind on this drifter is greatly reduced as it tracks closely with the other drogued drifters.

The wind correction for the downstream components increases the velocity difference between drifters for legs 1 and 2. However, for all legs this was the weakest wind component. Vertical and horizontal shears in the intense Loop Current, rather than the wind, probably were the most dominant causes of the deviation of the velocity records.

There are two important conclusions from that study. First, if accurate wind data are available, then windage corrections can be applied to surface current observations. But in the case of drogued drifters there is an additional unknown: the shear in the upper layer. Then this analysis is useful only as a diagnostic for identifying periods when the observations may not represent ocean conditions.

The experimental data also demonstrate another effect when there are large horizontal velocity gradients. A very small wind drag can move the drifter into a completely different current regime. This horizontal shear effect enhances the turbulent dispersive processes. This is demonstrated even more dramatically in the results of Cresswell and Garrett [3].

The separation of the drifters during leg 1 (fig. 3) exemplifies this type of effect. Here the northward motion of drifter 2 relative to 1 indicates that the current axis is to the west. Thus the east wind component caused the non-drogued drifters to move toward the axis. Although the east-west separation of the drogued and nondrogued drifters is a function of wind-induced motion, the north-south separation is caused by this horizontal current shear.

Because of this last effect, it is virtually impossible to correct a trajectory to obtain the true path of the water parcel originally tagged. Equation (3) requires as input the observed drifter velocity and wind. But the drifter integrates its velocity over time to obtain its trajectory. In regions of horizontal shear the wind will drive the drifter to a new parcel which in turn will have a different set of hydrodynamic forces. This will result in a drifter trajectory different from the original water parcel trajectory. See [3] for other examples of this effect.

Unfortunately it is not clear that windage corrections are useful for normal deployments where the wind data are indirect and available only at larger time intervals. KIRWAN *et al.* [4] obtained windage corrections for six drifters which had lost their drogues early in their deployment in the North Pacific.

The raw positions obtained from the NIMBUS G were interpolated to the synoptic time periods (0000, 0600, 1200, 1800 GMT) by cubic splines. The spline functions also provided velocity estimate of V at these times. The wind field at a particular drifter was obtained by a two-step process. The first was a planar interpolation through the three nearest grid points of the Fleet Numerical Weather Center (FNWC) surface wind analysis to each drifter position at the synoptic times. As this wind refers to a height of 19.5 m, it was necessary as a second step to reduce the wind to 1 m, the nominal height of the drifter above the sea surface.

This was accomplished through the following procedure. It was assumed that the friction velocity U_* is given by the drag coefficient C_D at 10 m as determined by SMITH and BANKE [5], *i.e.*

$$U_* = C_D |\boldsymbol{U}_{10}|^2, \tag{4}$$

where

$$C_D = 6.3 \cdot 10^{-4} + 6.6 \cdot 10^{-5} |\boldsymbol{U}_{10}| \ . \tag{5}$$

The relation between U_{10} and $U_{19.5}$ is assumed to be given by the logarithmic profile. This can be expressed as

$$U_* = k(|\boldsymbol{U}_{19.5}| - |\boldsymbol{U}_{10}|)/\ln 1.95 \ , \tag{6}$$

where $k = 0.35$ is von Karman's constant. Eliminating U_* between (4) and (6) yields

$$(k/\ln(1.95)^2 |\boldsymbol{U}_{19.5}| - |\boldsymbol{U}_{10}|)^2 = C_D |\boldsymbol{U}_{10}|^2. \tag{7}$$

This is a cubic in $|\boldsymbol{U}_{10}|$ whose roots are readily obtained by successive approximations. With $|\boldsymbol{U}_{10}|$ thus determined, the wind at 1 m was obtained from U_* and U_{10} by the logarithmic profile relation

$$|\boldsymbol{V}_a| = |\boldsymbol{U}_{10}| - (U_*/k) \ln 10 \ . \tag{8}$$

From (8) and the FNWC data, the components of $\boldsymbol{V}_a$ were calculated and then used in (3) to obtain the « corrected » surface current. Unexpectedly the correction was so large that at high winds the corrected surface current is opposed to the wind. Altering the drag coefficient within the wide range reported in the literature still produced unrealistic results.

One explanation for this discrepancy is that the FNWC wind field is wrong. But FRIEHE and PAZAN [6] in a case study from 1975-1976 have shown that this field is in good agreement with observations from Alpha buoy which was moored at 35° N, 155° W. Thus this explanation seems unlikely. Also it is easily shown that modifications of the logarithmic profile by stability cannot account for the discrepancy.

The only other explanations are that for high winds the logarithmic profile significantly overestimates the wind close to the sea surface and/or that the procedure suggested by [2] is not appropriate for this particular drifter hull. In this regard SHEMDIN [7] has observed in a laboratory study that there is a logarithmic velocity deficit in the vicinity of the free surface. In any event it is concluded that the uncorrected velocities are a better indicator of the true total surface current than are the velocities obtained from (5) and (10) for this type of buoy. Our experience with the NDBO buoy has been that, when it loses its drogue, the uncorrected surface velocity is readily correlated with the surface wind.

3. – Surface current response to wind.

3·1. *Background.* – The surface layer response of the ocean to the surface wind has been a fundamental problem in ocean dynamics ever since Ekman's [8] pioneering study. That model employed a constant eddy viscosity and assumed continuity of turbulent shear stress but not of velocity at the air-sea interface. It predicted that surface wind drift currents move 45° *cum sole* to the wind.

The extensive literature on the subject indicates a widespread uneasiness with this explanation of the surface layer response. As a result, the simple Ekman model has been extended to include variable eddy viscosities as well as boundary layers on either side of the air-sea interface. Then both shear stress and velocity may be continuous across the interface. Also, in these studies the angle between the wind drift current and the wind is determined by the boundary layer parameters. MADSEN [9] has given a lucid account of one such model as well as a review of previous results.

An alternate response mechanism has been suggested by LONGUET-HIG-

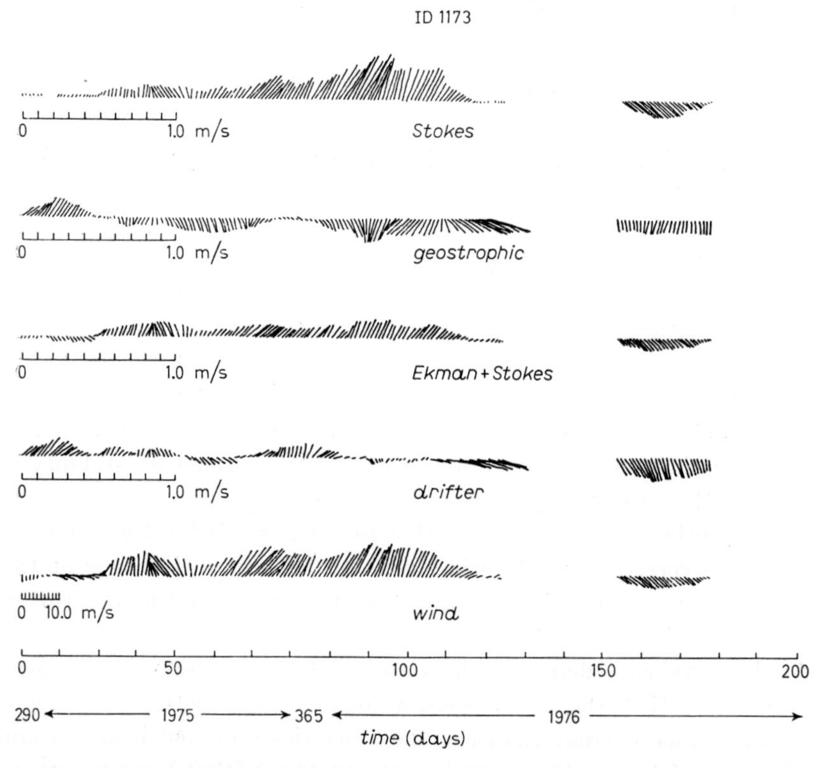

Fig. 6. – Stick diagrams of observed wind at 19.5 m, observed drifter velocities and the component velocities of the Ekman plus Stokes (case 2) model. The gap in the records is the result of missing wind data at FNWC.

GINS [10], KENYON [11] and IANNIELLO and GARVINE [12], who noted that much of the wind energy gets into the ocean through wave generation. The nonlinear character of the waves would then induce a Stokes drift which could be comparable to geostrophic or wind drift currents.

These mechanisms provide two possibilities for the response of the surface currents to the wind. For the Ekman theory, the wind drift surface current is proportional to the product of the wind magnitude and the vector wind. We call this the quadratic theory. On the other hand, a linear relation between the wind and surface current can arise from matching turbulent boundary layers on either side of the interface or from Stokes drift. Such a response is consistent with empirical studies cited by LANGE and HUHNERFUSS [13] and NEUMANN and PIERSON [14].

KIRWAN *et al.* [15] used the drifter data cited in [4], along with estimates of the surface wind, to test these mechanisms of surface layer response to the wind. Specifically, the following hypotheses were examined:

1) The wind drift surface current magnitude is a function of the surface wind speed squared as predicted by Ekman theory.

2) The « Ekman » angle of 45° is the best relation between the surface wind and the surface wind drift current.

3) Stokes drift is not a major component of the surface circulation.

The first two hypotheses test aspects in which the Ekman and recent boundary layer theories disagree. The last hypothesis addresses the question of the importance of Stokes drift in the wind response. The drifter data are

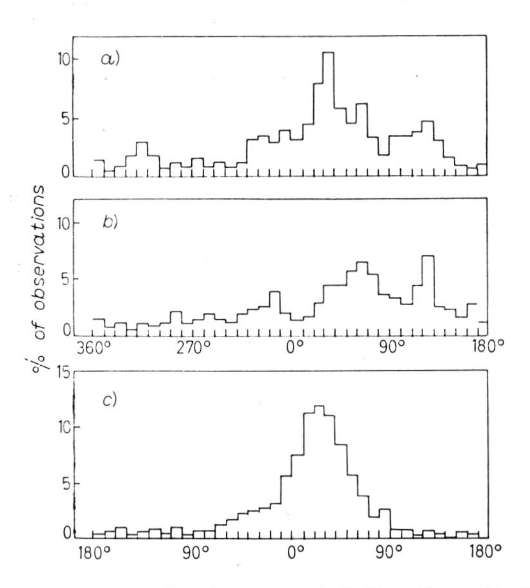

Fig. 7. – Composite histograms showing *a*) wind direction, *b*) drifter heading and *c*) difference angle. The directions are true.

especially suited for testing this hypothesis, as Stokes drift is seen only in Lagrangian measurements.

The two sources of data for the study are the velocities obtained from the drifter trajectories [4] and the FNWC surface wind objective analysis. The wind field evaluated at the longest running drifter is shown in fig. 6, taken from [15]. This stick diagram also summarizes the calculations discussed below. Focusing on just the winds and observed currents, the overall impression is that during steady wind periods the drifter velocities are generally to the right of the wind. Figure 7, also taken from [15], presents composite histograms of the wind azimuth, observed drifter azimuth and their difference for all six drifters. It is seen from this figure that on the average the drifter velocity is 25 to 30 degrees to the right of the wind with a standard deviation of 15°. The drifter azimuth, however, includes both wind-driven and geostrophic components.

3·2. *Analysis procedure.* – The estimates of the surface current from the drifters and the surface wind from FNWC make it possible to test the hypotheses stated in **3·1** by least-square methods. However, the errors in the data sets will adversely affect the correlations between the different fields. Thus there is a need for an *ad hoc* standard for the maximum amount of variance that can be explained by each hypothesis. For the quadratic and linear theories, the most general empirical form of the respective law is used as this indicator. The general form becomes the standard for objective tests of the relative effectiveness of each special case for that particular theory.

Specifically, Ekman and boundary layer theory suggests a standard statistical model for the quadratic and linear theories of the form

$$U = U_G + a_{11} F_1(U_a, V_a) + a_{12} F_2(U_a, V_a) , \tag{9}$$

$$V = V_G + a_{12} F_1(U_a, V_a) + a_{22} F_2(U_a, V_a) . \tag{10}$$

The subscripts G and a refer respectively to the surface geostrophic current and surface wind.

The form for F_1 and F_2 depends on which theory is specified. For the Ekman theory

$$F_1 = M_a U_a , \tag{11}$$

$$F_2 = M_a V_a , \tag{12}$$

$$M_a = (U_a^2 + V_a^2)^{\frac{1}{2}}, \tag{13}$$

and for the boundary layer theory

$$F_1 = U_a , \tag{14}$$

$$F_2 = V_a . \tag{15}$$

As will be seen shortly, this last model cannot discern between matched boundary layers and Stokes drift.

The U_G, V_G and a_{ij}'s are computed by least squares. As the coefficients appear in a vector equation, two regression equations must be solved. Not only are the disturbance terms in the two equations correlated, but, as indicated below, in some of the calculations constraints imposed on the coefficients couple the equations directly.

For the quadratic and linear models the following three cases were investigated. The first was the general case in which each of the a_{ij}'s was taken as independent. This established the maximum amount of variance that could be explained by that model class. Physically this case allowed for an arbitrary rotation and general deformation of the wind drift surface current by the surface wind. The second case allowed for only an arbitrary rotation and stretching deformation of the wind drift current with respect to the surface wind. This was accomplished by imposing the constraints $a_{11} = a_{22}$ and $a_{12} = -a_{21}$. The last case requires $a_{11} = a_{22} = a_{12} = -a_{21}$. This fixes the rotation at 45° and determines the stretching deformation or one drag coefficient.

The least-squares equations are obtained by minimizing

$$\overline{E}^2 = \overline{(\boldsymbol{V}_D - \boldsymbol{V}) \cdot (\boldsymbol{V}_D - \boldsymbol{V})} \tag{16}$$

with respect to U_G, V_G and the a_{ij}'s.

Here V_D is the observed drifter velocity vector. The overbar means a time average. For the contrained cases (16) was modified by the method of Lagrangian multipliers.

In order to assess the wave drift, an *ad hoc* Ekman plus Stokes (ES) model was assumed. This was a superposition of wind and wave drift. For this model (16) was replaced by

$$\overline{E}^2 = \overline{(\boldsymbol{V}_D - \boldsymbol{V} - \boldsymbol{V}_S) \cdot (\boldsymbol{V}_D - \boldsymbol{V} - \boldsymbol{V}_S)}\,. \tag{17}$$

Rather than use a theoretical value of the coefficient as given by [16] or [17], a general form for Stokes drift, $\boldsymbol{V}_S$, was used; namely, $\boldsymbol{V}_S = b\boldsymbol{V}_a$. The regression technique provided estimates of b along with the a_{ij}'s, U_G and V_G. Three models were investigated. For case 1 the a_{ij}'s were constrained just as in case 2 of the quadratic theory. The second imposed the case-3 constraint, and the last considered just Stokes drift, *i.e.* $a_{ij} = 0$.

The prescription of the averaging interval is a delicate balance between two factors. If the averaging period is too short, the number of data points is not much more than the number of parameters and there is the risk that predictive capability will be artificially enhanced. On the other hand, if the averaging period is too long, nonstationary and spatial-scale effects may become important. Consequently, the question of appropriate averaging periods was investigated in some detail. For this data set it was found that tests of the three basic hypotheses which used statistics from averaging periods ranging from 5 to 21 days yielded the same conclusions with only minor variations in the significance levels.

Each of the models described above was subjected to four types of tests. The first determined the levels of significance of the variance that each of the models explained. Then they were tested for homogeneity of the data base. This was accomplished by testing the differences of regression coefficients for each (nonoverlapping) averaging period with the combined regression coefficients. The third test determined the « best » special case for each theory. Finally, a subjective comparison was made of the best special cases of the three different theories.

In the analysis described above, the drifters move continuously, whereas the FNWC surface wind analysis prescribes the wind on a $2.5° \times 2.5°$ grid every synoptic period. It is necessary then to obtain records of the wind field and drifter velocities at the same point in time and space. A two-step procedure accomplished this. First, from the time series of locations, the drifter position at each synoptic period was interpolated by spline functions. These functions also produced smooth estimates of the drifter velocity. Second, at each synoptic period the wind components were interpolated to the drifter positions by passing a plane through the nearest grid points. In summary, the drifter positions and velocities were interpolated to each synoptic period and then the wind field was interpolated to each drifter. In essence this procedure applies a space and time filter to the drifter data which is roughly comparable to that used in obtaining the wind field.

3·3. *Tests of models.*

3·3.1. Quadratic models. Typically an analysis of variance partitions the total variance into the sum of the model variance and a residual or unexplained variance. A common measure of the goodness of fit is the ratio of the explained or model variance to the variance R^2 of the observations. Except for the record lengths, plots of R^2 for all drifters are very similar; however, values for ID 1275 are somewhat lower than the others. Table II summarizes the average R^2 for the entire record for each drifter.

The first test of the model determines the probability that these R^2's are are due to chance. This required testing the null hypothesis $H: a_{ij} = 0$ against not H, by the F distribution. The degrees of freedom are $N - q - 1$ and q, where N is the number of observations and q the number of independent a_{ij}'s. For the 5-day and 21-day averaging periods N is, respectively, 20 and 84. For cases 1, 2 and 3, q is, respectively, 4, 2 and 1.

Consider first the 5-day averages. It was found that H could be rejected with only a 5% chance of error if $R^2 > 0.45$, 0.30 and 0.20 for any 5-day average for the first, second and third cases, respectively. Except for case 1

TABLE II. – *Average R^2 for quadratic-law models.*

ID	Case 1		Case 2		Case 3	
	5 d	21 d	5 d	21 d	5 d	21 d
1102	0.5414	0.4342	0.4375	0.3924	0.3002	0.2780
1173	0.5026	0.4441	0.3903	0.3988	0.2684	0.2778
1204	0.5482	0.4096	0.4456	0.3679	0.3017	0.2532
1232	0.5331	0.4652	0.4237	0.4493	0.2833	0.2944
1243	0.5764	0.5130	0.4900	0.4964	0.3497	0.3849
1275	0.4420	0.3307	0.3373	0.3036	0.2291	0.2468

Case 1 is the general case. Each of the a_{ij}'s is independent.
Case 2 is the generalized Ekman case. Here $a_{11} = a_{22}$, $a_{12} = -a_{21}$.
Case 3 is the simple Ekman case. Here $a_{11} = a_{12} = a_{22} = -a_{21}$.

for drifter 1275, R^2's exceed these critical values for well over half the record. For the 21-day averages, any value of $R^2 > 0.2$ is significant at the 1 % level for all three cases. Only case 3 had any periods where R^2 was less than this critical value. Thus for most of the 5-day and 21-day averaging periods the R^2's are significant. The only exception was the period 355/75 to 20/76. A comparable decrease in R^2 was observed for all active drifters during that time. Here the level of significance of the results was reduced considerably. We have no explanation for this decrease. Nevertheless, this period was included in the combined overall averages.

The next concern is the homogeneity of the data base. Following WILLIAMS [18] this was tested by comparing the regression coefficients from case 1 for each nonoverlapping 21-day averaging period with that obtained from a combined or overall regression. These coefficients were tested by generalized Students' t. In the analysis of the variance the degrees of freedom for the combined regression is 4 (the number of coefficients), for the difference of regression they are $4(m-1)$, and for the combined residual they are $N-3m$. Here m is the number of nonoverlapping averaging periods. It is basically the greatest integer in the record length in days divided by the averaging period.

It was found that the difference of the within-group coefficients and overall coefficients was not significant at the 85 % or higher level for all drifters except 1275, where the significance level was about 75 %. The significance levels would be considerably higher if the period 355/75-20/76 was not included in the analysis. This means that the data from any particular averaging period are not significantly different at these levels. Thus the results from any particular averaging period are a resonable representation of the entire record.

Considering the 21-day cases, table II shows that case 2, where the data specify the angle of rotation and the stretching deformation or the drag coefficient, explains 90 % of the variance of the first case. However, the third case, in which the angle between the wind drift current and the wind is fixed at 45°, explains just over 60%. For the 5-day cases, the respective percentages are 80 and 53.

The question of whether the last two cases explain an adequate portion of the first case is addressed as follows: For case 2 the null hypothesis $H: a_{11} = a_{22},\ a_{12} = -a_{21}$ is tested against not H by a F statistic. Specifically, F is given by

$$F_2 = (R_1^2 - R_2^2)(N-5)/2(1-R_1^2)\,, \tag{18}$$

with degrees of freedom of 2 and $N-5$. The subscripts refer to the appropriate case. Calculated F values for the 5-day and 21-day periods were all less than 1.7 and 3.2, respectively. In order to minimize the chance of a type-II error (accepting a false hypothesis), H should be tested at its lowest level of significance. It is found that H cannot be rejected for any drifter at a probability level greater than 80% and 90% for the respective averaging periods.

The test for case 3 proceeds in the same manner except that (18) is replaced replaced by

$$F_3 = (R_1^2 - R_3^2)(N-5)/3(1-R_1^2)\,. \tag{19}$$

The degrees of freedom are 3 and $N-5$. For this case the null hypothesis is $H: a_{11} = a_{12},\ a_{22} = -a_{21}$. The values of F_3 range from 1.9 to 2.7 for the 5-day averages and 3.3 to 8.4 for the 21-day averages. Since these values are higher than the previous case there is a risk of a type-I error (rejecting a true hypothesis) here. This is minimized by testing H at a high level of significance. For the 21-day averages, H is rejected at the 99% level for every drifter except 1275, where the significance level is 95%. For the 5-day averages, H is rejected at the 90% level except 1275, where the significance level is 80%.

To summarize the results of these tests on the quadratic models it was concluded that each of the models explained a significant portion of the variance, that the data set was relatively homogeneous and that case 2 was the best special case. It explained nearly as much variance as the general case but with two less parameters.

Table III summarizes the combined average angle between the surface wind and the model wind drift component for case 2 for ID 1173 as well as

$$\gamma = (a_{11}^2 + a_{12}^2)^{\frac{1}{2}},$$

or case 2 for each drifter. The angle is of the order of 15° *cum sole* and the values are within the range given by accepted values of drag coefficient C_D and vertical eddy viscosity K.

TABLE III. – *Average angle between wind drift and wind and the drag coefficient for the quadratic case* 2.

ID	Angle (°)		$\gamma \times 10^{-3}$	
	5 d	21 d	5 d	21 d
1102	—11.4	—14.1	0.95	0.81
1173	— 9.4	— 8.8	0.86	0.76
1204	—13.0	—10.2	1.39	0.91
1232	—11.9	— 6.4	0.91	0.85
1243	—12.7	—17.8	0.92	0.86
1275	—22.2	—24.8	1.22	0.97

TABLE IV. – *Average R^2 for linear-law models.*

ID	Case 1		Case 2		Case 3	
	5 d	21 d	5 d	21 d	5 d	21 d
1102	0.5761	0.4700	0.4721	0.4298	0.3305	0.3120
1173	0.5361	0.4845	0.4174	0.4449	0.2985	0.3247
1204	0.5905	0.4741	0.4839	0.4414	0.3407	0.3105
1232	0.5632	0.5085	0.4536	0.4905	0.3078	0.3284
1243	0.6014	0.5694	0.5088	0.5467	0.3677	0.4333
1275	0.4896	0.3877	0.3860	0.3643	0.2770	0.3090

Case 1 is the general case. Each of the a_{ij}'s is independent.
Case 2 is the generalized Ekman case. Here $a_{11} = a_{22}$, $a_{12} = -a_{21}$.
Case 3 is the simple Ekman case. Here $a_{11} = a_{12} = a_{22} = -a_{21}$.

TABLE V. – *Average angle between wind drift and wind and the drag coefficient for the linear case* 2.

ID	Angle (°)		$(a_{11}^2 + a_{12}^2) \times 10$	
	5 d	21 d	5 d	21 d
1102	—12.1	—16.2	0.13	0.12
1173	—13.2	—12.3	0.12	0.13
1204	—14.4	—12.0	0.16	0.15
1232	—12.2	— 7.8	0.13	0.14
1243	—15.6	—19.1	0.12	0.13
1275	—26.5	—26.3	0.14	0.13

3˙3.2. Linear models. Figure 8 taken from [14] is a plot of R^2 for ID 1173 for the linear models and table IV summarizes the average R^2's for the different drifters. The conclusions reached concerning the averaging period, the homogeneity of the data and the selection of the best model (case 2) are the same as in the quadratic theory. Table V lists the average angle and $(a_{11}^2 + a_{12}^2)^{\frac{1}{2}}$ for the second case for each drifter. As before, it is seen that the

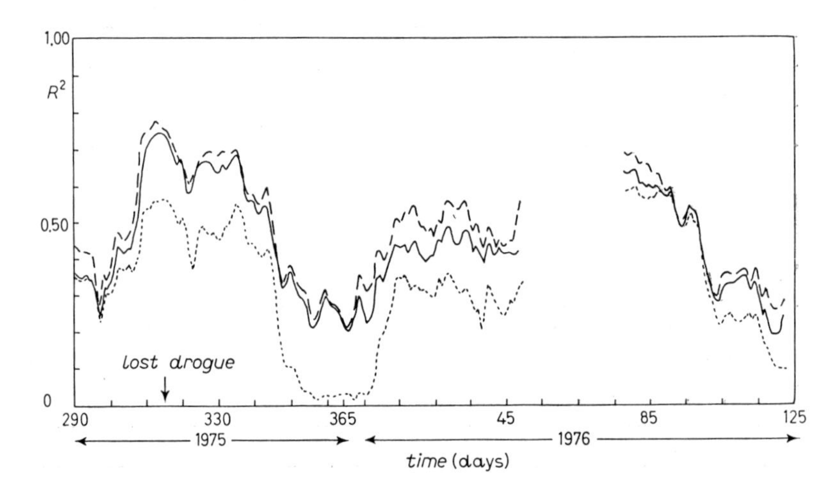

Fig. 8. – Ratio of model variance to observed variance for each case of the linear theory for the 21-day average (ID 1173): ——— case 1, ——— case 2, - - - case 3. The gap in the record is the result of missing wind data at FNWC.

angle is of the order of 15°. The scalar coefficient is within the range given in Madsen's study and is considerably less than the value used in attempting to correct for the wind drag on the buoy.

Comparing the R^2's of the linear and quadratic theories shows that the former explains more of the variance. Although the differences are not large, in view of the errors in the observations, they probably are sufficient to justify selecting the linear theory as the better representation of the data.

3˙3.3. Ekman plus Stokes drift models. Figure 9 shows the variance ratio for ID 1173 for the three cases of this theory, and table VI lists the average R^2's. Previous conclusions about the data set are valid here also. The F tests show that the hypothesis $H: a_{11} = a_{12} = a_{22} = -a_{21}$ cannot be rejected until the significance level is dropped to 80 % for both averaging periods. On the other hand, at the 95 % significance level, the hypothesis that all a_{ij}'s are zero is rejected for the 21-day averages. Thus the classic Ekman wind drift superposed with a Stokes drift component is the best case for this theory.

TABLE VI. – *Average R^2's for Ekman plus Stokes models.*

ID	Case 1		Case 2		Case 3	
	5 d	21 d	5 d	21 d	5 d	21 d
1102	0.4922	0.4294	0.4640	0.4232	0.3189	0.3916
1173	0.4526	0.4541	0.4171	0.4414	0.3352	0.3842
1204	0.5159	0.4552	0.4745	0.4349	0.3948	0.4112
1232	0.4922	0.5025	0.4576	0.4911	0.3612	0.4635
1243	0.5501	0.5575	0.5239	0.5409	0.3738	0.4840
1275	0.4096	0.3607	0.3625	0.3383	0.3192	0.2919

Case 1 is general Ekman plus Stokes drift. Here $a_{11} = a_{22}$, $a_{12} = -a_{21}$.
Case 2 is simple Ekman plus Stokes drift. Here $a_{11} = a_{12} = a_{22} = -a_{21}$.
Case 3 is only Stokes drift. Here $a_{11} = a_{12} = a_{22} = a_{12} = 0$.

Table VII shows the combined average angle for the first case. It is considerably greater than that given in the previous model studies. This points out a fundamental difficulty in comparing Eulerian and Lagrangian measurements of the surface current response to the wind. As the Stokes component is not present in Eulerian data, the angle between the wind and wind drift current in that system will be greater than that seen in the Lagrangian system.

Table VII also lists the b's for case 2. The values are considerably less than the theoretical value of 0.0158 [16]. However, there is no angular spreading in this model. There is some evidence [17] that the spreading function

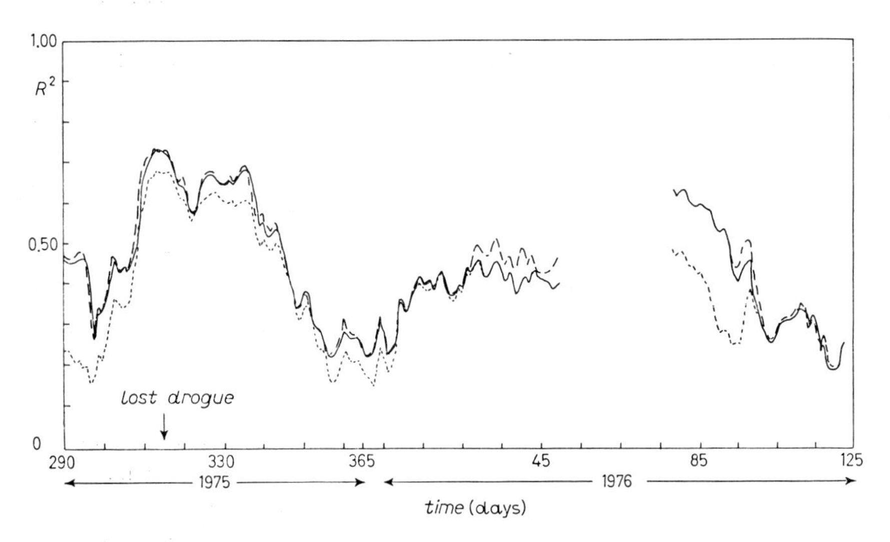

Fig. 9. – Ratio of model variance to observed variance for each case of the ES theory for the 21-day averages (ID 1173): – – – case 1, —— case 2, - - - case 3. The gap in the record is the result of missing data at FNWC.

TABLE VII. – *Average angle between wind drift and wind for Ekman plus Stokes drift case* 1 *and Stokes coefficient* (*b*) *for case* 2.

ID	Angle (°)		Stokes coefficient	
	5 d	21 d	5 d	21 d
1102	—51.0	— 74.6	0.0087	0.0093
1173	—32.3	— 65.2	0.0079	0.0096
1204	—31.9	— 84.2	0.0114	0.0129
1232	—49.0	— 35.6	0.0096	0.0115
1243	—59.6	—139.7	0.0081	0.0090
1273	—44.8	—116.2	0.0065	0.0078

obeys a $\cos^2$ law. Taking this into account reduces the theoretical value to 0.0134, still larger than that observed. Errors in the wind field will cause the least-squares estimates of b to be low. But, until these error variances are established, it is impossible to say if the discrepancy is the result of experimental error.

3'4. *Geostrophic calculations.* – In addition to establishing a model for predicting the wind-induced currents, the procedure adopted here also produces the surface geostrophic velocity. The stick diagram (fig. 6) shows these for the Ekman plus Stokes case. The combined average U_G and V_G are given in table VIII. It is very encouraging to find that the zonal component is of the order of 0.03 m s^{-1}, in excellent agreement with the mean annual dynamic

TABLE VIII. – *Geostrophic components, simple Ekman plus Stokes case* 21-*day average.*

ID	U_G	V_G
1102	0.03	—0.009
1173	0.05	—0.02
1204	0.02	—0.05
1232	0.03	—0.01
1243	0.02	—0.005
1275	0.03	0.02

topography [19, 20]. Except for ID 1275, there is a small southerly component to the surface geostrophic flow. This component is not inconsistent with southerly flow at the eastern end of the subtropical gyre.

Figure 6 also shows some evidence of a 20-day fluctuation in the geostrophic velocity. Similar fluctuations in this region have been associated with baroclinic mesoscale structures [21-23].

3·5. *Discussion.*

3·5.1. Linear and quadratic models. In regard to the tests of the linear and quadratic models, the results of this study support the following conclusions:

1) The two-parameter, or second cases are the most realistic. They explain nearly as much variance as the general cases but with two less parameters.

2) In comparing the linear and quadratic theories, it is clear that the linear theory is best.

3) The angles predicted by the general Ekman models are of the order of 15° and are in good agreement with the findings of Madsen [9].

Thus hypotheses 1) and 2) given in the introduction are rejected if Stokes drift is neglected.

3·5.2. Ekman plus Stokes model. The Ekman plus Stokes model calculations showed that case 2 worked as well as case 1. Moreover, the R^2's for the second case were indistinguishable from the linear case 2. There seems to be little doubt then that in the Lagrangian system there is a component of flow in the direction of the wind. How much of this is Stokes drift and how much is wind drag on the buoy is moot. In any event, windage is less than half that predicted.

This experiment, then, is insufficient to accept or reject hypothesis 3). Resolution of this issue will, no doubt, require better knowledge of the wind field, more detailed engineering studies of the drifter response to waves and wind than were available to us, as well as Eulerian measurements.

3·5.3. Geostrophic currents. The geostrophic velocities obtained from either Ekman plus Stokes or linear case 2 models are in excellent agreement with that calculated from the mean annual dynamic topography. This means that, in order to obtain the geostrophic component from drifter velocity records, it is necessary to evaluate the wind drift component as well.

4. – Circulation in the Eastern North Pacific.

4·1. *Background.* – There are a number of descriptions of the surface circulation in the North Pacific. Sverdrup *et al.* [24] identified two major eastward-flowing currents in the interior of the eastern Pacific. These are the Subarctic Current which they located at approwimately 45 °N and the North Pacific Current located approximately at 38 °N. The Defense Mapping Agency pilot charts show two anticyclonic gyres in the North Pacific [25]. The dynamic topography charts [19, 20] are generally consistent with these studies. However, their resolution is too coarse to resolve specific currents. TABATA [26, 27] has reported significant year-to-year variability.

Few studies have been made on the seasonal variability of the surface currents. Also, no studies have been made of the year-to-year surface current variability. However, NAMIAS [28] has shown that considerable interannual variability exists in the sea surface temperature field. WHITE and WALKER [29] indicate that this thermal variability may extend down to the main thermocline, thus suggesting the possibility of year-to-year fluctuations of the general circulation.

4·2. *Trajectories.* – The data set analyzed here comes from three drifter deployments which spanned the winters of 1976-1977 and 1977-1978. This has allowed us to look at both inter and intra annual variability of the circulation.

Figure 10 taken from [23] is a composite of the trajectories for 1976-1977. The most striking aspect of this figure is the large-scale eastward flow separating at the west coast of North America. Embedded in the main flow field are numerous mesoscale features. Figure 11 shows a typical mesoscale eddy. The size (~100 km) is roughly that reported for Gulf Stream rings. However, the period (three weeks) is about three times longer.

Figure 12 is a plot of some trajectories superimposed on Wyrtki's [19] mean annual 0/1000 db dynamic topography. Generally, there is good agreement between the trajectories and the dynamic topography. Both show the existence of the eastern portions of the subtropical and subarctic gyres, and both agree that the gyres split between 45 and 50° N at 140° W. Also, a close examination of the drifter data reveals that all drifters deployed in or to the north of the subarctic front stay in the subarctic gyre; those deployed south of the subarctic front stay in the subtropical gyre. The lone exception ends up in the stagnation point between the gyres.

Three trajectories of the drifters deployed along 43° N show the sharpness of the split between the gyres. The drifter deployed at 43° N, 162° W (ID 1562) ends up in the subarctic gyre, while ID 1525, deployed at 43° N, 154° W, stops in the subtropical gyre. ID 1513, deployed at 43° N, 158° W, between the first two, spends its last 110 days in aimless excursions centered about 51° N, 133° W, midway between the two gyres.

A curious feature of the subarctic gyre trajectories is the large anticyclonic meander centered at 57° N, 138° W. The meander was observed by three different drifters, the first being ID 1046 near the end of March 1977. The last drifter to detect this feature was ID 1376 in mid-May, about seven weeks later. Curiously, ID 1615, which was deployed 4° west of 1376, arrived at the meander three weeks before. Figure 13 shows the trajectory of ID 1615

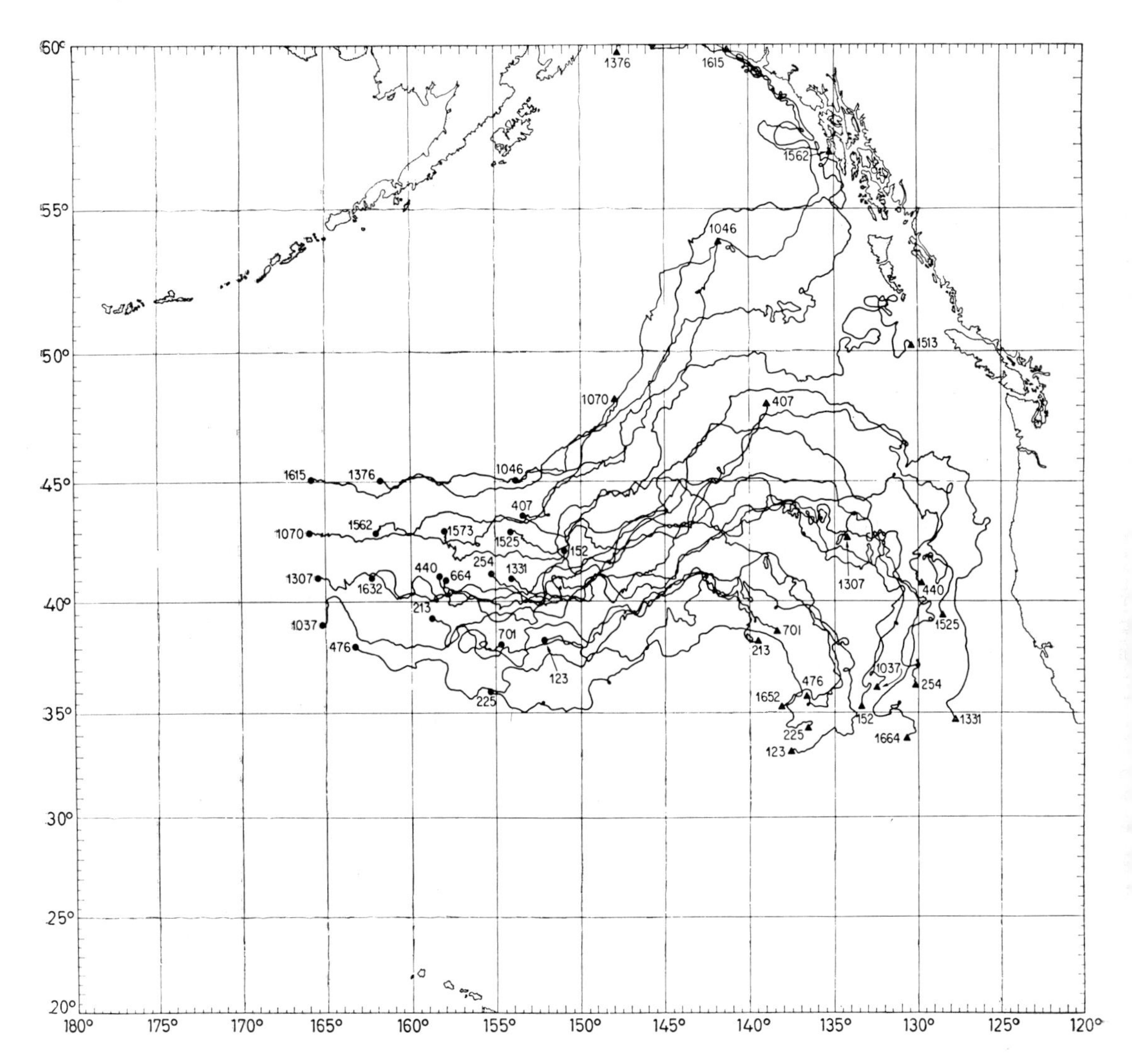

Fig. 10. – A composite of the trajectories of 22 drifters deployed during the period 10 September 1976 through 31 August 1977.

as it traversed the meander. Again the fact that the drifters which detected the meander were all deployed along 45° N is an indication of the smooth nature of the large-scale flow.

Another conspicuous feature of both the trajectories and the mean annual dynamic topography is the northward meander in the eastern portion of the subtropical gyre. However, the northward and westward extent of the meander in the mean annual dynamic topography is not nearly as great as it is in the trajectories. As the drifter observations were made during the development of the largest thermal anomaly ever observed in the North Pacific, it is likely

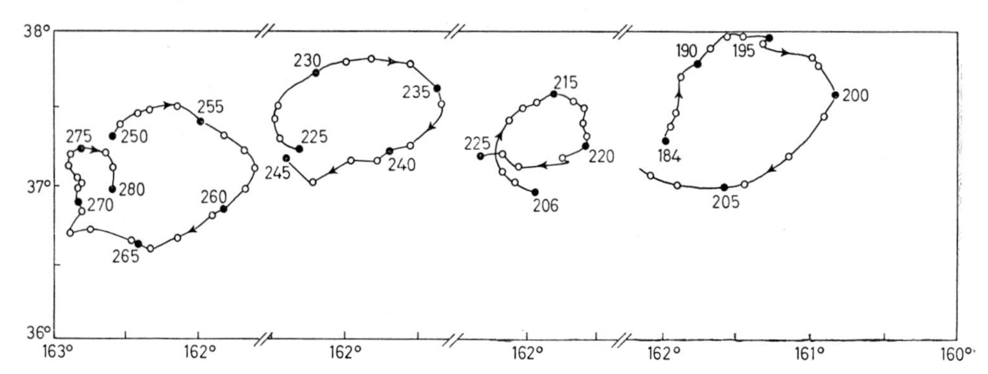

Fig. 11. – The trajectory of ID 0115 from Julian day 184 through 28 01 76. During this period the drifter completed four revolutions about an eddy which was migrating to the west-southwest at about 0.02 m s^{-1}.

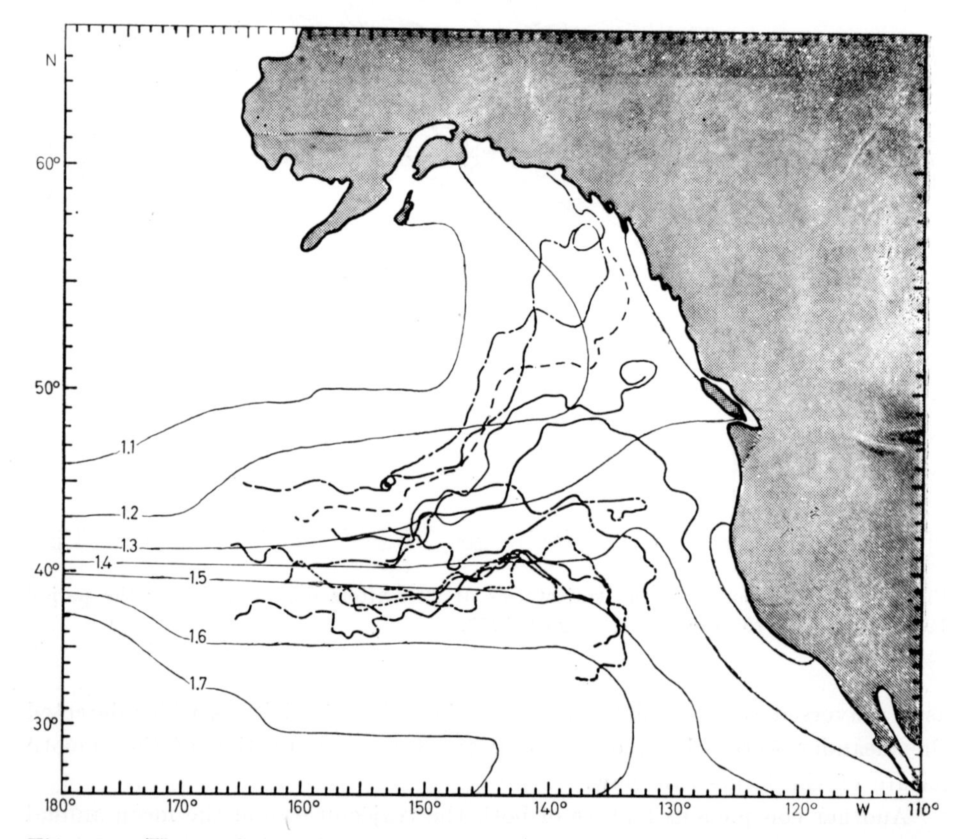

Fig. 12. – Eleven drifter trajectories superimposed on the 0/1000 db mean annual dynamic topography: —·—·— ID 1615, ID 1046; —··—··— ID 1307, ID 152; ····· ID 1562, ID 1513, ID 1525; — — — ID 476, ID 701; - - - ID 123, ID 213; ——— dinamic topography 0/1000. Note that all drifters deployed east and south of ID 1513 (deployed at 43 °N, 158 °W) were in the subtropical gyre. Those deployed west and north were in the subarctic gyre.

that this meander represents a significant intra-annual variation of the general circulation.

Because the drifters were deployed at different times and in different locations, a comparison of different trajectories can provide some information on the inter-annual variability of the general circulation. The trajectories indicate that the transition from season to season is not smooth.

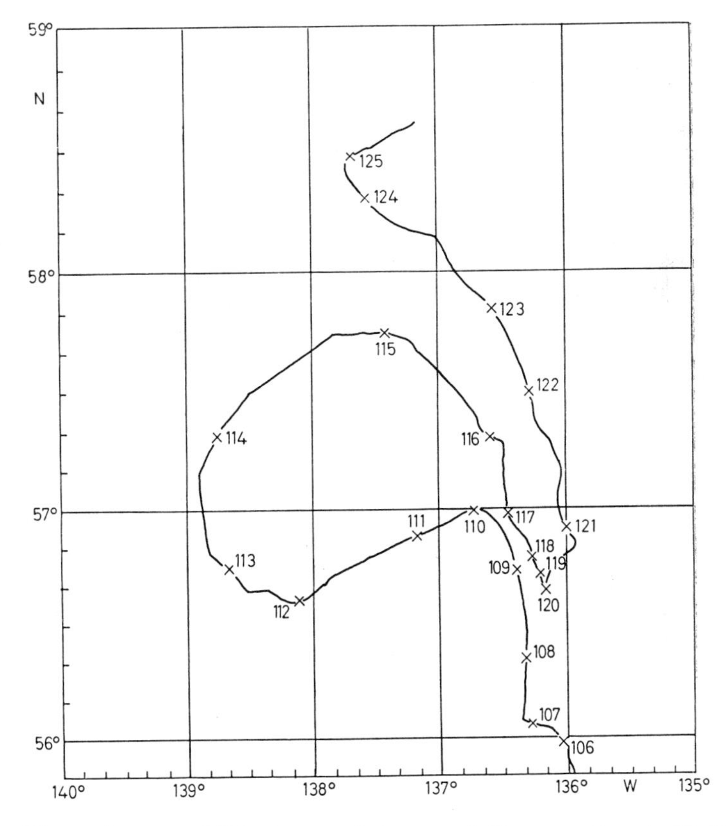

Fig. 13. – The trajectory of ID 1615 as it traversed the meander at 57° N, 138 °W. The ticks are the interpolated positions at 1200 GMT. Period covered was 16 April to 5 May 1977.

Over the period of November 1976-March 1977 two large-scale changes in the trajectories occurred. Both were associated with the surface winds. The trajectories in fig. 14 show that in November the four easternmost drifters developed a significant northward flow and that by January all the drifters were moving to the north. Figure 15 shows the monthly averaged trajectories superposed with the monthly surface pressure as obtained from FNWC. The trajectories demonstrate the dramatic response of the surface currents to the surface winds, even on a monthly averaged basis. Note also that the trajectories point to the right of the isobars. This is an indication of the Ekman response discussed in sect. **3**.

The other large-scale change is shown in the last panels of fig. 14 and 15. In March the drifters ceased their northward movement and turned to the east. This coincided with the wind field weakening and veering to the east.

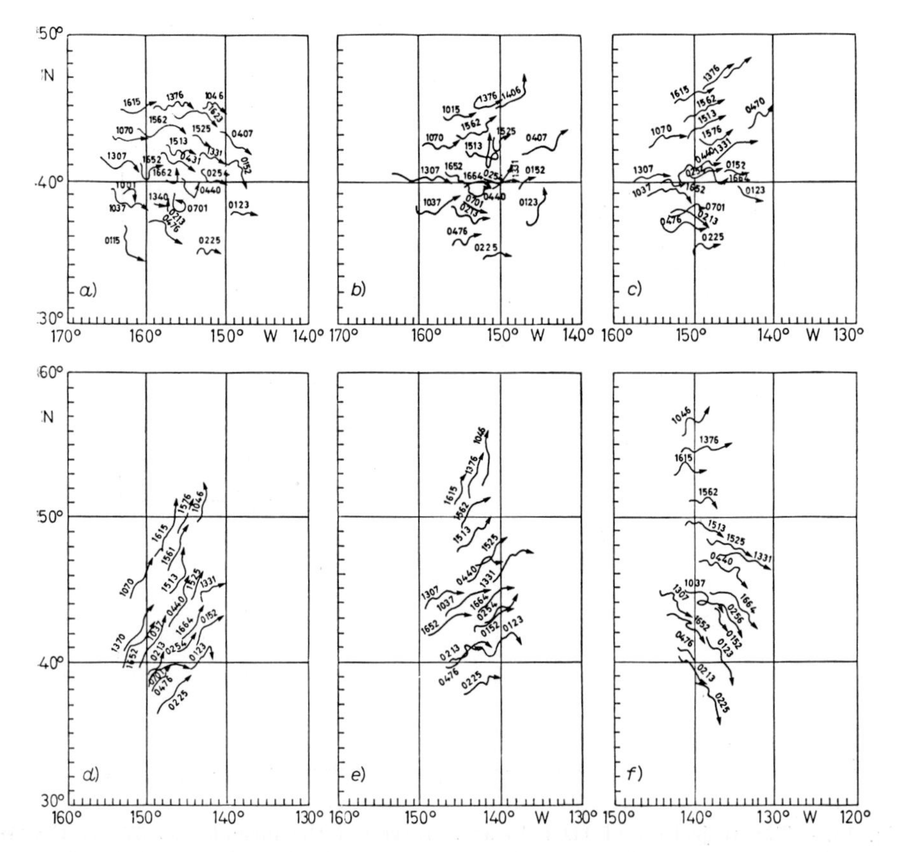

Fig. 14. – The trajectories by month for *a*) October 1976, *b*) November 1976, *c*) December 1976, *d*) January 1977, *e*) February 1977, *f*) March 1977. Only drifters which transmitted for the entire month are shown.

Figure 16 shows the trajectories of the third drifter deployment which spanned the winter of 1977-1978. The trajectories show many of the same features observed the previous year (fig. 10). These include the broad eastward flow with superposed mesoscale features. As before, the split between the subarctic and subtropical gyres seems to occur for drifters deployed along 43° N.

There are two curious features in this composite. First ID 0755 detected an intense anticyclonic circulation structure off the western coast of Canada. It was centered at 139° W, 51° N, somewhat south of the structure detected the previous year. The other curious feature is the trajectory of ID 1606

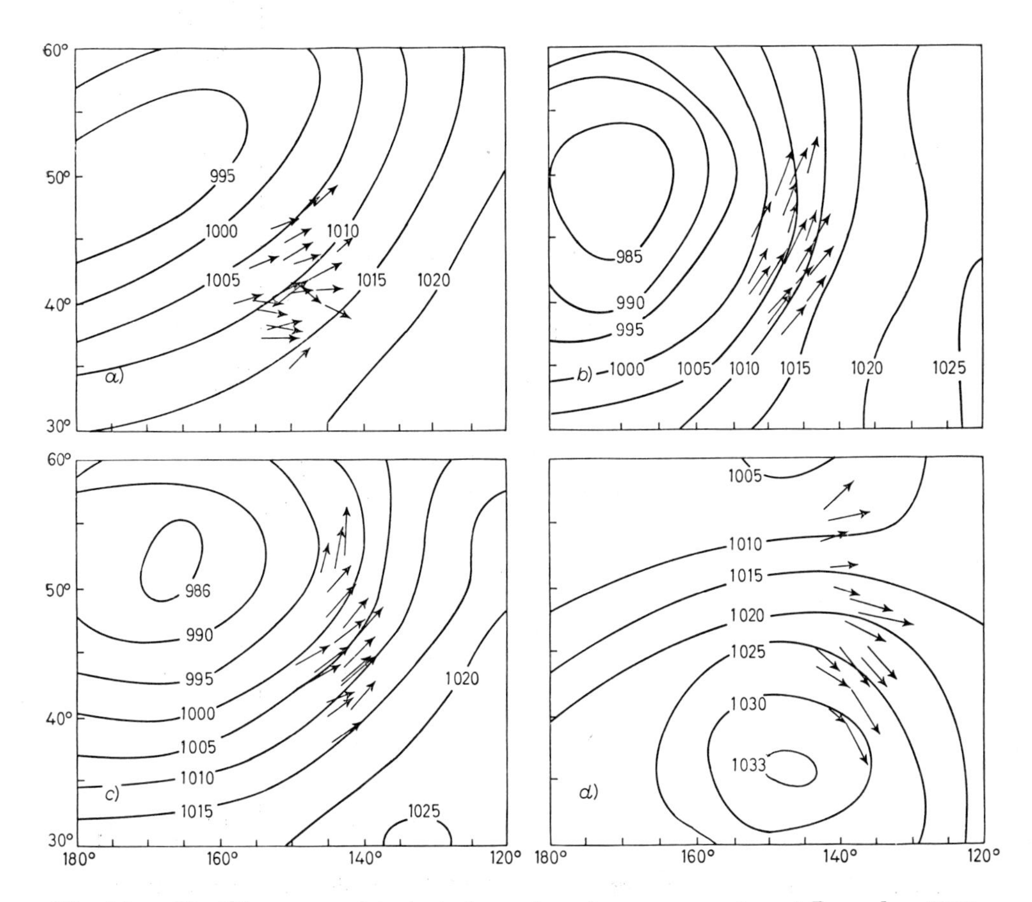

Fig. 15. – Monthly averaged trajectories and surface pressure for *a*) December 1976, *b*) January 1977, *c*) February 1977, *d*) March 1977.

which was deployed at 170° W, 30° N. This drifter spent 18 months meandering aimlessly about its deployment position, rarely wandering more than 200 km from the deployment site. Such a behavior has not been observed in any other deployment.

Conspicuously absent from this figure is the northward meander during the winter. This is because northward winds did not develop during this period.

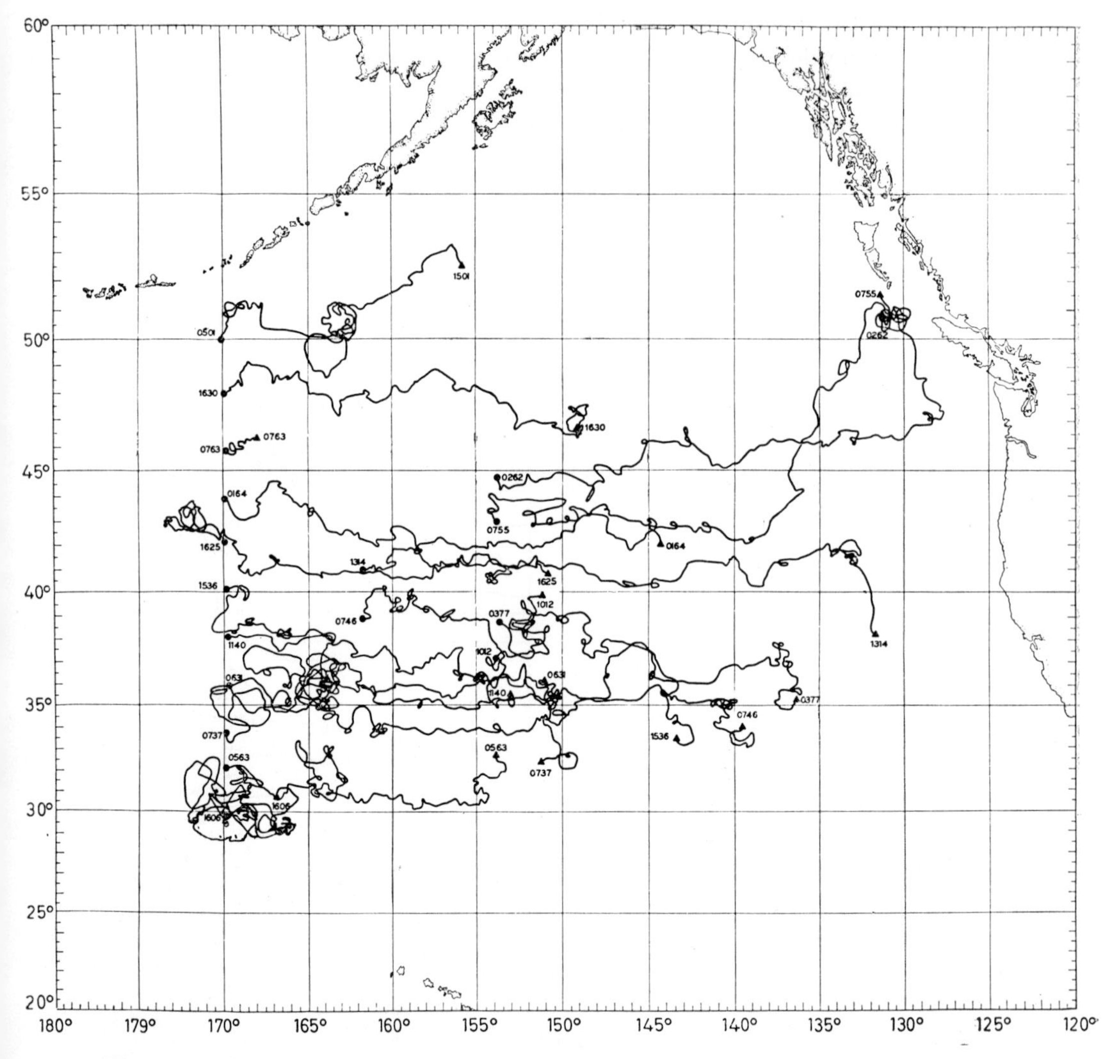

Fig. 16. – A composite of the trajectories of 17 drifters deployed during the period 1 June 1977 through 30 April 1978.

4'3. *Conclusions.* – Allowing for mesoscale variability, the trajectories for both years deployments are reasonably consistent with the mean annual dynamic topography. Both show a split between the subarctic and subtropical gyres at about 50° N. All trajectories show a northward meander in the eastern part of the Pacific. The flow around the meander is consistent with the strong northward winds observed in this region from September 1976 through February 1977. This was a period of anomalous conditions in both the atmosphere and sea surface temperature distribution; however, the connection between these is not yet well established. It is noteworthy that this feature was not observed in 1977-1978. We speculate that the anomalous conditions have occurred enough times to produce a vestige in the mean annual dynamic topography [19, 20].

The observations for both years show that all drifters deployed in the subarctic front or to the north stay in the subarctic gyre and those deployed in the subtropical front or to the south stay in the subtropical gyre. This seems to be the consequence of the smooth nature of the large-scale flow and the lack of dispersion by the mesoscale.

The drifter data show that the mesoscale produces very little dispersion in the eastern Pacific beyond a 300 km scale. This is in sharp contrast to drifter and hydrographic data from the western Pacific and Atlantic basins. The difter data do not support the concept of large-scale turbulent motion and random dispersion of clusters of parcels on time and space scales in excess of 30 days and 300 km, respectively. Finally, it seems that the drifters tend to find and follow the strongest currents.

5. – Tracking the Kuroshio.

5·1. *Background.* – As one of the major western boundary currents (WBC), the Kuroshio has long been an object of intense study. In many respects, the Kuroshio is quite similar to the Gulf Stream, another well-studied WBC. There is, however, one major difference. The Gulf Stream has only one preferred path as it flows along the eastern seaboard of the U.S., but, in the region south of the main island of Honshu, the Kuroshio follows one of two possible paths. One path follows quite closely the coastline of Honshu. In the other, a large meander develops south of Honshu. This meander is believed to extend as far south as 30° N or about 300 km from the other path. Examples of these two paths are shown in fig. 17, which was taken from bulletins that are distributed regularly by the Japan Hydrographic Department.

When the Kuroshio is in its meander mode, a cold-water mass develops between the meander and the coastline. South of the meander, a warm-core eddy is believed to exist. This meander circulates some of the Kuroshio water with the subtropical counter current system [30] and perhaps the Kuroshio itself south of Kyushu.

Because the strong currents of the Kuroshio have such a significant effect on the shipping industry, and the intrusion of the cold-water mass greatly affects the fishing industry and weather over the northern part of Honshu, Japan expends a considerable effort in monitoring the Kuroshio. This includes almost continuous ship surveys by such diverse agencies as Japan Fisheries Agency, Japan Meteorological Agency and the Hydrographic Department. These ship surveys include closely spaced expendable bathythermograph (XBT) stations, some hydrographic stations and continuous underway geo-

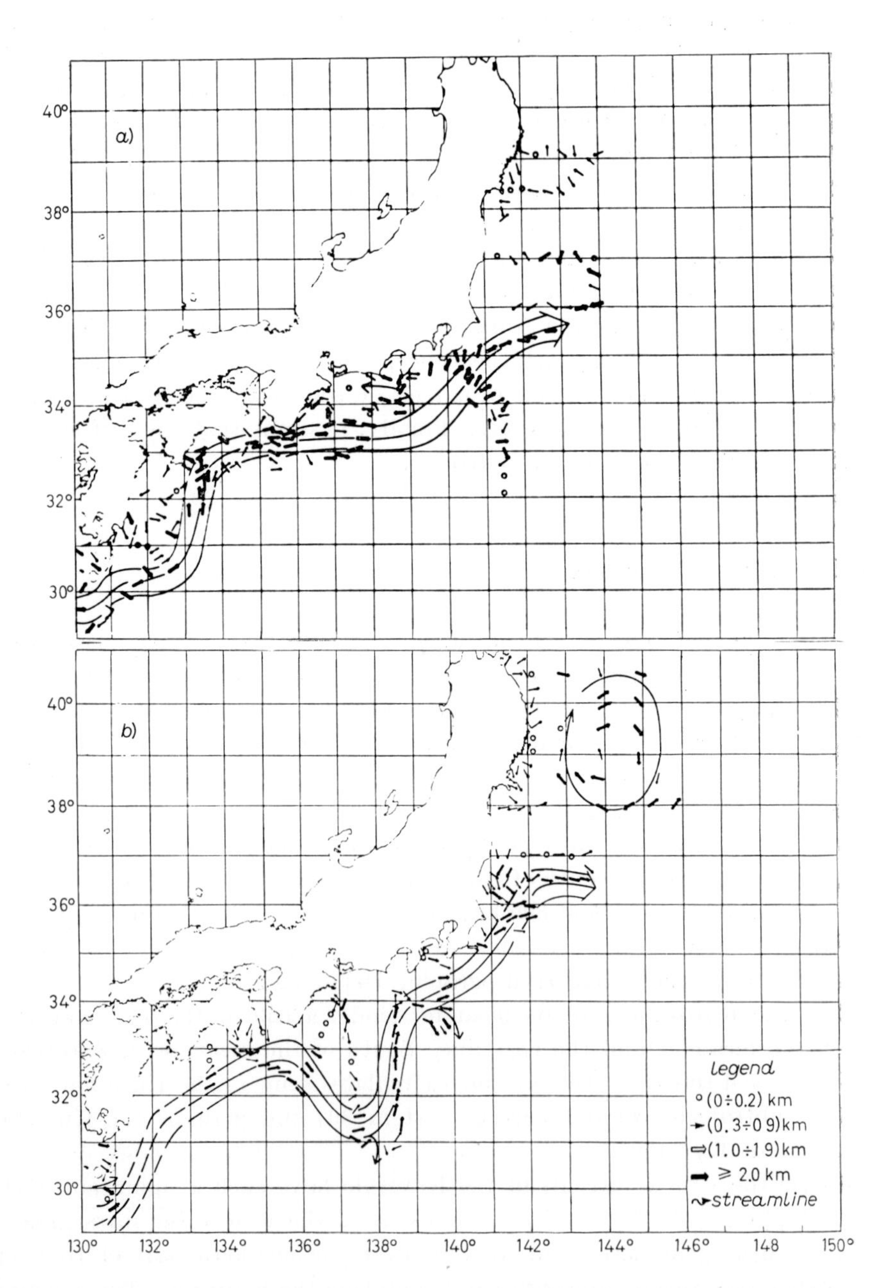

Fig. 17. – Two paths of the Kuroshio. The coast line path was observed from 1 to 17 June 1975. The meander path was observed from 1 to 15 December 1975.

magnetic electro kinetograph (GEK) measurements. The Hydrographic Department is responsible for assimilating all of the data from these diverse sources and producing a bulletin on the location and intensity of the Kuroshio. This bulletin comes out every other Saturday and includes observations made as late as the previous day. Figuring conservatively, over six hundred days of ship time per year are allocated by the Japanese government for monitoring the Kuroshio. Despite this effort, the Japanese feel that there are significant scales of motion of the Kuroshio which are not detected by either the frequency or the scale of the surveys.

Four drifters, each drogued at 100 m by a 9.2 m personnel parachute, were deployed off the island of Kyushu. The deployments were made from the R/V TAKUYO which was made available by the Hydrographic Department of the Marine Safety Agency of Japan.

This pilot experiment has a number of goals. First, we wished to compare the results of the drifter tracks and velocities with that obtained by GEK and the hydrographic surveys. The second goal was to verify the existence of the cold- and warm-water eddies. The third objective was to study the variability of the currents in the extension region.

5˙2. *Discussion of trajectories west of* 140° E. – The four drifters were deployed along a line normal to the axis of the Kuroshio just east of Honshu. The trajectories up to Julian day 59 are shown in fig. 18 along with results of the GEK survey. Drifter 106 was deployed just to the landward side of the high-speed core of the Kuroshio as determined by GEK. Drifter 130 was deployed just to the seaward side of this core. Drifters 307 and 341, respectively, were deployed out of the flank of the current. The deployment scheme was selected with the expectation that 106 would break off into the cold-core eddy and that 341 would peel off into the warm-core eddy.

Figure 18 shows that the drifters all moved at approximately the same velocity (~ 2 m/s) until Julian day 54. At that time, 341 moved off to the southeast along the meander axis. However, as indicated in fig. 18, it did not leave the Kuroshio as we had anticipated. At the other extreme was 106, which from day 56 to day 60 was entrained in a small anticyclonic eddy. It rejoined the main flow of the Kuroshio on day 62.

Figure 19 shows that the trajectories of all four drifters, along the eastern flank of the meander, are quite close, even though there is as much as eleven-day difference in the time of traverse. This supports the view of Japanese scientists that during days 47-59 the meander was in an extreme western position, and that it was beginning to move eastward. Thus in eleven days the western part of the current could have moved 20 km to the east. Such east-west migrations of the meander are well documented in the Japanese oceanographic literature.

Up to the time the drifters reached the Izu Ridge, which runs southward along 140° W, the 100 m temperature field, the current chart and trajectories

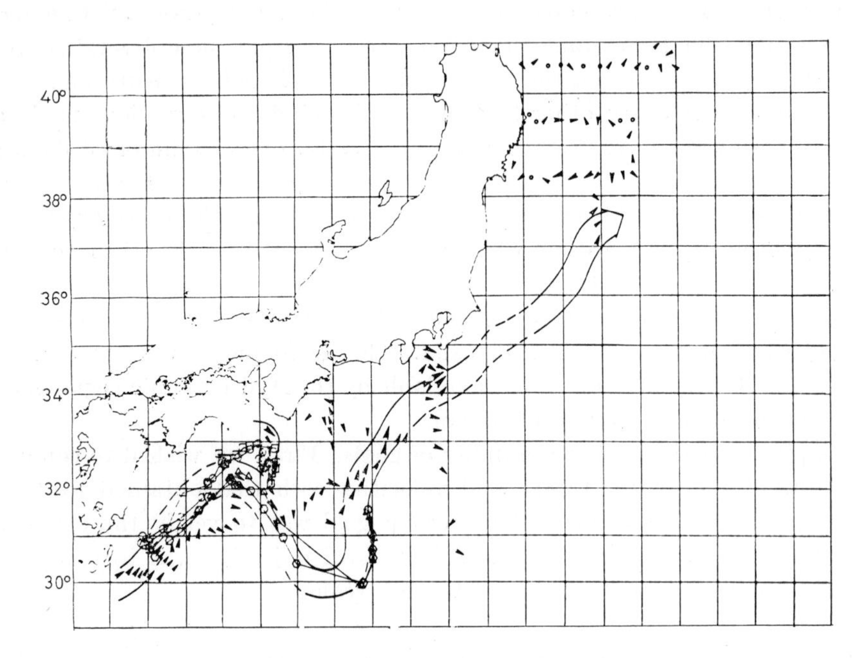

Fig. 18. – Summary of trajectories and GEK survey for Julian days 47-59.

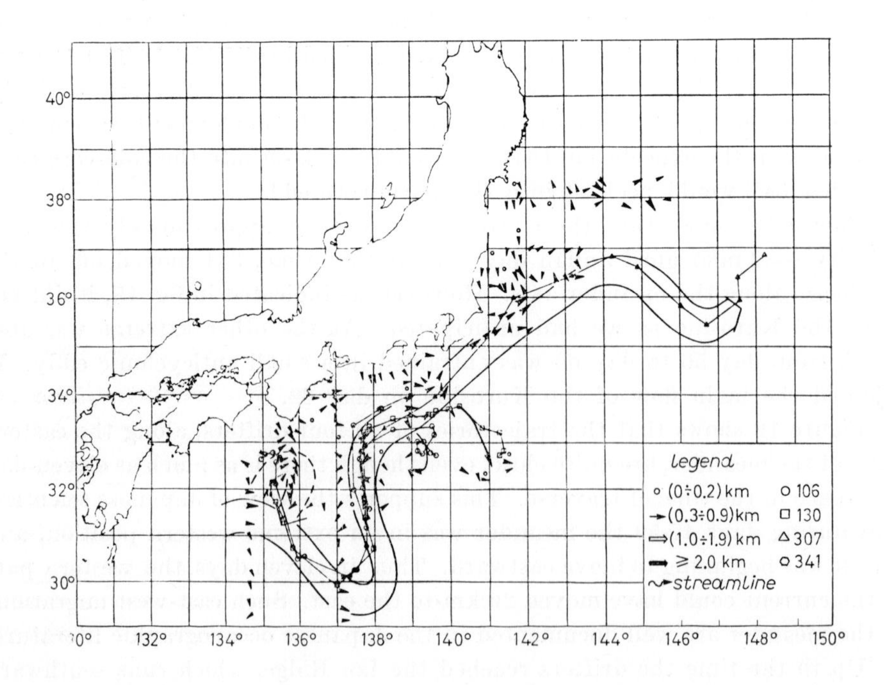

Fig. 19. – Summary of trajectories and GEK survey for Julian days 60-74.

are in excellent agreement. It is noted that 307 went straight through the Izu Ridge right in the core of the current. The track of 130 as it peels off from the main flow east of the ridge is also indicated in the current chart. Drifter 341, after it reaches the ridge, enters an anticyclonic eddy which is not inconsistent with the surface flow field as determined by GEK.

For days 75-90, only 106 and 341 remain in the survey area. The latter is seen in fig. 18 to be following the main current axis east of the Izu Ridge. Drifter 341 drifted slowly southward (about 0.2 m/s) until it reached 30° 30′. At this point it turned eastward and then northward. This track is still consistent with the GEK survey.

5'3. *Discussion of trajectories in Kuroshio extension.* – It is well known that the flow east of the Izu Ridge is quite complex. Data from numerous hydrographic surveys indicate a very complicated structure of the dynamic topography with rapid large-scale changes. Ocean fronts and large eddies are conspicuous features here.

The trajectories in this region (fig. 20 and 21) attest to this. Drifters 307 and 130 came through the Izu Ridge at the same point, but separated in time by about five days. Their initial tracks upon departing the ridge are nearly identical to 148° E. At this point, however, 130 broke off in a large warm-core eddy north of the Kuroshio. Drifter 307, on the other hand, followed a wavelike pattern to approximately 151° E, at which time it did an about-face and

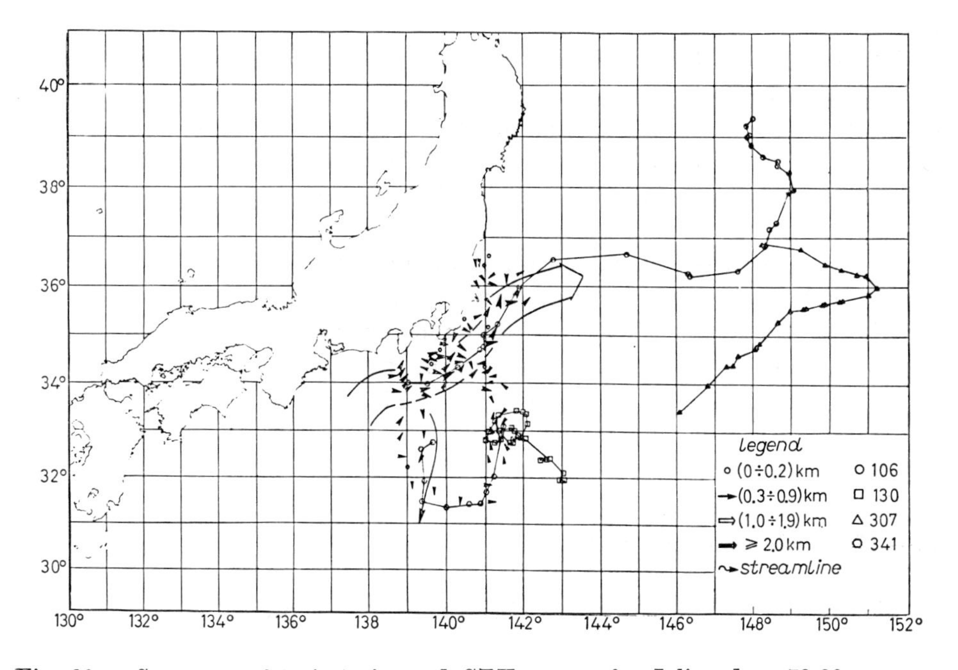

Fig. 20. – Summary of trajectories and GEK survey for Julian days 73-90.

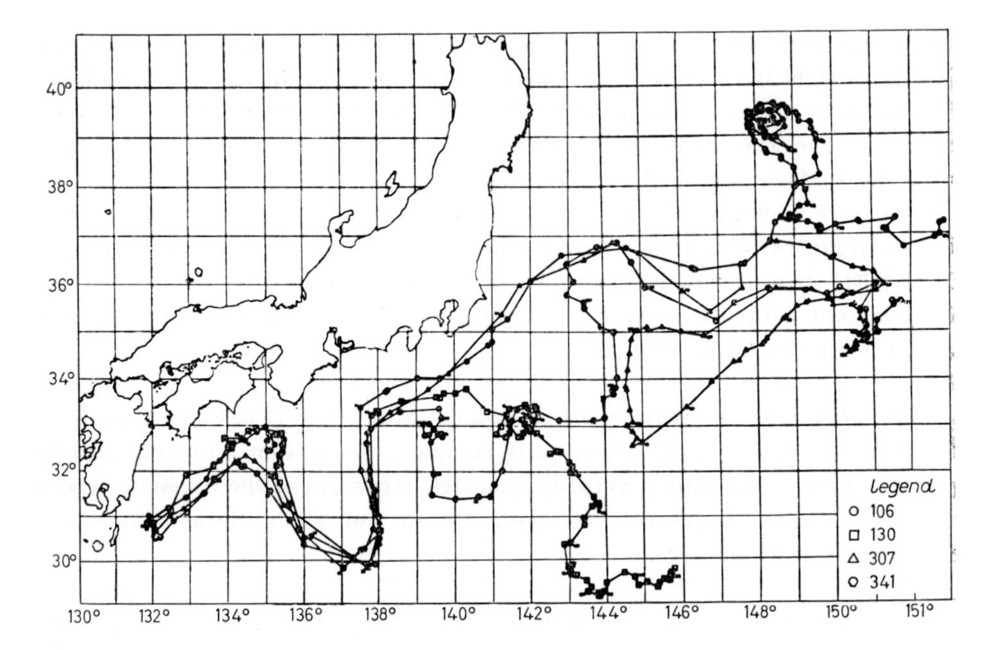

Fig. 21. – Summary of drifter trajectories from time of deployment to Julian day 125 of 1977.

started flowing to the southwest. On day 95 it turned abruptly northward and then on day 105 eastward.

Drifters 106 and 341 had been moving slowly just to the south of the mean axis of the Kuroshio. They arrived in this area by completely different routes; 341 having performed an end run around the ridge, while 106 came shooting through the gap at 33° 30′. Since day 109, however, 341 had been following closely the trajectory of 307, except that it looped much further to the north.

5·4. *Summary.* – To summarize, the agreement between the trajectories and the surveys is excellent. Main features of the Kuroshio such as path and velocity are duplicated in both the drifter and survey data sets. The trajectories, however, failed to detect either the warm- and cold-core eddies associated with the meander. Also, it was quite remarkable to find that 130 pinched off into a warm-core ring.

6. – Mesoscale eddies off Eastern Australia.

6·1. *Background.* – Eddies have been known to be present off Eastern Australia since the early 1960s. HAMON [31], in fact, raised the question: to what extent could the East Australian Current (EAC) be considered a system of eddies rather than a « current »? For a simple model he suggested that the EAC system consisted of a current coming southward adjacent to the continent and then swinging out to sea just north of Sydney (34° S). This current completed its U turn some 200 km offshore, thence to run north and east in the direction of New Zealand. At its southern end the EAC system included at least one anticyclonic (warm) eddy. HAMON suggested that eddies formed by a pinching-off of the northern current and that, once formed, they moved southward. The eddies were about 200 km across, had dynamic heights about 0.7 dyn·m above the surrounding water and surface currents up to 2.0 m s^{-1}. At a depth of 250 m the currents were reduced to half of their surface values. He found a useful empirical relation between dynamic height and the temperature measured at 240 m by mechanical bathythermographs (MBT). In wintertime he noticed that regions of high dynamic height were characterized by deep surface mixed layers. Subsequent work by HAMON, GODFREY and GREIG [32] with merchant ship drift data from near the shelf edge north of Sydney showed regions of strong current to move southward at several kilometers per day.

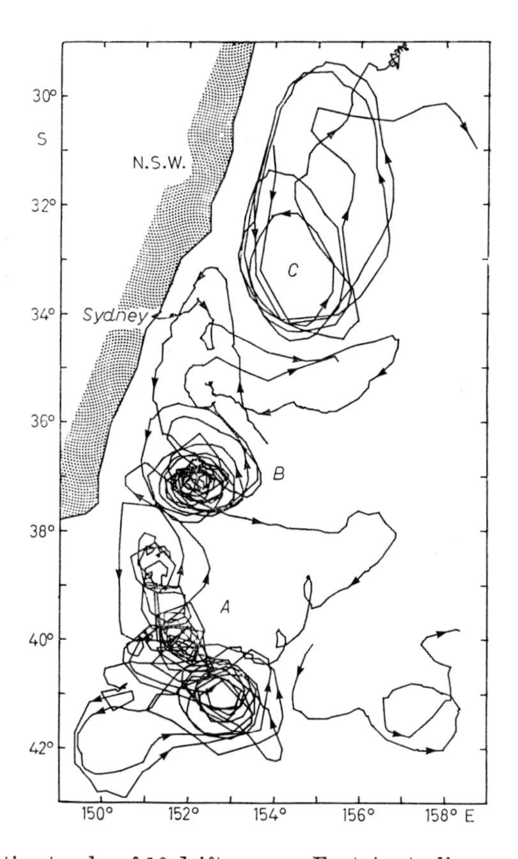

Fig. 22. – Cumulative tracks of 10 drifters near East Australian coast March-June 1977.

ANDREWS and SCULLY-POWER [33] and later NILSSON, ANDREWS and SCULLY-POWER [34] used expendable bathythermographs (XBTs) that enabled them to map eddies in more detail than had previously been possible. They found « winter » eddies to have a deep isothermal surface mixed layer and « summer » eddies to have a subsurface isothermal layer beneath a summer « cap ».

These researchers always faced a prolonged shipborne search, either when seeking new eddies, or when relocating old ones. In 1977 this was circumvented through the use of satellite-tracked drifters [35]. Twelve drifters were used, initially to locate three different anticyclonic eddies (fig. 22), and later to follow the evolution of one of these. The eddies were found to hold drifters for up to five months. Somewhat surprisingly, the eddies could be followed for up to 12 months through the joint efforts of ships and drifters. The eddies showed no consistent movement behavior. Sometimes they were seen to remain fixed for several-month long periods; at other times to move in near-straight lines, to reverse direction, and to traverse closed curves. Knowing the location of an eddy at all times from the drifters meant that even a short survey cruise could be effective. For example, a research trawler was directed to an eddy several times during the year for short periods to drop XBTs. This resulted in a better understanding of the evolution of EAC eddies. Armed with this new knowledge, but primarily without satellite-tracked drifters, researchers in 1979 were able to follow the physical, chemical and biological evolution of another eddy. In this section we cover some of the work done on EAC eddies since 1977.

6·2. *Methods.* – The drifters used in this study were 4 m long fibreglass-PVC spars or 2 m long « torpedos » with 4.5 m diameter parachutes at 20 m depth to serve as drogues. Balancing the wind drag on the exposed part of the spar with the drag of the parachute as discussed in sect. **2** reveals the slippage to be 1/200 of the wind speed; for the torpedo it is 1/500. Details of the transmitter, sea surface temperature sensor and solar charged power supply are given in [36]. In the present study the drifters were position fixed by the NASA NIMBUS-VI satellite.

6·3. *Results.*

Eddy B, 1977. The 1977 drifter track charts (fig. 22) both reveal the complexity of the circulation and enable the three anticyclonic eddies A, B and C to be followed. Eddy A moved down to the SE, taking two drifters with it, eddy B, also with two drifters, remained essentially stationary. Some of the complexity can be unravelled by preparing a latitude time series as shown in fig. 23. Eddy A's latitude increased through to May and then slowly decreased, while eddy B's latitude remained essentially unchanged until July, after which it showed an oscillation with period of approximately three months. The time series showed that, whenever two drifters were in eddy *B* together, their latitude oscillations due to rotation were commonly of the same period and with a consistent phase relationship. This suggests that at least an annulus of the eddy was rotating as a rigid body. Similar behavior was seen in an Indian

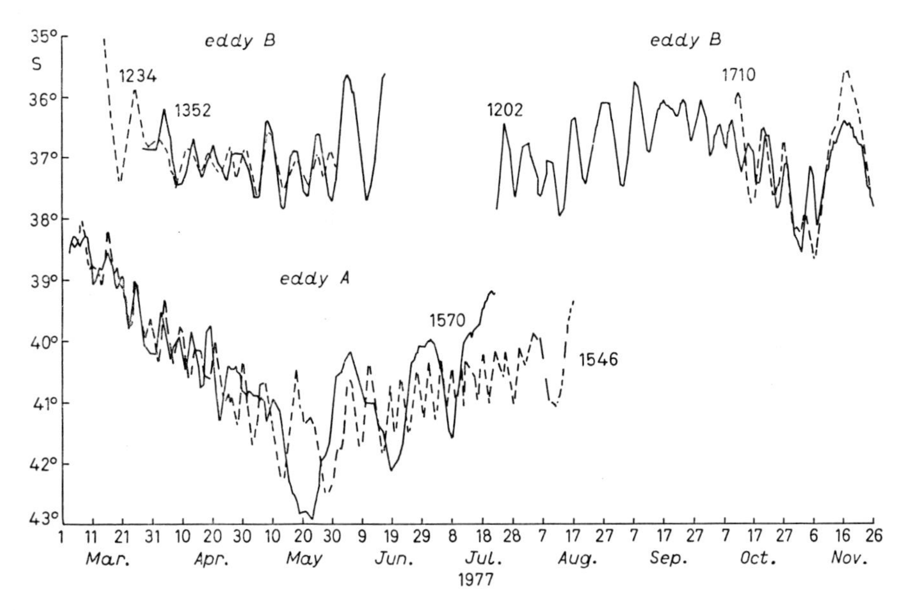

Fig. 23. – Latitude time series for drifters in eddies *A* and *B* during 1977.

Ocean eddy [37]. Exceptions to this occurred whenever one drifter was near the eddy center. The inner drifter rotated faster than the outer drifter. The two drifters in eddy *A* were released with parachute tether lines of 20 m and 200 m, respectively. From May the drifter with the shorter tether line rotated around the eddy four to five times faster than the other—in keeping with decreasing current speed with increasing depth reported by HAMON [31].

Latitude and longitude time series were used to determine the movements of the centers of eddies *A* and *B* and this information is given in fig. 24. Points are plotted at 10-day intervals: eddy *A* can be seen to have moved southward and then slowly northward; eddy *B* can be seen to have remained near 37° S, 152° E for the first few months and then to have described an anticlockwise arc. The dashed line gives a track inferred from ship data after both drifters in the eddy had escaped.

Figure 25 shows the tracks of two drifters, 1234 and 1352, in eddy *B* for a short period that included a two-day XBT survey. The dynamic height patterns determined from the XBT data and the drifter tracks show good agreement.

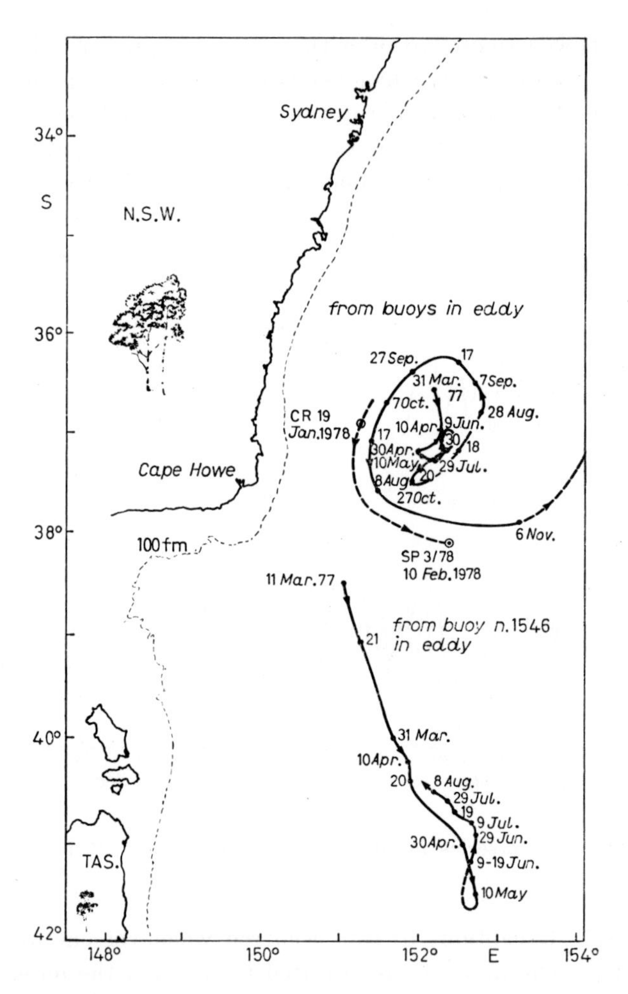

Fig. 24. – The movements of the centers of eddies A and B as tracked by drifters and observed by ship.

The complexity of the tracks of these drifters in eddy B can also be seen in fig. 26, which, for clarity, is broken into intervals April 26-May 5, May 5-22, May 22-31. In the second interval each drifter described two loops, a large loop containing a smaller one, while the two drifters maintained a consistent phase relationship. In the first and last intervals the drifter that appears to be nearer the eddy center, 1234, has described an extra loop.

A variety of factors can be involved in producing a drifter trajectory in an eddy: the rotation of the water in the eddy, the movement of the eddy, the shape of the eddy, the effect of the wind moving the upper layers of the ocean nearer or farther from the eddy center, the depth of the drogue of the drifter.

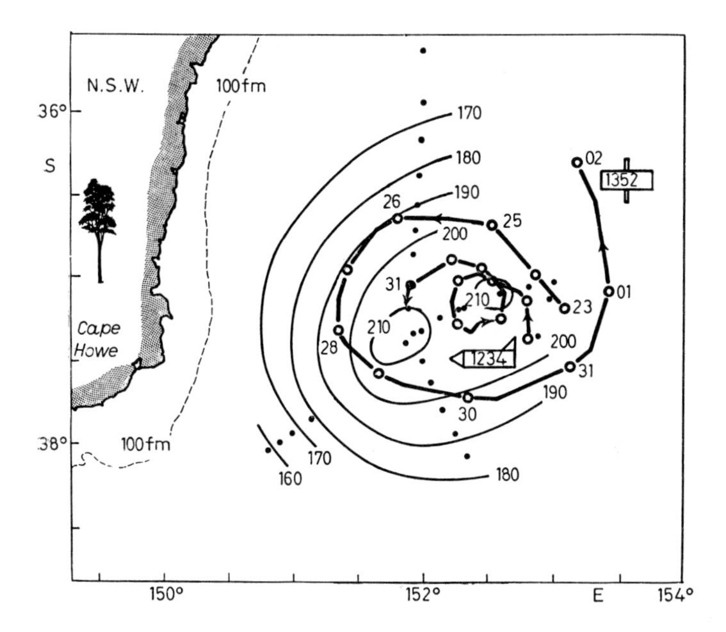

Fig. 25. – Cruise SPG/77 XBT track and surface dynamic topography $D(0/1300)$ dyn·cm for 26-29 May 1977 and drifter tracks with days of the month are superimposed.

To obtain a circular anticlockwise trajectory of radius A and angular velocity ω the x (east) and y (north) components can be expressed as

$$x = A \sin \omega t, \qquad y = -A \cos \omega t. \tag{20}$$

A linear time dependence added to the x component modifies (20) to

$$x = A \sin \omega t + Bt,$$

to produce a form looping to the east. Then, if there is the case of the eddy being carried in an anticlockwise circulat path of radius C and angular frequency ω_1, the x and y components become

$$x = A \sin \omega t + C \sin \omega_1 t. \tag{21}$$

$$y = -A \cos \omega t - C \cos \omega_1 t. \tag{22}$$

Radii and angular frequencies can be selected to produce trajectories similar to the ones in fig. 26b). For example, fig. 27 shows the trajectories that two drifters at different radii would follow in a circular eddy that itself was moving in a circle at half the water rotation rate. Perhaps, also, the shape of the eddy is an important factor: elliptical eddies have been observed with a current meter array [38] and with satellite imagery [39]. Complex looping trajectories could be produced by precessing elliptical eddies, or by circular eddies with an eccentric wobble.

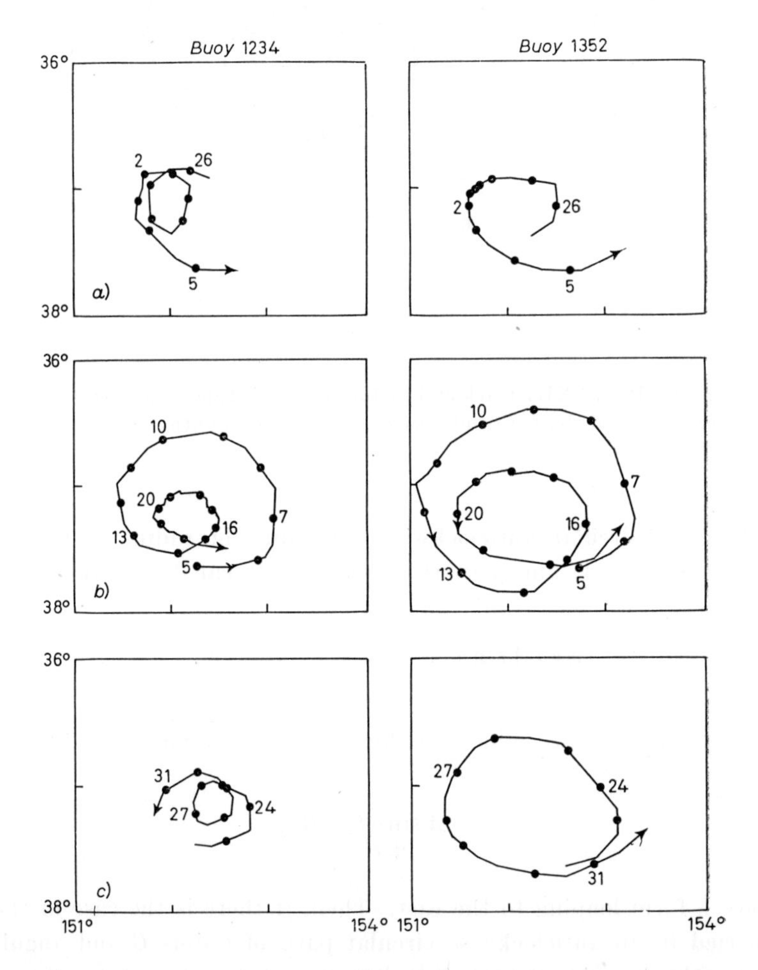

Fig. 26. – The trajectories of drifters 1234 and 1352 in eddy *B* April/May 1977: *a*) April 26-May 5, *b*) May 5-22 and *c*) May 22-31. The dots show estimated local noon positions.

The drifters in eddy *B* enabled its position to be known at all times and its evolution to be followed from ship surveys. The surface isothermal mixed layer cooled and deepened during the winter. At the end of September summer heating commenced and the isothermal layer was capped over by warmer water. It retained the temperature that was established during cooling and this then served as a « signature » for later identification ship surveys.

Eddy J, 1979. This eddy was first identified at 33° S, 154° E in March 1979 and it was possible to follow its evolution at least to December 1979. In March 1979 it was characterized by strong anticyclonic flow of up to 2.5 m s^{-1} and a dynamic relief of 1.0 dyn·m relative to the surrounding waters. The transports across the N and WSW radii were 35 and 50 Sverdrup, respectively, suggesting entrainment of water into the NW quadrant. The eddy had a « signature » in the form of a deep isothermal layer of temperature 19.7 °C and salinity 35.64 % between 220 and 380 m (fig. 28). This was probably caused by 1978 winter cooling and is identified as the « 1978 layer ».

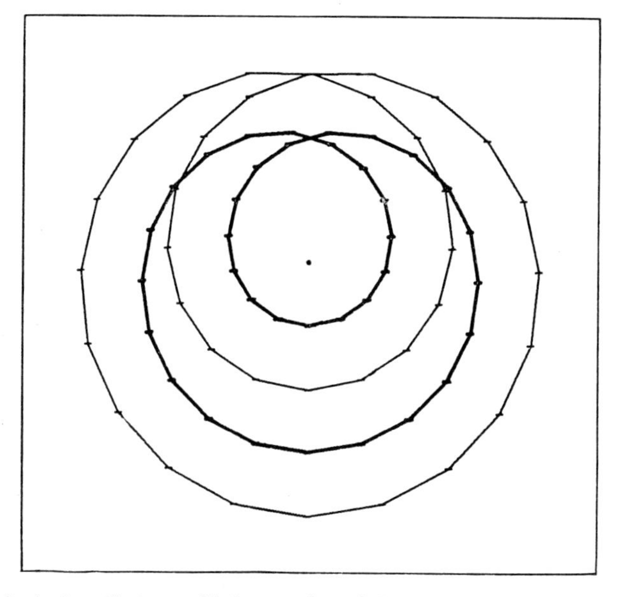

Fig. 27. – Trajectories that would be produced by two drifters at different radii in an eddy that itself was moving in a circle at half the rate that its water rotated. Horizontal scales are arbitrary.

As time proceeded, the 1978 layer was eroded way by unknown processes, while the 1979 layer cooled and mixed deeply. It reached down to join the 1978 layer and then down to a depth of 370 m at the eddy center. As in the case of eddy *B*, at the end of September the deep surface mixed layer became isolated from the surface by summer heating and perhaps by new warm EAC water from the north. This layer then retained its late-September temperature and salinity values as signatures and slowly decreased in vertical extent.

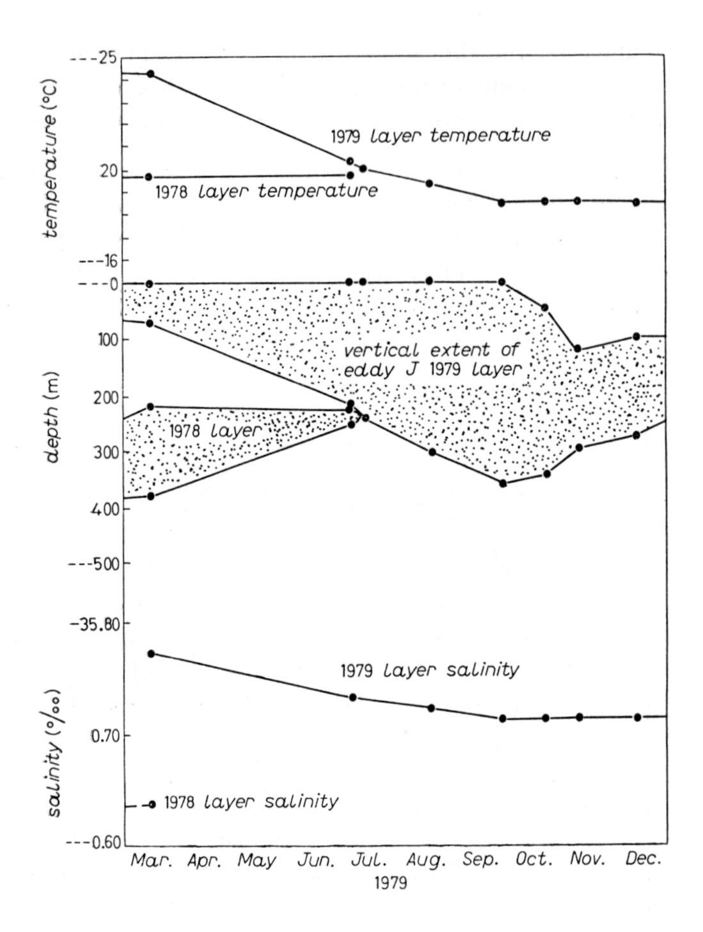

Fig. 28. – A time series to show the vertical extent, temperature and salinity of layers at the center of eddy *J* in 1979.

6·4. *Summary.* – Warm-core eddies are a major feature of the East Australian Current (EAC) system. Drifters enabled one eddy in particular to be carefully studied from the time that it pinched off from an EAC meander, through the winter cooling phase when a deep isothermal surface mixed layer 300 m formed within it, and then through to the « summer capping » of the temperature profile by solar heating. After this the isothermal layer became a signature for the eddy. The EAC eddies are now known to have associated with them a dynamic relief of as much as 1 m, current up to 2.5 m s^{-1} ($\sim$ 5 knots), rotation times generally around 1 week, diameters about 200 km, lifetimes of at least one year, movements that can be linear, curved, or, at times, capable of reversing. Some of the complex trajectories followed by drifters in the eddies could be due to the movements of the eddy centers.

7. – Envoi.

These lectures have documented some of the first uses of a fundamentally new experimental technique in oceanography. As with all new experimental approaches, an aspect of these investigations has been to establish the validity of the techniques. This is accomplished by error analysis, development of calibration procedures and repeated experiments with complementary data obtained from established observational techniques. The present status of error assessments and calibration procedures was discussed in sect. **2**. The discussion there indicates that future efforts likely will be directed towards improving the droguing techniques and quantifying the drogue depth and efficiency. This is essential if the physical processes affecting ocean surface layer motions and dispersion processes are to be elucidated.

Sections **3** through **6** presented results of some studies in which the experiments were repeated and/or complementary data were obtained which corroborated the drifter data. Section **3** presented a quantitative evaluation of competing models of surface current response to the wind. The study identified an important issue, namely the role of Stokes drift in the surface layer response.

Sections **4** and **5** presented a descriptive study of the circulation in the North Pacific. Three circulation regimes were depicted. In the Eastern Pacific the drifter data over a two-year period qualitatively agree with older descriptive studies which utilized field of mass observations averaged over many years. Broadly speaking, the circulation can be characterized by a east drift current on which are superposed some mesoscale features. In the Western Pacific the data show strong but very organized flow in the Kuroshio region west of 140° E. This picture is very consistent with contemporary data from GEK and the field of mass. The flow in the Kuroshio extension (east of 140° E to about 180°) is dominated by mesoscale variability. Superposed on this is a modest eastward drift. There is little support in these data for separate east and west subtropical gyres. Although one drifter ultimately migrated into the North Equatorial current and ended up aground in the Philippines, the others appeared to establish contiguous large-scale circulation between the Eastern and Western Pacific. Important unanswered questions are: what is the source of the intense mesoscale motions in the Western Pacific and what is the connection of the subtropical circulation with the equatorial circulation.

Section **6** presented detailed descriptions of the kinematics of intense and quasi-permanent mesoscale structures off the east coast of Australia. These features are quite similar to those described briefly in sect. **4** and **5**; in fact the dynamical processes are identical. The comparison of drifter data with repeated hydrographic observations gave a very consistent picture of the evolution of these features. Simple kinematic models of the trajectories compared well with the observations. This indicates a remarkable balance of forces acting within the eddies for extremely long periods. Unanswered as yet are the questions of what are the generating mechanisms for these structures and what is their long-term fate.

REFERENCES

[1] J. V. DENNIS and C. R. GUNN: *Econ. Biol.*, **25**, 407 (1971).
[2] A. D. KIRWAN jr., G. MCNALLY, M.-S. CHANG and R. MOLINARI: *J. Phys. Oceanogr.*, **5**, 361 (1975).
[3] G. R. CRESSWELL and J. GARRETT: Australia CSIRO Division of Fisheries and Oceanography, Report 115 (1980).
[4] A. D. KIRWAN jr., G. MCNALLY and S. PAZAN: *J. Phys. Oceanogr.*, **8**, 1146 (1978).
[5] S. D. SMITH and E. G. BANKE: *Q. J. R. Meteorol. Soc.*, **101**, 665 (1975).
[6] C. A. FRIEHE and S. E. PAZAN: *J. Appl. Meteorol.*, **17**, 1488 (1978).
[7] O. H. SHEMDIN: *J. Phys. Oceanogr.*, **2**, 411 (1972).
[8] V. W. EKMAN: *Ark. Mat. Astron. Pys.*, **2**, 1 (1905).
[9] O. S. MADSEN: *J. Phys. Oceanogr.*, **7**, 248 (1977).
[10] M. S. LONGUET-HIGGINS: *Deep-Sea Res.*, **16**, 431 (1969).
[11] K. E. KENYON: *J. Geophys. Res.*, **74**, 6991 (1969).
[12] J. P. IANNIELLO and R. W. GARVINE: *J. Phys. Oceanogr.*, **5**, 47 (1969).
[13] P. LANGE and M. HUHNERFUSS: *J. Phys. Oceanogr.*, **8**, 142 (1978).
[14] G. NEUMANN and W. J. PIERSON: *Principles of Physical Oceanography* (New York, N. Y., 1966), p. 315.
[15] A. D. KIRWAN jr., G. MCNALLY, S. PAZAN and R. WERT: *J. Phys. Oceanogr.*, **9**, 401 (1979).
[16] M. S. CHANG: *J. Geophys. Res.*, **74**, 1515 (1969).
[17] T. P. BARNETT and K. E. KENYON: *Rep. Prog. Phys.*, **38**, 667 (1975).
[18] E. J. WILLIAMS: *Regression Analysis* (New York, N. Y., 1959).
[19] K. WYRTKI: *J. Phys. Oceanogr.*, **5**, 450 (1975).
[20] J. L. REID and R. S. ARTHUR: *J. Mar. Res.*, **33**, *Suppl.*, 37 (1975).
[21] G. I. RODEN: *J. Phys. Oceanogr.*, **7**, 41 (1978).
[22] T. C. ROYER: *Science*, **199**, 1063 (1978).
[23] A. D. KIRWAN jr., G. MCNALLY, E. REYNA and W. J. MERRELL jr.: *J. Phys. Oceanogr.*, **8**, 937 (1978).
[24] H. W. SVERDRUP, M. W. JOHNSON and R. H. FLEMING: *The Oceans, Their Physics, Chemistry and General Biology*, Chapt. VII (New York, N. Y., 1942).
[25] S. TABATA: *Atmosphere*, **13**, 133 (1975).
[26] S. TABATA: *J. Fish. Res. Board Can.*, **18**, 1073 (1958).
[27] S. TABATA: *Trans. R. Soc. Can., Sect.* 3, *Ser.* IV, 367 (1965).
[28] J. NAMIAS: *J. Phys. Oceanogr.*, **1**, 65 (1971).
[29] W. B. WHITE and A. B. WALKER: *J. Geophys. Res.*, **79**, 4517 (1974).
[30] K. RIKIISHI and K. YOSHIDA: *Rec. Oceanogr. Works Jpn.*, **12**, 31 (1974).
[31] B. V. HAMON: *Deep-Sea Res.*, **12**, 899 (1965).
[32] B. V. HAMON, J. S. GODFREY and M. A. GREIG: *Aust. J. Mar. Freshwater Res.*, **26**, 389 (1975).

[33] J. C. ANDREWS and P. SCULLY-POWER: *J. Phys. Oceanogr.*, **6**, 756 (1976).
[34] C. S. NILSSON, J. C. ANDREWS and P. SCULLY-POWER: *J. Phys. Oceanogr.*, **7**, 659 (1977).
[35] C. S. NILSSON and G. R. CRESSWELL: *The formation and evolution of East Australia current eddies*, in press in *Prog. Oceanogr.* (1980).
[36] G. R. CRESSWELL, G. T. RICHARDSON, J. E. WOOD and R. WATTS: Australia CSIRO Division of Fisheries and Oceanography, Report 82 (1978).
[37] G. R. CRESSWELL: *Deep-Sea Res.*, **24**, 1203 (1977).
[38] M. N. KOSHLYAKOV and Y. M. GRACHOV: *Deep-Sea Res.*, **20**, 507 (1973).
[39] T. W. SPENCE and R. LEGECKIS: *J. Geophys. Res.*, **86**, 1945 (1981).

Planetary Solitary Waves and Their Existing Solutions in the Context of a Unified Approach.

P. Malanotte Rizzoli

Istituto per lo Studio della Dinamica delle Grandi Masse, CNR - Venezia, Italia
I.G.P.P., Scripps Institution of Oceanography - La Jolla, Cal.

1. – Introduction.

In this lecture a general review will be given illustrating the variety of circumstances in which solitary waves are found as solutions to the equations of motion valid for geophysical flow systems. In particular, we shall deal with those motions which are classified as planetary, that is with length scales of the order of the sizes of oceanic basins. Features like these include the longest waves in the atmosphere and the oceans. As, moreover, we shall deal with two (or three)-dimensional systems, our geophysical examples will have the appearance of large eddies.

The geophysical cases in which eddies can be observed experimentally range from cyclonic (or anticyclonic) shapes in the atmospheric isobaric maps, isolated or coupled in typical high-low pressure centers, eddy rings separating from major ocean currents, like the Gulf Stream in the Atlantic and the Kuroshio in the Pacific, and wandering through the ocean with long lifetimes; even to the giant eddies lasting for centuries like the Red Spot in the Jovian atmosphere.

Why are solitary-wave solutions important to model these planetary motions of geophysical systems? Most of our understanding of the oceanic or atmospheric planetary motions has been based either on analytic linear theories or on nonlinear numerical studies. Experimental data, however, taken during experiments like MODE (Mid Ocean Dynamics Experiment) and the following POLYMODE prove that linear theories are quite inadequate to model and predict these motions. Therefore, oceanographers dealing with the effects of nonlinearity on mesoscale ocean dynamics have usually approached the problem from the viewpoint that motion is turbulent. The effect of turbulent nonlinear interactions is to transfer energy among various spatial scales and to cause a loss of correlation between scales of motion which were initially correlated. Thus the system loses the memory of the initial state or, in other words, the flow loses predictability. Analytical turbulent theories, in fact, and related numerical experiments, bear upon the random-phase approximation for the Fourier—or normal—modes constituting the system.

However, there are fluid motions for which the nonlinearity plays a very different role, serving rather to preserve phase information and correlation against wave dispersion effects. These motions are those described by solitary-wave models. It is enough to think that an isolated initial disturbance evolving according to the Korteweg-de Vries (KdV) equation breaks up into a set of solitons to understand the organizing role which can be assumed by nonlinearity. Obviously, the predictability of such motions is greatly enhanced when solitons are generated.

For this reason, nonlinear solitary-wave models have received an increasing amount of attention and a good deal of the most recent literature has been devoted to find solitary solutions to the equations of motion governing large-scale geophysical flows under quite a variety of circumstances.

The first point of this lecture will be, therefore, to explore the variety of circumstances in which planetary waves may be found and to show corresponding experimental examples to which different authors have related their solitary eddies.

In the second part, a specific model will be chosen among those discussed in the first part, geophysically significant and particularly simple from the mathematical point of view. The procedure used in the first part will be applied to this specific model, thus allowing us to establish general criteria for the existence of solitary-wave solutions.

2. – A general potential-vorticity conservation equation for planetary motions.

The variety of existing solitary-wave solutions for mesoscale motions will be now investigated as well as their relationships according to general criteria.

To do this, a unified approach shall be presented, in the context of which all the existing solitary-wave models discovered by different authors can be derived and classified. This procedure is based upon a general, simplified dynamical equation which includes all important effects for mesoscale oceanic and atmospheric motions. This procedure has already been used in various papers cited in the following, particularly by FLIERL [1].

We consider a conserved quantity which can be thought of as a generalization of the potential vorticity for a stratified fluid. In formulating this general conservation law, we try to model the effects of a mean shear flow, stratification, topography, all in a simplified form. One must specify, however, that this model is not the exact law of conservation of potential vorticity for a stratified fluid. In its most general form, in fact, it cannot be derived exactly from the corresponding shallow-water momentum conservation equations, even though it is exactly correct in its quasi-geostrophic approximation. This general conservation model is, however, very useful, because from it all planetary solitary waves can be derived in different approximations and with a unified, very simple approach. For the single derivation of a specific solution the reader will be referred to the related author.

We consider, therefore, the following conservation equation:

$$\left(\frac{\partial}{\partial t} + U\frac{\partial}{\partial x} + \psi_x\frac{\partial}{\partial y} - \psi_y\frac{\partial}{\partial x}\right)\frac{\nabla^2\psi + f(y) - \partial U/\partial y}{1 - b(y) + \psi/f(y)\cdot R^2(y)} = 0\,, \tag{2.1a}$$

where subscripts indicate partial derivatives and

ψ = streamfunction of solitary wave, defined as

$$u = -\psi_y\,, \qquad v = +\psi_x\,;$$

$f(y) = f(0) + \beta y = f_0 + \beta y$ = Coriolis parameter in its β-plane approximation;

$U(y)$ = barotropic advective mean flow, latitude dependent;

$$\frac{D(y)}{D_0} = 1 - b(y) + \frac{\psi}{f(y)\cdot R^2(y)} \quad \text{normalized (generalized) depth}\,,$$

D_0 being the mean depth; $b(y)$ is, therefore, the topography variation around the mean depth.

The potential verticity, therefore, includes

$\omega = \nabla^2\psi = v_x - u_y$, relative vorticity of the solitary wave;

U_y = relative vorticity of the barotropic mean flow,

$f(y)$ = planetary vorticity (Earth rotation).

Dividing through the generalized depth gives the vorticity inputs due to the stretching or compressing of the water column determined by the topography variations. In particular, the term

$$\psi/f(y)\cdot R^2(y)$$

models the pycnocline displacement, that is a divergent term (equivalent to a surface elevation). Thinking, for instance, of a two-layer fluid, this term gives the variation in the height—and hence relative vorticity—of a fluid column limited at its lower end by a pycnocline interface. $R(y)$ is the Rossby radius of deformation, giving the scale of most energetic eddies in a stratified (or barotropic) fluid. The Rossby radius is, respectively,

$$R(y)=\frac{\sqrt{gH}}{f(y)} \qquad \text{for a barotropic flow at constant depth } H \text{ and}$$

$$R(y)=\frac{NH}{f(y)} \qquad \text{for a stratified fluid with } N^2=-\frac{g}{\varrho}\frac{\partial\varrho}{\partial z},$$

the Brünt-Vaisälä frequency.

The quasi-geostrophic approximation of (2.1a) is obtained in the limit

$$b(y)-\frac{\psi}{fR^2}\ll 1\,,$$

that is the topography variation plus pycnocline displacement are much less than the mean depth.

Then, Taylor expanding the denominator of (2.1a), one obtains the corresponding quasi-geostrophic model:

$$\left(\frac{\partial}{\partial t}+U\frac{\partial}{\partial x}+\psi_x\frac{\partial}{\partial y}-\psi_y\frac{\partial}{\partial x}\right)\left\{\nabla^2\psi+f(y)-\frac{\partial U}{\partial y}+f_0\cdot b(y)-\frac{\psi}{R_0^2}\right\}=0 \tag{2.1b}$$

with

$R=R(0)$ the now constant Rossby deformation radius at the average latitude.

We look for solitary-wave solutions steadily translating in the x-direction. Therefore, we pass to the co-ordinate system defined by

$$s=x-ct\,,\qquad \tau=t\,.$$

In this system, (2.1a) and (2.1b) become

$$J\left(\frac{\nabla^2\psi+f(y)-U_y}{1-b(y)+\psi/fR^2},\ \psi+cy-\int^y U\,\mathrm{d}y\right)=0 \tag{2.2a}$$

and

$$J\left(\nabla^2\psi+f(y)-U_y+f_0\cdot b(y)-\frac{\psi}{R_0^2},\ \psi+cy-\int^y U\,\mathrm{d}y\right)=0 \tag{2.2b}$$

with

$$J(a,b)=\frac{\partial a}{\partial x}\frac{\partial b}{\partial y}-\frac{\partial a}{\partial y}\frac{\partial b}{\partial x} \qquad \text{the Jacobian of } (a,b)\,.$$

This is equivalent to writing

$$\nabla^2\psi + f(y) - U_y = \left[1 - b(y) + \frac{\psi}{fR^2}\right] P\left(\psi + cy - \int^y U\,dy\right) \tag{2.3a}$$

and

$$\nabla^2\psi + \beta y - U_y + f_0\cdot b(y) - \frac{\psi}{R_0^2} = \hat{P}\left(\psi + cy - \int^y U\,dy\right); \tag{2.3b}$$

$P(\xi)$, and its quasi-geostrophic approximation $\hat{P} = P - f_0$, are the potential-vorticity functionals.

Let us make use of the fact that we are looking for solitary-wave solutions, that is

$$\psi \to 0 \qquad\qquad \text{as } x \to \pm\infty .$$

Then, in the asymptotic limit, following lines of constant

$$\xi = \psi + cy - \int^y U\,dy$$

extending from $-\infty$ to $+\infty$, we obtain the asymptotic forms of the functional P and, in the quasi-geostrophic approximation, $\hat{P}$:

$$P_\infty\left(cy - \int^y U\,dy\right) = \frac{f(y) - U_y(y)}{1 - b(y)}, \tag{2.4a}$$

$$\hat{P}_\infty\left(cy - \int^y U\,dy\right) = \beta y - U_y + f_0\cdot b(y). \tag{2.4b}$$

Using this asymptotic form for the functional, we can rewrite the model equations (2.3a), (2.3b) as:

general

$$\nabla^2\psi - \frac{1 - U_y/f(y)}{1 - b(y)}\frac{\psi}{R^2(y)} = \left[1 - b(y) + \frac{\psi}{fR^2(y)}\right]\left[P\left(\psi + cy - \int^y U\,dy\right) - P_\infty\left(cy - \int^y U\,dy\right)\right]; \tag{2.5a}$$

quasi-geostrophic

$$\nabla^2\psi - \frac{1}{R_0^2}\psi = \hat{P}\left(\psi + cy - \int^y U\,dy\right) - \hat{P}_\infty\left(cy - \int^y U\,dy\right). \tag{2.5b}$$

The definition used for P_∞ (or $\hat{P}_\infty$) is valid only for those isolines of $\xi = \psi + cy - \int^y U\,dy$ which extend continuously from $-\infty$ to $+\infty$. We shall call « streaklines » [1] lines of constant ξ.

In many solutions, however, it can happen that also closed streaklines exist. Thus we now encounter the fundamental distinction leading to the two basic categories of solitary-wave models.

I) P (or $\hat{P}$) is an analytic, single-valued functional. The same definition of P (or $\hat{P}$) can be used on open or closed streaklines.

II) P (or $\hat{P}$) is a multivalued functional. Different definitions of P (or $\hat{P}$) are used in the « interior » region (region of closed streaklines) and in the « exterior » region (region of streaklines extending from $-\infty$ to $+\infty$).

Figure 1 visually summarizes the two possibilities. Category II solutions constitute what in the literature have been called the « modon » solutions [2-4].

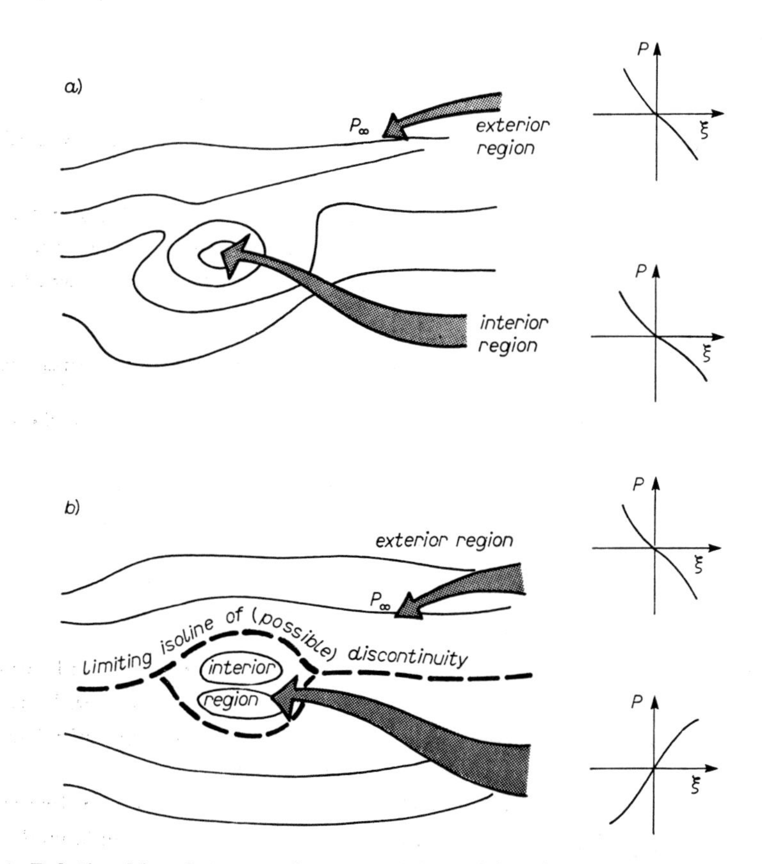

Fig. 1. – Relationships between the potential-vorticity functional P and the streak function $\xi = \psi + cy - \int^y U\,dy$ for *a*) analytic and *b*) modon solitary waves. From [1].

The fact that P (or $\hat{P}$) is multivalued implies that there is a « limiting » closed streakline $\xi = a$ upon which P will pass from the functional shape of the interior region to that of the exterior region. On this limiting streakline, the matching conditions will have to be imposed upon ψ and its derivatives. If P (or $\hat{P}$) is given a polynomial form, there will be a radial derivative of ψ, say $d^n\psi/dr^n$, which will be discontinuous on $\xi = a$ (see [3]).

A further very important remark to make is that in the general model equation (2.5a) there is the potentially nonlinear term

$$\psi \cdot P(\psi + \ldots)$$

which has disappeared in the quasi-geostrophic approximation (2.5b). This has the profound implication that quasi-geostrophic solitary waves to exist need steepening effects given by the interaction of the wave with some « external » factor, like topography or a mean shear flow, or otherwise need to have a modon structure. Nonquasi-geostrophic solitary waves, on the other hand, do not need any external feature to exist.

3. – The general classification of solitary-wave models: specific examples.

From this unified formulation, namely from eqs. (2.5a), (2.5b), one can derive all the solitary-wave solutions found in the literature by different authors.

Precisely, let us classify the solitary solutions according to the two following criteria:

I) P (or $\hat{P}$) is a) an analytic or b) a multivalued functional (modon).

II) In the category of *analytic functionals* solutions will be classified as

$$\text{weak:} \qquad v \ll c\,;$$

$$\text{strong:} \qquad v \gg c\,;$$

if c is the phase speed and v the particle speed of the wave.

Again, distinction will always be made between general and quasi-geostrophic models. The last difference is between barotropic and baroclinic solutions. As we shall see, there are solutions which can be only baroclinic, that is they need the divergence term ψ/R_0^2 to exist.

Table I shows all the existing solitary solutions which can be derived from eqs. (2.5a), (2.5b), classified according to the two above-mentioned criteria and related to the solutions previously found in the literature by different authors.

TABLE I. – *Planetary solitary-wave solutions.*

Authors	Analytic functional P		Multivalued P (modons) strong
	weak $v \ll c$	strong $v \gg c$	
LONG [5] (1964)		Q.G. Channel solutions. Barotropic and baroclinic. Steepening effects: interaction with shear flow $U(y)$.	
CLARKE [6] (1971)	non Q.G. β-plane barotropic channel solutions (allowing for divergence effects, no need of « external » features like shear flow, topography, etc.).		
CLARKE [6] (1971)	non Q.G. β-plane barotropic channel solutions, *weak* divergence with mean shear flow $U(y)$ *and* topography.		
MAXWORTHY and REDEKOPP [7-9] (1976-1977)	Q.G. Channel solutions. Barotropic (KdV) and baroclinic (mKdV). Steepening: interaction with $U(y)$.		
MALANOTTE RIZZOLI [10-12] (1978-1979)	Q.G. Channel solutions. Barotropic (but can include a divergent term ψ/R^2). Steepening: interaction with topography $h(y)$.	Q.G. Channel solutions. Barotropic and baroclinic. Steepening: interaction with topography $h(y)$.	
FLIERL [1, 13] (1979)		Q.G. Infinite β-plane. Radially symmetric. Steepening: interaction with $U(y)$ or $h(y)$. *Only baroclinic.*	

TABLE I (*continued*).

Authors	Analytic functional P		Multivalued P (modons) strong
	weak $v \ll c$	strong $v \gg c$	
STERN [2] (1975)			Stationary Q.G. Barotropic. Infinite β-plane. Limiting case ($c=0$) of
LARICHEV and REZNIK [3] (1976). FLIERL, LARICHEV, McWILLIAMS and REZNIK [4] (1979)			Q.G. Infinite β-plane. *Barotropic* and *baroclinic*.

We shall now show some examples of these solitary eddies, above all of those found in the most recent literature.

The first example is a geostrophic example referring to the atmosphere of Jupiter. The problem is that of an oval structure superimposed to a basic, zonal shear flow. This is the famous Giant Red Spot persisting for centuries. In fig. 2 we show an image of the Red Spot and of the circulation gyre in which it is included [7].

Experimental evidence shows that the Red Spot is situated in a region of anticyclonic shear (Jovian southern hemisphere), while, going towards the equator, the shear becomes cyclonic, until it matches with the so-called equatorial jet [14]. The circulation pattern is that shown in fig. 2. Jovian features persist for centuries. Observed during a period of weeks, the small spots around the Red Spot always move counterclockwise; therefore, the winds around it probably blow in that direction [14].

Observation of motion around the Red Spot suggests very interesting evidence. For instance, for many decades in this century a disturbance existed, which, starting at the southern edge, interacted several times with the Red Spot itself. During the interaction, observers noted a rapid acceleration of the disturbance and its re-emerging and reforming without any change of shape [14].

These properties suggested [7, 8] to model the Red Spot as a solitary-wave eddy. Specifically, REDEKOPP [9] formulates the theory of a solitary eddy superimposed on a mean shear flow, both in a barotropic and in a baroclinic atmosphere. For a barotropic atmosphere, the final equation is the KdV model. In a baroclinic atmosphere, with a constant Brünt-Vaisälä frequency, the modified KdV equation models the solitary eddy. This equation allows for both elevation (E) and depression (D) solitary-wave solutions.

In fig. 3 [7, 8] the mean shear flow is shown simulating the average, zonal circulation in the Jovian atmosphere. Also shown is the superposition of a D

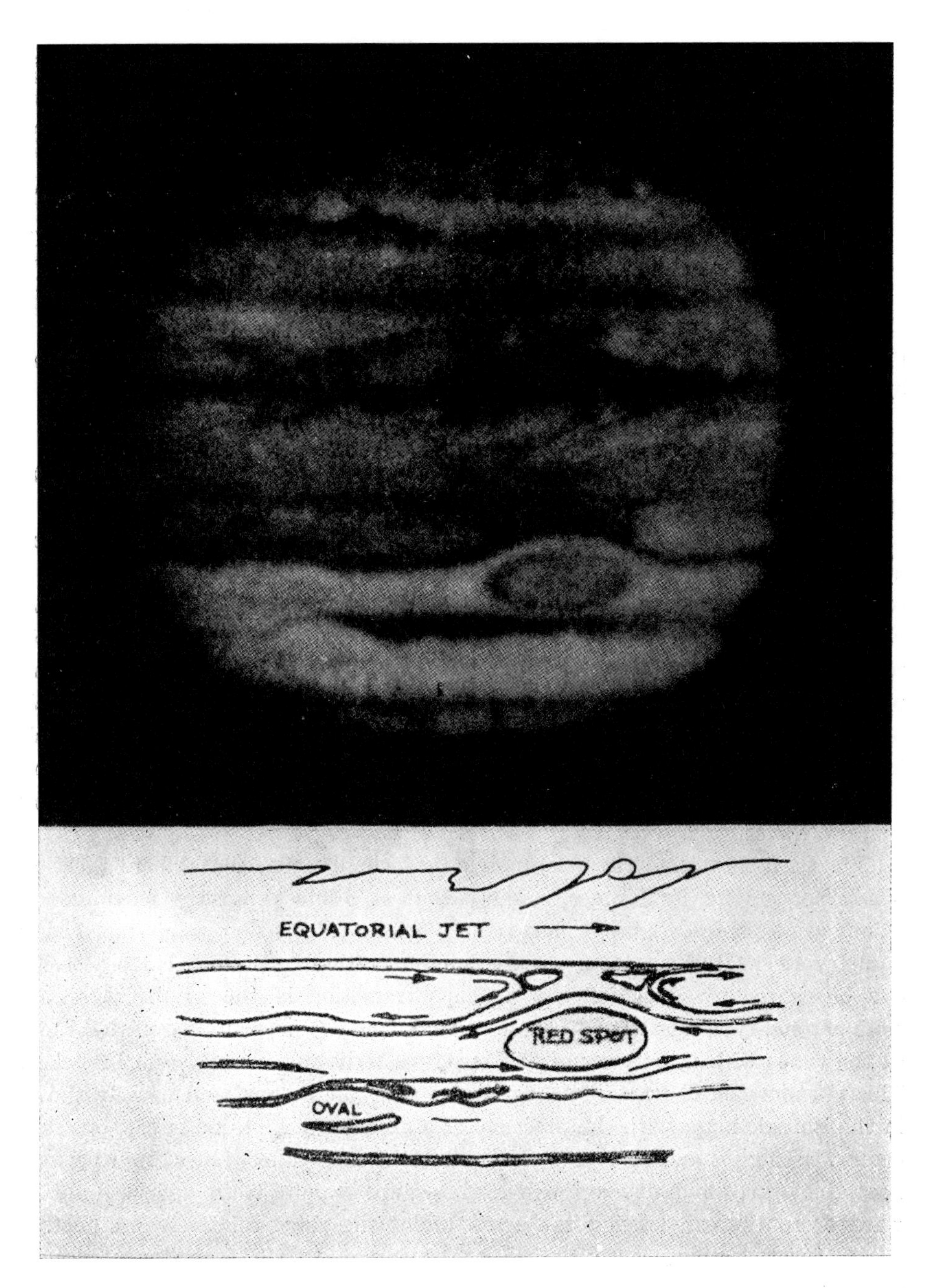

Fig. 2. – Picture of the giant Rep Spot of Jupiter and schematic sketch of the inferred circulation pattern in which it is embedded. From [7, 8].

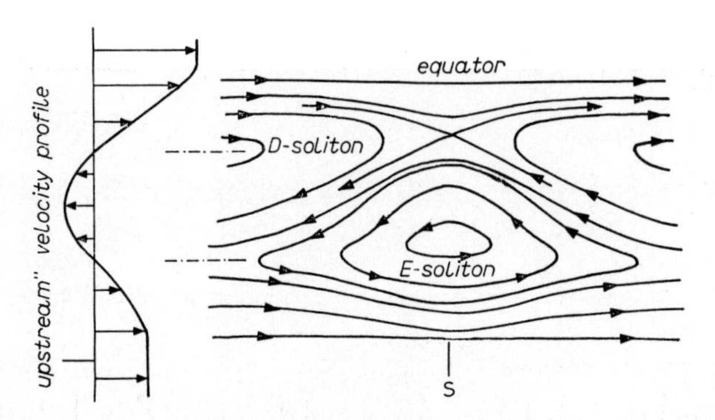

Fig. 3. – Theoretical mean shear flow profile and theoretical streamfunction composed by the superposition of depression (D) and elevation (E) solitary waves. From [7, 8].

and an E soliton solution. The theoretical pattern is strikingly resemblant to the Red Spot picture of fig. 2.

Here a point must be specified. It is the writer's opinion that it may be a rather strong and far-stretched statement to assert that the Red Spot « is » a solitary wave. Nevertheless, two statements can be made on safe grounds. First, there do exist in Nature experimental configurations characterized by remarkable permanence and stability, which last longer than a corresponding predictability time. Of these the Red Spot is an example. Second, for their properties these experimental configurations are strongly suggestive that delicate balances between nonlinearity and other physical effects like dispersion can be actually achieved in Nature, namely balances like those characterizing solitary waves.

We shall now give the example of a second experimental configuration which persists for long times, namely that of a blocking-ridge phenomenon in our atmosphere. Figure 4 shows the blocking-ridge situation characterizing January 1963 (McWilliams: private communication). Blocking-ridge activity can be characterized as follows: a sharp transition in the westerlies occurs from a zonal-type upstream to a meridional-type downstream, associated most of the times with a splitting of the basic westerly current into two branches. This phenomenon characteristically occurs over the oceans, and usually shows in the isobaric maps with the presence of a dipole of a high- and a low-pressure center, lasting for periods longer than a month. It is called « blocking » pattern because storms which develop westward and upstream of it are deflected northward or southward, leaving the areas under the ridge relatively unaffected.

Recently, blocking activity has been compared with a modon structure and consistency has been found between the experimental situation and the theoretical model [15]. The modon structure which can be associated with a blocking ridge is that of a simple dipole in the streamfunction (atmospheric pressure) as shown in fig. 5 ([3], the simplest case of a modon).

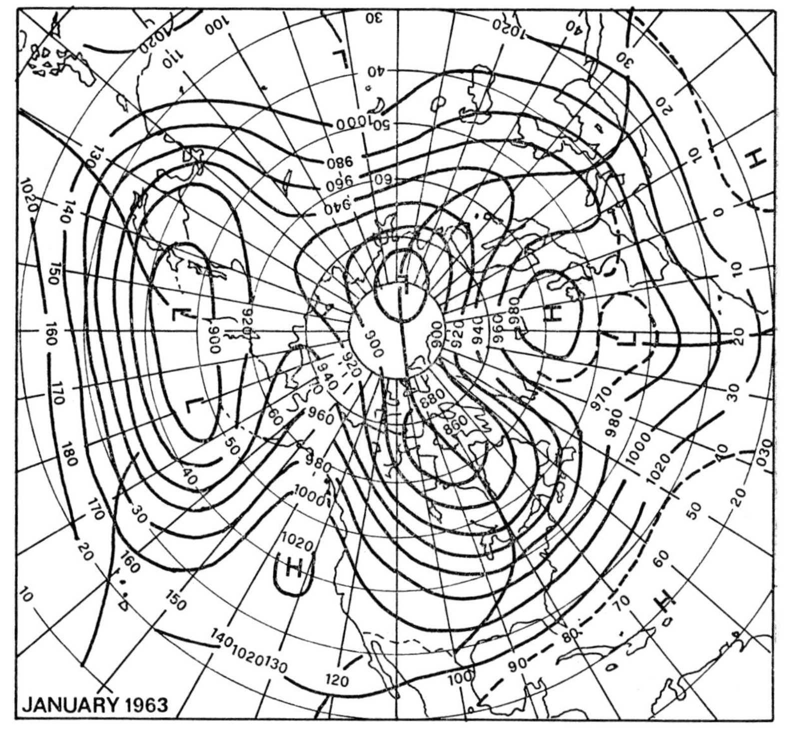

Fig. 4. – Blocking-ridge situation over the North Atlantic Ocean for January, 1963. McWilliams: private communication.

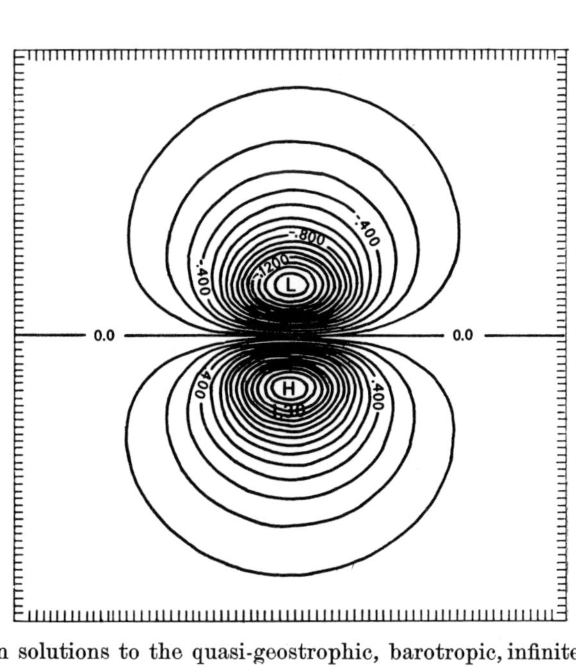

Fig. 5. – Modon solutions to the quasi-geostrophic, barotropic, infinite β-plane vorticity conservation equation. From [4].

In fig. 6 we show more complicated patterns of baroclinic modons, specifically for a two-layer model, and with different velocity distributions in the vertical. $\psi^{(1)}$ is the streamfunction for the upper layer, $\psi^{(2)}$ for the lower [4].

Figure 7 shows the example of an initial condition « close enough » to a permanent solution of the model equation. This is an asymmetric, quasi-geostrophic weak-wave type of solution in a zonal channel, over variable topography [10, 12]. This solution can be both barotropic and baroclinic. The shown example is the barotropic one. As is clear from fig. 7, the initial condition is not exactly the final, permanent-shape solution. To this, however, the initial eddy does evolve, through a steepening process, which makes the eddies also more symmetrical, thus increasing the dispersive effects. When

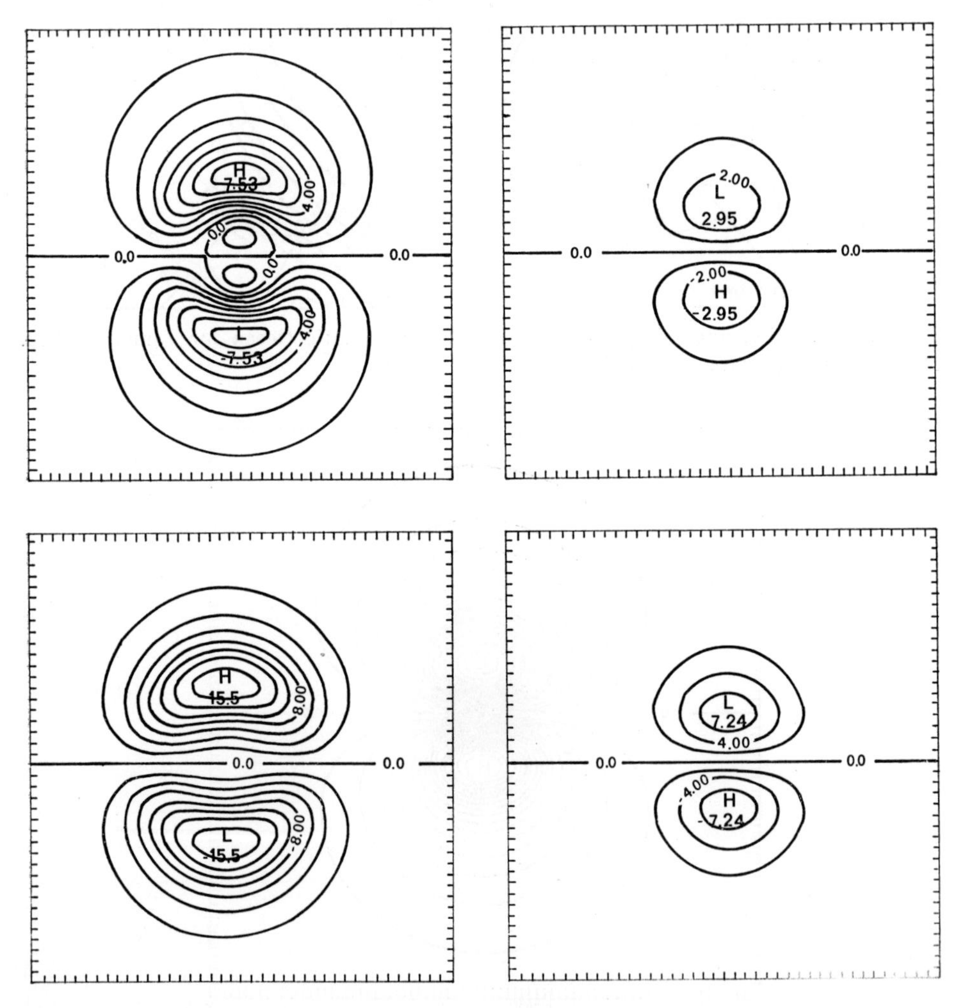

Fig. 6. – Different modon solutions, for a two-layer model in two different cases, for upper and lower layers, respectively. From [4].

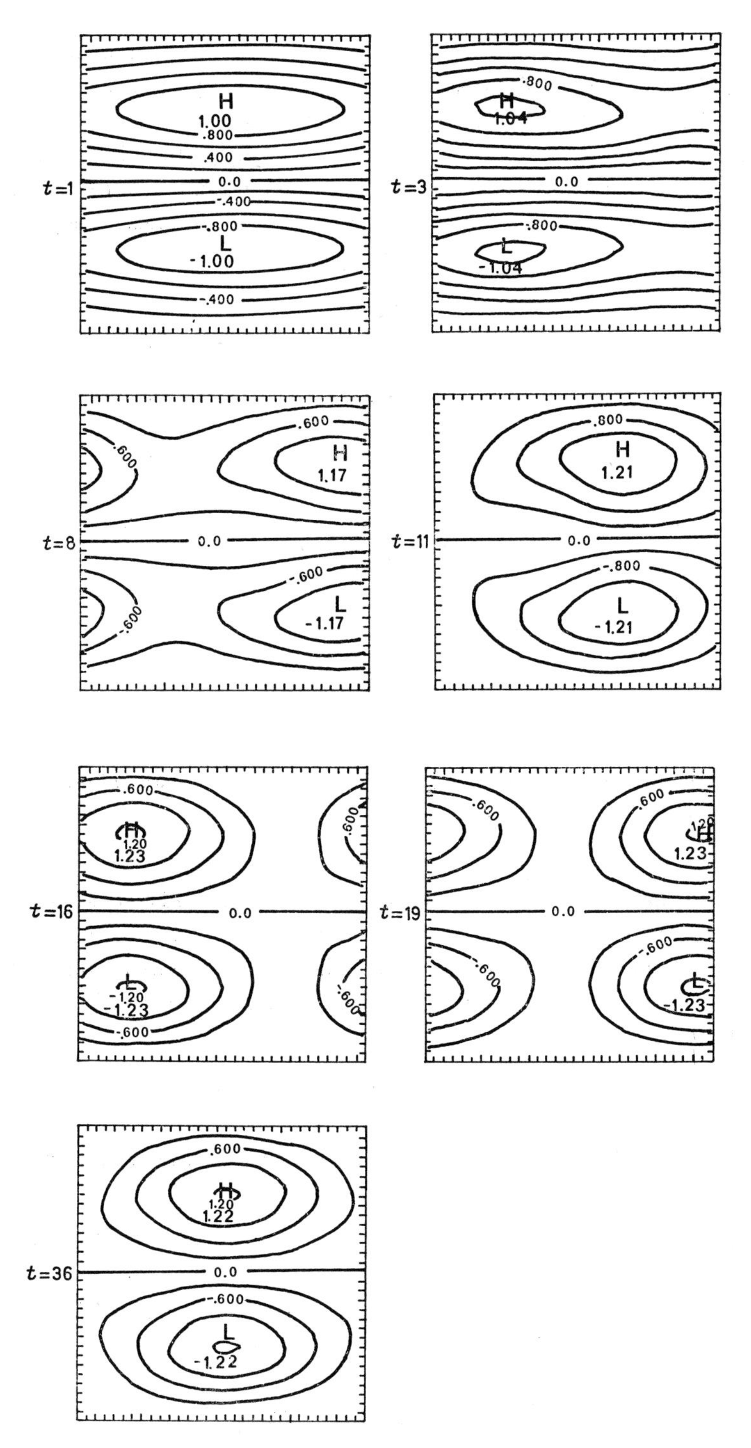

Fig. 7. – Evolution of a strong solitary-wave initial condition in the model equation (4.4) at the successive labeled dimensionless times. From [10, 12].

the balance between nonlinearity and dispersion is finally reached, the eddies evolve in the numerical basin without change of shape and speed over many crossings. This also is a dipole solitary structure.

As a final example in fig. 8 we show a radially symmetric isolated eddy, solution to the vorticity conservation equation on an infinite β-plane, over topography or a mean shear flow which provides the nonlinear steepening inter-

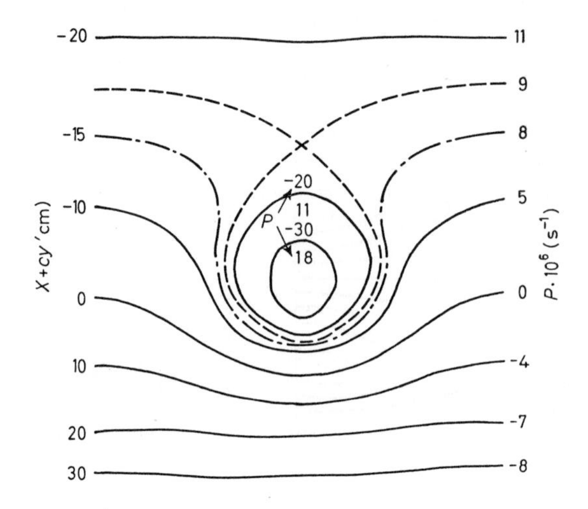

Fig. 8. – Baroclinic solitary eddy with radial symmetry on an infinite β-plane over a mean shear flow. From [13].

action [13]. For dispersion to be weak, however, the eddy scale must be large compared to the deformation radius, so that the baroclinicity is essential. The radial symmetry of the eddy makes the nonlinear effects also weak, even though the solution itself is strong, namely the particle speed v is large compared to the phase speed c.

4. – The « minimal » model equation: application of the unified approach.

We shall now consider a particular model, a simplification of the general conservation equation in its quasi-geostrophic version (2.5*b*). Specifically, this is the barotropic, quasi-geostrophic potential-vorticity conservation equation over variable topography:

$$\nabla^2 \psi_t + h_y \psi_x + J(\psi, \nabla^2 \psi) = 0 \tag{4.1}$$

with the usual meaning of symbols.

This will be called the « minimal » equation because it

includes the strictly minimum necessary effects to find solitary-wave solutions (steepening given by interaction with topography),

can include β (equivalent linear topography),

baroclinicity is very simply included adding a divergent term (the so-called « equivalent barotropic » equation),

is very simple to handle mathematically.

The application of the unified procedure described in the previous section will allow us

to answer the geophysically important question about which kind of topographic shapes can support solitary waves,

to establish general criteria of existence of the solitary-wave solutions.

Repeating the general procedure, namely passing to the reference frame moving with the wave speed c, we obtain

$$\nabla^2 \psi + h(y) = P\left(\frac{\psi}{c} + y\right). \tag{4.2}$$

The solitary-wave solution $\psi \to 0$ as $x \to \pm\infty$. This means that the asymptotic form of the functional P is in this case the topography itself, on those streaklines extending continuously from $-\infty$ to $+\infty$

$$P_\infty(y) = h(y). \tag{4.3}$$

In principle, it is possible for the functional P to be different from h on closed streaklines (the interior region). We shall here consider, however, the case in which $P(\xi) = h(\xi)$ everywhere. Then the equation can be written as

$$\nabla^2 \psi = P\left(\frac{\psi}{c} + y\right) - P_\infty(y) = h\left(\frac{\psi}{c} + y\right) - h(y). \tag{4.4}$$

Baroclinic effects are easily included adding a divergent term. Thus our final model is the equivalent barotropic equation over variable topography:

$$\nabla^2 \psi - \frac{1}{R_0^2}\psi = h\left(\frac{\psi}{c} + y\right) - h(y). \tag{4.5}$$

This equation is nothing else than (2.5b) when $U = 0$, $\beta = 0$ and $h(y) = f_0 \cdot b(y)$.

This equation allows for modon solutions. In fact, if we take $h(y)$ to be a multivalued functional, linear in y in the exterior region and different in the interior region, we obtain exactly the Flierl, Larichev, McWilliams and Reznik [4] equation, for the simplest modon solution of fig. 5.

We shall here limit ourselves to analytic functionals $h(\xi)$.

Accordingly, we shall explore and classify all existing solitary solutions according to the following three criteria:

I) form of the functional $P = h(y)$;

II) intensity of the solution:

$v \ll c$ weak,

$v \gg c$ strong;

III) symmetry of the solution:

asymmetric

$\delta^2 = L_1^2/L_2^2 \ll 1$ with $L_1 = y$-length scale, $L_2 = x$-length scale;

radially symmetric

$\delta^2 \sim O(1)$ barotropic,

$\delta^2 = R_0^2/l^2$ baroclinic, if l is the eddy radius.

Let us first make the following general consideration. We are looking for solitary-wave solutions which are weakly dispersive and weakly nonlinear and in which nonlinearity and dispersion balance each other. This means that the zeroth-order balance of the model equation must furnish linear, dispersionless waves as solutions. Including a « weak » nonlinearity and an « equally weak » dispersion, we shall obtain solitary waves traveling faster than the zeroth-order, linear phase speed.

There are only two ways in which nonlinearity can be made a higher-order effect.

The waves themselves are weak: $\psi/c \ll y$ $(v \ll c)$ and one can Taylor expand the right-hand side of (4.4) or (4.5) around ψ/c.

The waves are strong $(v \gg c)$, but the topography itself is weak enough for $h(Z)$ to be quasi-linear, so as all nonlinear terms are at higher order. In this case, one can Taylor expand around y.

Thus we shall consider the two cases for criterion I):

$h(Z)$ has a general shape,

$h(Z) = mZ + \varepsilon h^*(Z)$ is quasi-linear,

$h^*(Z)$ has itself a general shape.

TABLE II.

Case	Solitary-wave solutions
I	*Solitary solutions*, MALANOTTE RIZZOLI [10]
a1) *General* $h(Z)$	$\psi \to \varepsilon\psi$, $\delta^2 = L_1^2/L_2^2 = \varepsilon$, $\partial^2/\partial x^2 = \varepsilon\,\partial^2/\partial y^2$.
b1) *Weak* $v \ll c$	Barotropic and baroclinic.
c1) *Asymmetric*	Sturm-Liouville theory *ensures* existence for most general $h(Z)$.
II	
a1) *General* $h(Z)$	$\psi \to \varepsilon\psi$.
b1) *Weak* $v \ll c$	*Barotropic*: $\delta^2 = 1$, *no solitons*. Nonlinearity (weak) cannot balance dispersion $O(1)$.
c1) *Radially symmetric*	*Baroclinic*: $\delta^2 = R^2/l^2 = \varepsilon$, *no solitons*. In fact, zeroth order gives $h'(y) = \text{const} = \beta$; *quasi-linear topography*.
III	
a1) *General* $h(Z)$	$\delta^2 = L_1^2/L_2^2 = \varepsilon$, $\partial^2/\partial x^2 = \varepsilon\,\partial^2/\partial y^2$.
b2) *Strong* $v \gg c$	$c \to \varepsilon c$.
c1) *Asymmetric*	*No solitons*, dispersion (weak) cannot balance nonlinearity ($O(1)$).
IV	
a1) *General* $h(Z)$	*Solitons may exist*. But $\nabla^2\psi - (1/R^2)\psi \sim h(\psi/c)$.
b2) *Strong* $v \gg c$	A nonlinear eigenvalue problem for ψ, with $\psi \to 0$ as $r \to +\infty$, and c the eigenvalue.
c2) *Radially symmetric*	If solitary solutions exist, they do for *special* shapes of $h(Z)$. No general statement can be made as in case I.
V	
a2) *Quasi-linear* $h(Z) \approx mZ + \varepsilon h^*(Z)$	If $h(Z) = mZ + h^*(Z)$, then solitary solutions, *but* this is a particular case of I.
b1) *Weak* $v \ll c$	*No solitons* in general.
c1) *Asymmetric*	$\left\{\begin{array}{l} h(Z) = mZ + \varepsilon h^*(Z) \\ \psi \to \varepsilon\psi \end{array}\right\}$ make nonlinearity even weaker than dispersion (weak).
VI	
a2) *Quasi-linear* $h(Z) = mZ + \varepsilon h^*(Z)$	*No solitons*. $\psi \to \varepsilon\psi$.
b1) *Weak* $v \ll c$	*Barotropic* $\delta^2 \sim O(1)$. Nonlinearity (weak) cannot balance dispersion.
c2) *Radially symmetric*	*Baroclinic* $\delta^2 = R^2/l^2 = \varepsilon$. Order zero: $c_0 = -mR^2$, but *again* nonlinearity is *too weak* to balance dispersion. Linear waves are the solution.

TABLE II (*continued*).

Case	Solitary-wave solutions
VII	*Solitary solutions*, new, MALANOTTE RIZZOLI [16].
a2) *Quasi-linear* $h(Z) = mZ + \varepsilon h^*(Z)$	*Barotropic and baroclinic.*
b2) *Strong* $v \gg c$	$\delta^2 = L_1^2/L_2^2 = \varepsilon,\ c \to \varepsilon c,\ \partial^2/\partial x^2 = \varepsilon\,\partial^2/\partial y^2$.
c1) *Asymmetric*	*Channel solutions* (L_1 = channel width).
VIII	*Solitary solutions*, FLIERL [13]
a2) *Quasi-linear* $h(Z) = mZ + \varepsilon h^*(Z)$	*Only baroclinic.*
b2) *Strong* $v \gg c$	$\delta^2 = R^2/l^2 = \varepsilon$.
c2) *Radially symmetric*	$c_0 = -mR^2$ westward or eastward going according to topographic slope; *numerical* solutions.

Table II summarizes all the existing solitary solutions derivable from (4.5). We want here to show the detailed procedure for two cases of table II, namely case I and case VII.

For case I we shall use the barotropic equation (4.4). Under its assumption we can write

$$\varepsilon(\psi_{yy} + \varepsilon\psi_{xx}) = h\left(\varepsilon\frac{\psi}{c} + y\right) - h(y) = h(y) + h'(y)\frac{\varepsilon\psi}{c} + \frac{h''(y)}{2}\frac{\varepsilon^2\psi^2}{c^2} - h(y) + O(\varepsilon^3),$$

that is

$$\psi_{yy} + \varepsilon\psi_{xx} = h'(y)\frac{\psi}{c} + \varepsilon\frac{h''(y)}{2}\frac{\psi^2}{c^2} + O(\varepsilon^2), \tag{4.6}$$

Expanding, we obtain

$$\psi = \psi_0 + \varepsilon\psi_1, \qquad c = c_0 + \varepsilon c_1.$$

The order-zero system is

$$\psi_{0yy} - \frac{1}{c_0}h'(y)\psi_0 = 0 \tag{4.7}$$

with $\psi_0 = g(s)\cdot\Phi(y)$, $s = x - ct$.

The Sturm-Liouville system gives the zeroth-order eigenfunctions $\Phi_n(y)$ and eigenvalues c_{0n} in the longitudinal channel.

The order-ε system is

$$(4.8)\qquad \psi_{1yy}-\frac{1}{c_0}h'(y)\psi_1=-\frac{2c_1}{c_0}\psi_{0yy}+\frac{c_1}{c_0^2}h'(y)\psi_0-\psi_{0xx}+\frac{h''(y)}{2c_0^2}\psi_0^2=$$

$$=-\frac{c_1}{c_0^2}h'(y)\psi_0-\psi_{0xx}+\frac{h''(y)}{2c_0^2}\psi_0^2$$

using the zeroth-order equation (4.7).

Using the orthogonality condition of the left-hand side operator to avoid secular growth, that is multiplying through Φ and integrating over the channel width, we obtain

$$(4.9)\qquad \frac{c_1}{c_0^2}a_1 g+\frac{1}{2c_0^2}a_2 g^2+a_3 g_{xx}=0$$

with

$$a_1=+\int_0^1 h'\Phi^2\,\mathrm{d}y\,,\qquad a_2=-\int_0^1 h''\Phi^3\,\mathrm{d}y\,,\qquad a_3=+\int_0^1 \Phi^2\,\mathrm{d}y\,.$$

(4.9) is the KdV equation. This had been obtained in [11] with a different procedure, starting directly from the time-dependent equation. (4.9) gives the solitary-wave solution $g(x)$ and the velocity correction, amplitude-dependent $c_1(A)$. For all consequences of (4.9) and properties of the solitary solutions the reader is referred to the cited ref. [11].

For case VII we shall also use the barotropic model (4.4), which we can now write

$$(4.10)\qquad \psi_{yy}+\varepsilon\psi_{xx}=h\left(\frac{\psi}{c}+\varepsilon y\right)-h(\varepsilon y)=$$

$$=h\left(\frac{\psi}{c}\right)+\varepsilon y h'\left(\frac{\psi}{c}\right)-h(0)-\varepsilon y h'(0)+O(\varepsilon^2)\,.$$

h being quasi-linear,

$$h\left(\frac{\psi}{c}\right)=m\frac{\psi}{c}+\varepsilon h^*\left(\frac{\psi}{c}\right),\qquad h(0)=\varepsilon h^*(0)\,,$$

$$h'\left(\frac{\psi}{c}\right)=m+\varepsilon h^{*\prime}\left(\frac{\psi}{c}\right),\qquad h'(0)=m+\varepsilon h^{*\prime}(0)\,,$$

(4.10) becomes

$$(4.11)\qquad \psi_{yy}+\varepsilon\psi_{xx}=m\frac{\psi}{c}+\varepsilon\left[h^*\left(\frac{\psi}{c}\right)-h^*(0)\right]+\varepsilon^2 y\left[h^{*\prime}\left(\frac{\psi}{c}\right)-h^{*\prime}(0)\right]+O(\varepsilon^3)\,.$$

Again, expanding

$$\psi = \psi_0 + \varepsilon\psi_1 \ldots, \qquad c = c_0 + \varepsilon c_1 \ldots,$$

we obtain, for the zeroth-order solution in the zonal channel,

$$\psi_0 = G(x) \cdot \sin(n\pi y) . \tag{4.12}$$

If $m < 0$, $c_0 < 0$ westward going waves (equivalent β).
If $m < 0$, $c_0 > 0$ eastward going waves.
The order-ε equation, again applying the orthogonality condition, is

$$\frac{mc_1}{c_0^2}\left[\int_0^1 \Phi^2 \,\mathrm{d}y\right] G(x) + \left[\int_0^1 \Phi^2 \,\mathrm{d}y\right] G_{xx} - \int_0^1 \left[h^*\left(\frac{\psi_0}{c_0}\right) - h^*(0)\right] \Phi \,\mathrm{d}y = 0 . \tag{4.13}$$

Solitary-wave solutions can be obtained depending on the functional shape of h^*.

If $h(\psi_0/c_0) \propto \psi_0^2/c_0^2$, we get the KdV solitons, for which solutions exist only for odd n-modes. If $h^*(\psi_0/c_0) \propto \psi_0^3/c_0^3$, we obtain the mKdV model, and so forth. For further details upon this strong asymmetric solitary waves the reader is referred to [16].

5. – Conclusions.

Table II allows us to establish general criteria in the context of the minimal equation and thus to answer the question about which kind of topographic shapes support solitary-wave solutions.

The only case in which a general statement can be made in total generality is case I. Weak-amplitude, asymmetric solitons are supported in a channel by essentially any kind of topography.

On a β-plane, or conversely if the topography has a lowest-order term which is linear, strong solitary solutions are supported both in the channel (asymmetric) and in an infinite plane (radially symmetric). In the channel they can be either barotropic or baroclinic; in the infinite plane they are only baroclinic. In both cases, the topographic variations around the mean, linear slope can be quite general, essentially of all shapes which allow the final equation to evolve into a one-dimensional model supporting solitons. The variety of topographic shapes $h^*(\xi)$ is, however, not totally general for case VII as for case I of table II.

In any case, the results summarized in table II are wide enough to ensure that solitary waves can exist under very general circumstances, including very complex topographies realistically simulating those found in Nature.

REFERENCES

[1] G. FLIERL: *Polymode News*, **62**, 1 (1979).

[2] M. E. STERN: *J. Mar. Res.*, **33**, 1 (1975).

[3] V. D. LARICHEV and G. M. REZNIK: *Polymode News*, **19**, 1 (1976)

[4] G. FLIERL, V. LARICHEV, J. MCWILLIAMS and G. REZNIK: *Dyn. Atmos. Oceans*, **4**, 205 (1980).

[5] R. R. LONG: *J. Atmos. Sci.*, **21**, 197 (1964).

[6] R. A. CLARKE: *Geophys. Fluid Dyn.*, **2**, 343 (1971).

[7] T. MAXWORTHY and L. G. REDEKOPP: *Nature* (*London*), **260**, 509 (1976).

[8] T. MAXWORTHY and L. G. REDEKOPP: *Icarus*, **29**, 261 (1976).

[9] L. G. REDEKOPP: *J. Fluid Mech.*, **82**, 725 (1977).

[10] P. MALANOTTE RIZZOLI: Ph. D. Dissertation, Scripps Institute of Oceanography (1978).

[11] P. MALANOTTE RIZZOLI and M. C. HENDERSHOTT: *Dyn. Atmos. Oceans*, **4**, 247 (1980).

[12] P. MALANOTTE RIZZOLI: *Dyn. Atmos. Oceans*, **4**, 261 (1980).

[13] G. FLIERL: *Dyn. Atmos. Oceans*, **3**, 15 (1979).

[14] A. P. INGERSOLL: *Sci. Am.*, **95**, 46 (1976).

[15] J. MCWILLIAMS: *Dyn. Atmos. Oceans*, **4**, 225 (1980).

[16] P. MALANOTTE RIZZOLI: *Polymode News*, No. 75 (1980).

The Stability of Planetary Solitary Waves.

P. MALANOTTE RIZZOLI

Istituto per lo Studio della Dinamica delle Grandi Masse, CNR - Venezia, Italia
I.G.P.P., Scripps Institution of Oceanography - La Jolla, Cal.

1. – Introduction.

In recent years new interest has arisen in a very special type of solution to a wide variety of equations, a solution which models physical phenomena ranging from motions of one-dimensional nonlinear lattices to the propagation of eddies in geophysical fluids. These solutions have been labeled « solitary waves », or, to use a more recent terminology, « solitons », and were introduced to hydrodynamical science by RUSSELL [1] more than a century ago.

Mathematically, in 1895, KORTEWEG and DE VRIES [2] provided a simple analytic foundation for the study of solitary waves by developing an equation (KdV equation) for shallow-water waves which, though ignoring dissipation, included both nonlinear and dispersive effects. The solitary wave of Korteweg and de Vries was long considered a rather unimportant curiosity in the mathematical structure of nonlinear wave theory, it being thought that only very particular phenomena subject to only special initial conditions could be described by these solitons in their time evolution. It was supposed, furthermore, that the solitons, even when existing, would have been completely destroyed by collisions among themselves or by interaction with any other perturbation possible in the context of the considered physical phenomenon. In summary, they were thought to be unstable curiosities.

This has not proved to be the case. In the last decade, soliton solutions have been discovered not only for the KdV equation but also for many other nonlinear equations. They have, moreover, been proved to be remarkably stable in the one-dimensional case, so much that one could define solitons (following SCOTT [3]) by saying: « the soliton is a localized traveling wave which asymptotically preserves its shape and velocity upon collisions with other solitons ».

A great deal of the vast literature on solitons and other permanent-form nonlinear solutions deals with systems which have many degrees of freedom but only one space dimension; idealized strings with many molecules, narrow channels filled with fluid, such as the one along which SCOTT and RUSSELL first observed the solitary wave, etc. Corresponding studies of systems with two or three space dimensions are much fewer. Such two- or three-dimensional studies are, however, of paramount importance, for the existence of form-preserving nonlinear motions observed in solitons contradicts a fundamental randomness postulate of statistical mechanics and of turbulence theory. This postulate states that, in a system with many degrees of freedom, the interactions between different components have such a complex history that we can best describe the average properties of the system by assuming these interactions to be as random as is consistent with overall conservation of mass, energy, etc. This postulate successfully accounts for some of the macroscopic behavior of turbulent flows both in the ocean and in the atmosphere.

But this postulate is not universally valid. Recognition of this fact has, until recently, rested exclusively upon the behavior of a class of one-dimensional, nonlinear systems displaying solitons or other form-preserving nonlinear solutions. A by now classical work [4] has shown that a smooth initial condition, such as a single cosine, given to the KdV equation breaks into a series of well-localized pulselike shapes, solitons, freely streaming through one another at amplitude-dependent speeds and re-emerging from their collisions unaltered in shape and speed. At a certain time, the so-called recurrence time, all the solitons produced by the smooth initial condition focus together again, reproducing a state almost identical to the initial one. This correctly implies remarkable stability of the KdV solitons.

In summary, the behavior of these one-dimensional systems contradicts the randomness postulate. But, if the postulate is not universally valid for one-dimensional systems, it does not need to be universally valid for two- or three-dimensional systems, in particular for geophysical systems like the atmosphere and the ocean. In fact, in analogy with the one-dimensional case, the postulate is not obeyed in those situations for which the continuum models of the two-dimensional flows allow for stable permanent-form nonlinear wave solutions.

Thus in the last decade much work has been devoted to finding permanent-form solutions to nonlinear equations describing the most different geophysical phenomena. Apart from the now classical case of gravity water waves, for which envelope soliton solutions have also been discovered [5], solitary and cnoidal internal waves as well as solitary edge waves have been discussed. For model equations suitable to describe planetary motions in the oceans and in the atmosphere, solitary-wave solutions are nonlinear Rossby waves.

The first such studies concerning solitary Rossby waves were those by LONG [6] and by BENNEY [7]. They discussed long waves in a barotropic atmosphere having a horizontally sheared basic zonal flow and obtained the KdV equation for the nonlinear perturbations. Their analysis was limited to the case of a small-velocity shear superimposed on a basic order-one uniform flow, and they did not discuss the possible types of streamline flow patterns associated with these waves. Solitary Rossby waves were also studied by LARSEN [8] and CLARKE [9], but they also limited themselves to the analytical formulation for the solutions in different cases, again not giving any information about possible flow patterns and their specific properties. More recently, REDEKOPP [10] has given a very detailed discussion of zonally propagating long planetary waves in an atmosphere with a horizontally sheared zonal flow. The wave amplitude obeys the KdV or modified KdV equation, depending on the existence and nature of the atmospheric density stratification.

Other kinds of permanent-form Rossby waves have also been discovered. They are solutions to the potential-vorticity conservation equation on an infinite β-plane [11]. This work has recently been extended to a two-layer model, and a variety of cases for the solutions' behaviour has been examined [12]. These solutions, the « modons », are nonanalytical and require a finite discontinuity in the vorticity gradient at the eddy radius to satisfy the condition that the solution remains bounded as $r \to + \infty$. Also very recently, solutions simulating an isolated eddy ring have been discovered which are radially symmetrical in a baroclinic fluid [13].

It must be pointed out that in all other previously mentioned studies no discussion was given of the stability of these solutions. But stability is a fundamental property which must be thoroughly understood before the physical significance of these solutions becomes clear. In the following sections, we shall first derive the permanent-form solutions of our basic model equation, and develop an analytical theory to investigate their stability properties [14]. To our knowledge, no other analytical treatment has been given until now for the stability of two-dimensional solitary eddies. We shall then explore the stability of the solitary solutions with respect to finite-amplitude perturbations in the initial conditions [15] as well as upon collisions between themselves [16]. Finally, stability will be investigated with respect to perturbations in the solitary-wave Fourier phases and in the bottom relief itself [17].

2. – The basic model equation, its permanent-form solutions and the related analytical stability analysis.

In this section we choose a model equation which supports solitons. This model has been chosen because, although simple, its solutions range from linear waves at infinitesimal amplitude, to form-preserving nonlinear solutions at moderate amplitude and dispersion, to two-dimensional turbulence over bottom relief at high amplitude.

Consider divergenceless, quasi-geostrophic, barotropic flow in a zonal channel over variable relief. The relief is given by

$$D(x, y) = D_0\left[1 - \frac{|\Delta D|}{D_0} h(x, y)\right], \tag{2.1}$$

where $(|\Delta D|/D_0)\, h \ll 1$ is the fractional relief variation around the mean value D_0. Then the motion is described by the dimensionfull equation (primes denote dimensionfull quantities)

$$\nabla^2 \psi'_{t'} + J'(\psi', \nabla^2 \psi' + f_0^* h) = 0\,, \tag{2.2}$$

where

$$\nabla^2 = \frac{\partial^2}{\partial x^2} + \frac{\partial^2}{\partial y^2}$$

is the horizontal Laplacian and

$$J(a, b) = \frac{\partial a}{\partial x}\frac{\partial b}{\partial y} - \frac{\partial a}{\partial y}\frac{\partial b}{\partial x}$$

is the Jacobian of (a, b).

Subscripts indicate partial derivatives.

Here $f_0^* = f_0(|\Delta D|/D_0)$, f_0 being the Coriolis parameter; hence $h \sim O(1)$. Equation (2.2) is our model equation. Scaling it by

$$y' = L_1 y\,, \qquad x' = L_2 x = \frac{L_1}{\delta} x\,, \qquad t' = \frac{t}{f_0^* \delta}\,, \qquad \psi' = (UL_1)\psi\,,$$

where L_1 is the north-south (channel width) length scale and U the velocity scale, and considering only zonal relief $h = h(y')$, we get

$$\delta^3 \psi_{xxt} + \delta \psi_{yyt} + \delta \psi_x h_y + \varepsilon \delta J(\psi, \psi_{yy}) + \varepsilon \delta^3 J(\psi, \psi_{xx}) = 0\,.$$

Here $\delta = L_1/L_2$ is the aspect ratio of cross-channel to axial length scales, while $\varepsilon = U f_*^0 / L_1$ is the small-amplitude parameter, an equivalent Rossby number. Attention is focused on systems in which nonlinearity and dispersion are of the same order of magnitude. Therefore, putting $\varepsilon = \delta^2$, we get

$$\psi_{yyt} + h_y \psi_x + \varepsilon[\psi_{xxt} + J(\psi, \psi_{yy})] + \varepsilon^2 J(\psi, \psi_{xx}) = 0\,. \tag{2.3}$$

Upon imposing progressive wave solutions $s = x - ct$, the equation is transformed into a frame moving with the wave velocity c; in such a frame the motion is steady. Solutions of (2.3) may then be expanded in ε as

$$\begin{cases} \psi = \psi_0(s, y) + \varepsilon \psi_1(s, y) + \dots\,, \\ c = c_0 + \varepsilon c_1 + \dots\,. \end{cases} \tag{2.4}$$

At the channel borders $y = 0, 1$ we could either apply periodic boundary conditions on ψ or require the borders to be streamlines $\psi = 0$. We consider here the case of solutions periodic in $0 \leqslant y \leqslant 1$, the other case having the same formal mathematical treatment. Putting the expansion (2.4) into (2.3) yields the zeroth-order system

$$\psi_{0yys} - \frac{1}{c_0} h_y \psi_{0s} = 0 . \tag{2.5}$$

Upon separation of variables $\psi_0 = g(s) \cdot \Phi(y)$,

$$\Phi_{yy} - \frac{1}{c_0} h_y \Phi = 0 , \qquad \Phi(0) = \Phi(1) . \tag{2.6}$$

This is a standard Sturm-Liouville problem. An infinite denumerable sequence of eigensolutions Φ_n and eigenvalues c_{0n} must therefore exist [18]. The zeroth-order solution, therefore, fixes the velocity c_0, even though the function $g(s)$ is still undetermined. To this order, the solution is an arbitrary superposition of linear $(\varepsilon \to 0)$ dispersionless $(\delta^2 \to 0)$ topographic Rossby waves.

At the next order ε, the effects of dispersion and nonlinearity enter simultaneously:

$$\psi_{1yys} - \frac{1}{c_0} h_y \psi_{1s} = - \frac{c_1}{c_0} \psi_{0yys} - \psi_{0sss} + \frac{1}{c_0} (\psi_{0s} \psi_{0yyy} - \psi_{0y} \psi_{0syy}) \tag{2.7}$$

with ψ_1 subject to the same periodic boundary conditions as ψ_0. Substitute $\psi_0 = g(s) \cdot \Phi(y)$ with Φ given by (2.6), multiply through Φ and integrate in y over the channel width. The right-hand side of (2.7) must be identically zero to avoid secular growth of ψ_1; $g(s)$ is thus found to have to satisfy the KdV equation

$$\frac{c_1}{c_0^2} a_1 g_s + \frac{1}{c_0^2} a_2 g g_s + a_3 g_{sss} = 0 \tag{2.8}$$

with

$$a_1 = \int_0^1 h_y \Phi^2 \, \mathrm{d}y , \qquad a_2 = - \int_0^1 h_{yy} \Phi^3 \, \mathrm{d}y , \qquad a_3 = \int_0^1 \Phi^2 \, \mathrm{d}y . \tag{2.9}$$

The solution of (2.8) is the periodic cnoidal wave

$$g(s) = A \, \mathrm{cn}^2 (Bs/n^*) , \tag{2.10a}$$

where n^* is the parameter of the Jacobian elliptic function cn.

In the limit $n^* \to 1$ this becomes the solitary wave

$$g(s) = A \operatorname{sech}^2 (Bs)\,, \tag{2.10b}$$

which we will consider hereafter. Substitution of (2.10b) into (2.8) fixes both the adjustment c_1 of the velocity and the inverse width B of the solitary wave. Thus

$$c_1 = -\frac{a_2}{3a_1} A\,, \qquad B^2 = \frac{a_2}{12 a_3 c_0^2} A \tag{2.11}$$

for the solitary-wave solution ($n = 1$).

To obtain the soliton solutions, the adimensionalized equation of motion (2.3)

$$\psi^*_{yyt} + h_y \psi^*_x + \varepsilon[\psi^*_{xxt} + J(\psi^*, \psi^*_{yy})] + \varepsilon^2 J(\psi^*, \psi^*_{xx}) = 0$$

was transformed into a frame of reference moving with the soliton speed c (* denotes the full solution of this equation). Such a frame is defined by the change of co-ordinates

$$s = x - ct\,, \qquad \tau = t\,.$$

In it, the soliton is steady, while any other superimposed motion will be time dependent. The thus transformed equation is

$$\begin{aligned} -c\psi^*_{yys} + \psi^*_{yy\tau} + h_y \psi^*_s + \\ + \varepsilon[-c\psi^*_{sss} + \psi^*_{ss\tau} + \psi^*_s \psi^*_{yyy} - \psi^*_y \psi^*_{syy}] + \varepsilon^2[\psi^*_s \psi^*_{ssy} - \psi^*_y \psi^*_{sss}] = 0\,. \end{aligned} \tag{2.12}$$

Now put

$$\psi^* = \psi(s, y) + \delta\psi(s, y, \tau)\,, \tag{2.13}$$

where $\psi(s, y)$ is the basic soliton solution and $\delta\psi$ a time-dependent perturbation, which will be considered small with respect to the soliton itself:

$$\delta\psi \ll \psi\,.$$

Substituting (2.13) into (2.12), expanding again the solitary-wave solution as well as the wave speed c in powers of ε, as in (2.4), allowing for the basic-state resulting eqs. (2.5) and (2.7), finally linearizing in the perturbation amplitude, we obtain

$$\begin{aligned} \frac{1}{c_{0n}} \delta\psi_{yy\tau} - \delta\psi_{yys} + \frac{1}{c_{0n}} h_y \delta\psi_s + \varepsilon \Big\{ \frac{1}{c_{0n}} \delta\psi_{ss\tau} - \frac{c_{1n}}{c_{0n}} \delta\psi_{yys} - \delta\psi_{sss} + \\ + \frac{1}{c_{0n}} [\psi_{0ns} \delta\psi_{yyy} + \psi_{0nyyy} \delta\psi_s - \psi_{0ny} \delta\psi_{syy} - \psi_{0nsyy} \delta\psi_y] \Big\} + O(\varepsilon^2) = 0\,. \end{aligned} \tag{2.14}$$

Notice that $\delta\psi$ is not yet expanded in ε. Here we have chosen a specific basic state $\psi_{0n} = g_n(s)\cdot\Phi_n(y)$ with speed $c_n = c_{0n} + \varepsilon c_{1n} + \ldots$ defined by the zeroth-order eigenvalue problem for the soliton itself. Equation (2.14) allows for time separation

$$\delta\psi = \exp[\sigma\tau]\, G(s, y) = \exp[(\sigma_0 + \varepsilon\sigma_1 + \ldots)\tau][G_0(s, y) + \varepsilon G_1(s, y) + \ldots], \tag{2.15}$$

in which now both the perturbation amplitude $G(s, y)$ and the growth rate σ are expanded in powers of ε. Upon substitution of (2.15) into (2.14) we get at zeroth order

$$\frac{\sigma_0}{c_{0n}} G_{0yy} - G_{0yys} + \frac{1}{c_{0n}} h_y G_{0s} = 0 \tag{2.16}$$

and at order ε

$$\frac{\sigma_0}{c_{0n}} G_{1yy} - G_{1yys} + \frac{1}{c_{0n}} h_y G_{1s} + \frac{\sigma_1}{c_{0n}} G_{0yy} + \frac{\sigma_0}{c_{0n}} G_{0ss} - \frac{c_{1n}}{c_{0n}} G_{0yys} - \\ - G_{0sss} + \frac{1}{c_{0n}}[\psi_{0ns} G_{0yyy} + \psi_{0nyyy} G_{0s} - \psi_{0ny} G_{0syy} - \psi_{0nsyy} G_{0y}] = 0\,. \tag{2.17}$$

The zeroth-order equation (2.16) may be solved by separation of (s, y) variables only if we expand $G_0(s, y)$ in the orthogonal set of eigenmodes $\Phi_p(y)$ of the basic state. In fact, putting

$$G_0(s, y) = \sum_p F_{0p}(s)\cdot\Phi_p(y)$$

into the zeroth-order equation and remembering the eigenvalue equation

$$\Phi_{pyy} - \frac{1}{c_{0p}} h_y \Phi_p = 0\,,$$

we get

$$\sum_p \left\{\left(\frac{1}{c_{0n}} - \frac{1}{c_{0p}}\right) h_y \Phi_p F_{0ps} + \frac{\sigma_{0p}}{c_{0n} c_{0p}} h_y \Phi_p F_{0p}\right\} = 0\,.$$

Multiplying the equation through Φ_m $(m = 1, 2, \ldots, p, \ldots)$, integrating in y over the channel and using the orthogonality properties

$$\int_0^1 \Phi_m h_y \Phi_p \, dy = \delta_{p,m}$$

yields

$$\sum_p \left\{\left(\frac{1}{c_{0n}} - \frac{1}{c_{0p}}\right) \delta_{p,m} F_{0ps} + \frac{\sigma_{0p}}{c_{0n} c_{0p}} \delta_{p,m} F_{0p}\right\} = 0\,,$$

which gives, for every m,

$$\left(\frac{1}{c_{0n}}-\frac{1}{c_{0m}}\right)F_{0ms}+\frac{\sigma_{0m}}{c_{0n}c_{0m}}F_{0m}=0\,. \tag{2.18}$$

Two cases can now be distinguished for the perturbation amplitude at order zero.

a) The perturbation has the same cross-channel dependence as the basic state: $m = n$.

Equation (2.18) then gives

$$\sigma_{0n}\equiv 0\,. \tag{2.19}$$

The perturbation mode having the same y-dependence as the basic state is, at zeroth order, neutrally stable, and its s-dependence is yet undetermined.

b) The perturbation has different cross-channel dependence from that of the basic state: $m \neq n$.

Equation (2.18) can then be solved

$$F_{0m}(s)=M\exp\left[-\frac{\sigma_{0m}}{c_{0m}-c_{0n}}\,s\right]. \tag{2.20}$$

In the infinite channel, this solution must satify the boundary condition of being bounded as $s\to\pm\infty$. This implies that the necessary condition on σ_{0m} is

$$\operatorname{Re}\sigma_{0m}\equiv 0\,,\qquad\qquad \sigma_{0m}=i\gamma\,. \tag{2.21}$$

For the solution in a box of side 2π, with periodic boundary conditions, $F_{0m}(0) = F_{0m}(2\pi)$, this determines the eigenvalues $\sigma_{0m,l}$ for the zeroth-order growth rate

$$\begin{cases}\sigma_{0m,l}=-\,il(c_{0m}-c_{0n})\,,\\ F_{0m}\;\;=M\exp[ils]\,, & l=0,\ \pm1,\ \pm2,\ldots\end{cases} \tag{2.22}$$

In both these cases, the perturbation is thus neutrally stable at zeroth order.

To proceed in the analysis to order ε, one must consider eq. (2.17), valid at order ε. Then the only algebraically feasible case is that of a single-mode perturbation dependence at zeroth order, namely the zeroth order is taken to be

$$G_0(s,y)=F_{0m}(s)\cdot\Phi_m(y)$$

with growth rate $\sigma_m = \sigma_{0m} + \varepsilon\sigma_{1m}$. Even with this simplification, the mathematics is still rather cumbersome. Therefore, we shall limit ourselves to summarize the significant results, referring the interested reader to [14] for a detailed discussion.

Again, two cases of perturbation cross-channel dependence can be distinguished, the first with $m = n$ (the perturbation has the same cross-channel dependence as the basic state), the second with $m \neq n$. For $m = n$, it can be proved that the $O(\varepsilon)$ growth rate σ_{1n} is purely imaginary or identically zero. For $m \neq n$, in the above-mentioned reference it is shown that σ_{1m} is always purely imaginary. One can conclude, therefore, that, in the context of a linearized stability analysis, the permanent-form solution is neutrally stable to a single-mode perturbation in the initial condition up to $O(\varepsilon)$.

The zeroth-order results, summarized by (2.19) through (2.22), are, however, those strictly necessary to compare with them the results of the numerical stability experiments, exposed in the following section.

3. – Finite-amplitude perturbations in the solitary eddy initial conditions: numerical experiments.

It was shown in the previous section that disturbances having the same y-dependence as the basic state ($m = n$) have a zeroth-order growth rate $\sigma_{0n} \equiv 0$; they alter only the long-channel, *i.e.* s-structure of the basic state, thus actually acting as one-dimensional perturbations. The neutral stability of the solitary Rossby waves was also demonstrated up to $O(\varepsilon)$ for infinitesimal perturbations the functional shape of which is given by

$$\begin{aligned}(3.1)\qquad \delta\psi &= \exp[(\sigma_{0m} + \varepsilon\sigma_{1m} + \dots)\tau][F_{0m}(s) + F_{1m}(s) + \dots]\cdot\Phi_m(y) = \\ &= \exp[i(\gamma + \varepsilon\sigma_{1m} + \dots)\tau]\left[M\exp\left[-\frac{i\gamma}{c_{0m} - c_{0n}}s\right] + O(\varepsilon)\right]\cdot\Phi_m(y)\,,\end{aligned}$$

where $\sigma_m = (\sigma_{0m} \equiv i\gamma) + \varepsilon(\sigma_{1m} \equiv i\sigma^*_{1m}) + O(\varepsilon^2)$ is the growth rate of the perturbation, purely imaginary, and $s = s - c_n t$, where c_n is the speed of the soliton basic state $\psi_{0n} = g_n(s)\cdot\Phi_n(y)$, with $g_n(s)$ the cnoidal or solitary-wave solution to the KdV equation (2.8). (3.1) is a perturbation having a cross-channel (y) dependence different from that of the basic state ($m \neq n$), *i.e.* a truly two-dimensional infinitesimal perturbation. In the periodic box of size 2π, the single-mode perturbation (3.1) becomes

$$(3.2)\qquad \delta\psi = M\exp\left[i[ls - l(c_{0m} - c_{0n})\tau]\right]\cdot\Phi_m(y) + O(\varepsilon)$$

in the frame of reference moving with the basic-state speed c_n. In the stationary frame of reference, if we neglect terms of $O(\varepsilon)$, (3.2) can be written as

$$\delta\psi \simeq M \exp\left[il(x - c_{0m} t)\right] \cdot \Phi_m(y) . \tag{3.3}$$

Numerical experiments were performed to investigate the stability properties of the permanent-form solutions of the full model equation (2.3) with respect to perturbations with a functional shape given by (3.3) but with a finite amplitude M. We give the results of two numerical experiments, for two different cross-channel eigenmodes Φ_m. In all the following stability experiments the topography was taken to be

$$h(y) = -\sin 2y .$$

Then the zeroth-order eigenvalue problem (2.6) gives the Mathieu equation. As cross-channel structure for the basic state, the second lowest eigenmode was considered, corresponding to the second odd Mathieu function $\Phi_2(y) = se_2(y)$. All the following stability experiments were performed upon weak-amplitude permanent-form solutions with $A = -0.02$ and $B = 0.53$, as given by the second of (2.11).

In the first experiment, the lowest possible nonconstant eigenfunction $\Phi_1(y)$ for the relief $h(y) = -\sin 2y$ was considered. It corresponds to the first odd Mathieu function $se_1(y)$. For the x-dependence, the lowest wave number $l = 1$ was chosen. Then the perturbation was

$$\delta\psi = M \cos x \cdot \Phi_1(y) = M \cos x \cdot se_1(y) . \tag{3.4}$$

Perturbation (3.4) thus corresponds to the lowest two-dimensional linear periodic mode in the box. The experiment was carried out with $M = A$, perturbation and basic state having the same magnitude. In fig. 1 the physical-space patterns are given for the total streamfunction initial condition

$$\psi_{\text{total}} = \psi_{\text{sol}} + \delta\psi = A[\operatorname{sech}^2(Bx)\,\Phi_2(y) + \cos x \cdot \Phi_1(y)] . \tag{3.5}$$

The integration was continued up to $O(1/\varepsilon^2)$ dimensionless time, that is $T = 100$. In fig. 1 patterns are shown only through $T = 44$, the basin transversal time of the basic state. Subsequent evolution was not different. The field is clearly a superposition of two types of eddies, the faster—the disturbance—overtaking the slower many times, until at $T = 44$ a state almost identical to the initial one is reformed.

To observe more clearly the behavior of the perturbation itself, a widely used procedure has been followed [19]. Let us call run 1 the soliton basic-state evolution for the considered number of time steps N. Call run 2 the evolution of the perturbed field starting from (3.5). From the streamfunction

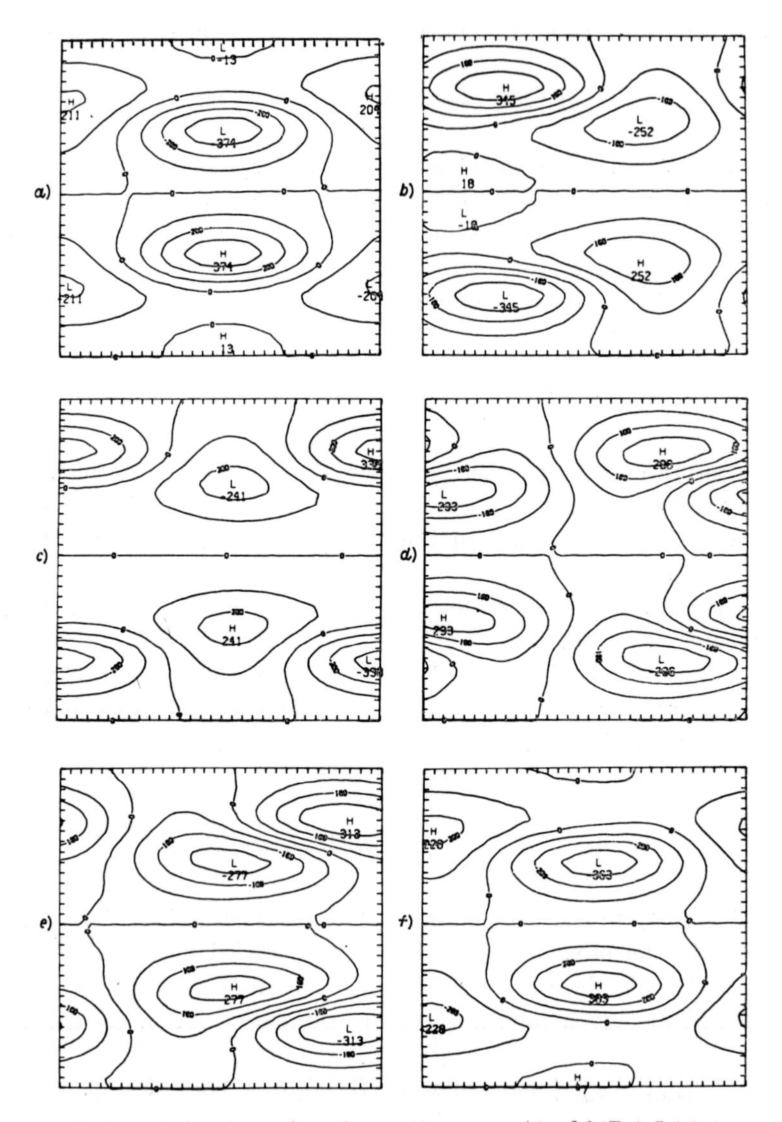

Fig. 1. – Evolution of the streamfunction pattern $\psi = A[\operatorname{sech}^2(Bx)\,\Phi_2(y) + \cos x \cdot \Phi_1(y)]$ at the following dimensionless times: *a*) $t = 0$, contour interval 0.01; b) $t = 9$, contour interval 0.009; *c*) $t = 22$, contour interval 0.01; *d*) $t = 26$, contour interval 0.008; *e*) $t = 34$, contour interval 0.008; *f*) $t = 44$, contour interval 0.01. $A = -0.02$, $B = -0.53$; Φ_1 and Φ_2 are, respectively, the first and second eigenmodes for the topography $h = -\sin 2y$. The field is scaled by 10^4.

results of each realization a difference streamfunction is then produced $\psi^{(3)} = \psi^{(2)} - \psi^{(1)}$ corresponding to the perturbation itself. This procedure is equivalent to the direct integration of the exact equation of motion for the finite-amplitude perturbation $\delta\psi$.

In fig. 2 the perturbation streamfunction thus obtained is shown from $T = 22$ to $T = 44$. Subsequent evolution was identical. The neutral stability of the perturbation is evident. The initial state, composed of four symmetrical eddies with alternating highs and lows both in x and in y directions, propagates westward with only an oscillation in amplitude around the initial value. We notice that the perturbation is by no means small. Its amplitude is the same as that of the basic state, therefore also its energy is of the same order

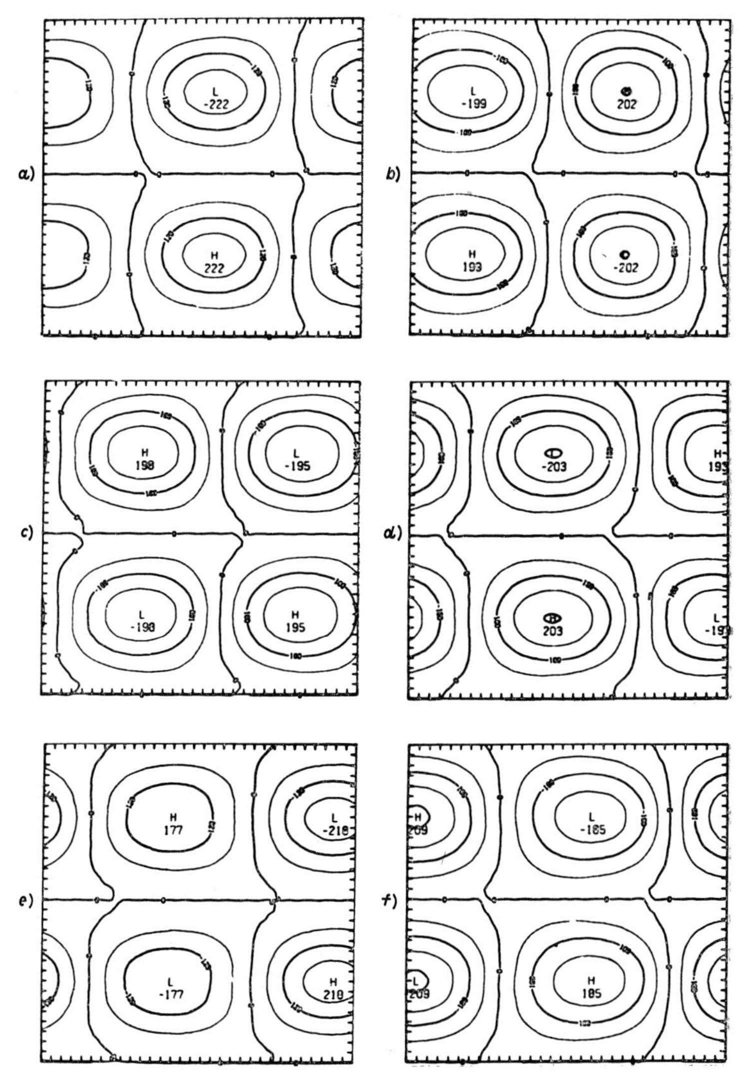

Fig. 2. – Evolution of the perturbation $\delta\psi = A \cos x \cdot \Phi_1(y)$ at the following dimensionless times: *a*) $t = 22$, contour interval 0.006; *b*) $t = 26$, contour interval 0.005; *c*) $t = 30$, contour interval 0.005; *d*) $t = 34$, contour interval 0.005; *e*) $t = 40$, contour interval 0.006; *f*) $t = 44$, contour interval 0.005. $A = -0.02$. The field is scaled by 10^4.

of magnitude, even though its root mean square vorticity ($\zeta_{\text{r.m.s.}}$) is one order of magnitude smaller. The perturbation evolves as a linear wave for which the interactions with itself are very small. This is because, its $\zeta_{\text{r.m.s.}}$ being one order of magnitude smaller than that of the soliton, self-advection of relative vorticity will, therefore, be small. An alternative way to see this is to notice that the lowest eigenmode $\Phi_1(y)$ has a shape very similar to $\sin y$, and nonlinear terms such as $J(\psi, \psi_{yy})$ vanish identically for purely sinusoidal functions. The neutral stability of the perturbation is evident also from the behavior of its energy spectrum. In fig. 3 the time evolution of the most energetic components of the isotropic spectrum is shown for the total field. Energy for the perturbation is concentrated at $K = 1$ and $K = 3$, and these two components exchange energy.

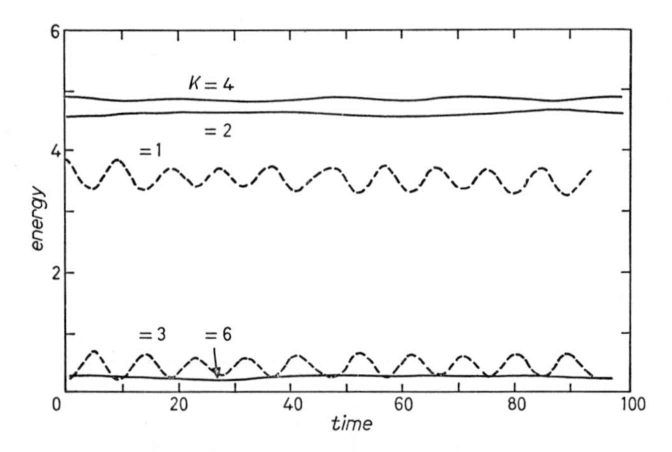

Fig. 3. – Time evolution of the modal energies of the soliton basic state ($K = 2, 4, 6$) and perturbation (3.4) ($K = 1, 3$) normalized with respect to the total soliton energy, in units 10^{-1}.

Perturbation (3.4) is a good first example with which to test extrapolation of the linearized stability analysis predictions to finite-amplitude perturbations of solutions of the complete model equation, because, as we have seen, the $O(\varepsilon)$ self-interaction terms $J(\delta\psi, \delta\psi_{yy})$ are negligible for (3.4). This property enables us to interpret simply the results of the linearized analysis from the point of view of triad interactions, because it means that the perturbation self-triads are small. Given this, the fact that the solitary-wave solution is neutrally stable to the single-mode perturbation up to $O(\varepsilon)$ implies that the triad interactions between perturbation and basic state are not resonant up to $O(\varepsilon)$.

This fact can be seen in the following way. At zeroth order, the dispersion relationship, respectively, for the soliton basic state and the single-mode perturbation (3.3) can be written as

$$\left\{\begin{aligned}\omega_{\text{sol}} &= |c_{0n}|k_x\,,\\ \omega_{\text{pert}} &= |c_{0m}|k_x\,.\end{aligned}\right.\tag{3.6}$$

The linear resonant triad interactions require the simultaneous fulfilling of the momentum and energy conservation conditions

$$\left\{\begin{aligned}&\boldsymbol{k}_1 + \boldsymbol{k}_2 = \boldsymbol{k}_3\,,\\ &\omega_1 + \omega_2 = \omega_3\,.\end{aligned}\right.\tag{3.7}$$

Dispersion relationships like (3.6) correspond to waves which are dispersionless in the x-direction. No resonant triads of the type perturbation-soliton-perturbation are then possible for $m \neq n$, that is for $|c_{0n}| \neq |c_{0m}|$. If $m = n$, we have seen that, for the perturbation growth rate at zeroth order, $\sigma_{0n} = 0$ and, therefore, no dispersion relationship like the second of (3.6) exists at zeroth order. Thus the results of the stability analysis imply that the $O(\varepsilon)$ triads are not resonant. For the perturbation given by (3.4), $|c_{0m}| \simeq 1.1$, while $|c_{0n}| \simeq$ $\simeq 0.132$, the perturbation and solitary-wave zeroth-order velocities differing by one order of magnitude. The nonresonant triad interactions with the basic state will have as their only effect a modulation of the perturbation itself. This is quite evident both in the physical-space patterns of fig. 2 and in the modal-energy behavior of fig. 3. For further details the reader is referred to [15].

The results of the stability analysis, that the triad interactions between perturbation and basic state are not resonant, are valid at zeroth order, namely for dispersion relationships exactly given by (3.6). Apart from the perturbation self-triads (which will not be negligible for perturbations having a cross-channel (y) dependence different from $\Phi_1(y) = se_1(y)$), the triad interactions between the perturbation and the basic state are modified by nonlinearity. The actual form of the dispersion relationships (3.6) is, therefore,

$$\left\{\begin{aligned}\omega_{\text{sol}} &= |c_{0n} + \varepsilon c_{1n}(A) + \ldots|k_x\,,\\ \omega_{\text{pert}} &= |c_{0m} + \varepsilon c_{1m}(A) + \ldots\,|k_x\,,\end{aligned}\right.\tag{3.8}$$

in which c_{1m} also depends upon the basic-state amplitude A. This implies that resonant triads could occur, although they were not possible when (3.8) were replaced by (3.6). But even so for perturbation (3.4) triads are not resonant, because the zeroth-order velocities c_{0m}, c_{0n} are widely different, by one order of magnitude. If for some other perturbation the zeroth-order perturbation and soliton speeds c_{0m}, c_{0n} were of the same order of magnitude, the nonlinearity could allow the possibility of resonances. The second single-mode finite-perturbation experiment was, therefore, performed choosing a cross-channel dependence for the perturbation for which i) self-interactions are not negligible and ii) the zeroth-order perturbation speed $|c_{0m}|$ is very near to $|c_{0n}|$. At $t = 0$ this second perturbation was given by

$$\delta\psi = M \cos 8x \cdot \Phi_6(y) , \tag{3.9}$$

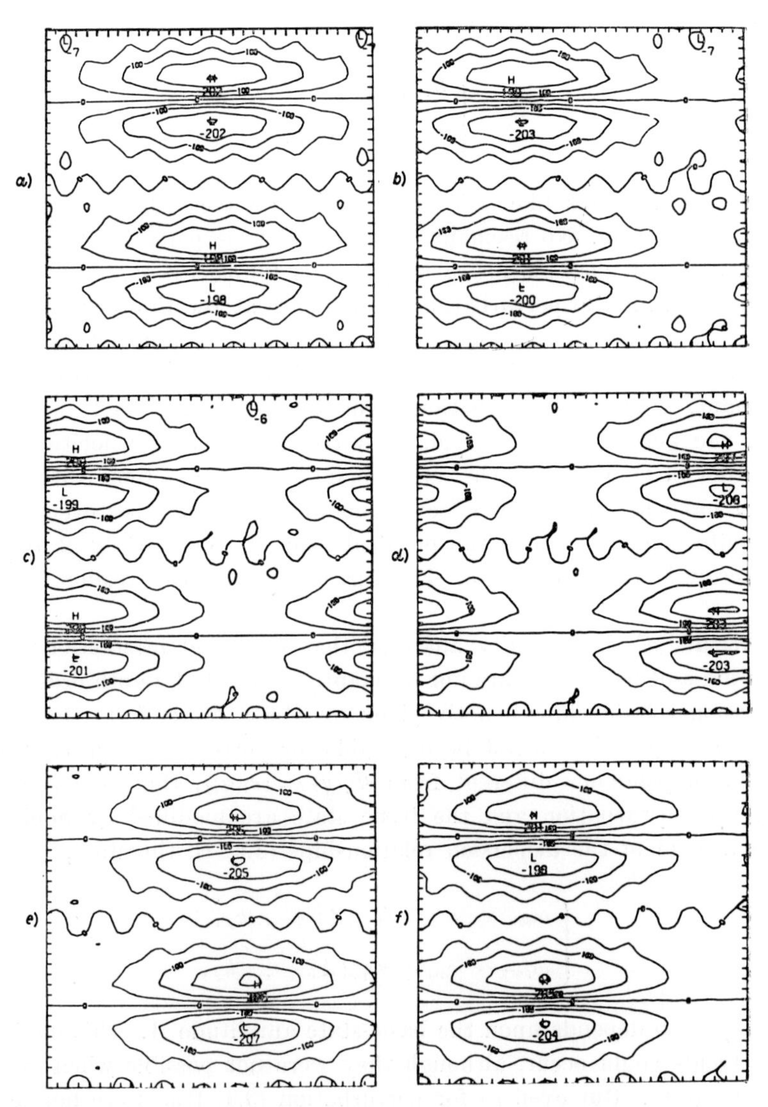

Fig. 4. – Evolution of the streamfunction pattern $\psi = A[\text{sech}^2 (Bx)\, \Phi_2(y) + A^* \cos 8x \cdot \Phi_6(y)]$ at the following dimensionless times: *a*) $t = 0$, *b*) $t = 8$, *c*) $t = 18$, *d*) $t = 25$, *e*) $t = 40$, *f*) $t = 50$. $A = -0.02$, $B = 0.53$, $A^* = 0.1$. Φ_2 and Φ_6 are, respectively, the first and sixth lowest eigenmodes for the topography $h = -\sin 2y$. The field is scaled by 10^4.

where $\Phi_6(y)$ is the eigenfunction having six zero crossings in the channel and $M = 10^{-1} A$ is the perturbation amplitude.

Figure 4 shows the perturbed solution during one transversal of the basin, up to $T = 50$. The experiment was continued up to dimensionless time $T = 150$ with unchanged subsequent evolution. Again, notice that (3.9) is not a small perturbation. Its r.m.s. particle velocity $u_{\text{r.m.s.}}$ and r.m.s. vorticity $\zeta_{\text{r.m.s.}}$ are

$$(u_{\text{r.m.s.}})_{\text{pert}} \sim \tfrac{1}{2}(u_{\text{r.m.s.}})_{\text{sol}}, \qquad (\zeta_{\text{r.m.s.}})_{\text{pert}} \sim (\zeta_{\text{r.m.s.}})_{\text{sol}}.$$

The self-interaction terms are not negligible and the basic requirement of the linearized stability analysis is thus not respected. Still, the physical-space pictures of fig. 4 show neutrally stable behavior persisting up to $T = 150$. The soliton progresses uniformly and a perturbation traveling at a different speed is clearly visible.

The $\Phi_6(y)$ eigenfunction has the eigenvalue $c_{0m} \simeq 0.05$. Equation (3.3) predicts that the perturbation should move eastward. This is indeed the case, as it is evident from fig. 5 in which the initial state is superimposed on the configuration at $T = 5$. This can also be see in fig. 6 giving the perturbation pattern itself. There the same eddy has been labeled at successive times. Figure 6 shows some coalescing of the smaller eddies into bigger ones. Still, the pattern remains remarkably coherent and conserves good enough memory of the initial state that we can track an original whorl without ambiguity. In the light of the previously given discussion on the possibility of resonant triads, note that now the soliton's most energetic modes are $k_{x1}(\text{sol}) = \pm 1$, while

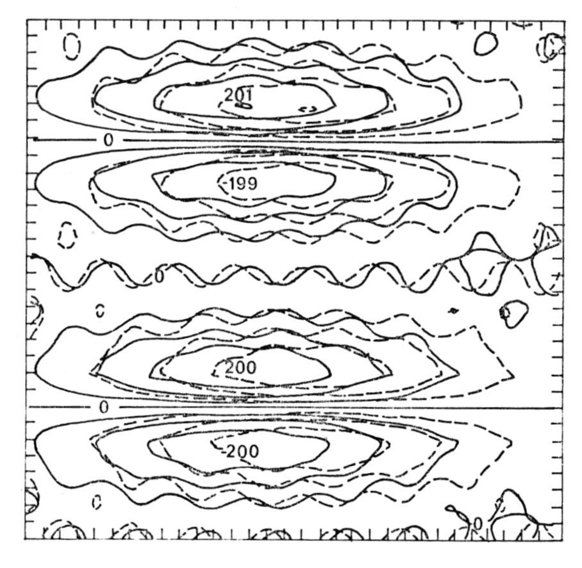

Fig. 5. – Initial state of the permanent-form solution plus the perturbation (3.9) (dashed line) superimposed to the configuration at dimensionless time $T = 5$ (solid line).

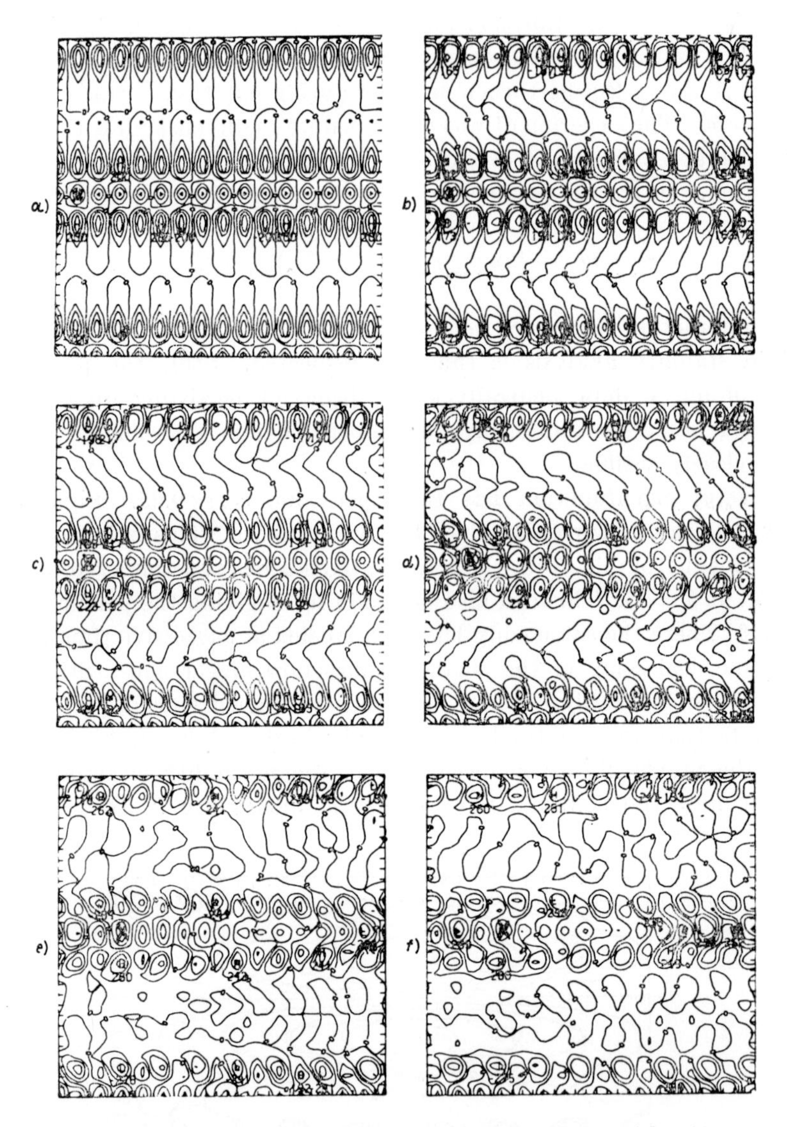

Fig. 6. – Evolution of the perturbation $\delta\psi = A^* \cos 8x \cdot \Phi_6(y)$ with $A^* = -0.002$ at the following dimensionless times: *a*) $t = 0$, contour interval 0.0005; *b*) $t = 5$, contour interval 0.0005; *c*) $t = 10$, contour interval 0.0006; *d*) $t = 20$, contour interval 0.0006; *e*) $t = 30$, contour interval 0.0007; *f*) $t = 40$ contour interval 0.0008. The field is scaled by 10^4.

initially $k_{x2}(\text{pert}) = 8$. Possible results of triad interactions are thus

$$k_{x3} = K_{x1} + K_{x2} = \begin{cases} 9\,, \\ 7\,. \end{cases}$$

If we consider only the linear contribution to the frequency, the corresponding condition $\omega_1 + \omega_2 = \omega_3(\text{pert})$ is

$$|c_{0n}|k_{x1} + |c_{0m}|k_{x2} = |c_{0m}|k_{x3}\,.$$

Since $|c_{0n}| \simeq 0.1$, $|c_{0m}| \simeq 0.05$, this gives

$$+\,|c_{0n}| + 8\,|c_{0m}| \simeq 9\,|c_{0m}| \qquad (0.5 \simeq 0.45)\,,$$

$$-\,|c_{0n}| + 8\,|c_{0m}| \simeq 7\,|c_{0m}| \qquad (0.3 \simeq 0.35)\,.$$

The resonant condition is thus approximately satisfied. This is true without taking into account the nonlinear contributions to the frequency. It is significant that the first new Fourier modes of the perturbation two-dimensional spectrum to appear are $a_{9,\pm1}$, $a_{7,\pm1}$.

The resonant-triad interpretation implies that the perturbation will now extract energy from the basic state before, and if, undergoing oscillations. The perturbation total energy, normalized with respect to the soliton energy, is shown in fig. 7. It grows initially. But it is noteworthy that, after the initial

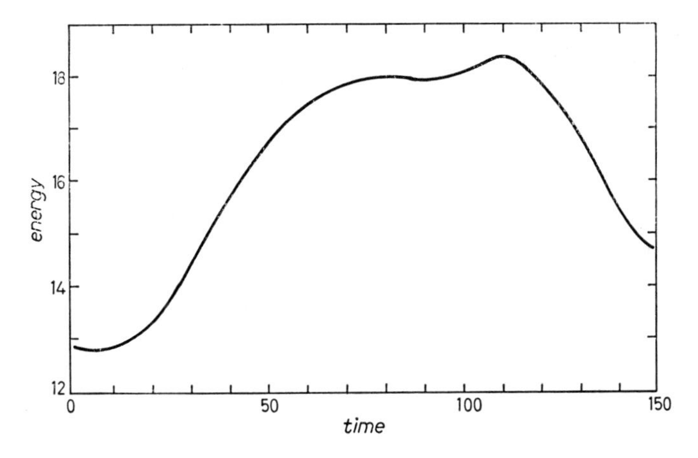

Fig. 7. – Time evolution of the total energy of the perturbation (3.9) normalized with respect to the total soliton energy. Vertical scale in units 10^{-2}.

growth (40 % of the initial amplitude in the first 80 dimensionless time units), the perturbation energy reaches a saturation level, oscillates around it and subsequently undergoes a rapid decrease. By $T = 150$ the rate of decrease seems to be slowing although the perturbation energy still has a higher value than the initial perturbation energy. This remarkable behavior suggests that, if the perturbation initial intensity does not exceed some critical value, resonant triads will be effective mainly in the initial stages of the evolution allowing the

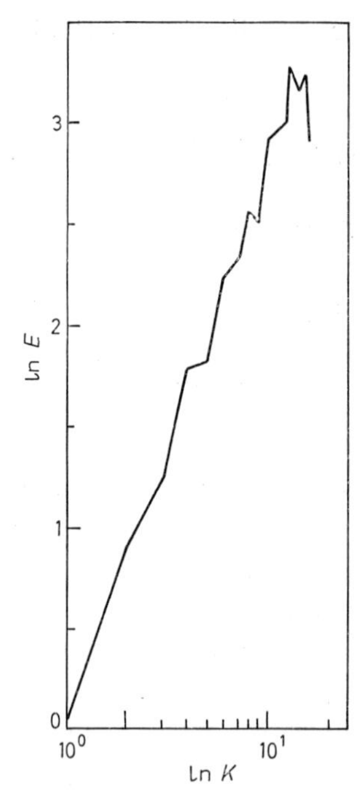

Fig. 8. – Isotropic energy spectrum for the random perturbation of experiments i), ii), iii) in logarithmic units. Horizontal co-ordinate: wave number.

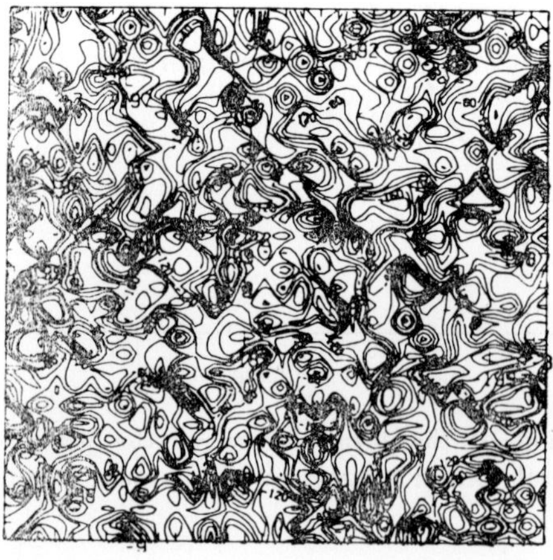

Fig. 9. – Streamfunction pattern for the random perturbation at $t = 0$. The field is scaled by 10^4.

perturbation total energy to grow at the most at a certain level, but not to the point of finally destroying the basic state. This behavior may be enhanced when considering general disturbances, for which many resonant triads are, therefore, possible.

The most general disturbance which can be imposed is a random perturbation of the streamfunction field at all mesh points, uncorrelated with the existing flow. Such a perturbation was obtained through a random-number generator. It was made to have a scalar energy spectrum proportional to k^3 (fig. 8) and had the initial physical-space pattern shown in fig. 9. Three numerical experiments were carried out, with three different amplitudes of the perturbation

$$\text{i)}\ M = 0.1A_{sol}, \qquad \text{ii)}\ M = 0.5A_{sol}, \qquad \text{iii)}\ M = A_{sol}.$$

Experiment i) was run up to $T = 150$, ii) and iii) up to $T = 100$.

Figure 10 shows the streamfunction patterns of the perturbed soliton for experiment i) over one transversal of the basin, from $T = 54$ to $T = 96$. Sub-

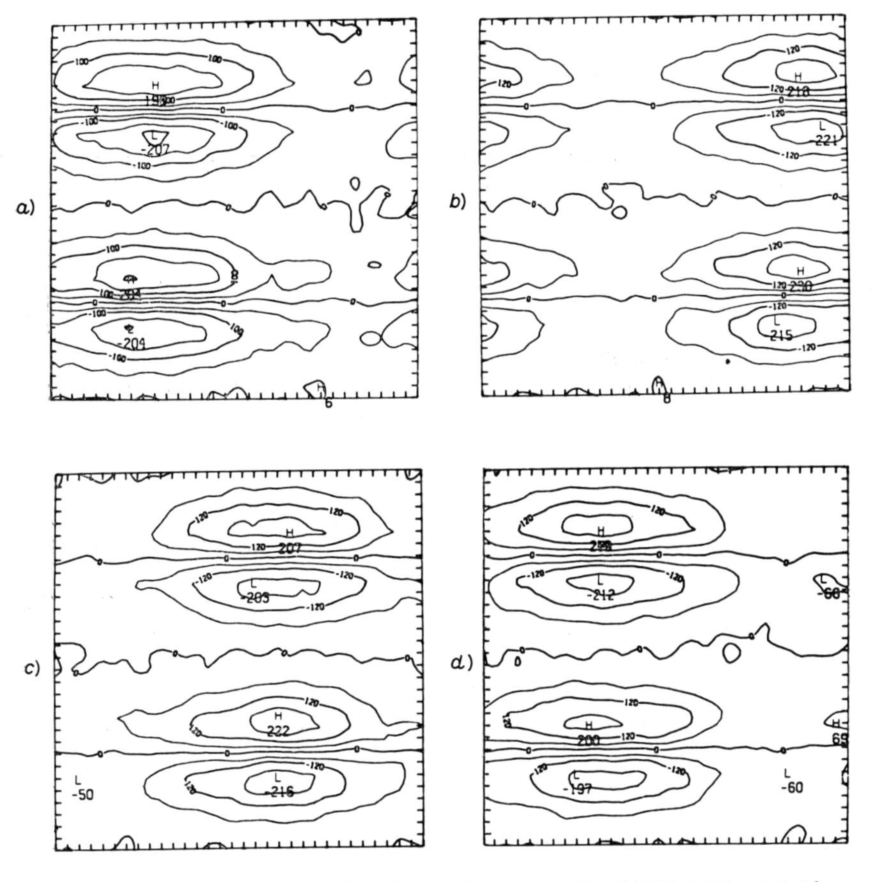

Fig. 10. – Evolution of the streamfunction pattern $\psi = A[\text{sech}^2(Bx)\,\Phi_2(y) + 0.1\,\delta\psi_{\text{random}}]$ at the following dimensionless times: *a)* $t = 54$, contour interval 0.005; *b)* $t = 72$, contour interval 0.006; *c)* $t = 84$, contour interval 0.006; *d)* $t = 96$, contour interval 0.006. $A = -0.002$, $B = 0.53$. The field is scaled by 10^4.

sequent evolution was identical up to $T = 150$. The physical-space representation indicates approximate neutral stability of the soliton. Still, the fact that the perturbation is extracting some energy from the mean field can be verified by looking at the perturbation itself. In fig. 11 this pattern is shown

Fig. 11. – Streamfunction pattern for the perturbation $0.1\ \delta\psi_{\text{random}}$ at dimensionless time $T = 50$. The field scaled by 10^4.

at $T = 50$. Comparison with the initial condition (fig. 9) clearly shows the clustering of small whorls into bigger ones, indicating an increase in large-scale perturbation energy.

Figure 12 shows the perturbed field for experiment ii) until $T = 30$. At this time the soliton is not easily recognizable, being far advanced in its way

TABLE I.

	Functional dependence	Amplitude	Total energy	$u_{\text{r.m.s.}}$	$\zeta_{\text{r.m.s.}}$	Energy of perturbed state $\psi_{\text{tot}} = \psi_{\text{sol}} + \delta\psi$
soliton basic state	$A\ \text{sech}^2(Bx)\,\Phi_2(y)$	$A = -0.02$	1	0.02	0.07	
perturbation						
a)	random field	$10^{-1}A$	$7.11\cdot 10^{-2}$ at $t = 0$	0.006	0.07	1.09
b)	$\cos 8x \cdot \Phi_6(y)$	$10^{-1}A$	$1.28\cdot 10^{-1}$ at $t = 0$	0.01	0.07	1.13
c)	random field	$0.5A$	1.78 at $t = 0$	0.03	0.35	2.86
d)	random field	A	7.11 at $t = 0$	0.06	0.7	8.27

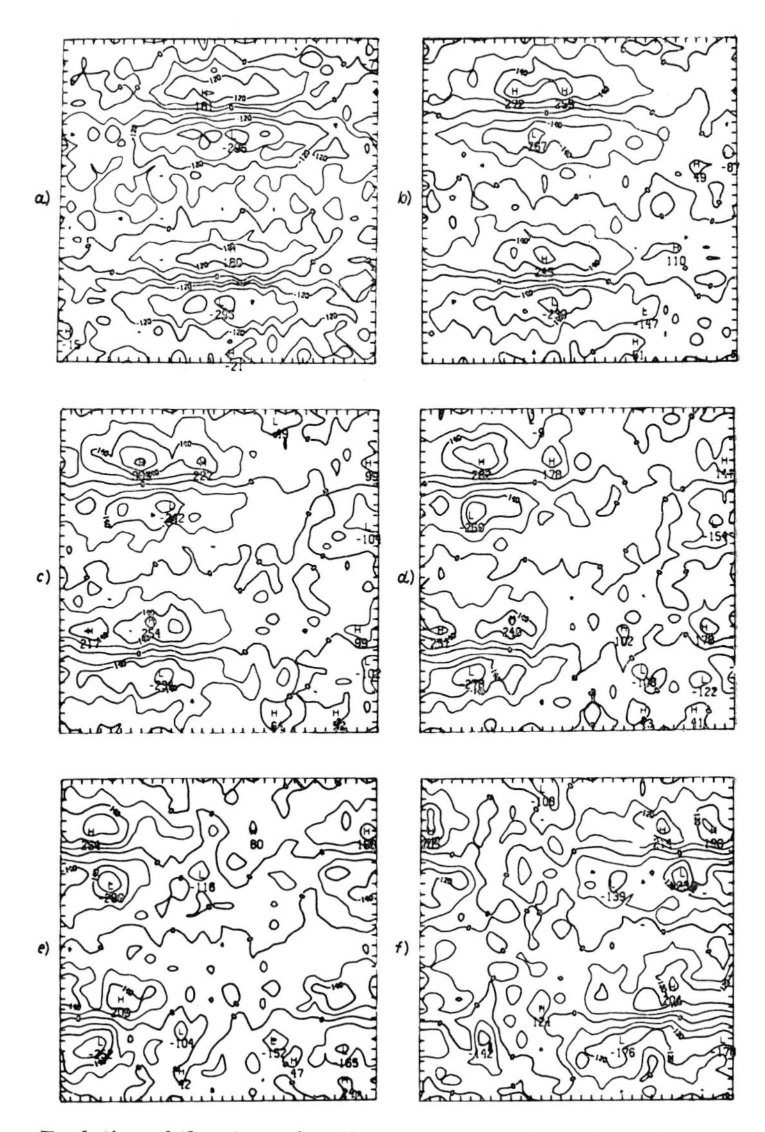

Fig. 12. – Evolution of the streamfunction pattern $\psi = A[\text{sech}^2(Bx)\,\Phi_2(y) + 0.5\,\delta\psi_{\text{random}}]$ at the following dimensionless times: *a*) $t = 0$, contour interval 0.006; *b*) $t = 5$, contour interval 0.007; *c*) $t = 10$, contour interval 0.007; *d*) $t = 15$, contour interval 0.007; *e*) $t = 20$, contour interval 0.007; *f*) $t = 30$, contour interval 0.006. $A = -0.02$, $B = 0.53$. The field is scaled by 10^4.

to destruction through interaction with the energetic perturbation. In experiment iii) the physical-space representation does not allow one to distinguish the soliton among the big random whorls of the disturbance. Experiment iii) is actually more a turbulence than a stability experiment.

Table I compares all these stability experiments, to which the experiment corresponding to the perturbation (3.9) has been added. The random field *a*) with an amplitude one order of magnitude smaller than the soliton is a moderate perturbation, as is evident when comparing the relative $u_{\text{r.m.s.}}$ and $\zeta_{\text{r.m.s.}}$ values. The previously considered single-mode disturbance *b*), (3.9), has av-

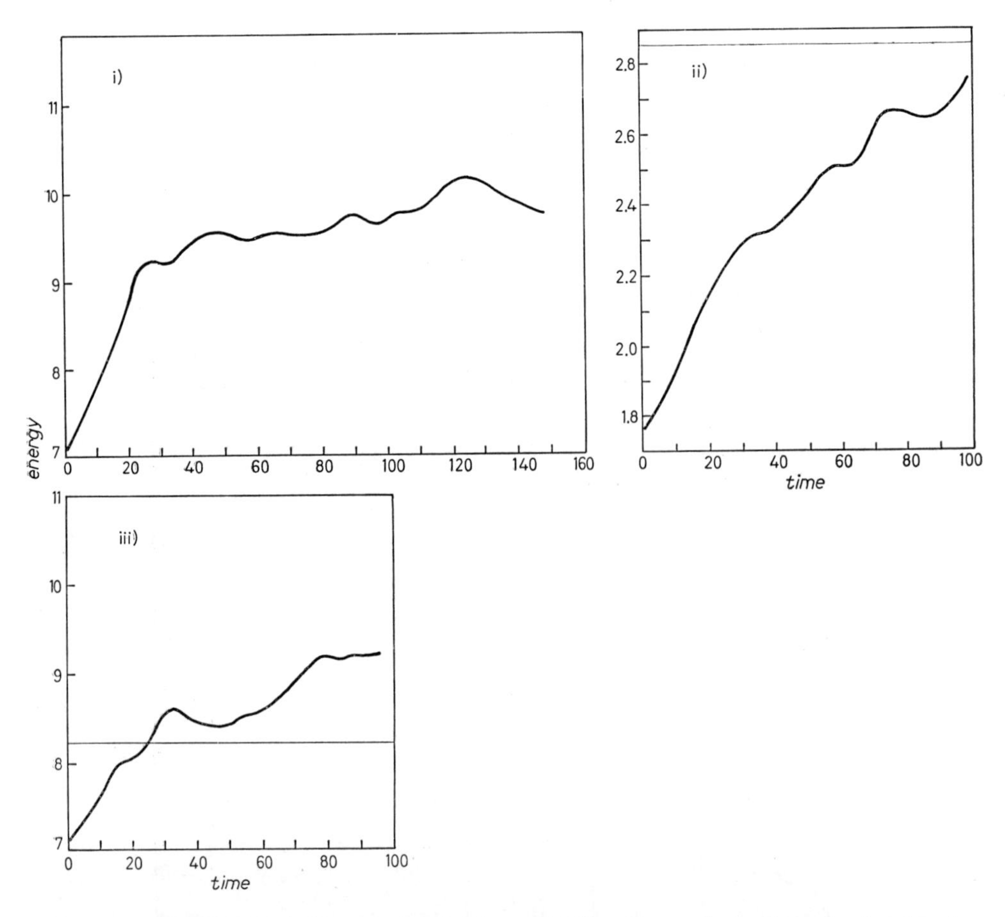

Fig. 13. – i) Time evolution of the total energy for the perturbation *a*) of table I, normalized with respect to the total soliton energy. Vertical scale in units 10^{-1}. ii) Time evolution of the total energy for the perturbation *c*) of table I normalized with respect to the total soliton energy. iii) Time evolution of the total soliton energy for the perturbation *d*) (turbulence experiment) of table I, normalized with respect to the total soliton energy.

erage physical properties the same order of magnitude as the soliton itself. Perturbations *c*) and *d*) are stronger than the basic state. Experiment *d*), as already mentioned, is a turbulence experiment and its significance for the soliton stability properties will be in the rate at which the initial soliton, distinguishable by examining the spectral properties, is destroyed.

Figures 13 i), ii), iii) show the perturbation energies for experiments i), ii), iii), cases *a*), *c*), *d*) of table I, normalized with respect to the total soliton energy. As in the previously discussed experiment *b*), the behavior for case *a*) is very remarkable. The energy increases until reaching a saturation value around which it oscillates and grows only very slowly. The total growth of the perturbation energy is of about 30 % of its initial value, as compared to the 40 % of the initial growth period in case *b*) (fig. 7). The oscillatory behavior begins at $T = 25$ in dimensionless units, earlier than for experiment *b*), $T = 80$.

Now let us consider a purely turbulent flow and two initial states differing by one another by an « error » energy. Their representative points in phase space will then be separated by a well-defined distance, very small if the error is small. The error energy is known to grow in time, the two point trajectories diverging in phase space until the two original states become as widely separated as two randomly chosen ones. The error energy itself will then approach a saturation limit given by the sum of the energies of the two initial states. For experiment *c*) the soliton is unstable to the random superimposed field which has an initial energy bigger than the soliton itself (1.78 relative to a normalized value 1 for the total soliton energy). In this case, in fact, in fig. 13 ii) it can be observed that the total perturbation energy grows in time by oscillating around a mean slope which decreases only slightly after the initial period of growth. At $T = 100$ the total perturbation energy has almost reached the value of the energy of the perturbed state (soliton + perturbation) given by the solid line, thus signaling the evolution of the field into turbulent flow. The corresponding line for the total energy of the perturbed state in cases *a*) and *b*) is well outside the scale of the drawing, fig. 13 i) and fig. 7, respectively. But, at $T = 150$, the total perturbation energy for case *a*) has reached a value of only 10 % the total energy of the perturbed state, and the total perturbation energy for case *b*), at the maximum level before the subsequent decrease, has reached a value of 20 % the total energy of the perturbed state. Experiment *d*) is a turbulence experiment. The initial state is effectively the random perturbation itself. In fact, at $T = 25$, the perturbation energy crosses the solid line representing the total energy of the perturbed state, clearly indicating that the motion is indeed completely turbulent. By $T = 100$, the energy has reached its saturation limit (fig. 13 iii)).

The observed passage from wavelike to turbulent behavior apparent from the previous series of numerical experiments should appear in the decorrelation of the locked soliton phases. Figures 14 *a*), *b*), *c*), *d*) show the evolution of the three phases $\Theta_{1,2}$, $\Theta_{1,4}$, $\Theta_{1,6}$ for experiments *a*) through *d*) of table I. As had to be expected, perturbation *a*) is the most stable. The locking is not only maintained, but becomes even better with the progressive reduction of $\Theta_{1,6}$ oscillations around $\Theta_{1,2}$. Perturbation *b*) is still stable. The phase behavior radically changes for cases *c*) and *d*). In case *c*), locking is first broken with the first sudden jump of phase $\Theta_{1,6}$, that of the least energetic mode among

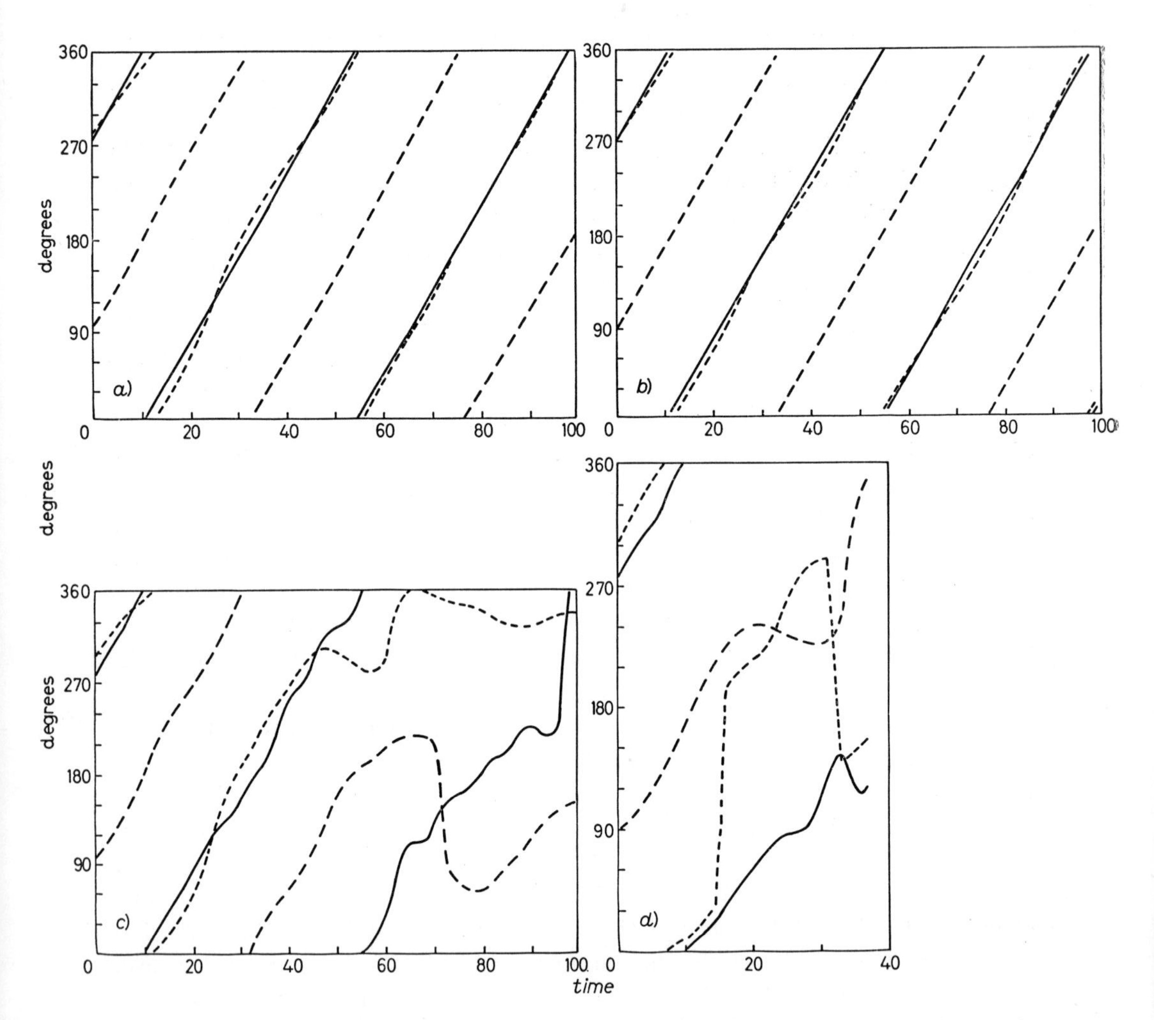

Fig. 14. – a) Time evolution of the Fourier phases of modes $a_{1,2}$ (———), $a_{1,4}$ (— — —), $a_{1,6}$ (- - -) of the soliton basic state when superimposed to perturbation a) of table I. b) As in a) but for perturbation b) of table I. c) As in a) but for perturbation c) of table I. d) As in a) but for case d) of table I.

the three, at $T = 45$. Successively, all phases are completely randomized. Case d) is, as noted, basically a turbulence experiment. The phase randomization occurs almost immediately at $T = 15$. This sudden mixing strongly suggests the existence of a sharp stochasticity border.

To conclude, we reiterate that the numerical experiments reported in b) and c) of table I show the neutral stability of the permanent-form solutions of the complete model equation to finite and general perturbations whose average properties like $u_{\text{r.m.s.}}$, $\zeta_{\text{r.m.s.}}$ may even be of the same order of magnitude as the basic state itself.

4. – Collision experiments.

Collision experiments are defined as those in which two—or more—different permanent-form solutions of the same model equation are given as initial condition to the numerical experiment and are let successively to evolve simultaneously, interacting nonlinearly between themselves. In the case in which the two permanent-form solutions have quite different intensities, the collision can be looked upon as a special case of stability experiment, in which the stronger solitary eddy is the basic state and the weaker one the superimposed perturbation.

Collision experiments between two different solitary-wave solutions are carried out in analogy to the one-dimensional case, to explore whether, and up to which point, two-dimensional solitary eddies maintain the same collision characteristics as one-dimensional solitons.

Let us first summarize the fundamental properties of these last ones from the general point of view of stability. We can define as one-dimensional solitons those nonlinear solutions of the considered model equation which

are stable under any superimposed perturbation,

are stable upon collisions among themselves, whatever the relative amplitudes of the colliding solitons. They re-emerge from their mutual interaction unaltered in shape and amplitude, that is speed, with a phase shift in their space-time trajectories as the only consequence of the interaction itself.

Analogous properties can be investigated for two-dimensional solitary eddies. We have already seen that the first property is not completely respected. In fact, two-dimensional solitary waves remain stable under superimposed perturbations—therefore maintaining their soliton nature—until the perturbation intensity does not exceed the intensity of the basic state itself. When this threshold is overcome, the soliton-locked Fourier phases are decorrelated, the perturbation energy grows without limitation, the solitary eddy is destroyed and turbulent behavior ensues.

As far as the second property is concerned, namely the behavior upon collision, two cases can immediately be distinguished, both for the weak- and the strong-wave case:

a) the two solutions have the same cross-channel (y) dependence,

b) the two solutions have different cross-channel (y) dependence.

For case *a*), when the two permanent-form solutions have the same modal y (along crest) variation, then the collision is really one-dimensional. The one-dimensional property of surviving upon the mutual interaction must be observed, both in the weak- and the strong-wave case. This is indeed the case.

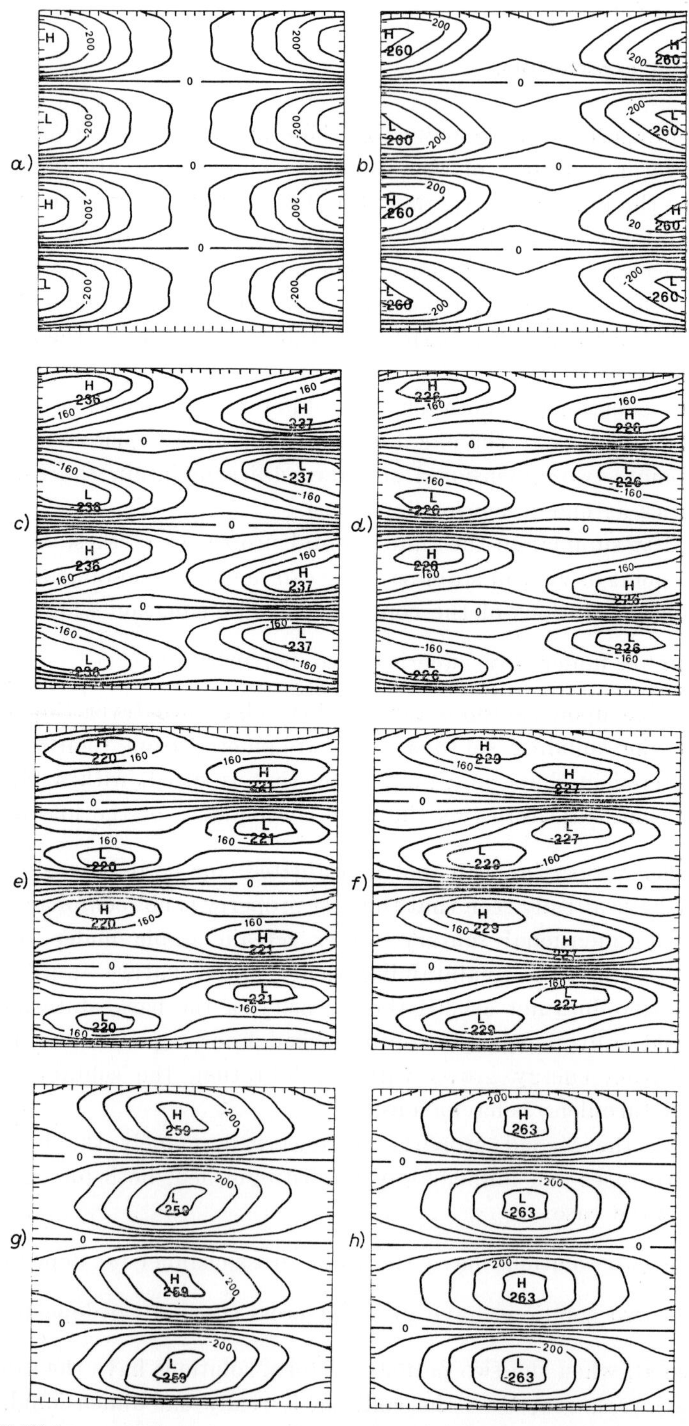

Fig. 15. – Collision experiment over $h = -\sin 2y$. Weak asymmetric solitary-wave case. Evolution of total field $\psi_{\text{total}} = A[\text{sech}^2(Bx)\,\Phi_2(y) + \text{sech}^2(Bx)\,\Phi_{-2}(y)]$, $A = 0.02$, $B = 0.53$. The field is scaled by 10^4. *a)* $t = 0$, *b)* $t = 4$, *c)* $t = 7$, *d)* $t = 9$, *e)* $t = 11$, *f)* $t = 15$, *g)* $t = 20$, *h)* $t = 22$.

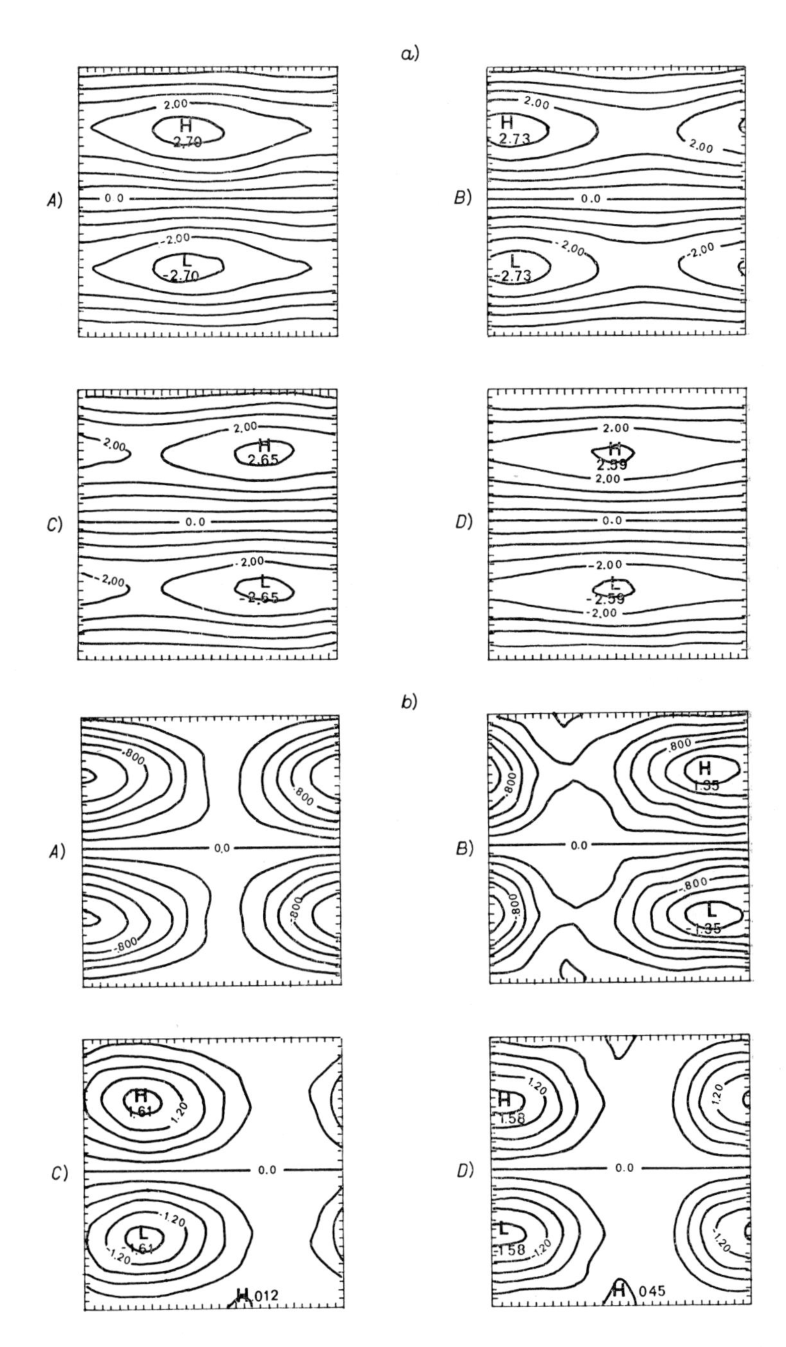

Fig. 16. – *a*) Collision experiment over $h = -\sin y - \sin 2y$. Strong asymmetric solitary-wave case. Evolution of total field $\psi_{\text{total}} = [A_1 \operatorname{sech}^2(B_1 x) + A_2 \operatorname{sech}^2(B_2 x)]\,\Phi_1(y)$, $A_1 \simeq -2.4$, $A_2 \simeq -1.2$: *A*) $t = 0$, *B*) $t = 4$, *C*) $t = 8$, *D*) $t = 11$. *b*) Evolution of subtracted permanent solution $\psi^{(2)} = \psi_{\text{total}} - A_1 \operatorname{sech}^2(B_1 x)\,\Phi_1(y) \simeq A_2 \operatorname{sech}^2(B_2 x)\,\Phi_1(y)$: *A*) $t = 0$, *B*) $t = 2$, *C*) $t = 8$, *D*) $t = 10$.

Figures 15*a*), *b*) show a collision experiment for the weak-wave case. The bottom relief is again $h(y) = -\sin 2y$, which allows both for westward and eastward going solutions, with the same zeroth-order speed (in absolute value) and the same cross-channel structure, only phase shifted. The two weak solutions are considered with the same amplitude $A = -0.02$ and taken for the initial state to be superimposed at one border of the periodic, numerical box. Figures 15*a*), *b*) show the two solutions emerging from each other, separating, moving towards their next interaction (collision) in the center of the box, and their merging again to reproduce a state identical to the initial one.

Figures 16*a*), *b*) show the corresponding collision for two solitary solutions having the same cross-channel dependence, now for the strong-wave case. For this experiment, the bottom relief was $h(y) = -\sin y - \sin 2y$. The same lowest eigenmode $\Phi_1(y)$ was considered, very similar to a sine dependence. The two strong solutions superimposed had amplitudes $A_1 \simeq -2.4$ and $A_2 \simeq \simeq -1.2$.

Figure 16*a*) shows the evolution of the total field. Initially, the solitary eddy with $A = -2.4$ is in the center of the box, while the other one is on its boundary. Being not only strong but also wide, the two permanent eddies now fill up the whole numerical basin. Therefore, to observe the collision properties, the eddy with $A \simeq -2.4$ was subtracted from the total field evolution. The remaining field will have to evolve as a permanent wave of amplitude roughly of -1.2, even though much more distorted than the single soliton due to the always intense nonlinear interactions. This is indeed the case, as shown in fig. 16*b*). The remainder of the evolution was identical over more than three basin transversals.

To complete the investigation of the collision properties, collision experiments must be performed for case *b*), namely between two solitary-wave solutions having a different cross-channel, modal structure. Again, two cases can be distinguished for what is now a really two-dimensional collision.

The two different permanent-form solutions have amplitudes of the same order of magnitude. Experiments for this case show that the permanent eddies preserve their own identity in the interaction, thus maintaining their soliton nature (P. MALANOTTE RIZZOLI: unpublished results).

The two permanent solutions have widely different amplitudes. Then the small-amplitude solitary eddy can be considered as the perturbation in the corresponding stability experiment. Figures 17*a*), *b*) show this last case [16].

The considered relief is $h(y) = -\sin y - \sin 2y$. On it, the two superimposed permanent solutions are

the lowest cross-channel mode $\Phi_1(y)$ with amplitude $A \simeq -2.4$,

the second eigenmode $\Phi_2(y)$ with $A = -0.02$.

Figure 17*a*) shows the total field, essentially identical to the strong eddy which evolves unaffected by the small superimposed wave, the perturbation.

Fig. 17. – *a*) Collision experiment over $h = -\sin y - \sin 2y$. Strong asymmetric solitary-wave case. Evolution of total field $\psi_{\text{total}} = A_1 \operatorname{sech}^2(B_1 x)\, \Phi_1(y) + A_2 \operatorname{sech}^2(B_2 x)\, \Phi_2(y)$, $A_1 \simeq -2.4$, $A_2 = -0.02$: *A*) $t = 9$, *B*) $t = 13$, *C*) $t = 16$. *b*) Evolution of subtracted field $\psi^{(2)} = \psi_{\text{total}} - A_1 \operatorname{sech}^2(B_1 x)\, \Phi_1(y)$: *A*) $t = 0$, *B*) $t = 1$, *C*) $t = 7$, *D*) $t = 18$. At $t = 0$, the initial condition is the permanent solution $A_2 \operatorname{sech}^2(B_2 x)\, \Phi_2(y)$. The field is scaled by 10^4.

Again subtracting the strong eddy from the total field, we obtain the remaining field, the weak solitary eddy shown in fig. 17*b*). The first pattern is the eddy at the initial time, $t = 0$. But already at $t = 1$ the eddy is completely distorted by the strong shear of the superimposed field. This breaks the weak eddy into an increasing number of smaller whorls, until—as is evident from the last picture of fig. 17*b*)—turbulent behavior results.

Thus an extra dimensionality really alters the soliton properties. Two-dimensional, asymmetric solitary waves are not the complete equivalent of one-dimensional solitons. For this last collision experiment, the stability properties established in the previous section are again observed.

Nevertheless, two-dimensional solitary waves are robust enough to be of significance and importance as initial states of enhanced predictability, in juxtaposition to flows characterized by turbulent behavior.

5. – The stability of the permanent-form solutions to perturbations of the initial phases and relief.

This section continues the study of the stability of the solitary solutions of the full model equation. The perturbations considered are i) perturbation in the initial phases of the Fourier components of the soliton, ii) perturbation of the topography field.

i) The soliton's most striking property is phase locking of the Fourier modes into which it can be decomposed. The opposite extreme, turbulent

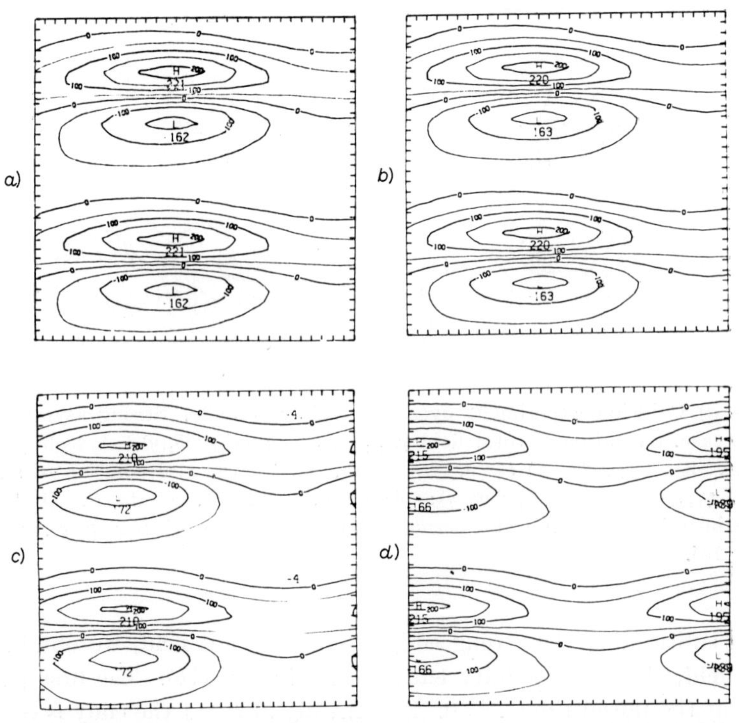

Fig. 18. – Evolution of the streamfunction pattern $\psi = A\,\mathrm{sech}^2(Bx)\,\Phi_2(y)$ with the Fourier phases randomly perturbed at $t = 0$, at the following dimensionless times: *a*) $t = 0$, *b*) $t = 1$, *c*) $t = 7$, *d*) $t = 12$. $A = -0.02$, $B = 0.53$. The field is scaled by 10^4.

motion, is instead characterized by the complete randomness of the phases. We can imagine decomposing any field into a series of orthogonal functions. Suppose a soliton is present in an observed field. The question immediately arises whether it will persist in an enhanced correlation of its spectral components. These considerations motivated a numerical experiment in which the soliton initial phases were randomly altered. Figure 18 shows the resulting physical-space patterns in the initial stages. The remainder of the evolution was completely analogous, with no change of shape for more than one transversal of the basin. The resulting eddies are neither identical with nor symmetrical to the unperturbed initial soliton; rather we seem to see a different persistent solution, again uniformly translating, as shown by the persistence of the new shape and the increased value of its numerical speed ($c \simeq -0.152$). Figure 19 shows the behavior of the phases for modes $a_{1,2}$, $a_{1,4}$, $a_{1,6}$. They are

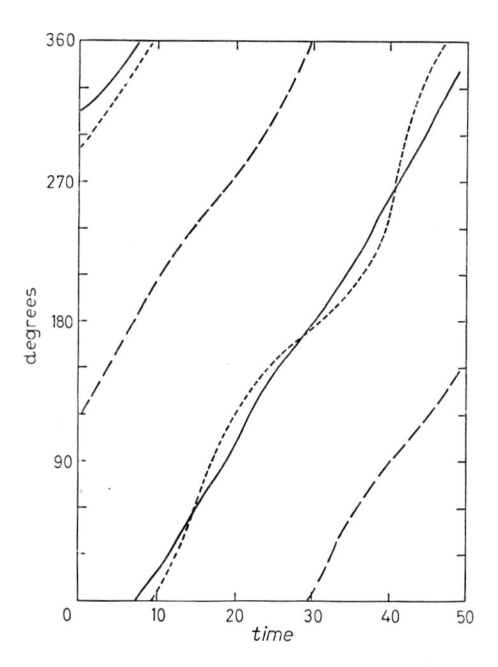

Fig. 19. – Time evolution of the perturbed Fourier phases of modes $a_{1,2}$ (———), $a_{1,4}$ (— — —), $a_{1,6}$ (- - -).

evidently locked; the only effect of the initial random perturbation is in their broader oscillations around a constant average slope. We suggest that we have a perturbed permanent-form solution different from the original soliton. We suggest that this solution, with perturbed initial phases, evolves as the superimposition of some of the basic permanent-form solutions allowed by the full model equation. It is evidently not straightforward to effect this decomposition. Nevertheless, the difference between this perturbed solution and the unperturbed soliton appears, over the limited time of the integration, to evolve as a permanent-shape eddy with a y-dependence very similar to another of the eigenfunctions allowed by eq. (2.6).

ii) In this experiment a small random perturbation has been added to the smooth basic topography:

$$h_{\text{total}} = -\sin 2y + 0.1 h_{\text{random}}(x, y)\,. \tag{5.1}$$

Here $h_{\text{random}}(x, y)$ is an $O(1)$ random function. For our full model equation it is believed that, under certain circumstances, the flow tends to evolve towards a final statistically stationary equilibrium state in which the average streamfunction is an energy-weighted version of the topographic field, with positive correlation $\langle \psi h \rangle$ between streamfunction ψ and relief h (anticyclonic flow over mountains [20]). The question thus arises whether the behavior of the permanent-form solutions described in this and the preceding sections can be ascribed to a manifestation of the properties of this equilibrium state. If we had started with all the energy concentrated in the topographic spectrum, which, anyway, is not the present case, this energy would be trapped there. Changing the bottom relief would allow energy to flow to the new topographic modes.

These considerations motivated a numerical experiment in which the relief was perturbed as in eq. (5.1). Now the topographic spectrum has energy at all scales, though the main mode is still $K = 2$. This crudely represents an ocean bottom with a mean slope over which rough, small features are imposed. Figure 20 gives the corresponding physical-space evolution of the flow from $T = 29$ to $T = 50$. The random relief is evident through the appearance of shorter wavelengths perturbing the basic streamfunction pattern. These higher modes, however, do not alter either the soliton permanence or its propagation speed, the value of which is still $c \simeq -0.140$. Figure 21 shows the behavior of the soliton Fourier modal energies. The phases, not shown, maintain their usual locking. In fig. 21, after an initial evolution in which the most energetic mode $K = 4$ gives up energy while $K = 2$ grows, these two most energetic modes undergo small oscillations around a common mean value. The initial decrease of $K = 4$, and, to a minor extent, of $K = 6$, is only partially compensated by the increase of $K = 2$. The rest of this energy flows to higher modes. These, not shown, reach a steady average value around which they subsequently oscillate, at the same time as that at which the dominant soliton modes 2 and 4 become stationary. Adding a bigger perturbation to the topography would probably disrupt this particular solution, because this solution would no longer be the correct one for the new topographic field, as derivable through the analytical procedure given in sect. **2**. Allowing for the relief to

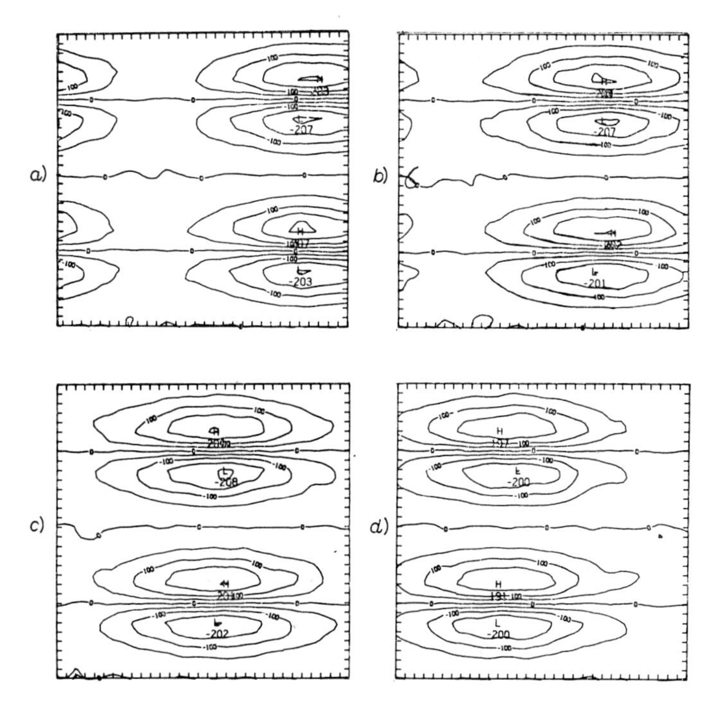

Fig. 20. – Evolution of the streamfunction pattern $\psi = A\,\mathrm{sech}^2(Bx)\,\Phi_2(y)$ over the perturbed relief $h_{\mathrm{total}} = -\sin 2y + 0.1 h_{\mathrm{random}}(x, y)$ at the following dimensionless times: *a*) $t = 29$, *b*) $t = 36$, *c*) $t = 46$, *d*) $t = 50$. $A = -0.02$, $B = 0.53$. The field is scaled by 10^4.

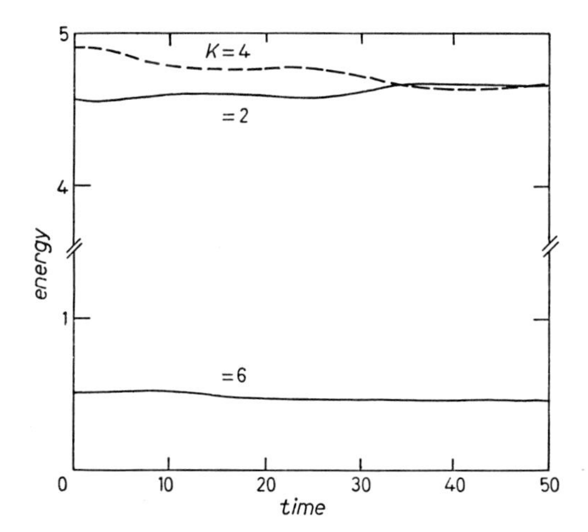

Fig. 21. – Time evolution of the soliton modal energies $K = 2, 4, 6$ normalized with respect to the total soliton energy when evolving over the perturbed relief (5.1). Vertical scale in units 10^{-1}.

be of the form

$$h_{\text{total}} = h(y) + \varepsilon h_p(x, y)\,, \tag{5.2}$$

where ε is the soliton expansion parameter, and substituting (5.2) into the complete model equation (2.3), we get

$$\psi_{yyt} + h_y \psi_x + \varepsilon[\psi_{xxt} + J(\psi, \psi_{yy}) + J(\psi, h_p)] + O(\varepsilon^2) = 0\,. \tag{5.3}$$

Upon repeating the procedure discussed in sect. **2** to obtain the solitary solutions, an identical zeroth-order equation results. At order ε, instead of the KdV equation, we get

$$\left[\frac{c_1}{c_0^2}a_1 - \frac{1}{c_0^2}\int_0^1 h_{py}\Phi^2\,\mathrm{d}y\right]g_x + \frac{1}{c_0^2}a_2 g g_x + a_3 g_{xxx} + \frac{1}{c_0^2}\left[\int_0^1 h_{px}\Phi\Phi_y\,\mathrm{d}y\right]g = 0\,, \tag{5.4}$$

where a_1, a_2, a_3 are the coefficients defined by (2.9a). Effects of the y- and x-dependent structure of the perturbation topography can now be distinguished. The y-dependent part simply modifies the value of the $O(\varepsilon)$ speed correction c_1. In the present case, this modification is evidently small, as the average numerical speed is the same as for the regular soliton ($c \simeq -0.140$). The real perturbing part is thus provided by the x-dependence of the perturbation relief. The effect of this last, however, is also small in the present case, as is shown by the numerical experiment.

6. – Conclusions.

In the introduction, the basic motivation of this research was established as the recognition that not all nonlinear systems with many degrees of freedom exhibit the usually postulated mixing of phases leading to some sort of statistical equilibrium. For two- and three-dimensional systems, specifically those of relevance for the study of the ocean and the atmosphere, this postulate successfully accounts for many aspects of the macroscopic behavior of turbulent flows. In analogy to the one-dimensional case, this lecture gives evidence that, for the considered model, the existence of the stable permanent-form solutions here presented appreciably prolongs the deterministic evolution of the flow from certain initial conditions. The postulate thus does not appear to universally govern two-dimensional systems. This is true in spite of the fact that two-dimensional solitary eddies are not as robust as the one-dimensional solitons. In fact, as shown in sect. **3** and **4**, the solitary eddy does not survive when perturbed by a disturbance having its same intensity or greater. This both in the case in which the superimposed perturbation is of general nature and in the case in which it is constituted by a different solitary solution of the same model equation, with a different along-crest structure. Thus, differently from the one-dimensional case, a stochasticity border can be envisioned; wavelike, soliton behavior is observed until, upon surpassing a critical perturbation value, the evolution becomes that of a turbulent flow. Numerical evidence in support of this picture shows a sudden loss of phase locking as high-amplitude flow evolves.

Nevertheless, the permanent-form solutions are strong enough to prolong considerably the flow predictability when evolving from the solitary-eddy initial conditions. The considered model equation has geophysical significance, but the importance of the given results lies more in their furnishing a two-dimensional counterexample to the randomness postulate than in their specific features.

REFERENCES

[1] J. S. Russell: *Rep. Meet. Brit. Assoc. Adv. Sci.*, **14**, 311 (1845).
[2] D. J. Korteweg and G. de Vries: *Philos. Mag.*, **39**, 422 (1895).
[3] A. C. Scott, F. Y. F. Chu and D. W. McLaughlin: *Proc. IEEE*, **61**, 1443 (1973).
[4] N. J. Zabusky and M. D. Kruskal: *Phys. Rev. Lett.*, **15**, 240 (1965).
[5] G. B. Whitham: *Linear and Nonlinear Waves* (New York, N. Y., 1974), p. 636.
[6] R. R. Long: *J. Atmos. Sci.*, **21**, 197 (1964).
[7] D. J. Benney: *J. Math. Phys. (Cambridge, Mass.)*, **45**, 52 (1966).
[8] L. H. Larsen: *J. Atmos. Sci.*, **22**, 222 (1965).
[9] R. A. Clarke: *Geophys. Fluid Dyn.*, **2**, 343 (1971).
[10] L. G. Redekopp: *J. Fluid Mech.*, **82**, 725 (1977).
[11] V. D. Larichev and G. M. Reznik: *Polymode News*, No. 19 (1976).
[12] G. Flierl, V. D. Larichev, J. McWilliams and G. Reznik: *Dyn. Atmos. Oceans*, **4**, 205 (1980).
[13] G. Flierl: *Dyn. Atmos. Oceans*, **3**, 15 (1979).
[14] P. Malanotte Rizzoli and M. C. Hendershott: *Dyn. Atmos. Oceans*, **4**, 247 (1980).
[15] P. Malanotte Rizzoli: *Dyn. Atmos. Oceans*, **4**, 261 (1980).
[16] P. Malanotte Rizzoli: *Polymode News*, No. 75 (1980).
[17] P. Malanotte Rizzoli: Philosophy Doctor Dissertation, Scripps Institution of Oceanography, p. 147 (1978).
[18] E. L. Ince: *Ordinary Differential Equations* (New York, N. Y., 1956).
[19] D. K. Lilly: *Geophys. Fluid Dyn.*, **4**, 1 (1972).
[20] R. L. Salmon, G. Holloway and M. C. Hendershott: *J. Fluid Mech.*, **75**, 691 (1976).

The Predictability Problem of Planetary Motions in the Atmosphere and the Ocean.

P. MALANOTTE RIZZOLI

Istituto per lo Studio della Dinamica delle Grandi Masse, CNR - Venezia, Italia
I.G.P.P., Scripps Institution of Oceanography - La Jolla, Cal.

1. – Introduction.

One of the problems which have received an increasing amount of attention in recent years is the problem of predicting the mesoscale motions of geophysical fluids, namely both in the ocean and in the atmosphere. In particular, the problem was first investigated in the specific context of atmospheric motions, evidently for the much more immediate impact which the weather prediction has on practical activities and utilization in the human society.

The predictability problem essentially arose with the advent of modern, high-speed computers. In the decade 1960-1970, what seemed a goal beyond any hope became a reality, and global atmospheric-circulation models began to be constructed. These models integrated numerically the full, complex, nonlinear set of partial differential equations of fluid dynamics (primitive equations), with complex boundary conditions, often involving the phenomenological specification of some field functions.

There was, however, a very important limitation in the specification of the initial state for these global circulation models. As in the atmosphere there is motion at all scales, from the planetary ones to the smallest eddies damped by viscosity, the initial state for these models could be based on the observation only of those scales of motion large enough to be resolved by the observing network of stations. Thus, apart from the experimental error implicit in every measurement, this meant that the scales smaller than the network grid spacing remained completely undetermined. In addition, these unmeasured smaller scales introduce, through aliasing, an error in the specification of the larger scales of motion.

This is the problem of the uncertainty of initial data. No matter how accurately a numerical model computes the evolution of the motion, the predictability of a flow will depend on the magnitude of the initial error as determined *a*) by the accuracy and spatial resolution of the observing system and *b*) on the growth rate of the error itself as determined by the equations of motion.

But, in the decade 1960-1970, it was felt that the problem was essentially of being capable of specifying accurately enough both the initial conditions and a suitable numerical algorithm. There was, therefore, the optimistic attitude that geophysical flow systems can be treated with a deterministic approach, that is at all future times the error can be kept arbitrarily small by making the initial error sufficiently small. It was not thought about the instability of these flow systems in phase space. With this, it is meant that, if the system is unstable to small perturbations and two very near initial conditions are chosen, the two points representing these two initial conditions in phase space will have diverging trajectories. These trajectories will diverge exponentially and the two initially near systems will evolve towards final states as widely separated and uncorrelated as two randomly chosen states.

The point, therefore, was made of an intrinsic predictability problem for which the observable behavior of a so-called deterministic system may not be distinguished from that of a stochastic, unpredictable one. The first to introduce these ideas was LORENZ [1], who studied in a by now classical paper the time evolution and predictability of a flow possessing many interacting scales of motion. LORENZ pointed out that the uncertainties of small-scale statistics have unknown effects on the larger-scale statistics. Even though the direct effect upon scales 10^3 times larger may be small, the effect cannot be so upon scales only twice as large as the smallest ones. The important question is then whether and how fast small errors in the small scales propagate and lead to big errors in the large scales, making them completely unpredictable after a finite amount of time, the predictability time.

These ideas proved so fundamental as to destroy the faith in the deterministic approach leading to the global circulation models of the 1960-1970. Therefore, Lorenz's work will be briefly summarized and explained as the classical turning point in the investigation of planetary motions and their predictability.

2. – The predictability of a flow with many interacting scales of motion.

Instead of dealing with the complicated, full set of primitive equations, LORENZ chose a different approach. He considered only simple model equations, important, however, in describing in an idealized way the dynamics of large-scale flows. For these simple equations initial and boundary conditions as well as the resolving numerical schemes can be specified quite well. Specifically, Lorenz's [1] model was

$$\frac{\partial}{\partial t}(\nabla^2\psi) + J(\psi, \nabla^2\psi) = 0 \tag{2.1}$$

with the following definition:

ψ = streamfunction of the horizontal (two-dimensional) motion related to the horizontal velocity components by the definition

$$u = -\psi_y\,, \qquad v = +\psi_x\,.$$

Subscripts indicate partial derivatives.

$$\nabla^2 = \frac{\partial^2}{\partial x^2} + \frac{\partial^2}{\partial y^2} \qquad \text{horizontal Laplacian},$$

$$J(a, b) = \frac{\partial a}{\partial x}\frac{\partial b}{\partial y} - \frac{\partial a}{\partial y}\frac{\partial b}{\partial x} \qquad \text{Jacobian of } (a, b)\,.$$

(2.1) is the so-called vorticity conservation equation, valid for mesoscale two-dimensional motions on a plane with constant rotation, in the nondivergent approximation. Under this hypothesis, it can be exactly derived from the horizontal momentum and continuity equations, where vorticity is defined as

$$\omega = \nabla^2 \psi\,.$$

To investigate how fast small errors in the small scales propagate and affect the larger scales, LORENZ [1] carried out a series of numerical predictability experiments. He defined two different realizations of the flow as described by the two streamfunctions ψ and $\psi' = \psi + \varepsilon$, where the difference ε is viewed upon as the initial error. ε is governed by the model equation

$$\frac{\partial}{\partial t}(\nabla^2\varepsilon) = -J(\psi, \nabla^2\varepsilon) - J(\varepsilon, \nabla^2\psi) - J(\varepsilon, \nabla^2\varepsilon)\,. \tag{2.2}$$

If ε is a small error, the term $J(\varepsilon, \nabla^2\varepsilon)$ is much smaller than the other nonlinear terms. LORENZ therefore linearizes (2.2) into

$$\frac{\partial}{\partial t}(\nabla^2\varepsilon) = -J(\psi, \nabla^2\varepsilon) - J(\varepsilon, \nabla^2\psi)\,. \tag{2.3}$$

(2.3) is Lorenz's model equation. If one chooses to follow the statistics of the flow fields, these are decomposed into Fourier components:

$$\psi = \sum_{\boldsymbol{k}} S_{\boldsymbol{k}} \exp[i\boldsymbol{k}\cdot\boldsymbol{r}]\,, \qquad \varepsilon = \sum_{\boldsymbol{k}} e_{\boldsymbol{k}} \exp[i\boldsymbol{k}\cdot\boldsymbol{r}]\,,$$

where $\boldsymbol{k}$ is the wave number vector with $|\boldsymbol{k}| = 2\pi/\lambda$, if λ is the wavelength of the considered scale of motion; $\boldsymbol{r}$ is the vector distance from the origin of the reference system; $S_{\boldsymbol{k}}$, $e_{\boldsymbol{k}}$ are the Fourier amplitudes, respectively, for the streamfunction ψ and the error ε. To follow the statistics of the system, under certain closure assumptions (the details of which can be found in Lorenz's paper), LORENZ studies the two quantities

$$E = \tfrac{1}{2}\,\overline{\nabla\psi\cdot\nabla\psi}\,,$$

the kinetic energy of the flow, which is a constant of the motion, and

$$F = \tfrac{1}{2}\,\overline{\nabla\varepsilon\cdot\nabla\varepsilon}\,,$$

the error energy. Overbars indicate ensemble averages.

To study the predictability of the motion, one must follow the time evolution of the error energy F, when the error ε evolves according to (2.3). If there is no predictability at sufficiently long times ($t\to+\infty$), then the two initial conditions ψ and $\psi'=\psi+\varepsilon$ will become uncorrelated as $t\to+\infty$, that is the two corresponding phase points will have diverging trajectories, that is the system loses the memory of its initial state and the error energy becomes like the vector difference between two randomly chosen ψ and ψ'. The only constraint is the constancy of the total flow energy E.

Then it can be shown that the only limitation imposed upon the error energy density $F(\boldsymbol{k})$ by the intrinsic nonlinearity of eq. (2.2) is that

$$\lim_{t\to+\infty} F(\boldsymbol{k}) \to 2E(\boldsymbol{k})\,, \tag{2.4}$$

if the system behaves stochastically as $t\to+\infty$. The energy density is defined as the energy of the flow per unit wave number. The stochastic behavior of the system—and the consequent loss of flow predictability—can thus be put into evidence by observing if the error energy F obeys the limit (2.4) as time evolves.

Therefore, a predictability time for the scale at wave number $\boldsymbol{k}$ can be defined as the time for the error energy density $F(\boldsymbol{k})$ to become equal to the basic-state energy density $E(\boldsymbol{k})$. LORENZ [1] chooses as an upper limit the energy density $E(\boldsymbol{k})$ itself—and not its double as given by (2.4)—because he works upon the linearized equation (2.3), for which the intrinsic limit (2.4) has been eliminated. For the basic state, LORENZ [1] chooses a scalar energy density spectrum typical of three-dimensional turbulence, namely

$$E(k) \propto k^{-\frac{5}{3}}\,. \tag{2.5}$$

LORENZ carries out a series of numerical experiments under the previous assumptions. A typical example is shown in fig. 1 (from [1]). The initial error energy density has the magnitude $2^{-16}E$ and is confined to the smallest scale of motion (highest wave number).

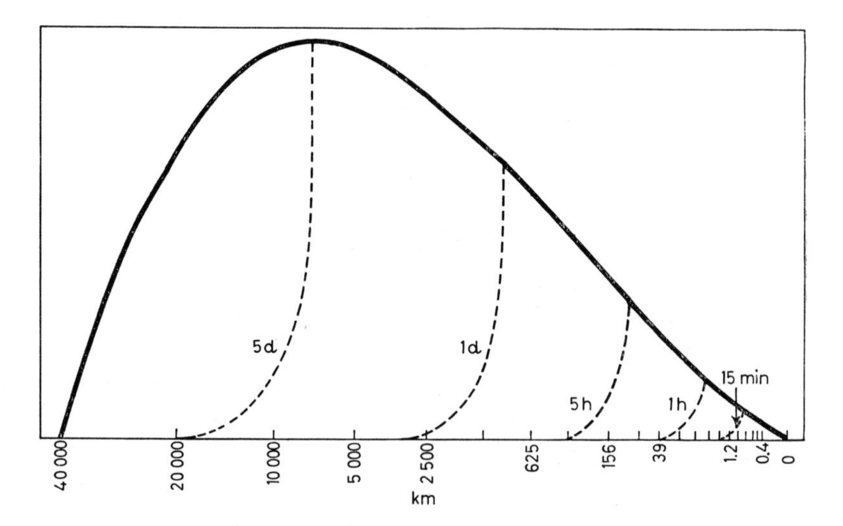

Fig. 1. – Basic energy spectrum (full curve) and error energy spectra (broken curves) at 15 min, 1 h, 5 h, 1 d and 5 d, as interpolated from numerical solution of eq. (2.3). Broken curves coincide with the full curve to the right of their intersections with the full curve. Horizontal co-ordinate is fourth root of wavelength $\sqrt[4]{\lambda}$, labeled according to wavelength. Vertical co-ordinate is energy per unit logarithm of wave length divided by fourth root of wavelength. Areas are proportional to energy. From [1].

As is evident from the figure, the error present only at the smallest scale at the initial time $t = 0$ propagates towards the larger scale gradually affecting all of them. After 15 minutes, the error energy spectrum (broken curves) coincides already with the basic energy spectrum (full curve) up to wavelengths of 1.2 km. After 1 day, wavelengths of 2500 km have already been affected. All wavelengths, up to the planetary one of $\lambda = 40\,000$ km, completely lose predictability in about two weeks, and this final predictability time cannot be improved by any reduction in the magnitude of the initial error.

Lorenz's classical study, therefore, destroyed the faith in the deterministic approach towards weather—and in general geophysical fluid motions—prediction. In the following years of the 70's research capitalized completely upon unpredictable stochastic systems. If, in fact, the effect of nonlinear interactions upon the flow evolution is to destroy its predictability, the best one can do is to study the flow statistics, and predict the evolution of its statistical properties, like spectral energy density, phase correlations, higher moments and so forth.

Bearing upon the stochastic approach, it was natural that research studies invoked turbulence theories to investigate the mesoscale oceanic and atmospheric motions.

3. – The planetary motions and their connection with two-dimensional turbulence.

A basic aspect of turbulent motions is their apparent total unpredictability. Turbulence consists of a continuing series of flow instabilities. Any introduced uncertainty continues to grow. Two initial states differing by some infinitesimal amount will diverge until they differ as much as two randomly chosen states having the same statistical properties.

Why were two-dimensional turbulence theories chosen to deal with planetary motions? It was recognized that both the atmosphere and the ocean behave roughly as barotropic, two-dimensional fluids at the large, planetary scales.

As far as the ocean is concerned, fig. 2 (from [2]) shows a very illustrative example. The figure shows current measurements in the MODE region (Mid Ocean Dynamic Experiment), at a more energetic part of the North Atlantic, just south of the Gulf of Mexico. The currents exhibit roughly a 50-day period and have penetrated right to the ocean bottom. The currents are thus essentially barotropic, with very little vertical dependence.

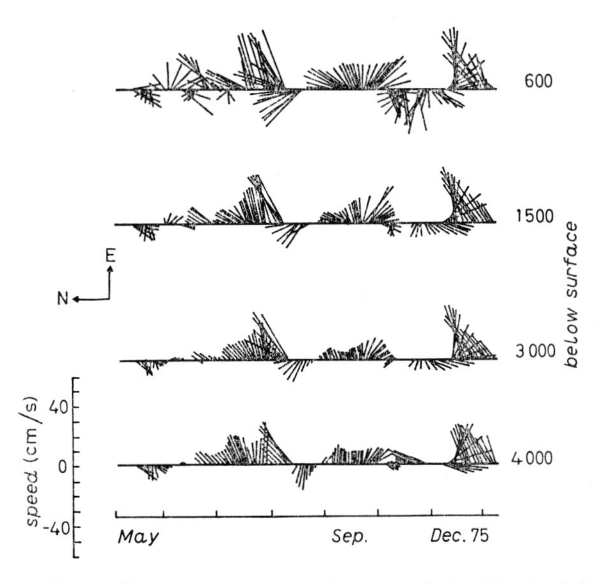

Fig. 2. – Currents observed at a more energetic part of the North Atlantic, just south of the Gulf Stream. The currents exhibit roughly a 50-day period and have penetrated right to the ocean bottom. From [2].

As far as the atmosphere is concerned, at the large scales, it behaves also as a two-dimensional, incompressible turbulent fluid. Figure 3 (from [3]) shows the eddy kinetic-energy spectrum in the atmosphere from the observations of several authors, for a variety of seasons, latitudes and pressures. All spectra at the wave numbers between 10 and 20 show a tendency towards a k^{-3} dependence, typical of the inertial range of two-dimensional turbulence.

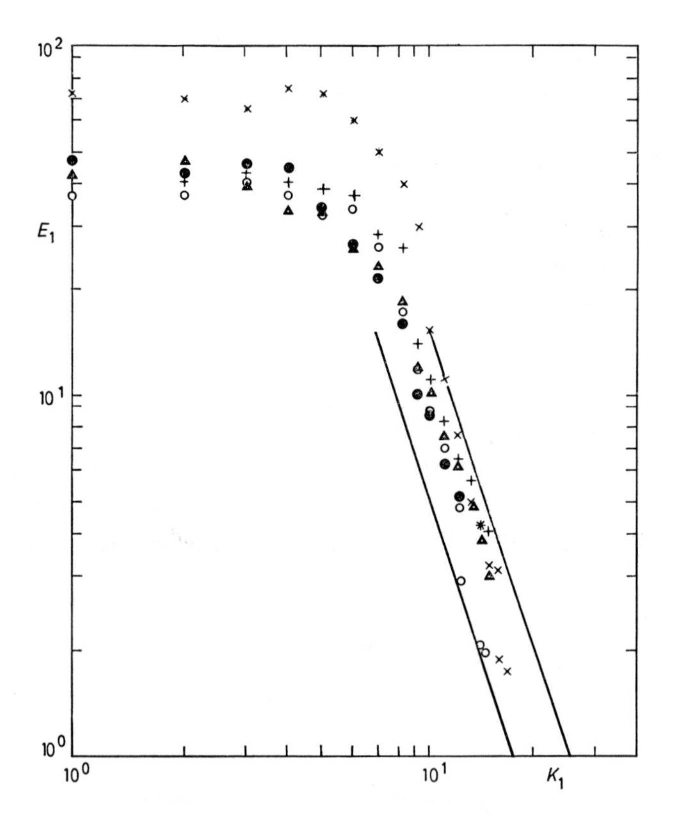

Fig. 3. – Observations of eddy kinetic scalar energy spectrum in the atmosphere from + Saltzman and Fleisher (1962), • Horn and Bryson (1963), ▴ Wiin and Nielsen (1967), ○ Julian *et al.* (1970), × Kao and Wendell (1970). Energy unit 10^{-4} rad^2 d^{-2}. From [3].

Two-dimensional turbulence is very different from the three-dimensional one. The lack of a variable topography prevents the stretching or compressing of water columns. This adds a second integral invariant of the motion to the integral invariant of energy. Thus, not only the total energy is conserved, but also the total vorticity itself ω is a constant of the motion, as well as all its powers. Of these, the most important is the enstrophy, defined as the square of the vorticity ω^2.

The presence of two integral invariants has profound consequences in the turbulent behavior of the fluid. Without entering into details, which can be found elsewhere [4, 5], it is enough to say that the energy inertial range of 3-D turbulence ($E(k) \propto k^{-\frac{5}{3}}$) is substituted by an enstrophy inertial range ($E(k) \propto k^{-3}$) in 2-D turbulence. In this last one, a further inertial range appears at the very long wavelengths (small k) in which energy flows from right to left, namely towards the largest λ.

As shown in fig. 2 and 3, it was natural to resort to 2-D turbulence theories to model mesoscale oceanic and atmospheric motions. A further example of a predictability study will be now shown, taken from [4, 5] and carried out upon a 2-D turbulence model. LILLY [4, 5] considers the following system:

$$\frac{\partial \omega}{\partial t} + u\frac{\partial \omega}{\partial x} + v\frac{\partial \omega}{\partial y} = F + \nu\nabla^2\omega - \bar{\boldsymbol{K}}\omega\,, \tag{3.1}$$

where $\omega = \nabla^2\psi$ is the previously defined vorticity of the system.

This model, even though similar, is more complex than the one considered by LORENZ. LILLY [4, 5], in fact, considers the following further terms:

F = a forcing function giving a vorticity input at intermediate wave numbers;

$\nu\nabla^2\omega$ = a viscositylike dissipation term for vorticity, particularly important at high k's;

$\bar{\boldsymbol{K}}\omega$ = a surface drag dissipation term (proportional to vorticity), particularly important at low k's.

LILLY considers scalar energy spectra for the initial state typical of 2-D turbulence, that is

$$E(k) \propto k^{-3}\,.$$

LILLY carries out the following numerical experiment. In the first run (run 1 with streamfunction ψ_1) the experiment is continued until a statistically steady state is reached, with an energy spectrum $\propto k^{-3}$ in the enstrophy inertial range. The experiment was thereafter the prosecution of the evolution for a certain, prechosen time. In run 2 (streamfunction ψ_2) the inital condition was altered adding a small white-noise perturbation to the streamfunction ψ_1 at all grid points, uncorrelated with it. The white-noise perturbation had a scalar error energy spectrum $\propto k^{+3}$. The difference between the two runs $\psi_2 - \psi_1$ was then evaluated and the error energy spectrum consequently computed. Figure 4 shows the growth of the difference energy spectrum from the initial shape $\propto k^3$, initially various orders of magnitude smaller than the basic-state energy spectrum. This last one is given by the series of dots and its limiting shape is finally attained by the error energy in about 4000 dimensionless time steps.

All these studies had as resulting effect the loss of faith in the deterministic approach. The opposite faith was built, namely that to solve the predictability problem for geophysical flow motions one had first to solve the closure problem of turbulence theories.

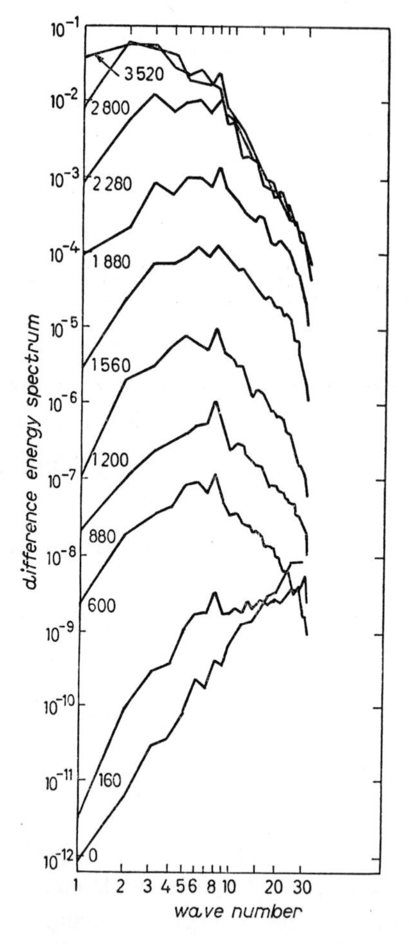

Fig. 4. – Growth of the difference energy spectrum in a numerical predictability experiment of eq. (3.1). The various spectra are labeled by the (dimensionless) time step. From [4].

As a matter of fact, for these systems, characterized by strong nonlinear dynamics, the previous results seem to indicate that the best we can do is to investigate the evolution of the ensemble averages of the system. The statistical approach, bearing upon unpredictability, assumes as valid *a priori* the so-called « random-phase postulate ». This fundamental randomness postulate states that, in a system with many degrees of freedom, the nonlinear interactions between different components have such a complex history that we can best describe the system average properties by assuming these interactions to be at random as is consistent with the conservation of the general invariants (mass, energy, etc.). This postulate thus establishes that any nonlinear system with many scales of motion will have a time evolution leading to the mixing of Fourier phases and, in the long range, to statistical equilibrium and equipartition of energy among the various Fourier modes (in the absence of sources or sinks).

But not all nonlinear systems lead to unpredictability and to energy equipartition among the degrees of freedom. This leads us to the famous example of the Fermi-Pasta-Ulam problem.

4. – The predictability problem in one-dimensional systems: the Fermi-Pasta-Ulam (FPU) model.

The fundamental randomness postulate valid for turbulence theories and statistical equilibrium is not universally obeyed. Recognition of this fact until recently rested exclusively on the behavior of a wide class of one-dimensional, nonlinear systems, of which the FPU model is the most famous example.

As early as 1955, FPU undertook a numerical study of the one-dimensional anharmonic lattice, that is a chain of mass point oscillators coupled by nonlinear springs. Specifically, they considered the following model:

$$\omega_0^{-2}\ddot{y}_i = (y_{i+1} - 2y_i + y_{i-1}) + \alpha[(y_{i+1} - y_i)^{p+1} - (y_i - y_{i-1})^{p+1}] \tag{4.1}$$

with

$i = 1, \dots, N$ number of the mass points,

$\omega_0^2 = k/m =$ frequency of the linear spring mass system $\equiv 1$.

Double dots indicate the second time derivative,

$p = 1$ defines a quadratic nonlinearity,

$p = 2$ defines a cubic nonlinearity,

α defines the strength of the nonlinearity.

The considered boundary conditions were zero or periodic:

$$y_0 = y_N\,(\equiv 0)\,.$$

The considered initial conditions were of a single wave form:

$$y_i|_{t=0} = a\sin\left(i\frac{\pi}{N}\right) \quad \text{and} \quad \dot{y}_i|_{t=0} = 0\,.$$

FPU [6] wanted to investigate the approach of (4.1) to thermal equilibrium. That is, starting from the previously given initial condition with all the energy concentrated in one single Fourier mode, they expected the nonlinear interactions to excite higher and higher modes, until, eventually, the system would evolve towards a state of equipartition among all the Fourier modes.

To their big surprise, the system did not approach equilibrium at all. Only a few « new » modes were excited and the system instead exhibited long time recurrences for which, after a certain time, the recurrence time, all the energy went back to the initially excited Fourier mode. Various experiments were carried out, varying the nonlinearity strength α, $\frac{1}{4} \leqslant \alpha \leqslant 1$, and the number of degrees of freedom, $N = 16, 32, 64$.

They all gave the previously described results, as shown in fig. 5 (from [6]), in which the plotted quantity, total energy (sum of kinetic + potential) initially concentrated in one single mode, goes back to it after a recurrence period.

This lack of randomness was explained by ZABUSKY and KRUSKAL [7] and ZABUSKY [8], who related the discretized lattice to a continuum model, precisely

$$\frac{\partial u}{\partial t} + u \frac{\partial u}{\partial x} + \delta^2 \frac{\partial^3 u}{\partial x^3} = 0 , \tag{4.2}$$

the Korteweg-de Vries (KdV) equation for the quadratic lattice ($p = 1$), and

$$\frac{\partial u}{\partial t} + u^2 \frac{\partial u}{\partial x} + \delta^2 \frac{\partial^3 u}{\partial x^3} = 0 , \tag{4.3}$$

the modified Korteweg-de Vries (mKdV) equation for the cubic lattice ($p = 2$). δ^2 measures the relative intensity of nonlinear and dispersive terms.

For the quadratic lattice, (4.2) holds. Its solutions are the well-known cnoidal and solitary waves. A smooth initial condition, namely a cosine wave, imposed to the KdV equation, broke up into a series of well-defined and localized pulselike shapes, freely streaming through one another, and re-emerging from the collisions unaltered in shapes and speeds, that is in amplitudes. Due to these peculiar collision properties, ZABUNSKY and KRUSKAL [7] coined the name « solitons » for these nonlinear wave forms, which they found emerging from quite general initial conditions.

At a certain time, the FPU recurrence time, all the solitons produced by the smooth initial conditions focused together, reproducing a state almost identical to the initial one. Thus the FPU cycles were explained as being echoes of the solitons of the equivalent continuum model, and hence due to the remarkable stability properties of the KdV solitons.

Figure 6 shows a typical result of Zabusky and Kruskal's experiments. In fig. 6a) the smooth initial condition

$$u\big|_{t=0} = \cos \pi x$$

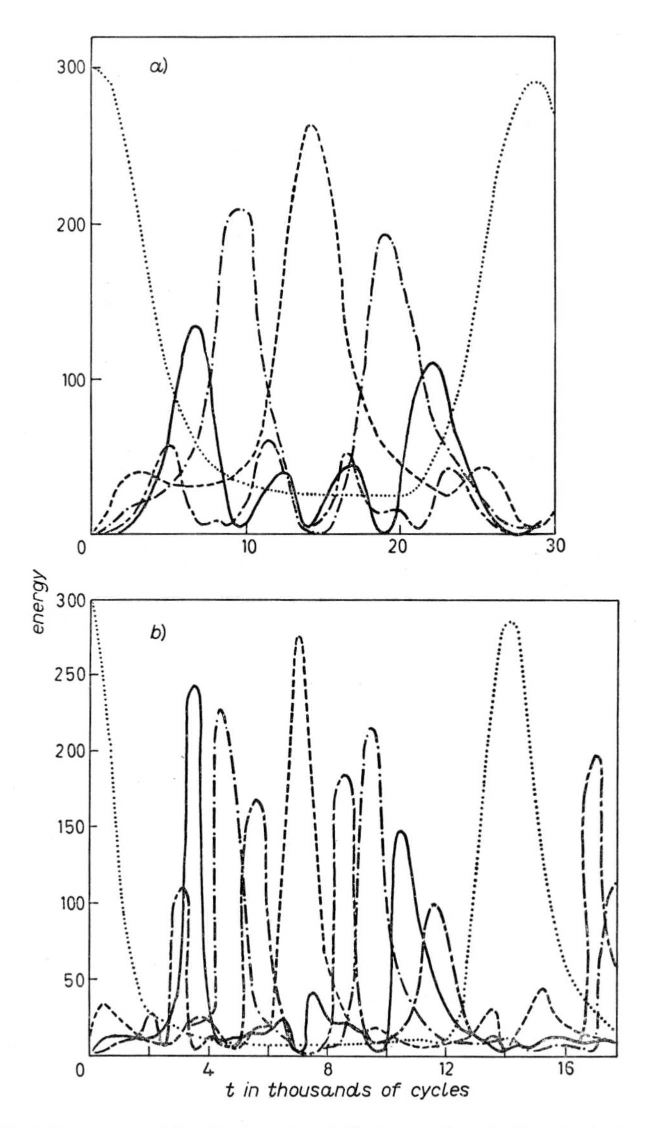

Fig. 5. – *a*) Total energy (kinetic + potential) in each of the first 5 modes for the lattice (4.1) with $p = 1$. Energy units are arbitrary. $N = 32$, $\alpha = \frac{1}{4}$. Initial state: single sine wave. About 30 000 computation cycles are computed. *b*) Same conditions as for *a*), but with $\alpha = 1$ (stronger nonlinearity), about 14 000 cycles were computed. From [6].

is shown as the dotted line (curve A)). Curve B) at time $t = t_{\text{breaking}}$ shows the first oscillating structure. Curve C) at $t = 3.6t_{\text{b}}$ shows a fully developed oscillatory structure with a collection of pulses. Each of these is a soliton; as bigger-amplitude solitons move faster, they overtake the smaller ones, merge together in the nonlinear interaction and re-emerge unaffected.

Figure 6*b*) gives the space-time trajectories of the produced solitons. Each of them moves along a straight-line trajectory, except when it interacts momentarily with other solitons. At this time they « accelerate » through one another, re-emerging unaffected with only a phase shift as a consequence of the collision.

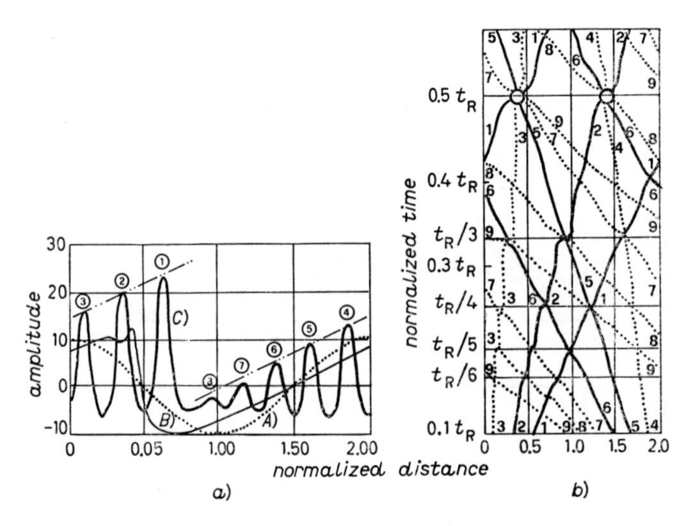

Fig. 6. – *a*) Solutions of the Korteweg-de Vries equation (4.2) at three different times: curve *A*) $t = 0$, curve *B*) $t = t_b$, curve *C*) $t = 36t_b$. $\delta = 0.022$. *b*) Soliton trajectories on a space-time diagram beginning at $t = 0.1t_R$. From [8].

At $t = t_R/2$ all odd solitons and even ones focus together.

At $t = t_R$ (not shown) all nine solitons focus together reconstructing the initial state.

This time t_R is just the FPU recurrence time for the periodic behavior of the anharmonic lattice.

The soliton solutions of the kdV equation and their properties can be summarized as follows:

$$u_t + \underbrace{uu_s}_{\text{nonlinearity}} + \underbrace{\mu u_{sss}}_{\text{dispersion}} = 0 . \tag{4.4}$$

Subscripts indicate, as usual, partial derivatives. We are in the reference frame moving with the wave:

$$s = x - ct .$$

If nonlinearity and dispersion balance each other, then the cnoidal wave solution is

$$u(s) = A\,\mathrm{cn}^2\frac{B\cdot s}{n}\,, \tag{4.5}$$

where n is the parameter of the Jacobian ellyptic function cn. In the limit $n \to 1$, this becomes the solitary wave:

$$u(s) = A\,\mathrm{sech}^2(Bs)\,. \tag{4.6}$$

Substitution of (4.6) into (4.4) gives

$$B^2 = \frac{A}{12\mu} \quad \text{and} \quad c = \frac{A}{3}\,. \tag{4.7}$$

The soliton properties in physical space are:

the speed c is amplitude dependent as from (4.7),

the KdV solitons are stable with respect to any superimposed perturbation,

they are stable with respect to collisions among themselves. The only collision effect is a phase shift in their space-time trajectories.

The soliton properties in Fourier space are:

the modal energies $E(k)$ are constant in time,

the phases of the Fourier modes composing the soliton structure keep their initial correlation, that is they are « locked » together.

Thus the soliton is a nonlinear entity which thoroughly contradicts the random-phase postulate of turbulence and statistical mechanics. As such, the soliton is a completely predictable nonlinear structure.

For one-dimensional systems, therefore, nonlinearities sometimes serve to preserve phase information rather than to destroy it. Those nonlinear systems which violate the random-phase postulate are also those capable of supporting as solutions these completely predictable entities, the solitons.

Does an analogous situation exist for systems with two or three dimensions? This question we shall explore in the next section.

5. – Two- and three-dimensional systems which support solitary-wave solutions.

Let us now examine model equations suitable to describe planetary motions in the ocean and the atmosphere, those same equations which are models for two-dimensional turbulence. Let us ask the question whether some of these models admit solitary-wave solutions, not obeying the randomness postulate and, therefore, endowed with extended predictability, and in which circumstance. The answer to this question is of fundamental importance to establish the direction in which efforts to predict geophysical flows will evolve. If it is negative, we must give up extended deterministic predictability as in principle impossible and content ourselves with purely statistical prediction. If it is positive, some extension of present estimates of predictability time may be sought, even though perhaps only from certain initial conditions.

The answer to the previous questions is positive, even though with stronger limitations for two- and three-dimensional systems than for the one-dimensional ones.

To give an example, let us consider the following simple model [9-11]:

$$\nabla^2 \psi + J(\psi, \nabla^2 \psi + f_0 h) = 0 \tag{5.1}$$

with the usual meaning of symbols. This is the vorticity conservation equation for a barotropic fluid, with constant rotation (f_0 = const = Coriolis parameter) over variable topography. The topography is given by

$$H(x, y) = D_0\left[1 - \frac{|\Delta D|}{D_0}\, h(x, y)\right] \tag{5.2}$$

with $|\Delta D| \ll D_0$, that is the topography variation is much smaller than its mean value. (5.1) admits as solutions

linear topographic Rossby waves at infinitesimal amplitude (nonlinearity is negligible);

solitary-wave solutions (KdV equation solutions) at finite amplitudes, when nonlinearity and dispersion balance each other;

two-dimensional turbulence over variable topography at high amplitudes.

In the turbulent limit, eq. (5.1) gives the same behavior for an error energy shown in fig. 1 (Lorenz's system) and 4 (Lilly's system). Two examples will now be shown of nonlinear solitary initial states which evolve in a predictable, deterministic manner in eq. (5.1).

Figures 7*a*), *b*) show a two-dimensional small-amplitude solitary solution constituted by four eddies, perturbed by a small superimposed disturbance. This notwithstanding, the solitary eddies evolve unperturbed and stable in their motion [11].

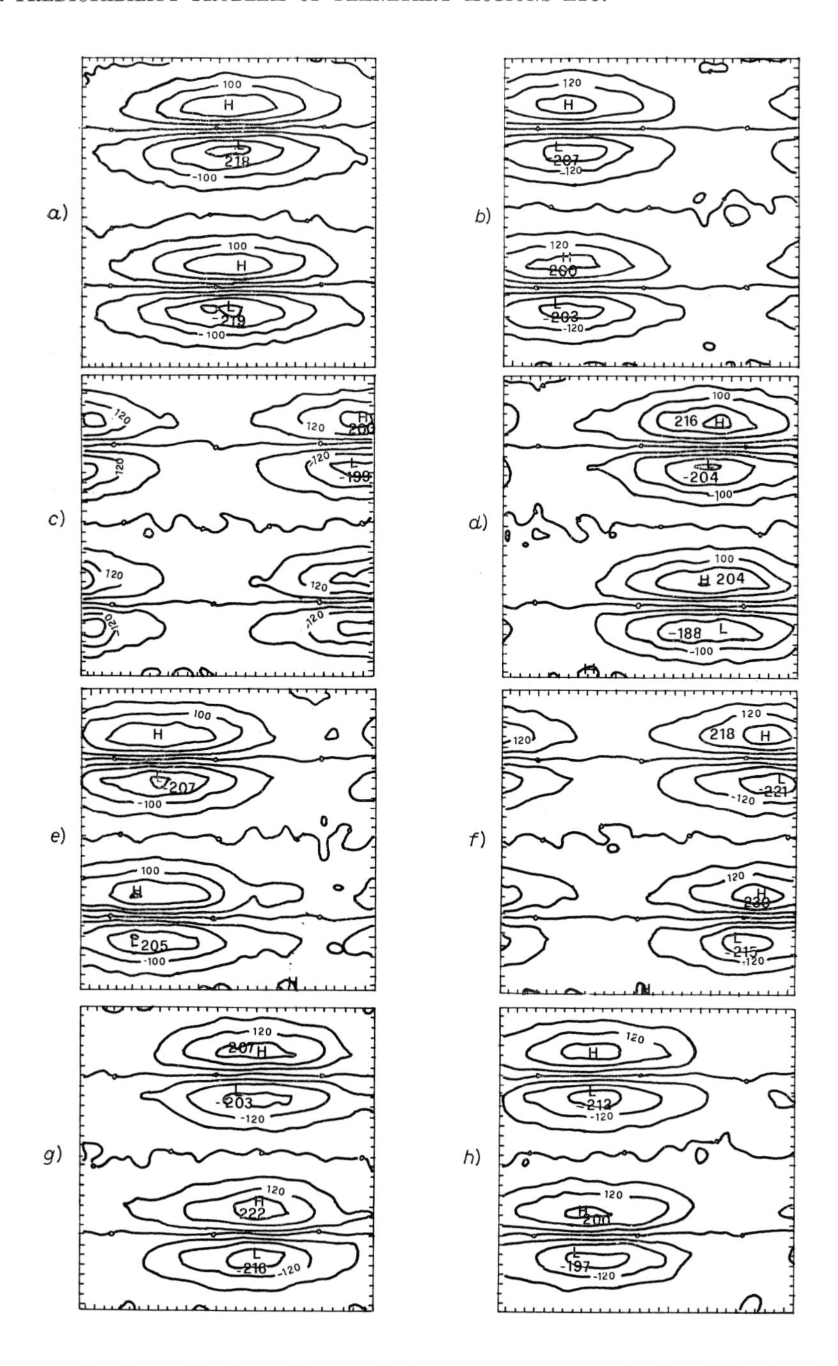

Fig. 7. – Small-amplitude soliton solution to eq. (5.1) at the following dimensionless times: *a*) $t = 0$, *b*) $t = 12$, *c*) $t = 24$, *d*) $t = 36$, *e*) $t = 54$, *f*) $t = 72$, *v*) $t = 84$, *h*) $t = 96$. Total streamfunction pattern $\psi = A[\text{sech}^2(Bx)\,\Phi_2(y) + 0.1\,\delta\psi_{\text{random}}]$, $A = -0.02$, $B = 0.53$, Φ_2 = second cross-channel eigenmode for $h(y) = -\sin 2y$, $\delta\psi_{\text{random}}$ = random superimposed perturbation. The field is scaled by 10^4. From [11].

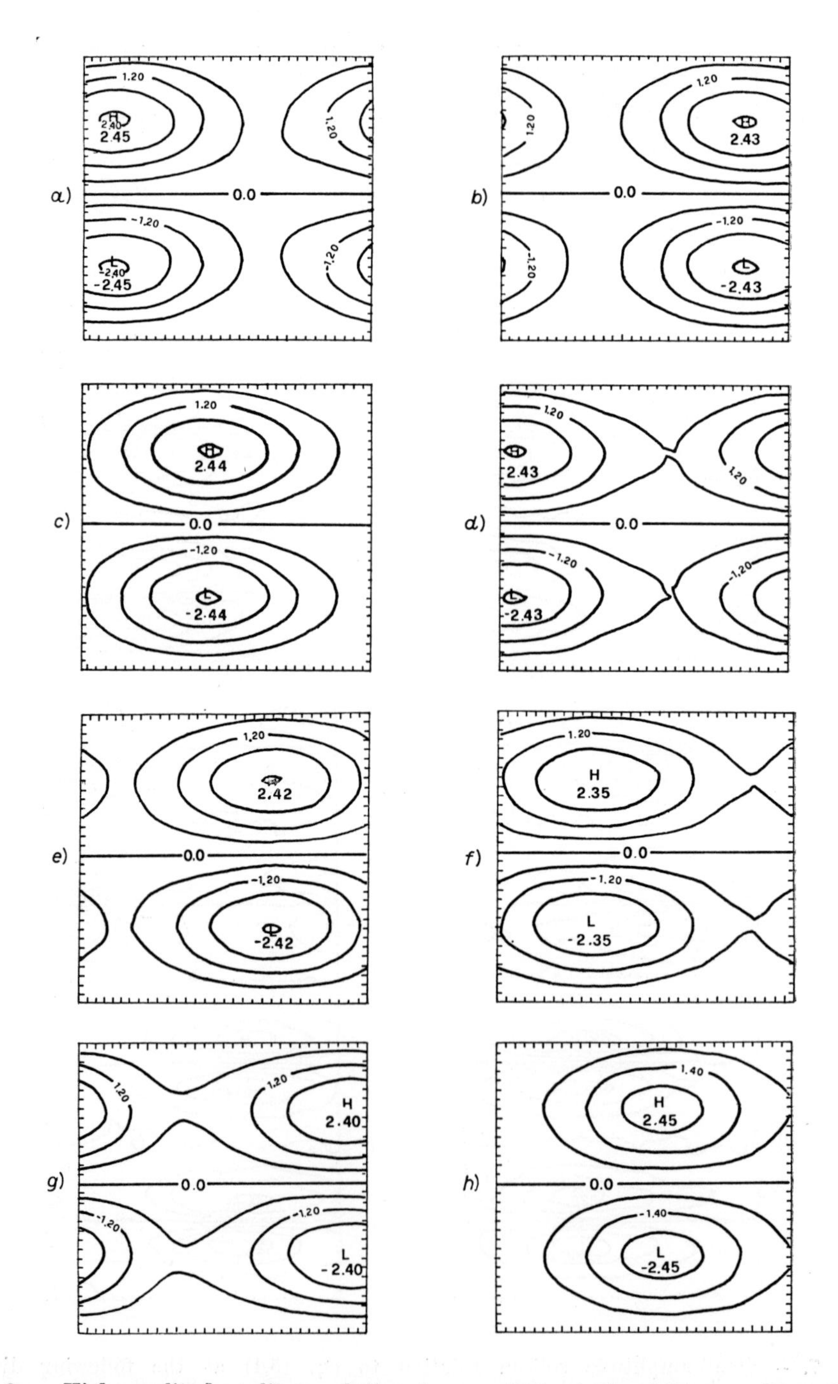

Fig. 8. – High-amplitude soliton solution to eq. (5.1) at the following dimensionless times: *a*) $t = 14$, *b*) $t = 17$, *c*) $t = 21$, *d*) $t = 25$, *e*) $t = 29$, *f*) $t = 32$, *g*) $t = 36$, *h*) $t = 39$. Total streamfunction pattern $\psi = A\,\mathrm{sech}^2(Bx)\,\Phi_1(y)$, $A = -2$, $B = 0.2$, Φ_1 = first cross-channel eigenmode for $h(y) = -\sin y - \sin 2y$. From [11].

Figures 8*a*), *b*) show the corresponding evolution of a high-amplitude solitary wave constituted now by two eddies with no superimposed perturbation. Again, the solitary eddies evolve in a completely stable, predictable manner [11].

6. – Conclusions.

Not all nonlinear systems in one, two, three dimensions obey the randomness postulate of turbulence and statistical mechanics which leads to the predictability problem. The existence of stable permanent-form solutions, the solitary waves of which two-dimensional examples are shown in fig. 7 and 8, appreciably prolongs the deterministic evolution and the consequent estimate of predictability, at least from certain initial conditions.

I would like to conclude with the following remarks, which clearly show the impact of the soliton concept upon planetary motions of geophysical flows, namely the ocean and the atmosphere, their modeling and their predictability:

« ...Is such a new approach in the offing for weather prediction? Perhaps not, but if it were, then conjecture about its possible form would be of obvious interest.

My own candidate for such an approach is the recognition, based on nonlinear wave theory, that many one-dimensional nonlinear systems with linear dispersion can have remarkably stable and completely predictable solutions called solitons....

The Earth's atmosphere is far more complicated than the examples studied so far, but certain intriguing qualitative questions arise. Is it possible that at times relatively stable structures can exist in the Earth's atmosphere whose predictability is far greater than that generally estimated? Do these arise from a balance between linear dispersion and nonlinear interactions leading to phase locking of waves of different wavelengths? If so, are our numerical methods competent to describe this balance?...

If these general ideas were to prove relevant to the problem of weather prediction, then the classical approaches would be replaced almost entirely by a new approach based on a quantum dynamics of the atmosphere, and present pessimistic conclusions would have to be re-examined » [12].

REFERENCES

[1] E. N. LORENZ: *Tellus*, **21**, 289 (1969).
[2] P. RHINES: *Annu. Rev. Fluid Mech.*, **11**, 401 (1979).
[3] C. E. LEITH: *J. Atmos. Sci.*, **28**, 145 (1971).
[4] D. K. LILLY: *Geophys. Fluid Dyn.*, **3**, 289 (1972).

[5] D. K. LILLY: *Geophys. Fluid Dyn.*, **4**, 1 (1972).
[6] E. FERMI, J. R. PASTA and S. M. ULAM: Los Alamos Sci. Lab. Rep., LA-1940 (1955).
[7] N. J. ZABUSKY and M. D. KRUSKAL: *Phys. Rev. Lett.*, **15**, 240 (1965).
[8] N. J. ZABUSKY: in *Nonlinear Partial Differential Equations*, edited by W. AMES (New York, N. Y., 1967), p. 223.
[9] P. MALANOTTE RIZZOLI: Philosophy Doctor Dissertation, Scripps Institution of Oceanography (1978), p. 147.
[10] P. MALANOTTE RIZZOLI and M. C. HENDERSHOTT: *Dyn. Atmos. Oceans*, **4**, 247 (1980).
[11] P. MALANOTTE RIZZOLI: *Dyn. Atmos. Oceans*, **4**, 261 (1980).
[12] C. E. LEITH: *Annu. Rev. Fluid Mech.*, **9**, 203 (1977).

Part II

NONLINEAR WAVE MECHANICS

Nonlinear Phenomena of Waves on Deep Water.

H. C. YUEN

Fluid Mechanics Department, TRW Defense and Space Systems Group, One Space Park Redondo Beach, Calif. 90278

1. – Introduction.

Water waves are probably the most familiar wave phenomenon found in Nature. However, common as the phenomenon may be, it possesses an extremely rich mathematical structure. In fact, it belongs to one of most difficult classes of wave mechanics, namely nonlinear, dispersive waves in three space and one time dimensions. Worse yet, the governing equations form a set of partial differential equations with nonlinear boundary conditions imposed on an unknown free surface. There are no known analytical methods to attack the full problem, and numerical techniques have only found recent success due to the advent of the large computers.

It may be somewhat discouraging to learn that even this set of monstrous equations represents only a highly idealized situation. We have already assumed that *a*) the waves are irrotational, *b*) the motion is inviscid, *c*) the fluid is incompressible, *d*) surface tension effects are negligible, *e*) the depth of the water is much greater than any relevant length scales in the problem, and *f*) the pressure over the free surface is a constant. In the course of this lecture series, I shall comment on the validity of some of these assumptions. For the most part, these assumptions still permit us to study a wide variety of wave phenomena on water with a good approximation.

Man's attraction to this fascinating problem developed early. Without going into even earlier calculations or observations of waves, we shall point out that water waves are probably one of the very first fluid-mechanical problems to be studied in a systematic fashion using the modern formulation of the Navier-Stokes type of equations.

The problem was first posed as a prize subject by the French Academy in 1816; specifically, it called for the wave pattern generated by a point source of disturbance, the so-called pebble-in-a-pond problem. The problem was solved independently by CAUCHY and POISSON, and it is now known by their names. The next giant step was taken by SCOTT RUSSELL in 1844 [1], who published an extremely comprehensive record of wave observations to the British

Association for the Advancement of Science. This *Report on Waves*, as it was entitled, was credited as the motivation for Stokes work and his ensuing treatise *Theory of Oscillatory Waves*, which appeared in 1847. In this monumental treatise, STOKES not only summarized the known results on linear wave theory, but introduced the famed Stokes expansion which later became the cornerstone of weakly nonlinear wave theory and the method of multiple scales.

Stokes' interest in the problem persisted. In 1880, he published a paper demonstrating that deep-water gravity waves have a limiting steepness. In other words, given the wavelength, the wave height cannot increase without bound even if there is continued increase in energy. When the wave height to wavelength ratio reaches 0.142, the wave attains a peak at the crest subtending an angle of 120°. Beyond this point, STOKES theorized that a smooth wave profile ceases to exist, and this limit marks the inception of wave breaking.

For years that followed, mathematicians and physicists worked on the extension of Stokes' findings. To some, the question of convergence and existence of the Stokes expansion presented the challenge. This problem was posed by BURNSIDE in 1916 [2], and settled by NEKRASOV in 1919 and LEVI-CIVITA in 1926. To others, the determination of finite-amplitude wave profiles was of interest. Notable work was performed by MICHELL [3], WILTON [4] and RAYLEIGH [5]. The question of wave evolution was virtually untouched except when the waves are infinitesimal. Throughout almost a century after Stokes' results were published, the stability of a train of steady, periodic, deep-water gravity waves was implicitly assumed.

It was against this background that the theoretical discovery of Sir James Lighthill using Whitham's theory [6] and the theoretical-experimental confirmation of Benjamin and Feir [7] that weakly nonlinear wave trains are unstable to long-wave modulational perturbations were received. The reaction was a surprise followed by great interest in the problem.

For the fifteen years following Lighthill's publication, advances in the study of deep-water gravity waves were made in rapid succession. The achievement came as a result of breakthroughs in analytical, computational and experimental techniques. In the center of the developments was the concept of solitons and wave coherence, but the new results soon extended far beyond these notions, and a new horizon was reached in the understanding of nonlinear wave phenomena found on deep water.

In this lecture series, we shall attempt to present a brief summary of some of the major contributions to the subject in the past fifteen years or so. Because of the limitations in knowledge of the author and the time allotted, the survey is by no means complete. It is hoped that the survey does give examples of interesting and challenging problems that can fully test the latest applied mathematical and fluid-mechanical research techniques, and provide motivation for scientists to further understand this familiar yet challenging phenomenon.

2. – The governing equations.

The equations taken for the study of gravity waves on deep water are the Euler equations for an incompressible, irrotational, inviscid fluid with a free surface:

$$\Delta\varphi = 0\,, \qquad -\infty < z < \eta(x, y, t)\,, \tag{1}$$

$$\begin{cases} \dfrac{\partial\varphi}{\partial t} + \dfrac{1}{2}(\nabla\varphi)^2 + gz = p\,, \\[2mm] \dfrac{\partial\eta}{\partial t} + \nabla\varphi\cdot\nabla\eta - \dfrac{\partial\varphi}{\partial z} = 0\,, \end{cases} \qquad z = \eta(x, y, t)\,, \tag{2}$$

$$\frac{\partial\varphi}{\partial z} \to 0\,, \qquad z \to -\infty\,, \tag{3}$$

where φ is the velocity potential, η is the free surface, g is the gravitational acceleration, p is an external surface pressure, the horizontal co-ordinates are (x, y) and the vertical co-ordinate is z, pointing upwards; $\Delta = \partial^2/\partial x^2 + \partial^2/\partial y^2 + \partial^2/\partial z^2$ is the Laplacian operator, and $\nabla = (\partial/\partial x, \partial/\partial y)$ is the horizontal gradient operator. The density has been normalized to unity with no loss of generality. For a wave to be considered a deep-water wave, the ratio of its wavelength λ to the water depth must be small. For surface tension to be negligible (*i.e.* for a wave to be considered a gravity wave), the wavelength must be substantially longer than 1.7 cm, the wavelength at which surface tension and gravitational effects are equal. Strictly speaking, motion in water is not inviscid. However, viscosity is most effective only for small-scale motions and is negligible for most of the gravity wave phenomena considered here. Therefore, potential flow provides a good approximation.

The dispersion relation for the linearized problem can be obtained by neglecting the nonlinear terms in (2) and applying the resulting linear boundary conditions at the undisturbed surface $z = 0$ with periodic conditions in the horizontal co-ordinates. We obtain a spatially periodic solution:

$$\eta(\boldsymbol{x}, t) = a\cos(\boldsymbol{k}\cdot\boldsymbol{x} - \omega t)\,, \tag{4}$$

$$\varphi = \frac{ga}{\omega}\sin(\boldsymbol{k}\cdot\boldsymbol{x} - \omega t)\exp\left[|\boldsymbol{k}|z\right]\,, \tag{5}$$

where $\boldsymbol{x} = (x, y)$ is the horizontal spatial vector, $\boldsymbol{k} = (k_x, k_y)$ is the wave vector, and ω the wave frequency in radians. The linear dispersion relation, which relates the frequency to the wave vector, is given as

$$\omega = \sqrt{g|\boldsymbol{k}|}\,. \tag{6}$$

The phase velocity, which gives the speed of advance of the individual crests, is

$$C = \frac{\omega}{|\boldsymbol{k}|^2}\boldsymbol{k} = \frac{\omega}{|\boldsymbol{k}|^2}(k_x, k_y) \tag{7}$$

and the group velocity, which describes the speed and direction of energy propagation, is given by

$$C_g = \frac{\partial \omega}{\partial \boldsymbol{k}} = \left(\frac{\partial \omega}{\partial k_x}, \frac{\partial \omega}{\partial k_y}\right). \tag{8}$$

The fact that $\boldsymbol{C}$ is a nontrivial function of $\boldsymbol{k}$ reflects the dispersive nature of the system.

Henceforth, unless stated otherwise, we shall be discussing one-dimensional waves which are independent of y. For simplicity of notation, we shall denote k_x by k in these one-dimensional problems.

3. – Concept of a wave train.

The expression for the free surface given by eq. (4) describes a linear wave train. Despite its obvious idealization, including perfect periodicity and infinite extent, the linear wave train occupies a uniquely important position in the study of properties of linear wave systems. Perturbations are first performed on the single wave train, and their effects on an entire wave system are determined by superposition. Unfortunately, the principle of linear superposition no longer holds in nonlinear systems. Even so, the wave train still provides the simplest entity that exhibits much of the richness of nonlinear wave systems.

A wave train can be characterized by three parameters: its amplitude a, wave number k and frequency ω. Generally, its evolution is described by equations for the conservation of wave number, the conservation of wave action and the dispersion relation. For weakly nonlinear waves, they are, respectively,

$$\frac{\partial k}{\partial t} + \frac{\partial \omega}{\partial k} = 0\,, \tag{9}$$

$$\frac{\partial a^2}{\partial t} + \frac{\partial}{\partial x}\left(\frac{\partial \omega}{\partial k}a^2\right) = 0\,, \tag{10}$$

$$\omega = \sqrt{gk}\left(1 + \frac{1}{2}k^2a^2\right). \tag{11}$$

These are the leading-order equations derivable from the averaged variational principle of Whitham [8-10] and are sometimes known as Whitham's conservation equations. Equivalent systems can be derived by a variety of methods, including the multiple-scale method [11].

Note that the dispersion relation contains a nonlinear correction which was first obtained by STOKES [12]. The absolute magnitude of the correction is small, and does not exceed 11 % even for the steepest possible waves. Nevertheless, its existence couples the three equations nonlinearly, and leads to some remarkable phenomena during the evolution of the weakly nonlinear wave train.

4. – Modulational instability.

Equations (9)-(11) were used by LIGHTHILL [6] to study the evolution of weakly nonlinear waves on deep water. Two sets of initial conditions were examined: a wave packet with a Gaussian envelope and a slightly modulated Stokes wave train. In the case of the wave packet, it was found that the envelope developed a cusp at its peak within a finite time. For the wave train, the modulation grew and again the envelope developed into a cusped shape. Smooth solutions ceased to exist beyond this stage of evolution, and the time taken to reach the singularities was found to be inversely proportional to the steepness of the waves.

These results were the first indications of modulational instabilities for weakly nonlinear waves on deep water. A detailed perturbation analysis of the uniform wave train performed on the Euler equations (1)-(3) was presented by BENJAMIN and FEIR [7], which confirmed the instability and extended the results to a wider range of perturbation wavelengths. It was found that a uniform wave train with amplitude a_0, wave number k_0 and frequency ω_0 was unstable to perturbations with wave number K in the range

$$0 < K < 2\sqrt{2}\, k_0^2 a_0 , \tag{12}$$

the maximum instability occurs when $K = K_{max}$, where

$$K_{max} = 2k_0^2 a_0 , \tag{13}$$

and the maximum growth rate $(\mathrm{Im}\,\Omega)_{max}$ is

$$(\mathrm{Im}\,\Omega)_{max} = \tfrac{1}{2}\omega_0 k_0^2 a_0^2 . \tag{14}$$

BENJAMIN and FEIR [7] also reported experimental data which are in fairly good agreement with the predictions regarding the wave number and growth rate of the instability. An example of such a modulation growing on an initially uniform wave train is shown in fig. 1. The wave forms in the figure correspond to the typical experimental situation, where the wave train evolves as it prop-

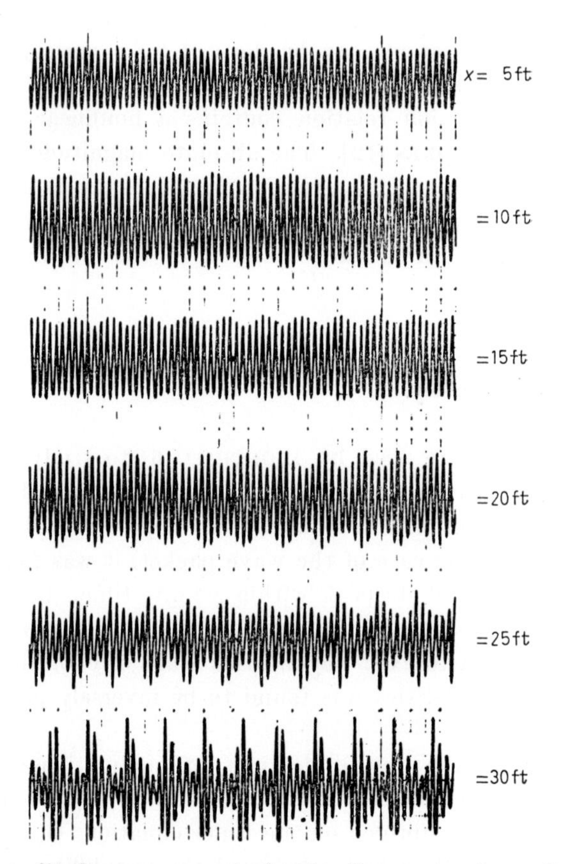

Fig. 1. – Wave amplitude measurements showing the onset and growth of the Benjamin-Feir modulational instability on an initially unmodulated nonlinear wave train. Each record shows the wave amplitude measured *vs.* time at fixed locations along a wave tank. The distance x is measured from the wavemaker. Although the output of every gauge was linearly proportional to wave amplitude, each gauge had a slightly different sensitivity. As a result, the oscillograph record of the output from any given gauge is an accurate representation of the wave form of the waves measured by that gauge, but the absolute magnitudes of the wave forms recorded by different gauges cannot be used to compare actual wave amplitudes at different stations unless differences in probe sensitivity are taken into account.

agates along a distance x, whereas the theory is usually formulated for evolution in time. The spatial and temporal cases can be related simply by transforming co-ordinates using the group velocity of the carrier wave.

5. – The nonlinear Schrödinger equation.

Identification of the instability led to the question of the evolution of the unstable nonlinear wave train. The analysis of Benjamin and Feir was valid only for the initial growth period; and Whitham's equations, as applied by LIGHTHILL, yielded singularities in the solutions within a finite time. Effort was, therefore, directed toward derivation of equations which would be valid for large times. CHU and MEI [11, 13] used the multiple-scale method to show that an additional correction term, effective when the curvature of the envelope becomes large, must be included in the dispersion relation. They showed further that the singularity LIGHTHILL encountered can be eliminated with the inclusion of the « curvature dispersion » term, but their computation met with another numerical difficulty at a later time, the physical significance of which was unclear. Equivalent correction terms in the dispersion relation were also obtained by BENNEY and ROSKES [14] and HAYES [15-17]. Meanwhile, HASIMOTO and ONO [18] used the derivative expansion method to show that the evolution of a nonlinear deep-water wave train obeys the nonlinear Schrödinger equation, and they recovered the results of Benjamin and Feir by a stability analysis of its uniform solution. YUEN and LAKE [19] showed that the nonlinear Schrödinger equation was a consequence of Whitham's equations when second variations are included in the dispersion relation. Soon it was realized that all methods, under the same set of assumptions, lead to the same nonlinear Schrödinger equation for the complex wave train envelope $A(x, t) = a(x, t) \exp [i\theta(x, t)]$:

$$i\left(\frac{\partial A}{\partial t} + \frac{\omega_0}{2k_0}\frac{\partial A}{\partial x}\right) - \frac{\omega_0}{8k_0^2}\frac{\partial^2 A}{\partial x^2} - \frac{1}{2}\omega_0 k_0^2 |A|^2 A = 0 , \tag{15}$$

where $\partial\theta/\partial t = \omega_0 - \omega$, $\partial\theta/\partial x = k - k_0$ and the free surface of water is $\eta(x, t) = \mathrm{Re}\,\{A \exp [i(k_0 x - \omega_0 t)]\}$.

This equation was first discussed in [20] in the general context, but, unknown to many workers in the field at the time, it (and its two-space-dimensional extension) was first derived for deep-water waves by ZAKHAROV [21] via a spectral method. Identification of the nonlinear Schrödinger equation as the correct equation for describing the time evolution of weakly nonlinear wave trains did not immediately resolve all the questions regarding the nature of wave train evolution, but it was a major step in that direction. For reasons of theoretical and experimental advantage, the next step actually involved consideration of finite wave packets rather than continuous wave trains.

6. – Envelope solitons.

One of the most important characteristics of the nonlinear Schrödinger equations is that it can be solved exactly for initial conditions which decay sufficiently rapidly as $|x| \to \infty$. This was done by ZAKHAROV and SHABAT [22] using the then newly discovered inverse-scattering method [23]. They showed

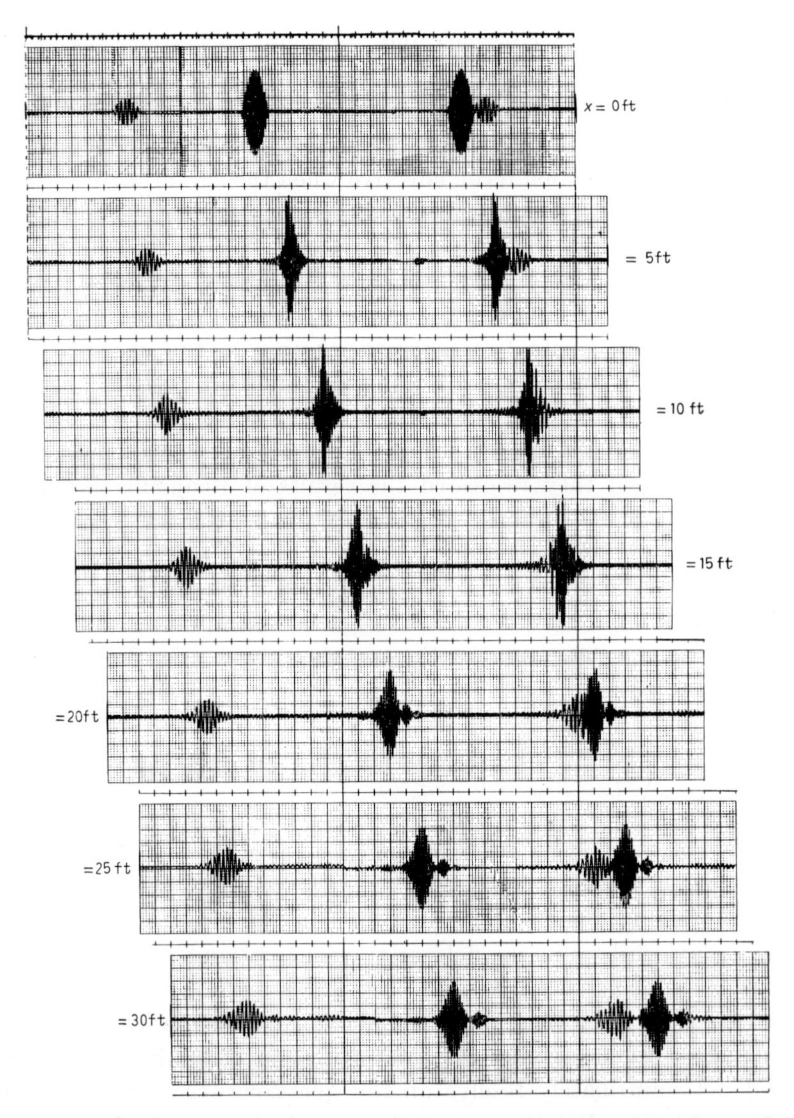

Fig. 2. – Example of one wave pulse overtaking and passing through another wave pulse. Laft-hand trace: First pulse alone $\omega_0 = 1.6$ Hz, initial $(ka)_{\max} \simeq 0.1$, 6-cycle pulse. Center trace: Second pulse alone $\omega_0 = 3$ Hz, initial $(ka)_{\max} \simeq 0.2$, 12-cycle pulse which disintegrates into two solitons. Right-hand traces: The two pulses together. At $x = 0$, the 3 Hz pulse is generated ahead of the 1.6 Hz pulse. As they propagate to $x = 30$ ft, the faster 1.6 Hz pulse passes through the 3 Hz pulse and emerges ahead of it, while the 3 Hz pulse is disintegrating into two solitons. Note that at 30 ft the pulses emerge from the interaction relatively unchanged from the forms they have at $x = 30$ ft for the no-interaction between pulses of different carrier wave frequencies cannot be described by a single equation such as (15). It has been shown, however, by OIKAWA and YAJIMA [24] that solitons with different carrier frequencies also survive interactions with no permanent change other than position and phase shifts. The governing equations for these multiwave systems can likewise be obtained by Whitham's methods by use of more than one phase function.

that any initial wave packet eventually evolves into a number of « envelope solitons » and a dispersive tail. The bulk of the energy is contained in the solitons, which propagate with permanent form once produced. Solitons also survive interactions with other solitons or wave packets. Since the nonlinear Schrödinger equation describes the envelope of deep-water waves with a carrier frequency, the theory predicts the existence of packets of deep-water waves with soliton properties. The existence of these envelope soliton properties would hardly have been expected on the basis of experience with linear deep-water wave systems in which wave components are uncoupled and highly dispersive.

The theoretical predictions of soliton properties were confirmed by experiments (an example is shown in fig. 2), and numerical solutions of the nonlinear Schrödinger equation were found to compare well with experimentally measured wave envelope profiles [19]. For the case of wave packets and solitons, the description of nonlinear deep-water wave dynamics provided by the nonlinear Schrödinger equation was found to be both qualitatively and quantitatively correct. Although the practical significance of these envelope solitons is questionable because of their one-dimensional nature, their mathematical and physical existence increased confidence in the nonlinear-Schrödinger-equation description and provided a dramatic demonstration of the potential significance of nonlinear effects on the dynamics of deep-water waves. The soliton investigations are also another example of the close relationship between experiment and theory that has become almost a tradition in this subject, and which has recently led to new insights into phenomena in more realistic wave systems, particularly phenomena influenced by the degree of coherence of the waves.

7. – Long-time evolution of an unstable wave train and recurrence.

There have been a variety of propositions regarding the possible end-state of a wave train undergoing modulational instability. These range from a complete breakdown of the wave train with equipartition of energy among all modes to a train of stable soliton envelopes (see, for example, [18, 25, 26]). The answer, however, turned out to be somewhat a surprise.

Experiments in the laboratory by LAKE *et al.* [27] indicated that the unstable modulations grew to a maximum and then subsided (fig. 3). Furthermore, at some stage of the evolution the wave train actually became nearly uniform again. Numerical computations using the nonlinear Schrödinger equation, which proved to be satisfactory for describing the long-time evolution of the wave packets, confirmed this interesting phenomenon. Thus, in the absence of viscous dissipation, there are no steady end-states, but a series of modulation and demodulation cycles, known as the Fermi-Pasta-Ulam recurrence phenomenon ([28], see also [29]).

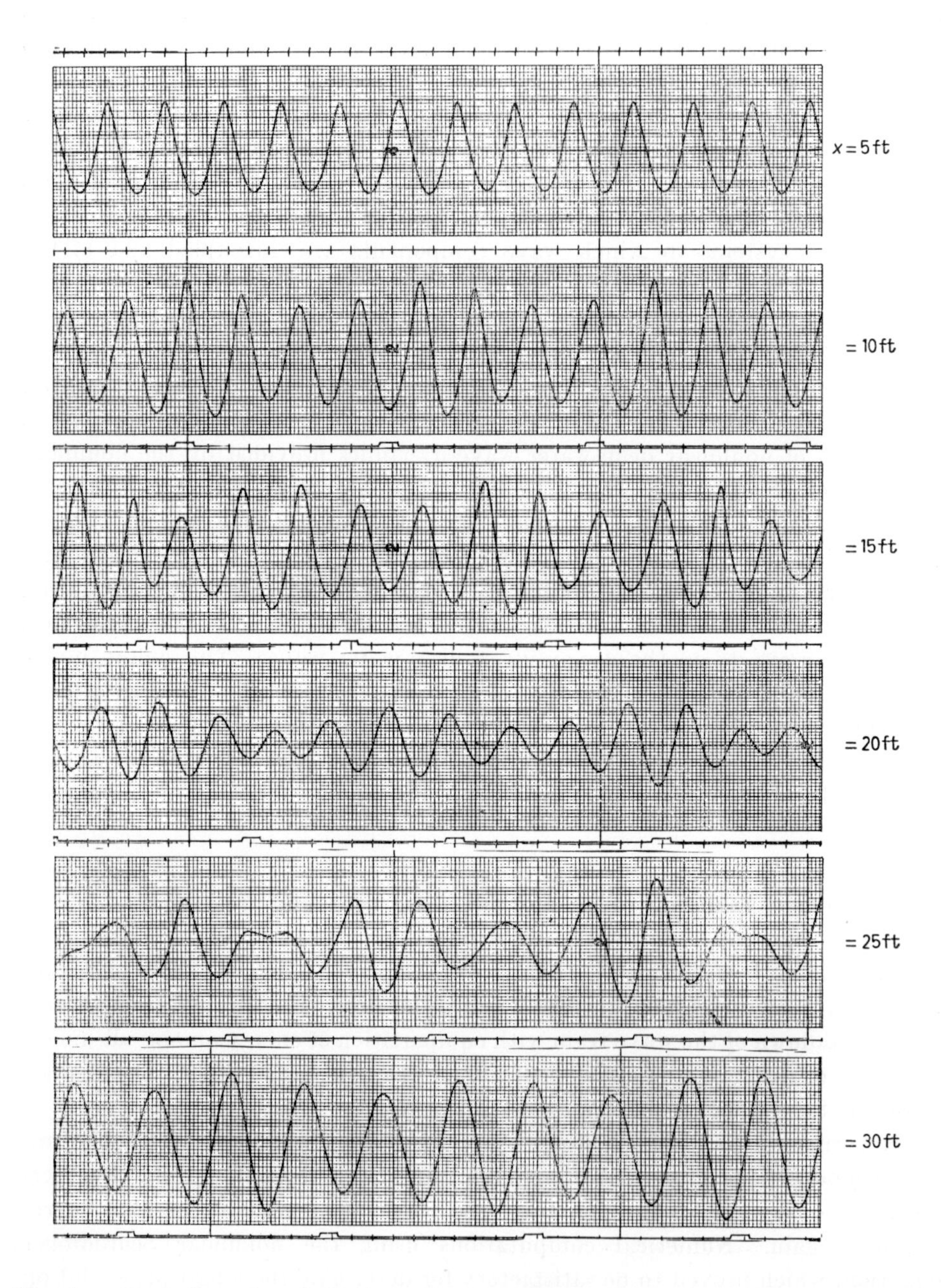

Fig. 3. – Evolution of an initially uniform weakly nonlinear wave train through one cycle of modulation and demodulation. Oscillograph records are shown on expanded time scale to display individual wave shapes; wave shapes in each modulation period are not exactly repetitious, because modulation period does not contain integral number of waves.

The existence of the Fermi-Pasta-Ulam recurrence has a special significance for the usefulness of the wave train in the study of deep-water waves. Had the end-state been one of complete breakdown (or thermalization), it would have indicated that the wave train is a totally idealized and artificial entity in that it would not occur naturally, and, even if it did, it could not sustain its identity as such. Moreover, the nonlinear Schrödinger equation, which was derived based on the concept of a wave train, would have truly limited applicability. Instead, because the nonlinear wave train maintains its coherence during its evolution, the wave train concept and the nonlinear Schrödinger equation are expected to be applicable to investigations of more realistic wave problems.

8. – Relation between initial conditions and long-time evolution.

The existence of the Fermi-Pasta-Ulam recurrence phenomenon also indicates that the solutions of the nonlinear Schrödinger equation have strong « memory » of their initial conditions. A positive link between the character of the long-time evolution and the initial conditions was reported by YUEN and FERGUSON [30, 31]. They showed by numerical experiment that the long-time evolution of an unstable wave train is governed by the unstable modes and their harmonics contained in the initial condition. The stable harmonics do receive energy, but they never appear as the dominant mode at any stage of the evolution. Their results are summarized in fig. 4.

A corollary to this observation is that, if the instability diagram which governs the eventual evolution has a high wave number cut-off, then energy in the solution for subsequent times must be effectively confined within the unstable modes. This rules out thermalization in the classical sense, which corresponds to equipartition of energy among all modes.

The nonthermalization of solutions of the nonlinear Schrödinger equation for smooth initial conditions was actually proved by THYAGARAJA [32], who constructed an algorithm for finding N, given any δ, so that the energy of the solution at all times is confined to the first N modes within an error of δ.

9. – Three-dimensional effects.

Thus far we have confined our discussion to one-space-dimensional problems. The equation for the complex envelope of a wave train propagating in two space dimensions is

$$i\left(\frac{\partial A}{\partial t}+\frac{\omega_0}{2k_0}\frac{\partial A}{\partial x}\right)-\frac{\omega_0}{8k_0^2}\frac{\partial^2 A}{\partial x^2}+\frac{\omega_0}{4k_0^2}\frac{\partial^2 A}{\partial y^2}-\frac{1}{2}\,\omega_0 k_0^2|A|^2A=0\,, \tag{16}$$

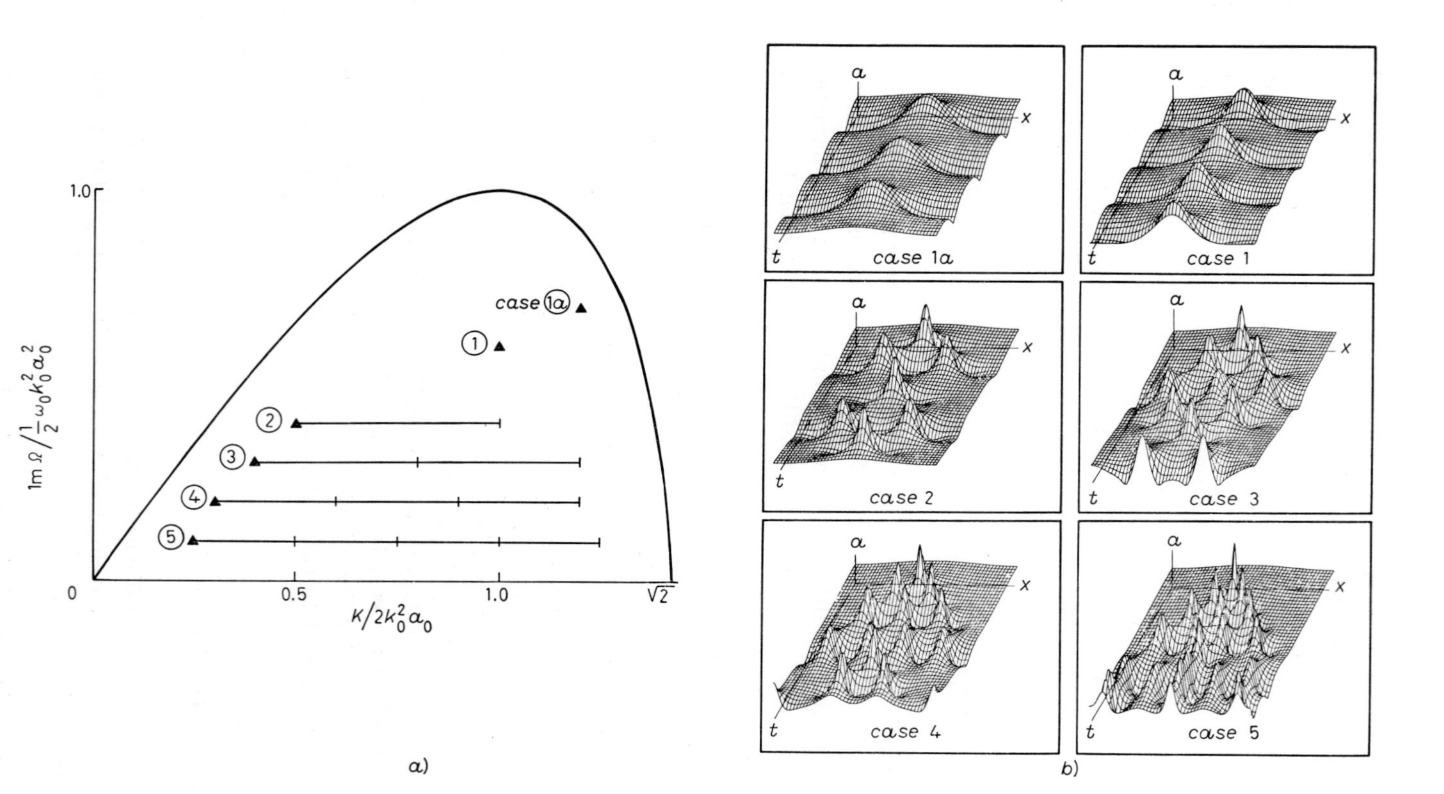

Fig. 4. – Relationship between initial conditions and long-time evolution of solutions of the nonlinear Schrödinger equation. *a*) gives the perturbation wave number K of the various cases and *b*) shows their corresponding time evolution. The full curve refers to the Benjamin-Feir analysis. The initial conditions consist of a uniform wave train with wave number k_0 and amplitude a_0 subject to a 1% perturbation. The numerals in circles in *a*) identify the cases in *b*), and also correspond to the number of harmonics of the perturbation (including the primary) which lie within the unstable regime according to the stability analysis. Note that the number of unstable harmonics corresponds exactly to the number of modes which dominate the evolution. For example, case 4 shows an evolution in which the 1st, 2nd, 3rd, 4th, 1st, 2nd harmonics, in that order, took turns in dominating the evolution as indicated by the number of peaks at various stages of evolution in the amplitude plot.

where the free surface of water is

$$\eta(x, y, t) = \text{Re}\,\{A(x, y, t) \exp\,[i(k_0 x - \omega_0 t)]\}\,. \tag{17}$$

When there is no y-dependence in the wave envelope, the equation reduces to the nonlinear Schrödinger equation (15). Note that the second-y-derivative term and the second-x-derivative term are of opposite sign, leading to a basically hyperbolic, rather than elliptic, equation. Also note that the carrier wave still propagates in the x-direction, but now its envelope entertains two-dimensional variations.

In the following we briefly review the concepts of envelope solitons, modulational instability and recurrence in the two-dimensional context.

9'1. *Envelope solitons.* – Plane envelope solitons exist for eq. (16) provided that the angle between the directions of propagation of the carrier wave and

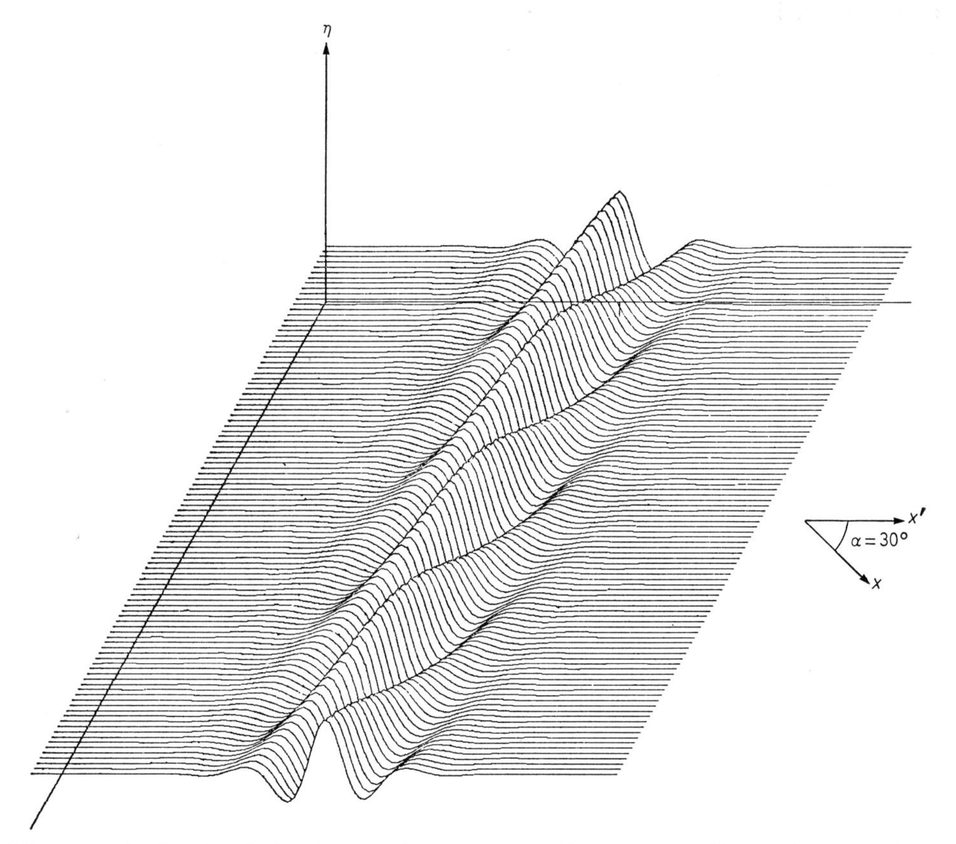

Fig. 5. – A sketch of the free surface corresponding to an oblique plane envelope soliton in two space dimensions (not to scale). The individual waves in the envelope propagate in the x-direction at the phase speed and the envelope propagates in the x'-direction at the group velocity.

of the envelope profile is less than 35.26°. An illustration of an oblique plane soliton is given in fig. 5. These two-dimensional plane solitons are infinite in extent. In addition, these solitons are unstable to two-dimensional disturbances [33, 34]. Thus far, no fully two-dimensional solutions exhibiting true soliton behavior have been found. Therefore, the possibility of applying the concept of solitons as fundamental entities to construct fully two-dimensional wave fields has not yet been realized, although there have been some attempts in this direction [35, 36].

9'2. *Modulational instability.* – Generalization of the stability analysis of a uniform wave train to two space dimensions yields the result illustrated in fig. 6. The Benjamin-Feir results are recovered by setting K_y to zero. Of particular significance is the fact that the two-dimensional instability does not have a high wave number cut-off. This result has major implications when one considers the long-time evolution of an unstable wave train in two dimensions, as discussed below.

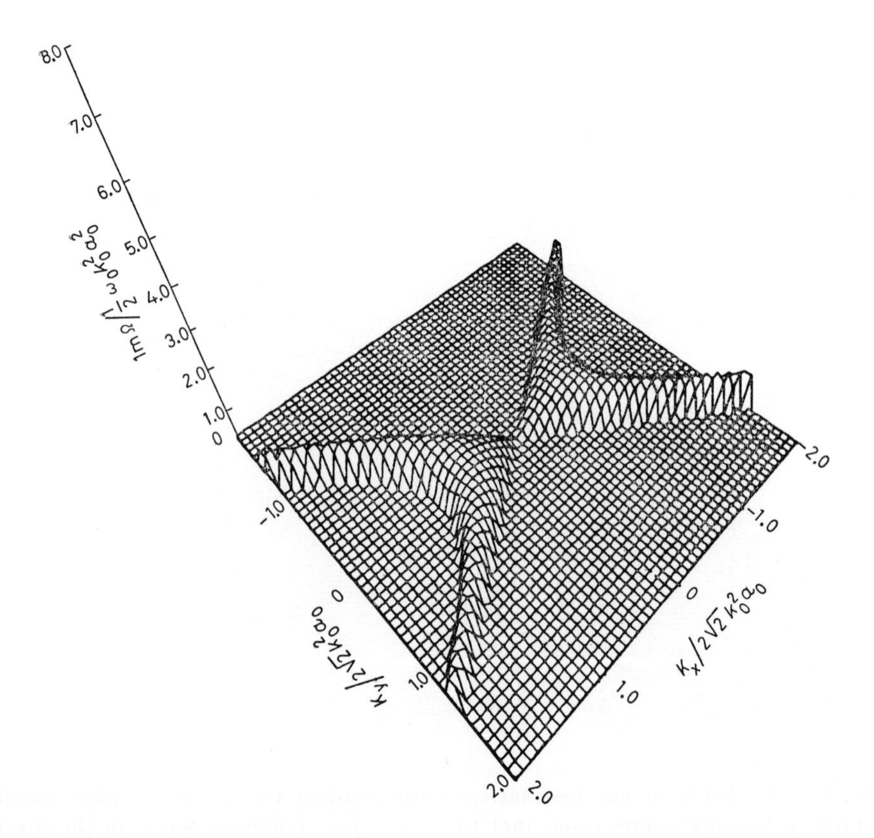

Fig. 6. – Stability diagram of a uniform wave train subject to infinitesimal two-dimensional plane-wave perturbations.

9'3. *Long-time evolution and recurrence.* – For a wide variety of initial conditions, the phenomenon of Fermi-Pasta-Ulam recurrence still exists for wave trains in two space dimensions [30, 31]. When one examines the relationship between initial conditions and long-time evolution for the two-dimensional nonlinear Schrödinger equation, however, one is led to the conclusion that, unlike the one-dimensional result, there exists for any fully two-dimensional unstable perturbation vector (K_x, K_y) an infinite set of integers (m, n) such that (mK_x, nK_y) lie in the unstable regime and are eligible to actively participate in the evolution. Furthermore, it has been confirmed by MARTIN and YUEN [37] that an unstable wave train does recur, but that during each new cycle unstable modes with larger values of m and n are excited. Thus the long-time evolution consists of a periodic return to the initial condition together with a gradual « leak » of energy to higher and higher modes. Although this may not be the classical definition of thermalization, such behavior clearly indicates that the validity of the two-dimensional equation ceases at a finite time, since it is derived on the presumption that most of the energy is confined within a narrow band around the carrier wave number.

10. – Higher-order effects.

Most of the desirable features of the solutions of the one-dimensional nonlinear Schrödinger equation are lost in the extension to two space dimensions. The inverse-scattering method can no longer be applied. No localized solitons exhibiting stability and immunity from interaction effects have been found. The concept of wave coherence, which was prominent and important in the one-dimensional case, appears to be of questionable relevance in light of the results in the two-dimensional case. Most perturbing of all, however, is the fact that certain solutions of the two-dimensional nonlinear Schrödinger equation experience a progressive leakage of energy to high modes, thus eventually extending beyond the domain of validity of the equation.

As these problems with the two-dimensional theoretical results were identified, the need for a more accurate description than that provided by the nonlinear Schrödinger equation became apparent. A nonlinear integro-differential equation, first derived by ZAKHAROV [21], was identified by CRAWFORD *et al.* [38] as being the desirable replacement. Zakharov's equation describes the time evolution of the complex envelope spectral function $\bar{A}(\boldsymbol{k}, t)$:

$$(18) \qquad i\frac{\partial}{\partial t}\bar{A}(\boldsymbol{k}, t) = \iiint_{-\infty}^{\infty} T(\boldsymbol{k}, \boldsymbol{k}_1, \boldsymbol{k}_2, \boldsymbol{k}_3)\,\delta(\boldsymbol{k} + \boldsymbol{k}_1 - \boldsymbol{k}_2 - \boldsymbol{k}_3)\cdot$$

$$\cdot \exp\left[i[\omega(\boldsymbol{k}) + \omega(\boldsymbol{k}_1) - \omega(\boldsymbol{k}_2) - \omega(\boldsymbol{k}_3)]t\right] \bar{A}^*(\boldsymbol{k}_1)\bar{A}(\boldsymbol{k}_2)\bar{A}(\boldsymbol{k}_3)\,\mathrm{d}\boldsymbol{k}_1\,\mathrm{d}\boldsymbol{k}_2\,\mathrm{d}\boldsymbol{k}_3\,,$$

where $\bar{A}(\boldsymbol{k}, t)$ is related to the free surface by the expression

$$\eta(\boldsymbol{x}, t) = \frac{1}{2\pi}\int \frac{|\boldsymbol{k}|^{\frac{1}{4}}}{\sqrt{2}g^{\frac{1}{4}}}\{\bar{A}(\boldsymbol{k}, t)\exp[i(\boldsymbol{k}\cdot\boldsymbol{x}-\omega t)] + \bar{A}^*(\boldsymbol{k}, t)\exp[-i(\boldsymbol{k}\cdot\boldsymbol{x}-\omega t)]\}\,\mathrm{d}\boldsymbol{k} \tag{19}$$

and $T(\boldsymbol{k}, \boldsymbol{k}_1, \boldsymbol{k}_2, \boldsymbol{k}_3)$ is a real interaction coefficient first calculated by ZAKHAROV [21] with minor algebraic corrections made by CRAWFORD *et al.* [38].

The nonlinear Schrödinger equation can be recovered, in one or two dimensions, by expanding the frequencies $\omega(\boldsymbol{k}_i)$ about a carrier wave vector to second order, and replacing the interaction coefficient $T(\boldsymbol{k}, \boldsymbol{k}_1, \boldsymbol{k}_2, \boldsymbol{k}_3)$ by its value when all four arguments are evaluated at the carrier wave vector $\boldsymbol{k}_0 = (k_0, 0)$. The resultant integral equation is then the Fourier-transformed equation for the complex amplitude $A(x, y, t)$ which satisfies the nonlinear Schrödinger equation. This simplification is equivalent to retaining only the leading-order terms in nonlinearity and dispersion.

Results obtained from the Zakharov integral equation were found to be very encouraging. The stability analysis for one-dimensional modulational perturbation yields an instability growth rate lower than that predicted by the nonlinear Schrödinger equation (and by the Benjamin and Feir analysis). The difference becomes appreciable for moderate wave steepness ($ka = 0.15$ and higher), and the growth rate predicted by the Zakharov equation agrees much better with the experimental data [38] (see fig. 7). For very large values

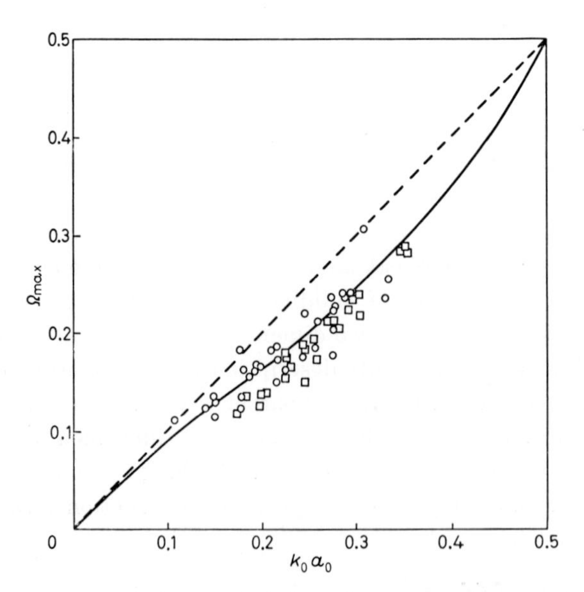

Fig. 7. – Plot of most unstable modulation frequency $\Omega_{\max}$ against carrier wave steepness $k_0 a_0$: ○, □ experimental data, —— results from stability analysis based on Zakharov's equation, – – – results from stability analysis based on nonlinear Schrödinger equation (or from Benjamin and Feir analysis).

of wave steepness, Zakharov's equation predicts a restabilization, which is in qualitative agreement with the exact numerical results of Longuet-Higgins [39] and Peregrine and Thomas [40].

Because the Zakharov equation does not require that there be a single constant wave vector, it can be used to investigate wave dynamics as a function of the band width of components in the spectrum as well as of wave steepness. As an example of the dependence of dynamics upon those parameters, one can calculate dispersive characteristics for such wave systems. An example of such a result taken from Crawford *et al.* [41] is shown in fig. 8, where com-

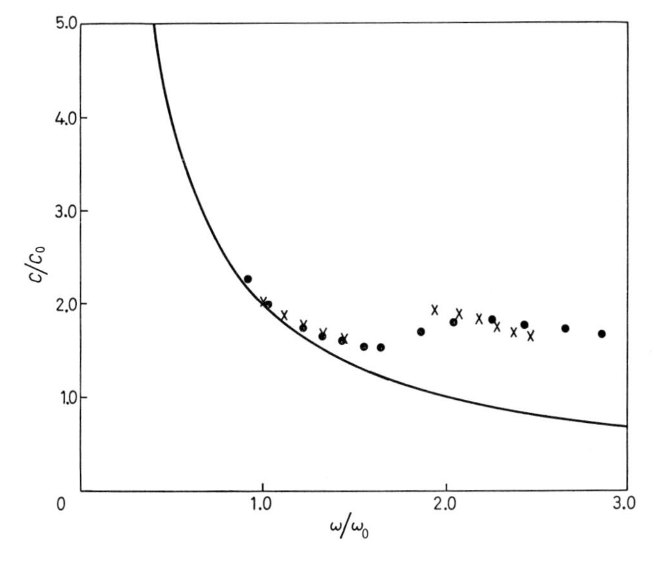

Fig. 8. – Plot of normalized component phase speed C/C_0 against normalized frequency ω/ω_0 for a spectrum of waves with characteristic band width (half width normalized by peak frequency) of 0.07, characteristic nonlinearity (root-mean-square wave slope) of 0.098. The phase speed and frequency are normalized by their values at the spectral peak. × numerical computation, • experimental data from laboratory (no wind), —— phase speed result from linear dispersion relation.

ponent phase speeds are plotted against component frequencies for a particular wave system. The experimental data are component phase speeds measured using a two-point filter and correlation technique for a wave system of finite band width and steepness. The theoretical results were obtained using the Zakharov equation for a spectrum of waves with the same band width and dominant frequency as those in the experiments. Depending upon band width and steepness, such wave systems may exhibit component dispersion ranging from that given by the linear dispersion relation to that of an effectively nondispersive phase-locked system in which components propagate at essentially a single speed. Most realistic wave systems would be expected to fall somewhere in between the two limits. An example is given in fig. 8. The dispersive properties of deep-water wave systems, therefore, vary continuously over this range of possibilities, with deviations from the linear results increasing with increasing steepness and decreasing band width.

The Zakharov equation can also be used to study the interaction of more than one wave component traveling in different directions. In particular, the problem of the generation of a third wave train by two resonantly interacting and intersecting wave trains was examined by CRAWFORD and YUEN [42], and the results compare very well with the experimental data of McGoldrick *et al.* [43] and Longuet-Higgins and Smith [44].

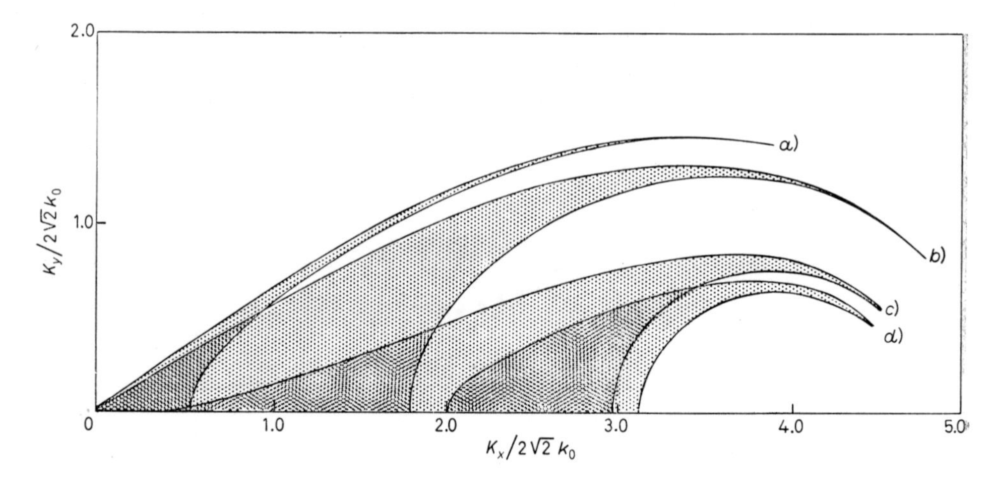

Fig. 9. – Stability boundary for a uniform wave train subject to infinitesimal two-dimensional plane-wave perturbations based on Zakharov's equation. Note that the stability boundary depends strongly on the carrier wave steepness k_0a_0: *a*) $k_0a_0 = 0.05$, *b*) $k_0a_0 = 0.2$, *c*) $k_0a_0 = 0.4$, *d*) $k_0a_0 = 0.44$. The shaded regions represent instability. The long waves (low wave number) have become stable for large values of k_0a_0 (0.4 and 0.44). For k_0a_0 greater than 0.495, the entire system is stable again. Whereas this value of k_0a_0 is unphysically large, the restabilization agrees with the trend exhibited by the exact calculations of Longuet-Higgins for the one-dimensional case.

The most important result, however, is the stability analysis in two space dimensions. CRAWFORD *et al.* [38] showed that, contrary to the results from the two-dimensional nonlinear Schrödinger equation, the instability diagram for the uniform wave train possesses a high wave number cut-off (see fig. 9). Therefore, the leakage to high modes experienced by certain solutions of the two-dimensional nonlinear Schrödinger equation need not occur in wave systems described by the Zakharov equation. The concept of wave coherence appears to be relevant to wave systems in two space dimensions after all, although positive confirmation has yet to be produced.

11. – Statistical theory.

The effects of randomness on the stability properties of a nonlinear wave train were examined by ALBER and SAFFMAN [45] and extended to two dimension by ALBER [46] using the nonlinear Schrödinger equation as a starting point. It was shown that the presence of phase and/or amplitude randomness reduces the growth rate of the modulational instability, and acts to suppress the growth of modulations having wavelength substantially shorter than the decorrelation length associated with the randomness. For sufficiently strong randomness, the modulational instability can be eliminated. More generally, ALBER and SAFFMAN [45] showed that the evolution of a random wave train obeys a transport equation similar to the Vlasov equation for weak turbulence in a plasma:

$$\frac{\partial F}{\partial t}+\left(\frac{\omega_0}{2k_0}-\frac{\omega_0}{2k_0^2}\,p\right)\frac{\partial F}{\partial x}-\omega_0 k_0^2\sin\left(\frac{1}{2}\frac{\partial}{\partial p}\frac{\partial}{\partial x'}\right)\left[F(p,x,t)\int\limits_{-\infty}^{\infty}F(p',x',t')\,\mathrm{d}p'\right]\Bigg|_{x'=x}=0\,,$$

where $F(p, x, t)$ is the Fourier transform of the two-point, one-time correlation function

$$F(p,x,t)=\frac{1}{2\pi}\int\limits_{-\infty}^{\infty}\exp[-ipr]\left\langle A\left(x+\frac{r}{2},t\right)A^*\left(x-\frac{r}{2},t\right)\right\rangle\mathrm{d}r$$

with $\langle\ \rangle$ denoting the ensemble average.

The point worth noticing is that this equation predicts an evolutionary time scale of $O(F)$ or $O(k^2a^2)$, which is a much faster time scale than the weak nonlinear energy transfer time scale of $O(F^2)$ or $O(k^4a^4)$ found by HASSELMANN [47, 48] for a random, homogeneous ocean. This apparent conflict was resolved by CRAWFORD, SAFFMAN and YUEN [49], who derived a more general equation for a random wave field based on the Zakharov equation, of which eq. (15) is a special case when constant peak frequency and narrowbandedness are assumed, and showed that the $O(k^2a^2)$ term represents effects of inhomogeneity and vanishes for a homogeneous ocean. The presence of the term does not lead to net energy transfer. In other words, the evolution of the random wave field consists of a relatively fast process of redistribution of inhomogeneity which is reversible, and a much slower process of net energy transfer among wave numbers which is irreversible. The former process is intimately related to the Fermi-Pasta-Ulam recurrence phenomenon associated with a single nonlinear wave train and it is responsible for the modulation scales and patterns of the wave field. The latter process is related to the much slower evolution of the overall spectral distribution of wave energy among components.

12. – The limiting wave.

The occurrence of instabilities and nonlinear effects in deep-water waves is associated with finite amplitude or, more accurately, with finite wave steepness, $k_0 a_0$. As amplitude or steepness increases from the infinitesimal-amplitude linear wave limit, the nature and variety of instabilities and nonlinear effects increase. In the following we describe some phenomena of wave trains and individual waves that occur as the wave amplitude or steepness becomes very large. First, however, we recall that STOKES [50] suggested that there is a maximum possible wave steepness for steady waves on deep water, so that « very large » refers to values of wave steepness approaching that of the limiting wave. The existence of a limiting wave and identification of its physical characteristics have been the subjects of numerous investigations ever since.

One line of investigation pertains to the calculation of profiles and flow fields of waves of finite and large amplitudes. STOKES [50] used an argument locally valid near the crest to demonstrate that the limiting-wave profile can have a 120° corner at its crest, which is a stagnation point. MICHELL [3] fitted a Fourier series to the leading-order singular term used by STOKES to arrive at the 120° result and determined that the height-to-wavelength ratio of the wave of greatest height is 0.142 ($ka = 0.446$). His analysis was refined and extended by WILTON [4] and HAVELOCK [51], and in a modified form by YAMADA [52], but with only slight changes in the value of the limiting steepness. These calculations were very laborious, and the complexity increased rapidly with the increasing number of terms retained in the expansion. Recently, GRANT [53] and NORMAN [54] reported difficulties in obtaining further terms in the expansion. Using the Padé approximants to accelerate convergence, SCHWARTZ [55] performed calculations which have accuracy equivalent to retaining 117 terms in the Stokes expansion and found that the nature of the singularity near the crest changes from square-root type to cube-root type when the steepness approaches the limiting value. Accurate profiles were also given for the entire range of wave steepness leading to the near-limiting wave. Similar calculations were performed by LONGUET-HIGGINS [56] and COKELET [57] with results in good agreement with those of SCHWARTZ [55]. The nature of the profile and the flow field near the crest of the near-limiting wave was examined analytically by LONGUET-HIGGINS and FOX [58], using matched asymptotic expansions. They showed that the maximum slope is slightly greater than the 30° implied by the Stokes conjecture, being 30.37°. Subsequently, LONGUET-HIGGINS and FOX [59] matched this inner solution to third order to an outer solution, providing the first definite link between the local corner flow at the crest as suggested by STOKES and a full-period wave profile.

The calculations of Schwartz [55] clearly identified an intriguing feature of large-amplitude waves: the phase speed, kinetic energy, potential energy and wave impulse each reach an absolute maximum before the wave reaches its maximum steepness. The more detailed investigation by LONGUET-HIGGINS and FOX [59] indicates that these quantities, after attaining their maxima, oscillate an infinite number of times before they reach the values for the steepest wave. The exact implication of this rather puzzling behavior has not yet been completely explored.

The other line of investigation concentrates on existence in the mathematical sense. The existence of small-amplitude, steady, periodic waves was proved by NEKRASOV [60] and later LEVI-CIVITA [61]. The existence of finite-amplitude and very steep waves was not established, however, until very recently. TOLAND [62], extending the work of Krasovskii [63] and Keady and Norbury [64], proved the existence of a « limiting wave » which has a stagnation point at the crest, and is smooth everywhere except at the crest. He showed that this wave is indeed the limit of a sequence of smooth, periodic steady waves, and that the maximum slope of some waves in the sequence must be at least 30°. He was unable, however, to ascertain the properties of the limiting wave near its crest. Depending on the type of singularity associated with the limiting-wave solution at the crest, the crest can either be a corner of 120° (and hence the Stokes corner flow), or an infinite number of ripples of large steepness.

13. – Restabilization.

The modulational instability for weakly nonlinear wave trains as calculated by BENJAMIN and FEIR [7], and from the nonlinear Schrödinger equation, is « similar » in the wave steepness (here taken to be $k_0 a_0 = \pi h/\lambda$), in the sense that when plotted in scaled variables, with the modulational wave number scaled by $k_0 a_0$ and the growth rate scaled by $k_0^2 a_0^2$, the results are no longer dependent on $k_0 a_0$. This was shown not to be so for the Zakharov equation, which has a stronger dependence on $k_0 a_0$. In fact, the Zakharov equation indicates that the modulational instability disappears for sufficiently large values of $k_0 a_0$. The Zakharov equation predicts wave train restabilization at $k_0 a_0 \doteq 0.5$. Since this is greater than the $k_0 a_0$ of the limiting wave, the prediction is only suggestive of the correct behavior. The value of $k_0 a_0$ at which restabilization actually occurs is around 0.34, as shown by the exact numerical calculations of Longuet-Higgins [39]. LONGUET-HIGGINS also showed that a second, more violent instability occurs at a larger value of $k_0 a_0$, but he was unable to obtain accurate results in the required range.

14. – Bifurcation of large-amplitude waves.

Compared to the question of existence, the uniqueness of steady periodic water waves has received much less attention. The main results are contained in a little-known paper by GARABEDIAN [65], who used topological methods to prove the nonexistence of asymmetric waves and the uniqueness of symmetric waves, provided that their crests and troughs are of the same height. In other words, perfectly uniform, symmetric wave trains are unique. His results apply to waves of arbitrary amplitudes.

The proviso that uniqueness is guaranteed only if the waves have uniform crests and troughs turns out to be critical. In fact, CHEN and SAFFMAN [66] have demonstrated numerically that, for sufficiently large amplitudes, the uniform wave train bifurcates into a train of waves with unequal crests or troughs, with the individual waves in the train remaining symmetric about their crests. These bifurcated wave trains acquire a periodicity with periods longer than that of the unbifurcated solutions, hence the bifurcation is subharmonic or fractional subharmonic. Consistent with the results of Garabedian [65], CHEN and SAFFMAN [66] reported no superharmonic bifurcations.

The critical amplitudes at which bifurcation occurs lie in the range of amplitudes where the wave train is stable to modulational perturbations. In fact, they coincide with points of neutral stability in the modulational-perturbation calculations. In the one-dimensional case, these points were calculated by LONGUET-HIGGINS [39] for several selected perturbational wavelengths and found to lie around $k_0 a_0 = 0.4$.

For the two-dimensional case, the neutral-stability curve in the limit of infinitely long perturbations can be obtained from the calculations of Peregrine and Thomas [40] to demonstrate that neutral stability occurs for smaller and smaller amplitudes for increasingly oblique perturbations. For finite perturbation wavelengths, the same trend is found from the Zakharov equations by CRAWFORD *et al.* [38]. Thus it should be possible for a uniform wave train to bifurcate into a steady, fully two-dimensional, periodic wave pattern even at small wave amplitudes. This question was pursued by MARTIN [67], who studied the bifurcation from the uniform solution of the two-dimensional nonlinear Schrödinger equation, which, despite its limitations at large amplitudes, should be adequate for small-amplitude situations. A two-parameter degenerate family of bifurcated solutions was found—one example of which is illustrated in fig. 10.

15. – Generation of capillary waves by steep gravity waves.

Thus far we have neglected the effects of surface tension on the premise that they are small compared to the effects of gravity for sufficiently long waves. Whereas this assumption holds for the bulk properties of the gravity waves, it breaks down locally near the crest of very steep gravity waves where the local curvature is large and the surface tension pressure strong. In fact, short capillary waves have been observed near the crests of steep gravity waves, with and without the presence of wind [68].

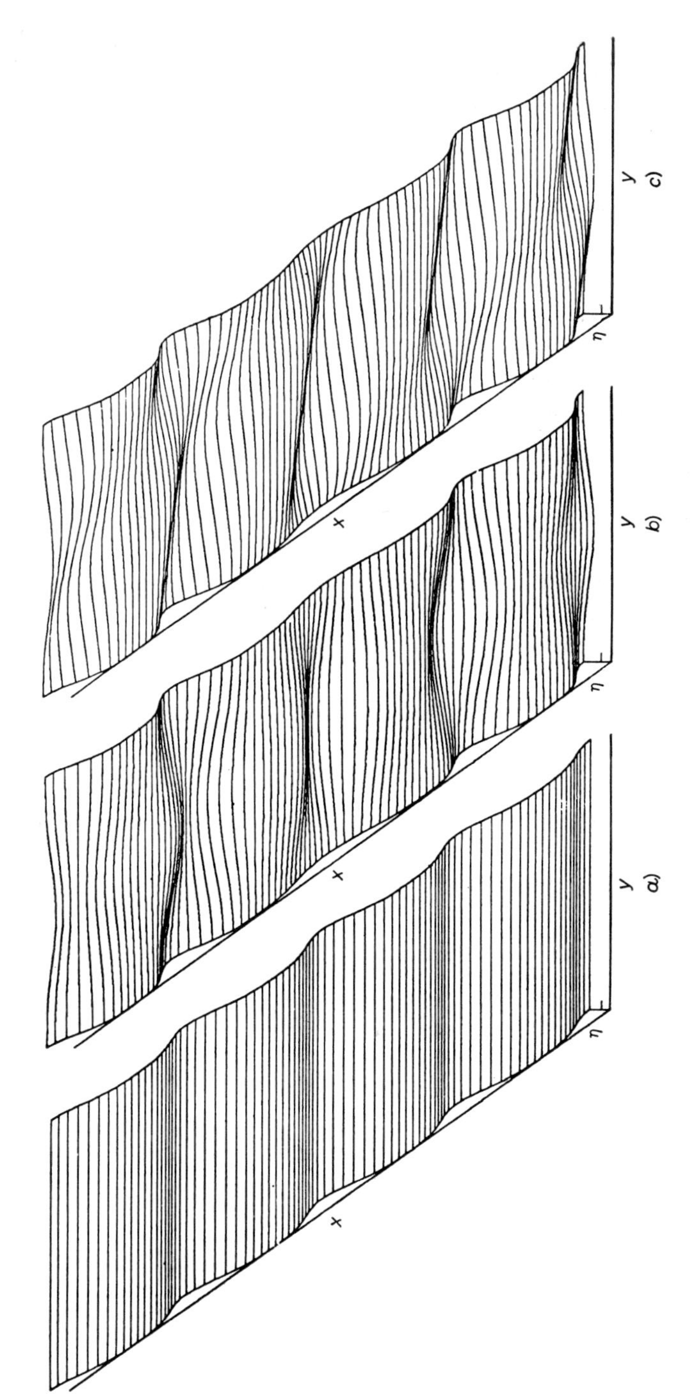

Fig. 10. – Example of steady two-dimensional bifurcated solutions from the uniform wave train solution of the nonlinear Schrödinger equation. Free surface $\eta(x, y)$ as a function of x and y. *a*) unbifurcated uniform solution, *b*) symmetric bifurcated solution, *c*) asymmetric bifurcated solution.

Theoretical considerations for the generation of short capillary waves by steep gravity waves have been given by LONGUET-HIGGINS [69]. Based on an argument by MUNK [70], LONGUET-HIGGINS modeled the crest of a steep, steady gravity wave by a localized pressure source. The situation is analogous to the fish-line problem considered by RAYLEIGH [71], in which capillary waves are found ahead of, and gravity waves behind, the steadily traveling pressure source. The relative locations of the capillary and gravity waves are caused by the fact that the group velocity is greater than the phase velocity in capillary waves and *vice versa* for gravity waves. The respective wavelengths of the capillary and gravity waves are determined by the requirement that the wave pattern remain steady with respect to the pressure source.

Applying this concept to the capillary-wave generation problem, LONGUET-HIGGINS concluded that 1) capillary waves are generated at the crest of the gravity wave and propagated down its front face; 2) the steepness of the capillary wave nearest to the crest is proportional to $\exp[-g/6T'\varkappa^2]$, where T' is the surface tension coefficient, and $\varkappa$ is the curvature of the underlying gravity wave at its crest; 3) the frequency and wavelength of the capillary waves are determined by the requirement that they remain steady with respect to the crest with the effect of the gravity wave orbital velocity taken into account; 4) viscous dissipation would act to kill off the capillary waves no more than a fraction of gravity wavelength down the front face of the gravity wave. These results were compared to observations by COX [68] and qualitative agreement was found except that there seemed to be an underprediction of the steepness of the first capillary wave.

The problem was re-examined by CRAPPER [72], who sought steady solutions of Whitham's equation describing capillary waves riding on a current field created by the steady gravity wave, with a source representing the surface tension pressure excess at the crest and with viscous dissipation. The result was substantially the same as in Longuet-Higgins' model, except that the steepness of the first capillary wave was overpredicted, rather than underpredicted. Some capillary waves were also permitted on the back side of the gravity wave.

Detailed measurements of capillary-wave generation by steep steady gravity waves were made with a high-resolution laser slope gauge under controlled conditions by CHANG, WAGNER and YUEN [73]. The results support the model of Longuet-Higgins [69], and suggest that the cause for underprediction of the lead capillary wave may be due to the fact that the crest curvature required in Longuet-Higgins' model is not given by the measured value, but is the idealized value in the absence of surface tension which is higher.

Note that the model of Longuet-Higgins represents a dynamic equilibrium reached when dissipation balances the production of capillary waves, and is not a steady situation in the usual sense. In fact, steady capillary-gravity wave solutions obtained as bifurcated states from pure gravity waves have capillary waves riding in the *troughs* of the gravity waves ([74], see also the experimental observations of Schooley [75]) and bear little resemblance to the Longuet-Higgins model. To maintain this dynamic equilibrium of the model, energy must be extracted from the gravity waves and effectively dissipated as capillary waves. LONGUET-HIGGINS [69] showed that through this mechanism the presence of capillary waves can result in a dissipation rate many times that caused by molecular viscosity acting on the pure gravity wave, and he proposed that this may be partly responsible for limiting the steepness that a real gravity wave can reach. In fact, FERGUSON, SAFFMAN and YUEN [76] showed by a model equation that the properties of capillary waves riding on gravity waves are strongly affected by the relative importance of viscosity and the degree of unsteadiness in each case.

16. – Wave breaking.

That deep-water waves « break » under certain circumstances has never been in dispute. The precise meaning of « wave breaking », however, is still unsettled. For example, one definition of wave breaking is the loss of a smooth wave profile, such as the attainment of a cornered crest in the Stokes limiting wave. Another definition is the development of a locally vertical slope, so that the wave profile becomes double-valued. Still another definition, more functional than mathematical, is the occurrence of a drastic, irreversible change in the important wave properties. This may include capillary-wave generation, although it more commonly refers to the appearance of turbulent patches with air entrainment.

In view of our lack of understanding of this complicated process, a precise, universally accepted definition of wave breaking may be too much to expect. For the present it may actually be best to choose to define wave breaking according to the particular requirements of the specific problem being addressed at the time, as long as the definition is reasonable and clearly stated.

The concept of « incipient wave breaking » was used by BANNER and PHILLIPS [77] to study the effects of wind drift on wave breaking. Incipient wave breaking is defined as occurring for a steady progressive wave when the particle velocity at the crest of the wave becomes equal to its phase velocity. Beyond this stage, it was argued, the water would outrun the wave, initiating wave breaking in the sense that water particles might run down the front face of the wave, or the wave profile might curl over. Note that this condition is also met by the Stokes limiting wave at the crest. BANNER and PHILLIPS [77] noted that, in the presence of wind, a surface drift layer is set up with thickness typically small compared to the wavelength of the steady wave under consideration. This layer has little effect on the phase velocity, but strongly affects the local particle velocity at the crest. Thus the presence of the wind drift layer can significantly increase the probability of incipient wave breaking. The same argument was applied to the situation of short steady waves in the presence of a long wave by PHILLIPS and BANNER [78]. In this case, the orbital velocity of the long wave plays the role of the wind drift.

In reality, however, deep-water gravity waves are seldom, if ever, steady. Even for the simplest situation of a single weakly nonlinear wave train, the existence of the Benjamin-Feir instability indicates that the wave train is normally in a modulated state. Since the phase velocity of the gravity wave is greater than its group velocity, the individual waves have continuously changing amplitude, slope and profile as they move and are, therefore, unsteady. In more complicated situations, where more than one wave system is involved, the concept of a steady wave system simply does not apply. It is under these inherently unsteady conditions that wave breaking is most likely to occur.

Experimental and theoretical investigations of unsteady wave breaking of waves on deep water are difficult and very few quantitative results are available. In order to study the details of unsteady wave breaking theoretically, the full set of equations (1)-(3) must be solved. This has been an impossible task analytically, and serious numerical attempts did not begin until the advent of large computers.

The first high-resolution calculations of unsteady deep-water waves of large amplitude were performed by LONGUET-HIGGINS and COKELET [79]. They took advantage of the fact that the velocity potential is completely determined by its values at the boundary, and rewrote the equations solely in terms of the free surface and tangential and normal derivatives of the velocity potential. All tangential derivatives could be evaluated by numerical differentiation, and the normal-derivative term determined by inverting an integral equation. As a consequence, only the boundary values of the velocity potential needed to be stored and updated. The savings in storage and operations permitted the use of a sufficient number of grid points to ensure high resolution while staying within the realm of economic reality. However, like other similar calculations [80-82], the time stepping led to grid scale oscillations which had to be suppressed by numerical smoothing.

LONGUET-HIGGINS and COKELET [79] presented computations of a steady wave evolving in response to an asymmetric pressure applied for half a wave period. After the wave was allowed to run free, the potential and kinetic energies oscillated strongly and the wave profile eventually achieved a vertical slope near the front of the crest and then curled over. The computation broke down shortly after this point. A subsequent paper by LONGUET-HIGGINS and COKELET [83] followed the development of individual waves in a wave train undergoing modulational instability to the point where some of them curled over. In all these calculations, the energy loss due to numerical smoothing remained acceptably small.

A generally accepted interpretation of these results is that the waves break because they are forced to contain more energy than would be contained in the steepest steady wave with the same wavelength. Calculations made by YUEN [84], however, using essentially the same method as that proposed by LONGUET-HIGGINS and COKELET [79] have shown that unsteady waves can break even when their steepness is much less than the limiting steepness and their energy much less than the corresponding limiting-wave energy. In particular, it was found that an initially symmetric sinusoidal profile with a steepness (h/λ) of 0.095 and an energy content approximately 60 % of the maximum allowable in a steady wave with the same wavelength « broke » (curled over) within one wave period. The proposition that waves can break at values of steepness and energy less than the limiting-wave levels is also supported by experiments [85] in which high-resolution laser slope gauges were used to measure unsteady waves breaking as they propagated through strong wave train modulations. These results indicate that the breaking of deep-water waves is strongly dependent upon the extent to which they are unsteady.

We must note that the numerical results described above do not take into account the effects of surface tension, which may be very important just as the wave begins to break (as described in the previous section). Since surface tension introduces a second length scale into the problem, one that is drastically different from the first, the inclusion of surface tension effects in the unsteady-wave-breaking problem will make an already difficult numerical task even more difficult.

17. – Conclusion.

In the relatively few years since the discovery of the modulational instability of weakly nonlinear, deep-water wave trains by LIGHTHILL [6] and BENJAMIN and FEIR [7], new phenomena associated with nonlinear deep-water waves have been identified at an accelerating rate. The first stage of investigations involved the properties of weakly nonlinear waves in one space dimension, and led to the discovery of the nonlinear Schrödinger equation, the envelope solitons, and the Fermi-Pasta-Ulam recurrence and its relationship to initial conditions. The second stage consisted of the attempts to extend to two space dimensions, resulting in exposing the limitations of the nonlinear Schrödinger equation and identifying Zakharov's equation. Concurrent with this line of investigations was the study of the properties of very steep waves. Much progress was made in this area, notably the stability calculations of large-amplitude waves by LONGUET-HIGGINS [39] in one dimension and by McLEAN *et al.* [86] in two dimensions, the calculation of unsteady evolution and wave breaking by LONGUET-HIGGINS and COKELET [79] and recent and stimulating topic of bifurcation of the uniform solution at large amplitude in one and two dimensions by CHEN and SAFFMAN [74] and SAFFMAN and YUEN [87, 88].

* * *

I am deeply indebted to Prof. P. G. SAFFMAN for this guidance and motivation in all aspects of this work. I am also grateful to Drs. J. CHANG, D. CRAWFORD, B. LAKE, Y. C. MA and D. U. MARTIN for invaluable discussions and contributions, and to Miss J. NAY for expert preparation of the manuscript. Much of this lecture has been adapted from [89].

REFERENCES

[1] J. SCOTT RUSSELL: *Rep. Brit. Assoc. Adv. Sci.*, **14**, 311 (1844).
[2] W. BURNSIDE: *Proc. London Math. Soc.*, **15**, 26 (1916).
[3] J. H. MICHELL: *Philos. Mag.*, **36**, 430 (1893).
[4] J. R. WILTON: *Philos. Mag.*, **26**, 1053 (1913).
[5] Lord RAYLEIGH: *Philos. Mag.*, **33**, 381 (1917).
[6] M. J. LIGHTHILL: *J. Inst. Math. Its Appl.*, **1**, 269 (1965).
[7] T. B. BENJAMIN and J. E. FEIR: *J. Fluid Mech.*, **27**, 417 (1967).
[8] G. B. WHITHAM: *J. Fluid Mech.*, **22**, 273 (1965).
[9] G. B. WHITHAM: *J. Fluid Mech.*, **27**, 399 (1967).
[10] G. B. WHITHAM: *J. Fluid Mech.*, **44**, 373 (1970).
[11] V. H. CHU and C. C. MEI: *J. Fluid Mech.*, **41**, 873 (1970).
[12] G. G. STOKES: *Trans. Cambridge Philos. Soc.*, **8**, 441 (1847).
[13] V. H. CHU and C. C. MEI: *J. Fluid Mech.*, **47**, 337 (1971).
[14] D. J. BENNEY and G. ROSKES: *Stud. Appl. Math.*, **48**, 377 (1969).
[15] W. D. HAYES: *Proc. R. Soc. London Ser. A*, **320**, 187 (1970).
[16] W. D. HAYES: *Proc. R. Soc. London Ser. A*, **320**, 209 (1970).
[17] W. D. HAYES: *Proc. R. Soc. London Ser. A*, **332**, 199 (1973).
[18] H. HASIMOTO and H. ONO: *J. Phys. Soc. Jpn.*, **33**, 805 (1972).
[19] H. C. YUEN and B. M. LAKE: *Phys. Fluids*, **18**, 956 (1975).
[20] D. J. BENNEY and A. C. NEWELL: *J. Math. Phys.* (*N. Y.*), **46**, 133 (1967).
[21] V. E. ZAKHAROV: *J. App. Mech. Tech. Phys.* (*USSR*), **2**, 190 (1968).
[22] V. E. ZAKHAROV and A. B. SHABAT: *Sov. Phys. JETP*, **65**, 997 (1972).
[23] C. S. GARDNER, J. M. GREENE, M. D. KRUSKAL and R. M. MIURA: *Phys. Rev. Lett.*, **19**, 1095 (1967).
[24] M. OIKAWA and N. YAJIMA: *J. Phys. Soc. Jpn.*, **37**, 486 (1974).
[25] T. B. BENJAMIN: *Proc. R. Soc. London. Ser. A*, **299**, 59 (1967).
[26] K. HASSELMANN: *Proc. R. Soc. London Ser. A*, **299**, 76 (1967).
[27] B. M. LAKE, H. C. YUEN, H. RUNGALDIER and W. E. FERGUSON jr.: *J. Fluid Mech.*, **83**, 49 (1977).
[28] E. FERMI, J. PASTA and J. ULAM: *Studies of nonlinear problems*, in *Collected Papers of Enrico Fermi*, Vol. **2** (Chicago, Ill., 1955), p. 978.
[29] A. C. SCOTT, F. Y. U. CHU and D. W. MCLAUGHLIN: *Proc. IEEE*, **61**, 1443 (1973).
[30] H. C. YUEN and W. E. FERGUSON jr.: *Phys. Fluids*, **21**, 1275 (1978).
[31] H. C. YUEN and W. E. FERGUSON jr.: *Phys. Fluids*, **21**, 2116 (1978).
[32] A. THYAGARAJA: *Phys. Fluids*, **22**, 2093 (1979).
[33] V. E. ZAKHAROV and A. M. RUBENCHIK: *Sov. Phys. JETP*, **38**, 494 (1974).
[34] P. G. SAFFMAN and H. C. YUEN: *Phys. Fluids*, **21**, 1450 (1978).
[35] B. I. COHEN, K. M. WATERSON and B. J. WEST: *Phys. Fluids*, **19**, 345 (1976).

[36] E. Mollo-Christensen and A. Ramamonjiarisoa: *J. Geophys. Res.*, **83**, 4117 (1978).
[37] D. U. Martin and H. C. Yuen: *Phys. Fluids*, **23**, 881 (1980).
[38] D. R. Crawford, B. M. Lake, P. G. Saffman and H. C. Yuen: *J. Fluid Mech.*, **105**, 177 (1981).
[39] M. S. Longuet-Higgins: *Proc. R. Soc. London Ser. A*, **360**, 489 (1978).
[40] D. H. Peregrine and G. P. Thomas: *Philos. Trans. R. Soc. London Ser. A*, **292**, 371 (1979).
[41] D. R. Crawford, B. M. Lake, P. G. Saffman and H. C. Yuen: *Effects of nonlinearity and spectral band width on the dispersion relation and component phase speed of surface gravity waves*, to appear in *J. Fluid Mech.* (1981).
[42] D. R. Crawford and H. C. Yuen: manuscript (1979).
[43] L. F. McGoldrick, O. M. Phillips, N. E. Huang and T. H. Hodgson: *J. Fluid Mech.*, **25**, 437 (1966).
[44] M. S. Longuet-Higgins and N. D. Smith: *J. Fluid Mech.*, **25**, 417 (1966).
[45] I. E. Alber and P. G. Saffman: *Stability of random nonlinear deep water waves with finite bandwidth spectra*, TRW Report No. 31326-6035-RU-00 (1978).
[46] I. E. Alber: *Proc. R. Soc. London Ser. A*, **363**, 525 (1978).
[47] K. Hasselmann: *J. Fluid Mech.*, **12**, 481 (1962).
[48] K. Hasselmann: *J. Fluid Mech.*, **15**, 273 (1963).
[49] D. R. Crawford, P. G. Saffman and H. C. Yuen: *Wave Motion*, **2**, 1 (1980).
[50] G. G. Stokes: *Supplement to a paper on the theory of oscillatory waves*, in *Mathematics and Physics Papers*, Vol. **1**, 314 (1880).
[51] T. H. Havelock: *Proc. R. Soc. London Ser. A*, **95**, 38 (1918).
[52] H. Yamada: *Rep. Res. Inst. Appl. Mech.*, **5**, 37 (1957).
[53] M. A. Grant: *J. Fluid Mech.*, **59**, 247 (1973).
[54] A. C. Norman: *J. Fluid Mech.*, **66**, 261 (1974).
[55] L. W. Schwartz: *J. Fluid Mech.*, **62**, 553 (1974).
[56] M. S. Longuet-Higgins: *Proc. R. Soc. London Ser. A*, **342**, 157 (1975).
[57] E. D. Cokelet: *Philos. Trans. R. Soc. London Ser. A*, **286**, 183 (1977).
[58] M. S. Longuet-Higgins and M. J. H. Fox: *J. Fluid Mech.*, **80**, 721 (1977).
[59] M. S. Longuet-Higgins and M. J. H. Fox: *J. Fluid Mech.*, **85**, 769 (1978).
[60] A. I. Nekrasov: *Izv. Ivanovo-Voznesensk. Politekh. Inst.*, 81 (1920).
[61] T. Levi-Civita: *Math. Ann.*, **93**, 264 (1925).
[62] J. F. Toland: *Proc. R. Soc. London Ser. A*, **363**, 469 (1978).
[63] Yu. P. Krasovskii: *USSR Comput. Math. Math. Phys.*, **1**, 996 (1961).
[64] G. Keady and J. Norbury: *Proc. Cambridge Philos. Soc.*, **76**, 345 (1978).
[65] P. R. Garabedian: *J. Anal. Math.*, **14**, 161 (1965).
[66] B. Chen and P. G. Saffman: *Stud. Appl. Math.*, **62**, 1 (1980).
[67] D. U. Martin: in preparation (1981).
[68] C. S. Cox: *J. Mar. Res.*, **16**, 199 (1958).
[69] M. S. Longuet-Higgins: *J. Fluid Mech.*, **16**, 138 (1963).
[70] W. Munk: *J. Mar. Res.*, **14**, 302 (1955).
[71] Lord Rayleigh: *Proc. London Math. Soc.*, **15**, 69 (1883).
[72] G. D. Crapper: *J. Fluid Mech.*, **40**, 149 (1970).
[73] J. H. Chang, R. N. Wagner and H. C. Yuen: *J. Fluid Mech.*, **86**, 401 (1978).
[74] B. Chen and P. G. Saffman: *Stud. Appl. Math.*, **60**, 183 (1979).
[75] A. H. Schooley: *J. Geophys. Res.*, **65**, 4075 (1960).
[76] W. E. Ferguson, P. G. Saffman and H. C. Yuen: *Stud. Appl. Math.*, **58**, 165 (1978).
[77] M. L. Banner and O. M. Phillips: *J. Fluid Mech.*, **65**, 647 (1974).
[78] O. M. Phillips and M. L. Banner: *J. Fluid Mech.*, **66**, 625 (1974).

[79] M. S. Longuet-Higgins and E. D. Cokelet: *Proc. R. Soc. London Ser. A*, **350**, 1 (1976).
[80] R. K. C. Chan and R. L. Street: *J. Comput. Phys.*, **6**, 68 (1970).
[81] R. L. Kulyaev: *Zh. Prikl. Mekh. Tekh. Fiz.*, **1**, 96 (1975).
[82] R. L. Kulyaev: *Zh. Prikl. Mekh. Tekh. Fiz.*, **6**, 82 (1977).
[83] M. S. Longuet-Higgins and E. D. Cokelet: *Proc. R. Soc. London Ser. A*, **364**, 1 (1978).
[84] H. C. Yuen: *Nonlinear deep-water waves. V: Unsteady wavebreaking*, TRW Report No. 31326-6013-RU-00 (1977).
[85] B. M. Lake and H. Rungaldier: in preparation (1981).
[86] J. W. McLean, Y. C. Ma, D. U. Martin, P. G. Saffman and H. C. Yuen: *Phys. Rev. Lett.*, **46**, 817 (1981).
[87] P. G. Saffman and H. C. Yuen: *Phys. Rev. Lett.*, **44**, 1097 (1980).
[88] P. G. Saffman and H. C. Yuen: *J. Fluid Mech.*, **101**, 797 (1980).
[89] H. C. Yuen and B. M. Lake: *Annu. Rev. Fluid Mech.*, **12**, 303 (1980).

Solitons and the Inverse Scattering Transform.

H. Segur

Aeronautical Research Associates of Princeton, Inc.
50 Washington Road, P.O. Box 2229, Princeton, N. J. 08540

Part I

The Physical Meaning of Equations with Solitons.

One of the important recent advances in mathematical physics has been the discovery that certain nonlinear evolution equations can be solved exactly as initial-value problems, using a method that may be called the inverse scattering transform (IST). One of the remarkable features of many of these special equations is that they admit as solutions extremely stable objects called solitons. For convenience, we may use « soliton theories » loosely to describe all of these equations solvable by IST.

It may be regarded as something of a miracle that there are nonlinear evolution equations that are completely integrable. It is a second miracle that several of these equations arise naturally as models of various physical systems, including aspects of ocean waves. My lectures at this School will describe some of the mathematical theory that has been developed in the last fifteen years to solve these special equations. Before discussing the theory, however, it may be useful to give some idea of the sense in which these soliton theories model ocean waves. Consequently, in this first part, I will try to describe the physical meaning of equations with solitons.

The spectrum of ocean waves is large and diverse, and soliton theories describe a rather small part of it. In order to see the context in which soliton theories arise, we may attempt a crude classification of ocean waves, admitting in advance that the classification probably will be incomplete.

Let us first classify ocean waves on the basis of whether the wave amplitudes are large or small. Large-amplitude waves may break, among other things. Fully nonlinear theories are needed to describe them, and will be discussed by other speakers at this School. I will restrict my attention to small-amplitude waves, because soliton theories arise in the context of small-amplitude waves. These waves are not necessarily infinitesimal, but they are small enough that we may develop weakly nonlinear theories to describe them. All of the soliton theories of water waves that have been developed to date have been weakly nonlinear theories.

Within the context of small-amplitude waves, we may classify the theories that have been developed as either stochastic or deterministic. Note that this is a classification of the theories (*i.e.* mathematics) rather than of the waves (*i.e.* physics). One of the interesting open questions in this subject is to identify the physical criteria that determine whether or not a given system of small-amplitude waves may exhibit chaotic behavior. I will return to this point in the fifth part. In the meantime, let us restrict ourselves to strictly deterministic theories.

By the time we have restricted ourselves to deterministic theories of small-amplitude waves, it begins to sound as if we will end up with the linear theory of infinitesimal waves, described in detail by LAMB [1], STOKER [2] and WEHAUSEN and LAITONE [3]. That turns out to be almost right. Soliton theories arise as (singular) perturbations of linear wave systems, and there are relations between the linear theory and the soliton theories in terms of their physical derivation, in terms of their methods of solution and in terms of their solutions themselves. Soliton theories are nonlinear, but its useful to keep re-iterating the question « How does this relate to linear theory? ».

The physical relation between linear theory and soliton theories is this. If the waves are infinitesimal, the linear theory gives a complete description of their evolution. If the waves have small but finite amplitude, then the linear theory breaks down after a finite time, and nonlinear corrections are needed to extend the range of validity of the theory to a longer time scale. Typically, soliton theories provide the nonlinear corrections to render the linear theory valid on a longer time scale. There is a short time scale on which the linear theory applies, followed by a longer time scale on which the soliton theory applies, perhaps followed by an even longer time scale on which something else applies.

What are these soliton theories? Here is a list of some of the equations that can be solved exactly by IST, and also model ocean waves.

1. – Small-amplitude waves propagating in only one spatial dimension.

a) The Korteweg-deVries (KdV) equation

$$u_t + 6uu_x + u_{xxx} = 0 \tag{1.1}$$

governs long surface (or internal) waves. (Here subscripts denote partial derivatives.) See [4-6].

b) Under other circumstances, the evolution of « long » internal waves may be governed instead by the modified Korteweg-deVries (mKdV) equation

$$u_t - u^2 u_x + u_{xxx} = 0 , \tag{1.2}$$

an equation due to BENJAMIN [7] and ONO [8]

$$u_t + u u_x + \partial_x^2 \frac{1}{\pi} ⨍ \frac{u(y, t)}{x - y} \, dy = 0 , \tag{1.3}$$

or by other models.

c) Nearly monochromatic (*i.e.* narrow-band) surface waves are governed by the nonlinear Schrödinger equation [9-11]

$$i\varphi_t + \psi_{xx} + \sigma |\psi|^2 \psi = 0 , \qquad \sigma = \pm 1 . \tag{1.4}$$

d) The sine-Gordon equation

$$\varphi_{xt} = \sigma \sin \varphi , \qquad \sigma = \pm 1 , \tag{1.5}$$

governs the waves that are nearly neutrally stable in a baroclinically unstable system on a beta-plane earth [12].

2. – Small-amplitude waves in more dimensions.

a) The equation of Kadomtsev and Petviashvili [13]

$$(u_t + u u_x + \sigma u_{xxx})_x + u_{yy} = 0 , \qquad \sigma = \pm 1 , \tag{1.6}$$

is a two-dimensional generalization of the KdV equation. Another generalization of the KdV equation (in a different limit) is the cylindrical KdV equation

$$q_t + (2t)^{-1} q - 6 q q_x + q_{xxx} = 0 . \tag{1.7}$$

b) The resonant interaction of three (narrow-band) packets of nearly resonant internal waves satisfies

$$\begin{cases} D_1 a_1 = i\gamma_1 \overset{*}{a}_2 \overset{*}{a}_3 , \\ D_2 a_2 = i\gamma_2 \overset{*}{a}_3 \overset{*}{a}_1 , \\ D_3 a_3 = i\gamma_3 \overset{*}{a}_1 \overset{*}{a}_2 , \end{cases} \tag{1.8}$$

where $D_j = \partial_t + (\boldsymbol{C}_j \cdot \nabla)$.

3. – Derivation of the KdV equation.

Clearly there is no point in trying to derive all of these equations. Rather, let us derive the KdV equation fairly carefully by a multiple time scale argument, and I simply will assert that the other equations may be derived by multiple time scale arguments as well. We begin with the following « exact » problem. We assume that the fluid is homogeneous, incompressible and inviscid. It is subject to a constant, vertical gravitational force (g), and rests on a horizontal impermeable bed at $z = -h$ (see fig. 1). The pressure vanishes at the free surface, defined by $z = \eta(x, y, t)$, where surface tension may be acting or not. (In this derivation we will omit it, but almost no qualitative changes are needed if it is included.) The motion of the fluid under these forces is assumed to be irrotational, two-dimensional ($\partial_y \equiv 0$), and either to vanish as $|x| \to \infty$ or to be periodic in x for all time.

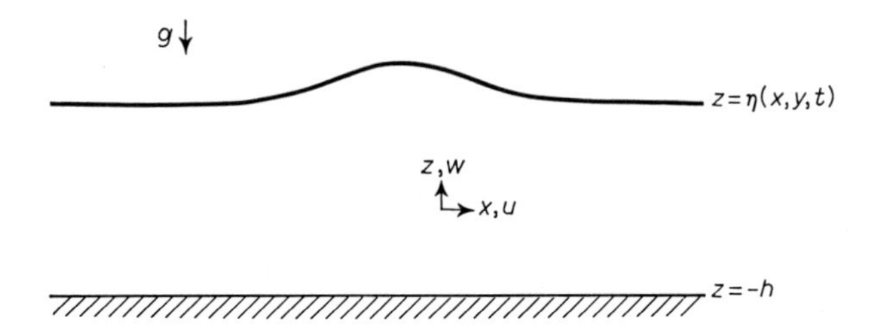

Fig. 1. – Physical configuration, showing notation for (1.9).

The governing equations under these conditions are well known (*e.g.*, [2]):

$$\nabla^2 \varphi = 0 , \qquad -h < z < \eta(x, t), \tag{1.9a}$$

$$w = \partial\varphi/\partial z = 0 , \qquad z = -h, \tag{1.9b}$$

$$\left.\begin{aligned} &\eta_t + u\eta_x = w , &(1.9c)\\ &\varphi_t + \tfrac{1}{2}|\nabla\varphi|^2 + g\eta = \text{const}, &(1.9d) \end{aligned}\right\} \qquad z = \eta(x, t),$$

and either

$$|\nabla\varphi|, \qquad \eta \to 0 , \qquad |x| \to \infty, \tag{1.9e}$$

or

$$|\nabla\varphi|,\ \eta \text{ periodic in } x \text{ with period } L .$$

These equations, along with appropriate initial conditions, uniquely determine the fluid motion, at least for some finite time. A consequence of these equations is that there are three globally conserved quantities:

$$(1.10)\qquad \begin{cases} M = \varrho\int \eta\,\mathrm{d}x\,, & \text{net mass of wave,} \\ m_x = \varrho\int\left\{\int_{-h}^{\eta} u\,\mathrm{d}z\right\}\mathrm{d}x\,, & \text{horizontal momentum,} \\ E = \varrho\left\{\int \tfrac{1}{2}g\eta^2 + \int_{-h}^{\eta}\tfrac{1}{2}|\nabla\varphi|^2\,\mathrm{d}z\right\}\mathrm{d}x\,, & \text{energy.} \end{cases}$$

To derive the KdV equation from (1.9), we must make more assumptions, to restrict the possible solutions. For this restricted class of solutions, we may replace (1.9) by a simpler problem, whose solutions will approximate some of the solutions of (1.9). In particular, the KdV equation follows by making the following assumptions.

i) Small-amplitude waves. If $\bar{\eta}$ represents a « typical » wave amplitude,

$$(1.11a)\qquad \varepsilon = \bar{\eta}/h \ll 1\,.$$

ii) Long waves. If k is a typical horizontal wave number,

$$(1.11b)\qquad (kh)^2 \ll 1\,.$$

iii) These two effects approximately balance,

$$(1.11c)\qquad (kh)^2 = O(\varepsilon)\,.$$

Two remarks are in order here. The first is that, because (1.9) uniquely determines its solution, it is not obvious that any additional assumptions are permitted. We must verify *a posteriori* that our final solution is consistent with the assumptions that led to it. Second, kh and ε are not well defined. We may use them as convenient computational tools only because they eventually drop out of the analysis.

Consistent with (1.11), we may scale the independent variables as

$$(1.12a)\qquad z^* = z/h\,, \qquad x^* = \sqrt{\varepsilon}\,x/h\,, \qquad t^* = \sqrt{\varepsilon}\,ct/h\,,$$

where $c^2 = gh$ (the only speed available in the linearized, long-wave limit). We also introduce a slow time,

$$(1.12b)\qquad \tau = \varepsilon t^*\,,$$

so that

$$\frac{\partial}{\partial t} = \sqrt{\varepsilon}\frac{c}{h}\left[\frac{\partial}{\partial t^*} + \varepsilon\frac{\partial}{\partial \tau}\right]. \tag{1.12c}$$

RAYLEIGH [14] noted that, if φ is analytic at $z = -h$, then we may expand φ in a power series there. The final result is that the solution of (1.9a), (1.9b) is

$$\varphi = \sum_{n=0}^{\infty} \frac{[-\varepsilon(1+z^*)^2]^n}{(2n)!}\left(\frac{\partial}{\partial x^*}\right)^{2n} \varphi_0(x^*, t^*, \tau). \tag{1.13}$$

If φ is analytic at $z = -h$, this series is convergent. If all of the derivatives of φ_0 are bounded, it is also asymptotic (in ε).

Next we expand the unknowns:

$$\begin{cases} \eta(x, t; \varepsilon) = h[\varepsilon\eta_1(x^*, t^*, \tau) + \varepsilon^2\eta_2 + \ldots], \\ u|_{z=-h} = \left.\dfrac{\partial\varphi}{\partial x}\right|_{z=-h} = \varepsilon\sqrt{gh}\,[u_1(x^*, t^*, \tau) + \varepsilon u_2 + \ldots]. \end{cases} \tag{1.14}$$

It follows that at the free surface, $z = \eta$,

$$\begin{cases} u = \varepsilon\sqrt{gh}\left[u_1 + \varepsilon\left(u_2 - \dfrac{1}{2}\dfrac{\partial^2 u_1}{\partial x^{*2}}\right) + O(\varepsilon^2)\right], \\ w = -\varepsilon\sqrt{\varepsilon gh}\left[\dfrac{\partial u_1}{\partial x^*} + \varepsilon\left(\dfrac{\partial u_2}{\partial x^*} + \eta_1\dfrac{\partial u_1}{\partial x^*} - \dfrac{1}{6}\dfrac{\partial^3 u_1}{\partial x^{*3}}\right) + O(\varepsilon^2)\right]. \end{cases} \tag{1.15}$$

Now we simply substitute this into (1.9c) and the tangential derivative of (1.9d), and collect terms. The result at leading order is

$$\begin{cases} \dfrac{\partial\eta_1}{\partial t^*} + \dfrac{\partial u_1}{\partial x^*} = 0, \\ \dfrac{\partial u_1}{\partial t^*} + \dfrac{\partial\eta_1}{\partial x^*} = 0. \end{cases} \tag{1.16}$$

The unique solution is

$$\begin{cases} \eta_1(x^*, t^*, \tau) = f(r, \tau) + g(l, \tau), \\ u_1(x^*, t^*, \tau) = f(r, \tau) - g(l, \tau), \end{cases} \tag{1.17}$$

where $r = x^* - t^*$, $l = x^* + t^*$. The initial data for (η, u) define (f, g), which inherit their boundedness, smoothness, etc. At this order, the solution consists of a left- and a right-traveling wave. There is no interaction and no evolution, so all waves are permanent waves.

At the next order, we obtain

$$\left\{\begin{aligned}&\frac{\partial \eta_2}{\partial t^*}+\frac{\partial u_2}{\partial x^*}+\frac{\partial \eta_1}{\partial \tau}+\frac{\partial}{\partial x^*}(u_1\eta_1)-\frac{1}{6}\frac{\partial^3 u_1}{\partial x^{*3}}=0\,,\\&\frac{\partial u_2}{\partial t^*}+\frac{\partial \eta_2}{\partial x^*}+\frac{\partial u_1}{\partial \tau}+u_1\frac{\partial u_1}{\partial x^*}-\frac{1}{2}\frac{\partial^3 u_1}{\partial x^{*2}\partial t^*}=0\,,\end{aligned}\right. \tag{1.18}$$

where (u_1, η_1) are defined by (1.17). The next step is more transparent if we eliminate u_2 (or η_2) and write the result in characteristic co-ordinates:

$$-4\frac{\partial^2\eta_2}{\partial r\,\partial l}-\frac{\partial}{\partial r}\left[2f_\tau+3ff_r+\frac{1}{3}f_{rrr}\right]+[2g_l f_r+gf_{rr}+g_{ll}f]-\\-\frac{\partial}{\partial l}\left[-2g_\tau+3gg_l+\frac{1}{3}g_{lll}\right]=0\,. \tag{1.19}$$

The dependence of each term on (r, l) is explicit, so (1.19) can be integrated easily. It is apparent that η_2 will grow linearly both in r and in l unless we require

$$\left\{\begin{aligned}&2f_\tau+3ff_r+\tfrac{1}{3}f_{rrr}=0\,,\\&-2g_\tau+3gg_l+\tfrac{1}{3}g_{lll}=0\,,\end{aligned}\right. \tag{1.20}$$

which can be reascaled to the form (1.1) if desired. We must also require that

$$f\,,\quad f_r\,,\quad \int_{-\infty}^{\infty} f\,\mathrm{d}r\,,\quad g\,,\quad g_l\,,\quad \int_{-\infty}^{\infty} g\,\mathrm{d}l \tag{1.21a}$$

all be bounded if $-\infty<x<\infty$, and that

$$\oint f\,\mathrm{d}r=0=\oint g\,\mathrm{d}l \tag{1.21b}$$

in the periodic problem. Then η_2 is bounded:

$$\eta_2(r, l, \tau)=\tfrac{1}{4}\left[g_l\int^r f\,\mathrm{d}\hat{r}+f_r\int^l g\,\mathrm{d}\hat{l}+2fg\right]+f_2(r,\tau)+g_2(l,\tau)\,. \tag{1.22a}$$

Similarly,

$$u_2=-\tfrac{1}{4}\left[g_l\int^r f\,\mathrm{d}\hat{r}-f_r\int^l g\,\mathrm{d}\hat{l}\right]-\tfrac{1}{4}(f^2-g^2)+\tfrac{1}{3}(f_{rr}-g_{ll})+f_2-g_2\,. \tag{1.22b}$$

The conditions (1.21) guarantee that the left- and right-going waves do not affect each other long enough to interact strongly on this time scale $(\tau = O(1))$. However, each of the two wave trains does interact with itself for a long time, and the two KdV equations govern the evolution of each wave train on this longer time scale. This is the main point of the derivation. The KdV equation governs the evolution on a slow time scale of a small-amplitude wave that satisfied the linear wave equation on a fast time scale.

To stop at $O(\varepsilon^2)$, one may set $f_2 = g_2 = 0$, and (1.22) gives the second-order corrections of the unknowns. To go on to third order, it is necessary to use f_2 and g_2 to eliminate secular terms at the next order. In principle, the expansion can be carried to arbitrarily high order, although this is rarely done in practice.

We will discuss how to solve the KdV equation in the second part. For the moment, we consider only some of the implications of this derivation. The first is that both the wave equation (1.16) and the KdV equation (1.20) are ε-independent. The expansion is intended to be valid in the limit $\varepsilon \to 0$, and we may take this limit without emasculating the governing equations. That is indicative that the KdV equation is a true asymptotic equation.

A second consequence of this derivation concerns the conservation laws. It follows from (1.20) that, if f vanishes rapidly as $|r| \to \infty$, or is periodic, then

$$I_1 = \int f \, dr , \qquad I_2 = \int f^2 \, dr , \qquad I_3 = \int [f^3 - \tfrac{1}{3}(f_r)^2] \, dr \tag{1.23}$$

all are conserved. In fact, the KdV equation has an infinite number of conservation laws. These may be reconciled with the three physical conservation laws in the following way. We obtained (1.20) by expanding the dependent variables, as in (1.14). If we also expand the mass integral (for example) in powers of ε, each coefficient in that expansion must also be conserved. The infinity of conserved quantities for the KdV equation are related to these coefficients. For example, we may easily verify that for the right-going waves (*i.e.* $g \equiv 0$), to $O(\varepsilon^2)$,

$$\left\{ \begin{aligned} M &= \frac{\varrho h^2}{\sqrt{\varepsilon}} [\varepsilon I_1 + O(\varepsilon^3)] , \\ m_x &= \varrho h^2 \sqrt{gh/\varepsilon} \left[\varepsilon I_1 + \frac{3}{4} \varepsilon^2 I_2 + O(\varepsilon^3) \right] , \\ \mathrm{KE} &= \frac{\varrho g h^3}{2\sqrt{\varepsilon}} [\varepsilon^2 I_2 + \varepsilon^3 I_3 + O(\varepsilon^4)] , \\ \mathrm{PE} &= \frac{\varrho g h^3}{2\sqrt{\varepsilon}} [\varepsilon^2 I_2 + O(\varepsilon^4)] , \\ E &= \mathrm{KE} + \mathrm{PE} . \end{aligned} \right. \tag{1.24}$$

PART II

Introduction to the Inverse Scattering Transform.

The first part was devoted to the physical derivation of equations that admit solitons. We saw that, even though the equations themselves are fully nonlinear, they typically arise in physical problems by eliminating secular terms in a weakly nonlinear theory.

This part is devoted to the method of solution of these special equations. This method goes under a variety of names, including the inverse scattering transform. It turns out that this method can be viewed as a generalization of Fourier analysis to certain nonlinear problems. It provides the exact solution to certain nonlinear evolution equations, just as the Fourier transform does for certain linear evolution equations.

The outline of this part is first to review briefly the method of Fourier transforms for linear problems, then to sketch how the IST works for certain nonlinear problems, and to show that it is a generalization of Fourier analysis. In the process we can say something about the solutions of these equations, and the class of equations to which the method applies. This approach is essentially that of [15].

1. – Linear evolution equations.

Let us consider three examples of linear problems on the infinite interval $(-\infty < x < \infty)$

$$a) \quad u_t + u_{xxx} = 0\,, \tag{2.1a}$$

$$b) \quad iu_t + u_{xx} = 0\,, \tag{2.1b}$$

$$c) \quad u_{TT} - u_{XX} + u = 0 \qquad \text{or} \qquad u_{xt} = u\,. \tag{2.1c}$$

Equation (2.1a) may be called the Airy equation; it arises in certain problems in optics. Equation (2.1b) is the time-dependent Schrödinger equation, with no potential. Equation (2.1c) is the Klein-Gordon equation, written both in laboratory and in characteristic co-ordinates. In all three cases, the equation holds for all real x, initial data also must be given at $t = 0$, and we will require that the initial data be smooth and decay rapidly as $|x| \to \infty$.

These equations all can be solved by Fourier-transform methods. The first step in that approach is to map the initial data into its Fourier transform:

$$\hat{u}(k) = \int_{-\infty}^{\infty} u(x, 0) \exp[-ikx]\,dx . \tag{2.2}$$

As t changes, $u(x, t)$ evolves according to a *partial* differential equation, but $\hat{u}(k, t)$ satisfies an *ordinary* differential equation. (This is precisely the advantage of Fourier transforms.) The equation is so simple that we often skip that step and simply look for solutions in the form

$$u(x, t) = \frac{1}{2\pi} \int \hat{u}(k) \exp[ikx - i\omega t]\,dk . \tag{2.3}$$

Substituting (2.3) into the (linear) evolution equation yields the (linear) *dispersion* relation, $\omega(k)$. In particular, for our example problems,

$$\begin{cases} a) & \omega = -k^3 , \\ b) & \omega = k^2 , \\ c) & \omega = 1/k . \end{cases} \tag{2.4}$$

If the problem has a dispersion relation, it contains all of the information that was in the original partial differential equation. For example:

i) For a first-order (in time) equation, if $\omega(k)$ is real for real k, then the original problem has an « energy » integral that is conserved. In our example problems,

$$\begin{cases} a) & \partial_t \int_{-\infty}^{\infty} |u|^2\,dx = 0 , \\ b) & \partial_t \int |u|^2\,dx = 0 , \\ c) & \partial_t \int |u_x|^2\,dx = 0 , \end{cases} \tag{2.5}$$

ii) If $d^2\omega/dk^2 \neq 0$, then the problem is « dispersive ». Each wave number travels with its own speed, and the waves sort themselves out in time. (These concepts are discussed in detail by STOKER [2] and by WHITHAM [16], among others.)

The net effect is that, in the long-time limit, the solutions of each of these equations have rather characteristic features determined largely by the group velocity $d\omega/dk$. For example, a typical solution of (2.1a) for large times is shown in fig. 2. The waves are spreading slowly to the left, because $d\omega/dk \leqslant 0$. Energy is conserved, because $\omega(k)$ is real, so, as the waves spread out, the amplitude tends to zero (for (2.1a), as $t^{-\frac{1}{3}}$ or faster).

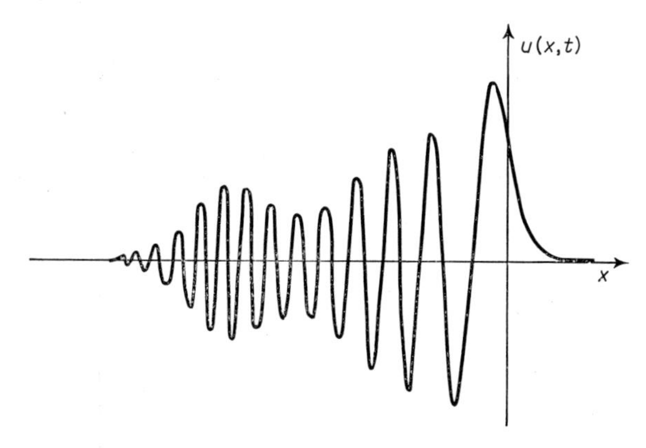

Fig. 2. – Typical long-time solution of (2.1), with the wave train spreading to the left. Modulations of the wave train are determined by the initial data.

Thus the solutions of these conservative, dispersive problems are comparatively simple for large times. They are characterized by the waves dispersing over larger and larger regions of space, with the amplitudes decaying as required by energy conservation. The method to obtain these solutions depends on the existence of two functions: $\hat{u}(k)$ represents the initial data, and $\omega(k)$ represents the evolution equation. This method of solution (Fourier transforms) may be represented schematically as follows:

$$u(x, 0) \to \hat{u}(k, 0) \xrightarrow{\omega(k)} \hat{u}(k, t) \to u(x, t) . \tag{2.6}$$

2. – Nonlinear evolution equations.

Consider next three nonlinear generalizations of the equations in (2.1):

a) the Korteweg-deVries equation

$$u_t + 6uu_x + u_{xxx} = 0 , \tag{2.7}$$

b) the nonlinear Schrödinger equation

$$iu_t + u_{xx} + 2|u|^2 u = 0 , \tag{2.8}$$

c) the sine-Gordon equation

$$u_{xt} = \sin u . \tag{2.9}$$

Why these examples? First, each of these equations linearizes to one of the linear equations in (2.1). Second, they have the same energy integrals as their linear counterparts. Third, each arises as a model of some aspect of ocean waves. Fourth, they all possess solitary-wave solutions. For the KdV equation

$$u(x, t) = 2\varkappa^2 \operatorname{sech}^2 [\varkappa(x - x_0 - 4\varkappa^2 t)], \tag{2.10}$$

where $\varkappa$ and x_0 are free constants (see fig. 3). Finally, and most importantly, each of these equations can be solved by the inverse scattering transform (IST).

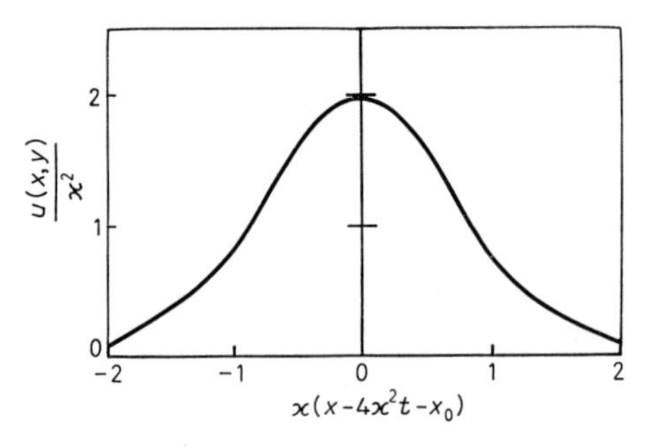

Fig. 3. – Soliton solution of the KdV equation (2.7).

Understanding how the IST works requires some knowledge of direct and inverse scattering theory. Here is an extremely superficial introduction for the Schrödinger scattering problem. For a very thorough treatment, see [17]. Let $q(x)$ be given, satisfying

$$\int_{-\infty}^{\infty} (1 + x^2)|q| \, dx < \infty . \tag{2.11}$$

The direct scattering problem is to find pairs $[\lambda, \psi(x; \lambda)]$ such that

$$\psi_{xx} + [\lambda + q(x)]\psi = 0 , \qquad -\infty < x < \infty. \tag{2.12}$$

Here λ is a real number, an « eigenvalue », and ψ is required to be a bounded function satisfying certain boundary conditions. For $\lambda > 0$, we may set $\lambda = k^2$, and require that

$$\begin{cases} \psi \sim a(k) \exp[-ikx] & \text{as } x \to -\infty, \\ \psi \sim \exp[-ikx] + b(k) \exp[ikx] & \text{as } x \to +\infty. \end{cases} \tag{2.13}$$

These are related by $|a|^2 + |b|^2 = 1$ (a Wronskian relation). (In the scattering context, these solutions are called « radiation », $a(k)$ is called the « transmission coefficient » and $b(k)$ the « reflection coefficient ».) For $\lambda < 0$, there are only a finite number of discrete eigenvalues (« bound states »). We may set $\lambda = -\varkappa_n^2$, and normalize $\int |\psi_n|^2 \, dx = 1$. The boundary conditions are

$$\begin{cases} \psi_n \sim d_n \exp[\varkappa_n x] & \text{as } x \to -\infty, \\ \psi_n \sim c_n \exp[-\varkappa_n x] & \text{as } x \to +\infty. \end{cases} \tag{2.14}$$

A given « potential », $q(x)$, generates certain scattering data $[a(k), b(k); c_n, \varkappa_n]$. These may be collected into a single function,

$$B(x) = \frac{1}{2\pi} \int b(k) \exp[ikx] \, dk + \sum_{n=1}^{N} c_n^2 \exp[-\varkappa_n x] . \tag{2.15}$$

The direct scattering problem is to find $B(x)$ for a given $q(x)$.

The inverse scattering problem is, given $B(x)$, to find the potential, $q(x)$, that generated it. This is an interesting mathematical question that was solved in a slightly different form in a famous paper by GEL'FAND and LEVITAN [18]. If all intermediate details are left out, the final result is that one must solve a linear integral equation

$$K(x, y) + B(x + y) + \int_x^{\infty} K(x, z) B(z + y) \, dz = 0 , \qquad y > x. \tag{2.16}$$

(An equation of this form now is called a Gel'fand-Levitan equation.) Then the solution of the inverse scattering problem is

$$q(x) = 2 \frac{d}{dx} K(x, x) . \tag{2.17}$$

Thus we may represent the direct and inverse scattering problems by

$$q(x) \to B(x + y) \to K(x, y) \to K(x, x) \to q(x) . \tag{2.18}$$

After that rather long detour, let us return to nonlinear evolution equations, and come to the main point. In 1967, GARDNER, GREENE, KRUSKAL and MIURA [19] made a remarkable discovery about the KdV equation (2.7). Denote its initial data by $u(x, 0)$ and consider

$$\psi_{xx} + [\lambda + u(x, 0)] \psi = 0 . \tag{2.19}$$

In this way, $u(x, 0)$ is mapped into scattering data, summarized by $B(2x, 0)$. As t changes, $u(x, t)$ evolves according to (2.7), and, of course, the scattering data change as well. The remarkable fact is that, if u satisfies (2.7), then

$$\partial_t \lambda = 0 . \tag{2.20}$$

In other words, $u(x, t)$ evolves through a family of potentials, all of which have exactly the same spectrum! (For this reason, CALOGERO and DEGASPERIS [20] prefer to let IST abbreviate the « iso-spectral transform ».) The rest of the scattering data also evolves simply:

$$\begin{cases} \dfrac{\partial}{\partial t} a(k, t) = 0 , \\ \dfrac{\partial}{\partial t} b(k, t) = 8ik^3 b , \\ \dfrac{\partial}{\partial t} c_n(t) \quad = 4\varkappa_n^3 c_n , \end{cases} \tag{2.21}$$

so that

$$B(2x, t) = \frac{1}{2\pi}\int_{-\infty}^{\infty} b(k) \exp[2ikx + 8ik^3 t]\, dk + \sum_{n=1}^{N} c_n^2 \exp[8\varkappa_n^3 t - 2\varkappa_n x] . \tag{2.22}$$

Therefore, $u(x, t)$ satisfies a nonlinear evolution equation (KdV), but $B(2x, t)$ satisfies a linear evolution equation whose dispersion relation is that of the linearized KdV equation,

$$\omega(2k) = -8k^3 . \tag{2.23}$$

But now we know $B(2x, t)$ for any t, and we may reconstruct the new potential, $u(x, t)$, *via* inverse scattering. To summarize, we have a method to solve the KdV equation:

$$u(x, 0) \to B(x+y; 0) \xrightarrow{\omega(2k)} B(x+y; t) \to K(x, y; t) \to u(x, t) . \tag{2.24}$$

In fact, this scheme represents the method of solution for all problems solvable by IST. The steps are:

i) Map the initial data of the nonlinear evolution equation into the scattering data, using the linear scattering problem.

ii) Let the scattering data evolve, in accord with the dispersion relation of the linearized evolution equation.

iii) Reconstruct the solution of the nonlinear evolution equation at a later time by solving a linear integral equation.

Notice that each step in this method is linear, and that the whole procedure parallels the method of Fourier transforms for linear problems. In this sense, the IST is a generalization of Fourier analysis to certain nonlinear problems.

What about the solutions of an equation solvable by IST on $-\infty < x < \infty$? The scattering data consist of a discrete spectrum (bound states) and a continuous spectrum (radiation), and these represent different kinds of solutions. Each discrete eigenvalue represents one solitary wave (or « soliton », since they are no longer required to be solitary):

$$B \sim c_n^2 \exp[-\varkappa_n x] \Rightarrow u \sim 2\varkappa_n^2 \operatorname{sech}^2\{\varkappa_n(x - x_n - 4\varkappa_n^2 t)\}, \tag{2.25}$$

where $x_n = x_n(c_n)$. The permanence of these waves is ensured by the fact that the eigenvalues $(-\varkappa_n^2)$ are time independent. Notice that each of these waves has positive speed, and that the bigger waves move faster.

The continuous spectrum requires a little more care, but in a crude sense it represents a part of the solution that behaves almost as if the problem were linear. More precisely, it represents a dispersive wave train in which the linearized group velocity (which is negative for the KdV equation) plays an important role. The net result is that the long-time solution of the KdV equation is comparatively simple, as shown in fig. 4. It may be worth emphasizing that, even though the equation is fully nonlinear, its long-time behavior may be predicted to any desired accuracy. The problem is stable, and there are no chaotic solutions.

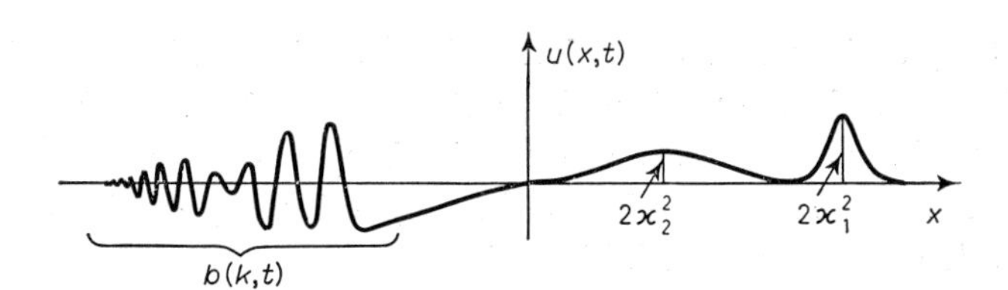

Fig. 4. – Typical long-time KdV solution. The solitons are determined by the discrete spectrum ($\varkappa_1^2$ and $\varkappa_2^2$ in this case), while the radiation is associated with the continuous spectrum ($b(k)$).

3. – Generalizations.

The reader may have the impression by now that IST is a miracle that works. To a certain extent that is true, but that is no reason not to use it wherever possible. The question is « To what problems does IST apply? » Next we show that there are infinitely many problems solvable by IST. (So even though it is a miracle, it may not be uncommon.) Let us consider a different scattering problem on $-\infty < x < \infty$:

$$\begin{cases} \partial_x V_1 + i\zeta V_1 = q V_2, \\ \partial_x V_2 - i\zeta V_2 = r V_1, \end{cases} \tag{2.26}$$

with q, $r \to 0$ as $|x| \to \infty$. This problem was first analyzed by ZAKHAROV and SHABAT [21] for $r = -q^*$ (* denotes complex conjugate). Note that, if $r = 1$, (2.26) can be reduced to (2.12). Note further that, if $r = 0$ and $V_1 \exp[i\zeta x] \to 1$ as $x \to -\infty$, then $\lim_{x\to+\infty} V_1 \exp[i\zeta x]$ is just the Fourier transform of $q(x)$. In this sense (2.26) is already a generalization of Fourier transforms.

To construct an IST out of (2.26), we need $[q(x, t), r(x, t)]$ to evolve so that $\partial_t \zeta = 0$, *i.e.* we will force the eigenvalue to be time independent. To do this, we allow the eigenfunctions to evolve according to linear equations:

$$\left\{\begin{aligned} \partial_t V_1 &= A V_1 + B V_2 , \\ \partial_t V_2 &= C V_1 - A V_2 , \end{aligned}\right. \tag{2.27}$$

where $A = A(q, r; \zeta)$, etc. Compatibility of (2.26) and (2.27) requires that $(V)_{xt} = (V)_{tx}$. Demanding $\partial_t \zeta = 0$ yields

$$\left\{\begin{aligned} A_x \qquad\quad &= qC - rB , \\ B_x + 2i\zeta B &= q_t - 2Aq , \\ C_x - 2i\zeta C &= r_t + 2Ar . \end{aligned}\right. \tag{2.28}$$

If (A, B, C) satisfy these coupled ordinary differential equations (in x), then the eigenvalue is constant, and we can construct an IST based on (2.26). Boundary conditions for (2.28) are obtained by comparing (2.26), (2.27) as $|x| \to \infty$. Because $(q, r) \to 0$, we know that $V_1 \sim C_1 \exp[-i\zeta x]$, $V_2 \sim C_2 \exp[i\zeta x]$ from (2.26). To ensure compatibility with (2.27), we require

$$A \to A_0(\zeta) , \qquad B, C \to 0 \qquad \text{as } |x| \to \infty. \tag{2.29}$$

This gives six boundary conditions for (2.28), which is now over-determined. Therefore, (2.28), (2.29) will have no solution except in special cases.

Given (2.26), the choice of nonlinear evolution equations is determined by $A_0(\zeta)$. Here we use the linearized dispersion relation

$$\omega(2\zeta) = 2iA_0(-\zeta) . \tag{2.30}$$

For example, if $A_0(\zeta) = -2i\zeta^2$, then (2.28) has a solution only if

$$\left\{\begin{aligned} iq_t + q_{xx} - 2q^2 r &= 0 , \\ ir_t - r_{xx} + 2qr^2 &= 0 . \end{aligned}\right. \tag{2.31}$$

If $r = -q^*$, each of these reduce to the nonlinear Schrödinger equation (2.8).

Theorem [15]. Let $\omega(k)$ be a ratio of entire functions, and be real for real k. Then (2.26) generates a nonlinear evolution equation that is solvable by IST, and whose linearized dispersion relation is $\omega(k)$. If $\omega(k)$ also is an odd function of k, then (2.12) generates a different nonlinear equation solvable by IST. Its linearized dispersion relation also is $\omega(k)$.

If your objective is to generate nonlinear equations solvable by IST, you need two ingredients:

i) a scattering problem and

ii) a linearized dispersion relation.

Each such pair generates one nonlinear problem solvable by IST. If your objective is to solve a particular equation, this is almost no help at all. In the fifth part, I will discuss how to determine whether a given equation can be solved by IST.

Part III

More Inverse Scattering on the Infinite Interval.

In the second part we saw that the inverse scattering transform works, and that it may be regarded as a generalization of the Fourier transform to certain nonlinear problems. For the KdV equation (2.7), IST may be represented schematically as follows:

$$\left\{\begin{array}{ccc} u(x,0) & \xrightarrow{\text{direct scattering}} & \{a(k,0),\, b(k,0),\, \varkappa_n,\, c_n(t)\} \\ & & \downarrow \\ & & \left[\begin{array}{ll} \dfrac{\partial a}{\partial t}=0\,, & \dfrac{1}{b}\dfrac{\partial b}{\partial t}=8ik^3 \\ \dfrac{\partial \varkappa_n}{\partial t}=0\,, & \dfrac{1}{c_n^2}\dfrac{\partial c_n^2}{\partial t}=8\varkappa_n^3 \end{array}\right] \\ & & \downarrow \\ u(x,t) & \xleftarrow[\text{inverse scattering}]{} & \{a(k,t),\, b(k,t),\, \varkappa_n,\, c_n(t)\}\,. \end{array}\right. \tag{3.1}$$

In this part we will examine IST on the infinite line in more detail. The extra information may help to explain why IST works and what sort of solutions it admits.

1. – Hamiltonian mechanics.

ZAKHAROV and FADDEEV [22] pioneered a description of IST as a canonical transformation of a Hamiltonian system to action-angle variables. This description is an alternative to that of IST as Fourier analysis for nonlinear problems. Both are legitimate; which is preferable is a matter of taste.

The reader may recall that Hamiltonian mechanics is simply a variational description of certain dynamical systems. (Basic references here are [23, 24].) In this formulation, one identifies generalized « co-ordinates » (q) and « momenta » (p), which describe completely the state of the system. If the problem has N degrees of freedom, then p and q are each N-dimensional vectors. For the partial differential equations of interest here, p and q are infinite-dimensional. One introduces a Hamiltonian, $H(p, q)$, which must have the property that Hamilton's equations

$$\dot{q} = \frac{\delta H}{\delta p}, \qquad \dot{p} = -\frac{\delta H}{\delta q} \tag{3.2}$$

are equivalent to the equations of motion of the system. Here $\dot{} = \partial_t(\;)$, and the derivatives of H in (3.2) are functional derivatives. In general H may depend on time, but for any of the conservative systems under consideration it does not. Any system that has such a variational formulation is said to be Hamiltonian.

It happens that problems solvable by IST are Hamiltonian. As examples, consider

$$\begin{cases} H_1 = -\int \{pq_x^2 + p^2 q_x - p_x q_{xx}\}\, dx\,, \\[2ex] H_2 = -\, i\int \{q_x p_x + p^2 q^2\}\, dx\,. \end{cases} \tag{3.3}$$

H_1 yields two evolution equations, which admit the identification $p = q_x$. Each equation reduces to the KdV equation under this identification. Similarly, H_2 gives the nonlinear Schrödinger equation if we identify $p = \pm q^*$ (complex conjugate).

Canonical transformations play an important role in Hamiltonian mechanics. Roughly speaking, a canonical transformation is simply a change of variables,

$$(p, q) \to (P, Q)\,, \tag{3.4}$$

that preserves certain projections of the volume of phase space (this corresponds to a mapping that is 1-1 and onto). One may check this property either by means of Poisson brackets (old-fashioned) or a symplectic form (newfangled). Necessarily, a canonical transformation does not affect the form of Hamilton's equations:

$$\mathcal{H}(P, Q): \; \dot{Q} = \frac{\delta \mathcal{H}}{\delta P}, \qquad \dot{P} = -\frac{\delta \mathcal{H}}{\delta Q}. \tag{3.5}$$

Out of all possible canonical transformations, an especially desirable one is one in which the new Hamiltonian is independent of Q. These are called action (P) and angle (Q) variables. Obviously, if $\mathcal{H} = \mathcal{H}(P)$, then from (3.5)

$$\dot{P} = 0, \qquad \dot{Q} = \frac{\delta \mathcal{H}}{\delta p} = \text{const}, \tag{3.6}$$

so that integration of the equations is as simple as possible. Another way to say this is that, if a Hamiltonian system has action-angle variables, then its motion is basically very simple, when viewed appropriately. Unfortunately, there is no general method known to determine whether a system has action-angle variables, or to find them if they exist.

The valuable insight of Zakharow and Faddeev [22] was that (3.1), the equations that describe the evolution of the scattering data for the KdV equation, was in the form of (3.6).

Theorem [22]. Let $u(x, t)$ represent a KdV solution on $-\infty < x < \infty$. The IST mapping

$$u(x, t) \to \left\{ \begin{array}{ll} P(k) = (k/\pi) \ln |a(k)|^2, & Q(k) = \operatorname{Im} \ln b(k) \\ P_n \;\;\; = -2\varkappa_n^2, & Q_n \;\;\; = \ln c_n \end{array} \right\} \tag{3.7}$$

is a canonical transformation to action-angle variables.

From this standpoint, it is not surprising that the KdV equation has an infinite set of constants of the motion. They are a representation of the (time-independent) action variables. Moreover, given that the equation on $-\infty < x < \infty$ has an asymptotic state, it is not surprising that the asymptotic behavior is relatively simple. The existence of action-angle variables means that motion is basically simple (when viewed appropriately); the simple asymptotic behavior reveals the basically simple motion.

2. – Scattering theory.

Another apparent miracle in IST is the fact that scattering theory works so well. Even after accepting that the potential in the scattering problem (*e.g.*, $u(x)$ in (2.12)) is determined by the scattering data, one is still surprised at the simplicity of the inversion procedure. In fact, scattering theory works as well as it does because it uses the powerful theory of analytic functions of a complex variable.

We may illustrate this close connection to the theory of analytic functions by considering a scattering problem due to ZAKHAROV and SHABAT [21] and generalized by ABLOWITZ, KAUP, NEWELL and SEGUR [15]:

$$\text{(3.8)} \qquad \begin{cases} \partial_x V_1 + i\zeta V_1 = q(x)\, V_2\,, \\ \partial_x V_2 - i\zeta V_2 = r(x)\, V_1\,. \end{cases}$$

Here $-\infty < x < \infty$, and we assume that $q(x)$, $r(x)$ vanish rapidly as $|x| \to \infty$. This scattering problem is appropriate for the nonlinear Schrödinger equation, the sine-Gordon equation and infinitely many other problems. However, as time dependence is not germane to scattering theory *per se*, we will hold time fixed for the current discussion. Thus (q, r) may be considered known functions that are absolutely integrable.

Solutions of (3.8) may be identified by the boundary conditions they satisfy, and we define for real ζ

$$\text{(3.9)} \qquad \begin{cases} \varphi = \begin{pmatrix} \varphi_1(x, \zeta) \\ \varphi_2(x, \zeta) \end{pmatrix} \to \begin{pmatrix} 1 \\ 0 \end{pmatrix} \exp[-i\zeta x] & \text{as } x \to -\infty\,, \\ \bar{\varphi} \to \begin{pmatrix} 0 \\ -1 \end{pmatrix} \exp[i\zeta x] & \text{as } x \to -\infty\,, \\ \psi \to \begin{pmatrix} 0 \\ 1 \end{pmatrix} \exp[i\zeta x] & \text{as } x \to +\infty\,, \\ \bar{\psi} \to \begin{pmatrix} 1 \\ 0 \end{pmatrix} \exp[-i\zeta x] & \text{as } x \to +\infty\,. \end{cases}$$

For real ζ, φ and $\bar{\varphi}$ are linearly independent for all real x, as are ψ and $\bar{\psi}$. Because (3.8) has only two linearly independent solutions, we may define $\{a(\zeta), \bar{a}(\zeta), b(\zeta), \bar{b}(\zeta)\}$ by

$$\text{(3.10)} \qquad \begin{cases} \varphi = a\bar{\psi} + b\psi\,, \\ \bar{\varphi} = \bar{b}\bar{\psi} + \bar{a}\psi\,. \end{cases}$$

These are related by a Wronskian relation, $a\bar{a} + b\bar{b} = 1$. From (3.9) and (3.10)

$$\text{(3.11)} \qquad \varphi \to \begin{pmatrix} a(\zeta) \exp[-i\zeta x] \\ b(\zeta) \exp[i\zeta x] \end{pmatrix} \qquad \text{as } x \to +\infty\,.$$

The set $\{a, \bar{a}, b, \bar{b}\}$ for real ζ makes up part of the scattering data. These functions may be viewed as representing the asymptotic behavior of certain solutions of (3.8) as in (3.11), or simply as Wronskians of solutions from (3.10).

Once these functions have been defined for real ζ, they may be extended into the complex ζ-plane. ABLOWITZ, KAUP, NEWELL and SEGUR [15] proved that

i) $\varphi(x, \zeta) \exp[i\zeta x]$ and $\psi(x, \zeta) \exp[-i\zeta x]$ are analytic functions of ζ for all real x if $\operatorname{Im} \zeta > 0$;

ii) $\bar{\varphi} \exp[-i\zeta x]$ and $\bar{\psi} \exp[i\zeta x]$ are analytic in ζ if $\operatorname{Im} \zeta < 0$;

iii) $a(\zeta) = \lim_{x\to\infty} \varphi_1 \exp[i\zeta x]$ is analytic for $\operatorname{Im} \zeta > 0$, and $\bar{a}(\zeta)$ is analytic for $\operatorname{Im} \zeta < 0$;

iv) the discrete eigenvalues of (3.8) are zeros of $a(\zeta)$ or $\bar{a}(\zeta)$ in their regions of analyticity;

v) as $|\zeta| \to \infty$, $\operatorname{Im} \zeta > 0$, $a(\zeta) \to 1 + O(\zeta^{-1})$,

$$\varphi_2 \exp[i\zeta x] \to -r(x)/2i\zeta + O(\zeta^{-2}),$$

$$\psi_1 \exp[-i\zeta x] \to q(x)/2i\zeta + O(\zeta^{-2}),$$

with similar results for $\operatorname{Im} \zeta < 0$.

Most of the results of IST are a consequence of these relations and the time dependence of the scattering data. In particular, one finds that $a(\zeta)$ and $\bar{a}(\zeta)$ are time independent. Then it follows from iv) that the discrete eigenvalues are time independent as well (the isospectral property). It follows from v) that $\ln a(\zeta)$ has an asymptotic expansion as $|\zeta| \to \infty$ for $\operatorname{Im} \zeta > 0$:

$$\ln a(\zeta) = \sum_{n=1}^{\infty} I_n/(2i\zeta)^n . \tag{3.12}$$

Because $a(\zeta)$ is time independent, the coefficients in this expansion must be time independent as well. These are the infinite set of conserved integrals of any of the evolution equations solved by (3.8). It also follows from v) that, if we can reconstruct $\varphi_2 \exp[i\zeta x]$ and $\psi_1 \exp[-i\zeta x]$ from the scattering data, then $q(x)$ and $r(x)$ may be obtained by taking a limit. We show next that inverse scattering theory follows just that strategy.

To obtain the linear integral equations that are the heart of inverse scattering theory, write (3.10a) as

$$\frac{\varphi \exp[i\zeta x]}{a} = \bar{\psi} \exp[i\zeta x] + \frac{b}{a} \psi \exp[i\zeta x], \qquad \zeta \text{ real}. \tag{3.13}$$

For simplicity, let us assume that $a(\zeta)$ has no zeros for $\mathrm{Im}\,\zeta \geqslant 0$. Then $\varphi \exp[i\zeta x]/a$ is analytic for $\mathrm{Im}\,\zeta > 0$ and vanishes as $|\zeta| \to \infty$ there, while $\bar{\psi} \exp[i\zeta x]$ has similar properties for $\mathrm{Im}\,\zeta < 0$. Now the problem of reconstructing these analytic functions from (3.13) is very similar to a famous problem posed by HILBERT (cf. [25], Chap. 5). That problem may be stated as follows. $F_+(z)$ is analytic for $\mathrm{Im}\,\zeta > 0$ and vanishes as $|z| \to \infty$ there. $F_-(z)$ is analytic for $\mathrm{Im}\,\zeta < 0$ and vanishes as $|z| \to \infty$ there. They are related on the real axis by

$$F_+(x) - F_-(x) = f(x) \qquad \text{on } z = x, \tag{3.14}$$

where $f(x)$ is a given function. The « Hilbert problem » is to construct both F_+ and F_- from $f(z)$. If $f(x)$ is absolutely integrable, the solution of the problem is given by

$$\mathscr{F}(z) = \frac{1}{2\pi i}\int \frac{f(x)\,\mathrm{d}x}{x - z} = \begin{cases} F_+(z)\,, & \mathrm{Im}\,z > 0\,, \\ F_-(z)\,, & \mathrm{Im}\,z < 0\,. \end{cases} \tag{3.15}$$

Comparing (3.13) and (3.14), it is evident that, because b/a is known for real ζ, if $\psi \exp[i\zeta x]$ were known for real ζ, then (3.13) would be solved by a formula like (3.15). However, because $\psi \exp[i\zeta x]$ is unknown, we obtain instead a linear integral equation, with a Cauchy-type singularity. Another singular integral equation follows from (3.10*b*). The usual Gel'fand-Levitan-type of integral equations for (3.8) are essentially the Fourier transform of these coupled singular integral equations.

The original work by GEL'FAND and LEVITAN [18] and others on inverse scattering theory did not make this connection to the Hilbert problem, which has been developed more recently (*e.g.*, [26]). The advantage of this approach is that it emphasizes the fundamental role played by analytic functions in the theory of inverse scattering.

3. – Solutions of the nonlinear Schrödinger equation.

The practical consequence of all of this remarkable structure is that the solutions of these special nonlinear evolution equations are quite predictable, and can be computed explicitly (especially for large times) once the initial data have been mapped into scattering data. We may illustrate this by focussing our attention on the nonlinear Schrödinger equation

$$iq_t + q_{xx} + 2|q|^2 q = 0\,, \tag{3.16}$$

which describes the nonlinear instability of a packet of nearly monochromatic water waves of small amplitude in one dimension.

The appropriate scattering problem is (3.8), with $r = -q^*$. The discrete spectrum in this case is represented by N discrete eigenvalues, $\zeta_j = (\xi + i\eta)_j$ with $\eta_j \geqslant 0$, along with N parameters, c_j. The continuous spectrum is represented by $b/a(\xi)$, on $\eta = 0$. (The equation

$$iq_t + q_{xx} - 2|q|^2 q = 0 \tag{3.17}$$

also has physical interest. In this case, $r = +q^*$ in (3.8), and there are no discrete eigenvalues.)

A single eigenvalue with $\eta > 0$ corresponds to a single-soliton solution of (3.16):

$$q(x, t) = 2\eta \operatorname{sech}\{2\eta(x + 4\xi t + x_0)\} \exp\left[-2i[\xi x + 2(\xi^2 - \eta^2)t + \varphi_0]\right]. \tag{3.18}$$

This is an « envelope soliton » (as opposed to a KdV-type soliton). It represents a one-dimensional wave packet that is stable with respect to one-dimensional perturbations, and is shown in fig. 5. For a one-dimensional packet of water waves that are nearly monochromatic and of small amplitude, (3.16) describes the Benjamin-Feir [27] instability, and one could think of an envelope soliton as a stable-equilibrium state for that process.

N discrete eigenvalues generate N solitons. If they have N different speeds ($\xi_j \neq \xi_k$ for $j \neq k$), they separate in space as $t \to \infty$, so that the long-time solution of (3.16) is simply a sum of N individual solitons. We should emphasize that solitons are intrinsically nonlinear objects, that disappear in the linear limit.

On the other hand, the « radiation », which corresponds to the continuous spectrum, is qualitatively quite similar to the solution of the linearized problem. For comparison we note that

$$iy_t + y_{xx} = 0 \tag{3.19}$$

has a family of self-similar solutions, including

$$y = t^{-\frac{1}{2}} A \exp[ix^2/4t + i\varphi]. \tag{3.20}$$

One may also solve (3.19) as an initial-value problem on $-\infty < x < \infty$, and evaluate the solution in the long-time limit. The result is that along $x/t = 2k$ (the group velocity)

$$y(x, t) \sim t^{-\frac{1}{2}} \hat{Y}(k) \exp[ix^2/4t - i\pi/4], \tag{3.21}$$

where $\hat{Y}(k)$ is the Fourier transform of the initial data. This has the form of a slowly varying similarity solution.

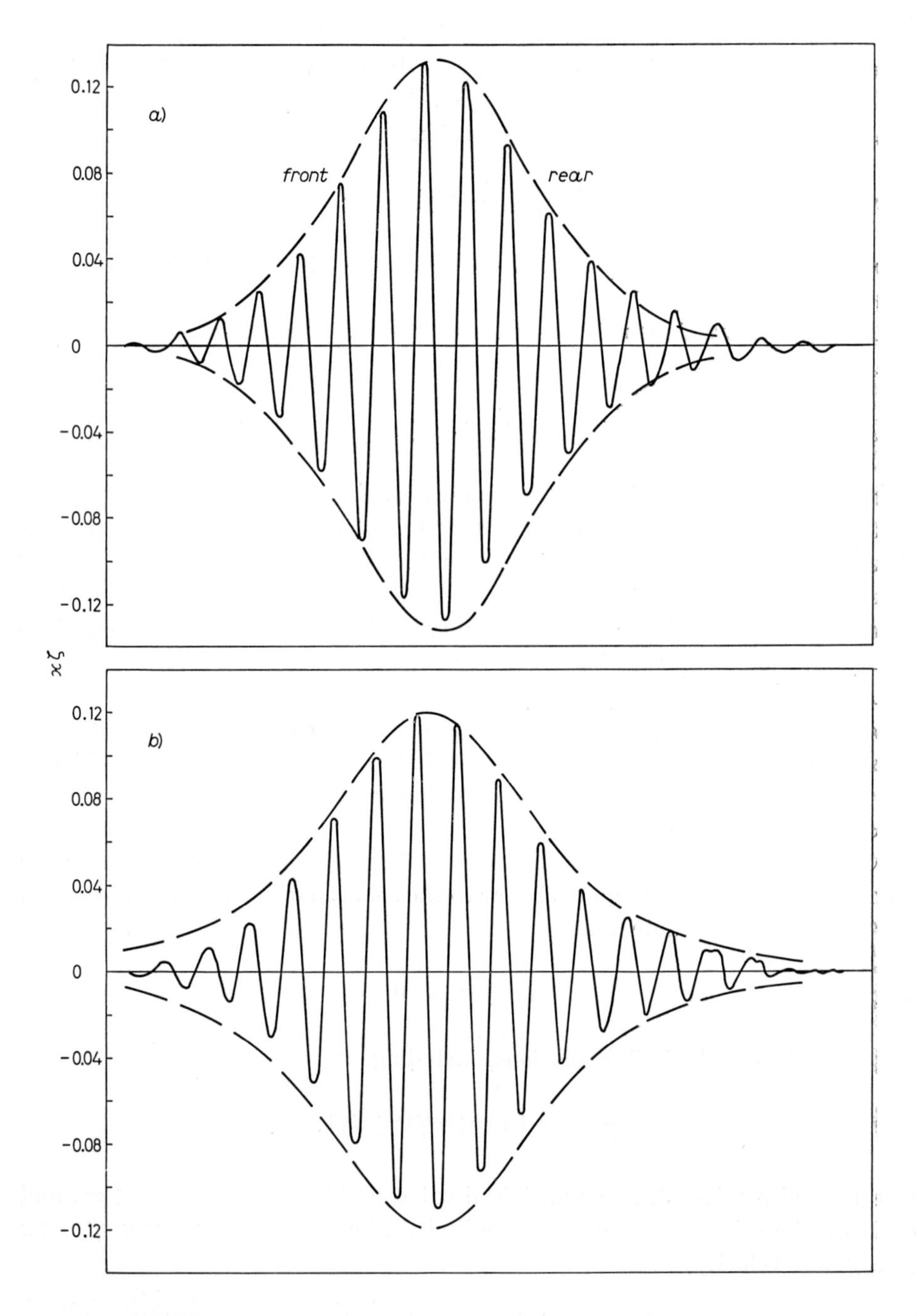

Fig. 5. – Measured packet of nearly monochromatic (frequency = 1 Hz) surface waves of small amplitude. The theoretical shape of the appropriate envelope soliton solution of (3.16) is given by the dashed line. *a*) 6 m downstream of wavemaker, *b*) 30 m downstream. From [28].

There is a nearly identical similarity solution of (3.16),

$$q = t^{-\frac{1}{2}} A \exp\left[i(x^2/4t + 2A^2 \ln t + \varphi)\right]. \tag{3.22}$$

In the absence of solitons, the asymptotic solution of (3.21) that evolves from appropriate initial data also takes the form of a slowly varying similarity solution. Along $x/t = -4\xi$,

$$q \sim t^{-\frac{1}{2}} f(x/t) \exp\left[i[x^2/4t + 2f^2 \ln t + g(x/t)]\right], \tag{3.23}$$

where

$$f^2(x/t) = -\frac{1}{4\pi} \ln\left[1 - |b/a(\xi)|^2\right], \tag{3.24}$$

and $g(x/t)$ also is determined by $b/a(\xi)$. (See [29-31].)

The general solution of (3.16) on $-\infty < x < \infty$ involves both solitons and radiation. For the KdV equation these two components separate in space, but for (3.16) they coexist. In the long-time limit, the solution consists of N envelope solitons riding on a sea of radiation. The solitons are permanent wave packets, while the radiation decays as $t^{-\frac{1}{2}}$.

PART IV

The Korteweg-deVries Equation with Periodic Boundary Conditions.

To this point we have discussed solitons and IST only on the infinite interval. However, the KdV equation with periodic boundary conditions also has applications in water waves. Moreover, the original discovery of solitons by ZABUSKY and KRUSKAL [32] was based on numerical experiments on the periodic KdV problem.

There is a theory of inverse scattering transforms for the KdV equation with periodic boundary conditions. The most complete version is due to MCKEAN and TRUBOWITZ [33]. In this part, we will follow the less general but simpler version of Dubrovin and Novikov [34]. In contrast to the theory of IST on the infinite interval, however, the theory for the periodic problem cannot yet be considered a practical tool for use in applications. After presenting the theory in its current form, we will identify some practical questions that have not yet been answered satisfactorily.

The first work on the periodic KdV problem

$$\begin{cases} u_t + 6uu_x + u_{xxx} = 0\,, \\ u(x, t) = u(x + L, t) \end{cases} \tag{4.1}$$

was done by KORTEWEG and DEVRIES [4], who found a periodic, traveling-wave solution of (4.1):

$$u(x, t) = 2p^2k^2 \operatorname{cn}^2 [p(x - ct + \bar{x}); k] + \beta\,. \tag{4.2}$$

Here $\operatorname{cn}(\varphi; k)$ is the Jacobian elliptic function with modulus k $(0 \leqslant k^2 \leqslant 1)$,

$$\begin{cases} c = 6\beta - 4p^2(1 - 2k)\,, \\ pL = 2K(k)\,, \end{cases} \tag{4.3}$$

where $K(k)$ is the complete elliptic integral of the first kind (cf. [35]). If we require $\oint u\, dx = 0$, as required by (1.21b), then

$$\beta = -2p^2 \left[\frac{E(k)}{K(k)} - 1 + k^2\right], \tag{4.4}$$

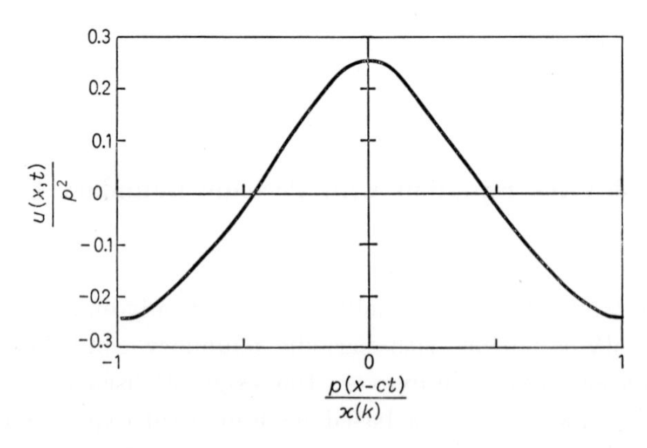

Fig. 6. – Cnoidal-wave solution of the KdV equation (4.1) with $k = \frac{1}{2}$ and $\oint u\, dx = 0$.

where $E(k)$ is the complete elliptic integral of the second kind. KORTEWEG and DEVRIES called these « cnoidal waves », by analogy with sinusoidal waves. A typical solution is shown in fig. 6, for $k = \frac{1}{2}$. These nonlinear, periodic, traveling waves reduce to infinitesimal sinusoidal waves if $k \to 0$,

$$u(x, t) \to p^2k^2 \cos 2p(x + 4p^2t + \bar{x})\,, \tag{4.5}$$

with $pL \to \pi$. At the other extreme, if $k \to 1$,

$$u(x, t) \to 2p^2 \operatorname{sech}^2 p(x - 4p^2 t + \bar{x}) . \tag{4.6}$$

Aside from these special solutions, however, almost nothing was known about the periodic KdV problem until the important numerical work of Zabusky and Kruskal [32], who were motivated by the earlier work of Fermi, Pasta and Ulam [36]. In both of these studies, the authors observed « recurrence of initial states » after a relatively short time. In other words, rather than showing any tendency to an equipartition of energy among all of the degrees of freedom of the system, the solution of the initial-value problem almost returned to its initial configuration repeatedly. The time required for this recurrence was short enough to suggest that only a few of the possible degrees of freedom of the system actually were participating in the process described by (4.1). But how could the solution of (4.1) be so constrained unless (4.1) itself carried additional constraints that had not been discovered? These extra constraints turned out to be the infinite set of conservation laws of the KdV equation [37]. The discovery of these conservation laws led in turn to the development of IST [19]. We should emphasize the historical importance of the careful numerical studies of Zabusky and Kruskal [32], which indicated that the KdV equation possessed additional mathematical structure.

Let us now outline the theory of IST for the KdV problem. The reader will observe that some of this theory is analogous to that on the infinite interval, but some of it is not. The scattering problem is

$$V_{xx} + [\lambda + u(x)]V = 0 , \tag{4.7}$$

just as it was on the infinite interval, but in this case $u(x)$ is a periodic function of x. Because scattering theory requires no knowledge of the time dependence of $u(x, t)$, we may consider (4.7) at a fixed time.

One of the difficult conceptual questions about the periodic KdV problem is to find what plays the role of the scattering data here, since there is no « point at infinity » where the solution simplifies. Without such a special point, we simply choose an arbitrary x_0 in order to begin the analysis. With x_0 fixed (and time fixed), we may identify two linearly independent solutions of (4.7), $\varphi(x; x_0, \lambda)$ and its complex conjugate φ^*, by imposing boundary conditions for real λ at $x = x_0$:

$$\begin{cases} \varphi(x_0; x_0, \lambda) = 1 , & \varphi^*(x_0; x_0, \lambda) = 1 , \\ \varphi_x(x_0; x_0, \lambda) = ik = i\sqrt{\lambda} , & \varphi^*_x(x_0; x_0, \lambda) = - ik . \end{cases} \tag{4.8}$$

One period to the right ($x \to x + L$), these two functions satisfy the same differential equation again, so they must be a linear combination of the φ and φ^*:

$$\begin{bmatrix} \varphi(x+L; x_0, \lambda) \\ \varphi^*(x+L; x_0, \lambda) \end{bmatrix} = \begin{bmatrix} a(x_0, \lambda) & b(x_0, \lambda) \\ b^*(x_0, \lambda) & a^*(x_0, \lambda) \end{bmatrix} \begin{bmatrix} \varphi(x; x_0, \lambda) \\ \varphi^*(x; x_0, \lambda) \end{bmatrix}. \tag{4.9}$$

The matrix connecting these two sets of solutions of (4.7) is called the « monodromy matrix ». In some ways it plays the role of the scattering data. Its coefficients are related by a Wronskian relation

$$|a|^2 - |b|^2 = 1. \tag{4.10}$$

The fact that the coefficients of (4.7) are periodic does not necessarily mean that its solutions are periodic. In fact, most of them are not. The solutions of (4.7) that are periodic play a fundamental role in the theory of inverse scattering. We define next the « Bloch eigenfunctions » to be solutions of (4.7) that satisfy

$$\begin{cases} \psi(x_0; x_0, \lambda) = 1, \\ \psi(x+L; x_0, \lambda) = \mu\psi(x; x_0, \lambda). \end{cases} \tag{4.11}$$

Because these must also be a linear combination of φ and φ^*, one may show that for each (x_0, λ)

$$\mu^2 - 2a_r\mu + 1 = 0, \tag{4.12}$$

where

$$a = a_r + ia_i. \tag{4.13}$$

Equation (4.12) admits three possibilities.

i) If $|a_r| > 1$, one root of (4.12) is larger than one in magnitude, and the other is smaller. From (4.11), these correspond to Bloch eigenfunctions that grow without bound, either as $x \to +\infty$ or as $x \to -\infty$. These are said to be « unstable ».

ii) If $|a_r| < 1$, we may define $a_r(\lambda) = \cos p(\lambda)$, and show that

$$\mu = \exp[ip]. \tag{4.14}$$

These represent « stable » eigenfunctions.

iii) If $|a_r| = 1$, *i.e.*

$$a_r^2 = 1, \tag{4.15}$$

then $\mu = \pm 1$ and the eigenfunctions are either periodic or antiperiodic functions of x. Thus (4.15) identifies the periodic solutions of (4.7).

These possibilities may be summarized on a « Floquet diagram », in which a_r is plotted as a function of λ for fixed x_0, as shown in fig. 7. The regions in which $|a_r| > 1$ are called « unstable bands ». Necessarily, each unstable band lies between two successive roots of (4.15), numbered λ_{2n} and λ_{2n+1}. One may show (by oscillation theorems, cf. [38]) that $\lambda_{2n} > \lambda_{2n-1}$ and that $\lambda_{2n+1} \geqslant \lambda_{2n}$. Moreover, one stationary point of a_r occurs in each unstable band, and none occur elsewhere. Unstable bands that consist of a single point (if $\lambda_{2n} = \lambda_{2n+1}$) are called « degenerate ».

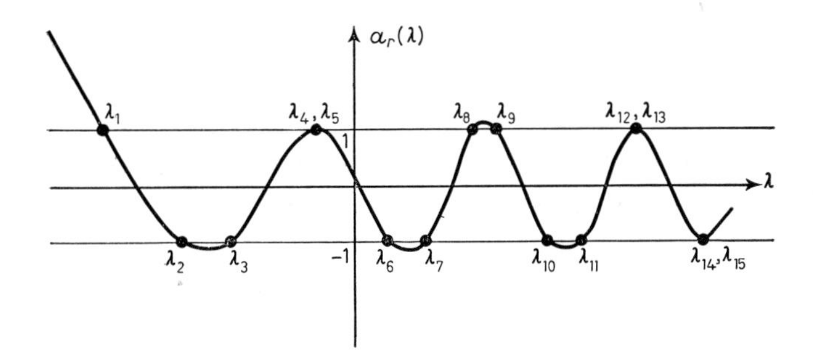

Fig. 7. – Floquet diagram for a particular $u(x)$ in (4.7). In this example there are four unstable bands.

Following DUBROVIN and NOVIKOV [34], we now assume that $u(x)$ generates $a_r(x_0, \lambda)$ with only a finite number (N) of nondegenerate bands. (Roughly, this corresponds to approximating a periodic function with a finite Fourier series.)

Next we define two spectra for (4.7). These may be defined either by attaching boundary conditions to (4.7), or by imposing conditions on the monodromy matrix.

i) The main spectrum is defined by (4.15) or by requiring that (4.7) admit periodic solutions. Points of this spectrum ($\lambda_1, \lambda_2, \ldots$) occur at the edges of the unstable bands. Note that periodic solutions of (4.7) are independent of x_0.

ii) The auxiliary spectrum ($\gamma_1, \gamma_2, \ldots$) is defined by

$$a_i + b_i = 0 , \tag{4.16}$$

or by requiring that there be a solution of (4.7) that satisfies

$$y(x_0; x_0, \lambda) = 0 = y_0(x_0 + L; x_0, \lambda) . \tag{4.17a}$$

It follows from (4.10) and (4.16) that $a_r^2 \geqslant 1$ at each of these points, which must, therefore, lie in the unstable bands. One may show that each unstable band contains exactly one point of the auxiliary spectrum. By assumption, at most N of these points of the auxiliary spectrum do not coincide with points of the main spectrum.

Given $u(x)$ and x_0, determination of $\{\lambda_j\}$ and $\{\gamma_j\}$ completes the direct scattering problem. The inverse mapping requires knowledge of the analytic properties of the monodromy matrix, as one might expect by analogy with the inverse problem on the infinite interval. The final result is miraculously simple:

$$u(x_0) = -\sum_{j=1}^{2N+1} \lambda_j + 2\sum_{j=1}^{N} \gamma_j . \tag{4.17b}$$

Now it remains only to find how the two spectra depend on x_0 and on time.

The boundary conditions that define $\{\lambda_j\}$ and $\{\gamma_j\}$ suggest that the main spectrum should be independent of x_0, but that the auxiliary spectrum should not be. This turns out to be the case. One shows that

$$\frac{\partial \lambda_j}{\partial x_0} = 0 , \tag{4.18a}$$

$$\frac{\partial \gamma_j}{\partial x_0} = 2i\sigma_j R^{\frac{1}{2}}(\gamma_j) / \prod_{k\neq j}^{N} (\gamma_j - \gamma_k) , \qquad j = 1, \dots, N , \tag{4.18b}$$

where $\sigma_j = \pm 1$ and

$$R(\lambda) = \prod_{j=1}^{2N+1} (\lambda - \lambda_j) . \tag{4.18c}$$

The x_0-dependence of $\{\gamma_j\}$ is as frightening as (4.17) was appealing. One is faced in (4.18b) with the integration of N coupled, nonlinear, ordinary differential equations, even before time dependence is brought into the picture! Fortunately, another miracle occurs, and there is a change of variables (involving hyperelliptic functions) that permits one to integrate (4.18b) by quadrature. Thus, although (4.18b) looks foreboding, it represents a straight-line motion when viewed appropriately. The cost of acquiring this simple picture is that one must introduce hyperelliptic functions.

The time dependence of the two spectra parallels their x_0-dependence. For the KdV equation,

$$\frac{\partial \lambda_j}{\partial t} = 0 , \tag{4.19a}$$

$$\frac{\partial \gamma_j}{\partial t} = 8i\hat{\sigma}_j \left[\prod_{k\neq j}^{N} (\gamma_j - \gamma_k)\right]^{-1} \left[\sum_{k\neq j}^{N} \gamma_k - \frac{1}{2}\sum_{k}^{2N+1} \lambda_k\right] R^{\frac{1}{2}}(\gamma_j) , \quad j = 1, \dots, N , \tag{4.19b}$$

where $\hat{\sigma}_j = \pm 1$, and $R(\lambda)$ was defined by (4.18c). Again, the integration in (4.19b) may be reduced to quadrature by introducing hyperelliptic functions.

What are the consequences of introducing hyperelliptic functions? One may show that for a N-band solution of the KdV equation (*i.e.* whose auxiliary spectrum contains only N nondegenerate points)

$$u(x, t) = 2 \frac{\partial^2}{\partial x^2} \ln \theta_N(\eta_1, \eta_2, \dots, \eta_N) + \text{const}\,, \tag{4.20a}$$

where

$$\eta_j = i(k_j x - \omega_j t)\,, \tag{4.20b}$$

each k_j and ω_j is constant, and θ_N is the theta-function, an analytic function that is periodic in each of its N variables separately (cf. [39]). It follows that the motion described by a N-band solution of the KdV equation may be thought of as uniform translation of a point along a straight-line path on a N-dimensional torus. Because any such solution has only N degrees of freedom, rather than infinitely many as (4.1) suggests *a priori*, it has a relatively short recurrence time that may be estimated from a knowledge of $\{\omega_j\}_{j=1}^N$ in (4.20b). This is apparently the explanation of the relatively short recurrence time observed numerically by ZABUSKY and KRUSKAL [32].

In an abstract sense, this theory of the periodic KdV problem (or its generalization by MCKEAN and TRUBOWITZ [33]) is complete. In a practical sense, it is difficult to extract from the theory the numbers that one needs to make comparison with experiments. (The difficulty here is simply that hyperelliptic functions have not yet been reduced to a practical engineering tool.) Here are some of the practical problems for which satisfactory answers are not yet available:

1) Given a N-band solution of (4.1), what is its recurrence time?
2) Given appropriate initial data for (4.1), estimate its recurrence time.
3) ZABUSKY [40] defined T_b to be the time of breakdown of the solution of

$$u_t + u u_x = 0 \tag{4.21}$$

that evolves from $u(x, 0) = \sin(2\pi x/L)$. He found empirically that the recurrence time of the corresponding solution of

$$u_t + u u_x + \delta^2 u_{xxx} = 0 \tag{4.22}$$

was

$$T_\text{r} = (0.71/\delta)\, T_\text{b}\,. \tag{4.23}$$

Can this formula be derived theoretically?

4) « Solitons » have come to mean special solutions of problems solvable by IST on the infinite interval. Each soliton is associated with a discrete eigenvalue in that formulation. However, the word was coined by ZABUSKY and KRUSKAL [32] to describe phenomena observed in the periodic problem. What were the « solitons » they observed? How does the band structure of a solution of the periodic problem relate to the number of solitons that one would observe numerically?

PART V

Deterministic and Chaotic Models.

In the course of these lectures, I have tried to survey the theory of nonlinear evolution equations (« soliton theories ») that can be solved by IST. If one were asked to identify the single feature of these equations that identifies them as special, one might choose the predictability of their solutions as their common identifying feature. This property may be stated in several ways:

i) These equations are computationally deterministic, not only for some finite time, but for all time.

ii) Their solutions are (neutrally) stable with respect to perturbations in the initial data.

iii) Their solutions are predictable.

All of these notions are closely related to each other, and they amount to saying that, for these special equations, the initial-value problem makes sense in a practical, real-world way. Because of their predictability, these completely integrable equations may be regarded as prototypes of *deterministic problems*.

As models of physical systems, they contrast sharply with stochastic models, which require a certain amount of unpredictability. This unpredictability may come from a variety of sources; one possibility is that the dynamics themselves may be unpredictable. An example of this inherent unpredictability may be seen in a system of coupled ordinary differential equations known as the « Lorenz model »:

$$\left\{\begin{aligned} \dot{x} \quad &= \sigma(y - x)\,, \\ \dot{y} + y &= -\,x(z - r)\,, \\ \dot{z} \quad &= xy - bz\,, \end{aligned}\right. \tag{5.1}$$

where $\dot{}$ $= \mathrm{d}/\mathrm{d}t$, and σ, r, b are parameters. In their original context these equations modelled a problem in convection, and σ represented the Prandtl number, r the Rayleigh number, and b was a measure of a length scale. For this discussion, however, we may regard them simply as a dynamical system with parameters.

LORENZ [41] chose certain values of the parameters ($\sigma = 10$, $b = \frac{8}{3}$, $r = 28$) and integrated the equation numerically. His results are described in detail in his original work, and may be summarized by saying that the solutions he computed were unpredictable. These equations have become a popular model of a dynamical system exhibiting chaotic behavior.

We may regard the Lorenz model with the particular parameters chosen by LORENZ as a prototype of a « chaotic model ». Its features may be characterized in the following ways:

i) The equations are deterministic for a finite time (*i.e.*, given finite initial data, a unique solution exists). However, unless the initial data are known with infinite accuracy, the solution becomes less and less determined by the initial data as time increases without bound.

ii) The solution is unstable with respect to perturbations of the initial data.

iii) Given any initial data, the solution is unpredictable over a long time scale in any practical sense.

Now we have two kinds of dynamical systems. The Lorenz model is chaotic, and one might imagine using statistical methods to analyze the problem further. Soliton theories are deterministic, and statistical methods would only obscure matters there. Thus a fundamental question is « How does one know whether a given model will be chaotic or predictable? »

This same question arises in the foundations of statistical mechanics and of the theory of turbulence. Perhaps the first context in which the question arose was theology. In that context, deterministic theory was called « predestination », while evidence of unpredictable behavior was attributed to « free will ». A corresponding debate, with somewhat different names, is now in progress in psychology.

No attempt will be made here to classify all dynamical systems on the basis of how chaotic their solutions are. We will examine a narrower question, which may be regarded as a first step in constructing such a grand classification scheme.

Q: What determines whether a given partial differential equation can be solved by IST?

This question has been the motivation of recent work by ABLOWITZ, RAMANI and SEGUR [42-44]. The answer seems to be related to what we may call the « Painlevé property » (which will be defined shortly).

After this rather long introduction, we may finally outline this part. First, we must define the Painlevé property for ordinary differential equations (ODEs), because everything else follows from it. Next we may show that ODEs with the Painlevé property are related to evolution equations solvable by IST. Our conjecture about characterizing these nonlinear partial differential equations (PDEs) then is almost obvious: they must reduce to ODEs of Painlevé type. What is less obvious is how to prove the conjecture, but a partial proof is available. The notion that ODEs of P-type are closely related to IST and complete integrability may be used in a variety of ways. We will look at two: i) to test whether a given PDE can be solved by IST and ii) to find conditions under which the Lorentz model is completely integrable.

1. – The Painlevé property.

Consider first a linear ordinary differential equation, say of second order:

$$\frac{\mathrm{d}^2 w}{\mathrm{d}z^2} + p(z)\frac{\mathrm{d}w}{\mathrm{d}z} + q(z)w = 0 . \tag{5.2}$$

For suitable $p(z)$ and $q(z)$, this equation may be viewed in the complex plane, and the singularities of the solution of (5.2) are found by examining $p(z)$ and $q(z)$ (*e.g.*, [45], Chap. 15). In particular, the general solution has two constants of integration,

$$w(z; A, B) = Aw_1(z) + Bw_2(z) , \tag{5.3}$$

and the location in the complex plane of the singularities of $w(z)$ does *not* depend on A or B. The singularities of a linear differential equation are said to be *fixed*, because they do not depend on the constants of integration.

Nonlinear differential equations lose this property. A very simple example of a nonlinear ODE is

$$\frac{\mathrm{d}w}{\mathrm{d}z} + w^2 = 0 ; \tag{5.4}$$

its general solution is

$$w(z; z_0) = \frac{1}{z - z_0} . \tag{5.5}$$

Here z_0 is the constant of integration, and it also defines the location of the singularity. This singularity is *movable*, because its location depends on the constant of integration.

Thus linear differential equations have only fixed singularities, while nonlinear equations can have both fixed and movable singularities. About 100 years ago, mathematicians asked the following question:

Q: Which nonlinear ODEs admit no movable branch points or essential singularities?

Movable poles are allowed, as are fixed singularities of any kind. We will refer to this property as the Painlevé-property, and equations that possess it will be said to be of *Painlevé-type*, or P-*type*.

It turns out that the only first-order equations with the Painlevé-property are generalized Riccati equations

$$\frac{\mathrm{d}w}{\mathrm{d}z} = p_0(z) + p_1(z)w + p_2(z)w^2 . \tag{5.6}$$

(A complete review of the nineteenth-century work in this field may be found in [45], Chap. 12-14.)

PAINLEVÉ and his co-workers were able to answer the question comprehensively for second-order equations of the form

$$\frac{\mathrm{d}^2w}{\mathrm{d}z^2} = F\left(\frac{\mathrm{d}w}{\mathrm{d}z}, w, z\right), \tag{5.7}$$

where F is rational in $\mathrm{d}w/\mathrm{d}z$ and w and analytic in z. They showed that out of all possible equations of the form (5.7) only 50 canonical equations have the Painlevé-property of admitting only poles as movable singularities. Further, they showed that 44 of these equations can be reduced to something already known, such as elliptic functions. That left six equations that defined new transcendental functions, called the Painlevé transcendents. The first three of these are

$$\frac{\mathrm{d}^2w}{\mathrm{d}z^2} = 6w^2 + z \qquad \mathrm{P_I},$$

$$\frac{\mathrm{d}^2w}{\mathrm{d}z^2} = 2w^3 + zw + \alpha \qquad \mathrm{P_{II}},$$

$$\frac{\mathrm{d}^2w}{\mathrm{d}z^2} = \frac{1}{w}\left(\frac{\mathrm{d}w}{\mathrm{d}z}\right)^2 - \frac{1}{z}\frac{\mathrm{d}w}{\mathrm{d}z} + \frac{1}{z}(\alpha w^2 + \beta) + \gamma w^3 + \frac{\delta}{w} \qquad \mathrm{P_{III}}.$$

There are three more.

The question of which equations have the Painlevé-property is appropriate at any order, but comprehensive results are available only at the first and second order.

2. – Relation to IST.

A relation between ODEs of P-type and the inverse scattering transform is formulated in the following

Conjecture [42]. Every nonlinear ODE obtained by an exact reduction of a nonlinear PDE solvable by some inverse scattering transform has the Painlevé-property.

Here are some examples. The Boussinesq equation

$$u_{tt} = u_{xx} + \left(\frac{u^2}{2}\right)_{xx} + \frac{1}{4}u_{xxxx} \tag{5.8}$$

is a nonlinear PDE solvable by IST [46]. An exact reduction to an ODE may be obtained by looking for a traveling-wave solution:

$$u(x, t) = w(x - ct) = w(z) . \tag{5.9}$$

Then (5.8) becomes

$$(1 - c^2)w'' + \left(\frac{w^2}{2}\right)'' + \frac{1}{4}w'''' = 0 , \tag{5.10}$$

which can be integrated twice. Depending on the constants of integration, the result after rescaling is either

$$w'' + 2w^2 + a = 0 \quad \text{or} \quad w'' + 2w^2 + z = 0 . \tag{5.11}$$

The first possibility defines an elliptic function, whose only singularities are poles. The second possibility is the equation for P_I. In either case, the ODE has the Painlevé-property. So the PDE solvable by inverse scattering reduces to an ODE of P-type.

Another example is the modified KdV equation

$$u_t - 6u^2u_x + u_{xxx} = 0 , \tag{5.12}$$

which can be solved by IST [47]. An exact reduction to an ODE may be obtained by looking for a self-similar solution:

$$\begin{cases} u(x, t) = (3t)^{-\frac{2}{3}}w(z) , \qquad z = x/(3t)^{\frac{1}{3}} , \\ \Rightarrow v''' - 6w^2w' - (zw)' = 0 . \end{cases} \tag{5.13}$$

This can be integrated once

$$w'' = 2w^3 + zw + \alpha \qquad P_{II}.$$

Again, the ODE is of P-type.

The sine-Gordon equation

$$u_{xt} = \sin u \tag{5.14}$$

can be solved by IST [48]. It has a self-similar solution

$$u(x, t) = f(z), \qquad z = xt. \tag{5.15}$$

If we set $w(z) = \exp[if]$, then

$$w'' = \frac{1}{w}(w')^2 - \frac{1}{z}w' + \frac{1}{2z}(w^2 - 1) \qquad P_{III}.$$

Again, the ODE is of P-type.

By now the pattern is evident, and I may simply state that there is a nonlinear evolution equation solvable by IST that reduces to P_{IV}, and another that reduces to P_V. The point is not that the evolution equation must reduce to one of the six Painlevé transcendents, which are all of second order, but that it must reduce to an ODE of P-type. We have checked an enormous number of examples. In every case checked, PDEs that can be solved by IST reduce to ODEs of P-type and PDEs that are not solvable by IST (*e.g.*, this may be determined by observing numerically that two solitary waves do not interact like solitons) reduce to ODEs that are *not* of P-type.

Thus there is some kind of relation between partial differential equations solvable by IST and ordinary differential equations of P-type. This relation can be used to examine either the ODEs or the PDEs. To see how it helps in the study of the ODEs, consider the mKdV equation and P_{II}. Recall that the last step of IST, the inverse scattering part, goes as follows (cf. [15]). $F(x, t)$ satisfies a linear partial differential equation

$$F_t + F_{xxx} = 0 \tag{5.16}$$

subject to some boundary and initial conditions. Then $K(x, y; t)$ satisfies a linear integral equation of the Gel'fand-Levitan-Marchenko type,

$$K(x, y) = F(x + y) + \int_x^\infty \int_x^\infty K(x, z) F(z + s) F(s + y)\, dz\, ds, \qquad y \geqslant x. \tag{5.17}$$

Once K is known, then $q(x, t) = K(x, x; t)$ satisfies the mKdV equation

$$q_t - 6q^2 q_x + q_{xxx} = 0 . \tag{5.18}$$

In the full IST treatment, F depends on the initial data of $q(x, 0)$ through the direct scattering problem. Here we simply start with F, and force everything to be self-similar:

$$\begin{cases} \xi = x/(3t)^{\frac{1}{3}}, & \eta = y/(3t)^{\frac{1}{3}}, \\ F(x, t) = (3t)^{-\frac{1}{3}} \mathscr{F}(\xi), & K(x, y; t) = (3t)^{-\frac{1}{3}} \mathscr{K}(\xi, \eta) . \end{cases} \tag{5.19}$$

Then (5.16) becomes a linear ODE

$$\mathscr{F}'''(\xi) - (\xi \mathscr{F})' = 0 \tag{5.20}$$

and a one-parameter family of solutions is

$$\mathscr{F}(\xi + \eta) = r \operatorname{Ai}\left(\frac{\xi + \eta}{2}\right), \tag{5.21}$$

where $\operatorname{Ai}(\xi)$ is the Airy function. The integral equation (5.17) becomes

$$\mathscr{K}(\xi, \eta) = r \operatorname{Ai}\left(\frac{\xi + \eta}{2}\right) + \frac{r^2}{4} \int_\xi^\infty \int_\xi^\infty \mathscr{K}(\xi, \zeta) \operatorname{Ai}\left(\frac{\zeta + \theta}{2}\right) \operatorname{Ai}\left(\frac{\theta + \eta}{2}\right) d\zeta \, d\theta , \quad n \geqslant \xi . \tag{5.22}$$

The Airy function decreases rapidly as its argument becomes large, so the integral term in (5.22) is very well behaved. Therefore, it is relatively easy to solve (5.22) for $n \geqslant \xi$. On $\eta = \xi$, the solution of (5.22) satisfies the self-similar form of the mKdV equation, *viz.* P_{II}:

$$\frac{d^2}{d\xi^2} \mathscr{K}(\xi, \xi) = 2\mathscr{K}^3(\xi, \xi) + \xi \mathscr{K}(\xi, \xi). \tag{5.23}$$

(Two different proofs of this fact are given by ABLOWITZ, RAMANI and SEGUR [42, 43].) The point here is that (5.22) is an exact linearization of P_{II}: every solution of the linear integral equation also solves P_{II}. The general solution of (5.23) involves two arbitrary constants; the linear integral equation gives a one-parameter (r) family, which includes all of the bounded real solutions of (5.23).

Next let us sketch a partial proof of why this test actually works. Consider a linear equation of the form

$$K(x, y) = F(x + y) + \int_x^\infty K(x, z) N(x; z, y) \, dz, \qquad y \geqslant x, \tag{5.24}$$

where F vanishes rapidly for large values of the argument and N depends on F. For example, in (2.16), $N(x, z, y) = F(z + y)$. In (5.22), we had

$$N(x, z, y) = \int_x^\infty F(z + s) F(s + y) \, \mathrm{d}s . \tag{5.25}$$

Other choices are also possible. We want to show that every solution of a linear integral equation like (5.24) must have the Painlevé-property. Then if K also satisfies an ODE, the family of solutions of the ODE obtained *via* (5.24) necessarily has the Painlevé-property as well. Thus the Painlevé-property is not out of the blue, it is a consequence of the linear integral equation.

Very roughly, the proof goes like this (for details, see [43], also [49]):

i) F satisfies a linear ODE, and, therefore, has no movable singularities at all.

ii) If F vanishes rapidly enough, then the Fredholm theory of linear integral equations applies. It follows that (5.24) has a unique solution in the form

$$K(x, y) = F(x + y) + \int_x^\infty F(x + z) \frac{D_1(x, z, y)}{D_2(x)} \, \mathrm{d}z , \tag{5.26}$$

where D_1 and D_2 are entire functions of their arguments. Then the singularities of K can only come from the fixed singularities of F, or the movable zeros of D_2. But D_2 is analytic, so these movable singularities must be poles.

3. – Applications.

Here are two examples of how the conjecture may be used.

Example. In $1 + 1$ dimensions, the nonlinear Schrödinger equation is

$$iu_t = u_{xx} + a|u|^2 u . \tag{5.27}$$

It can be solved by IST [21]. A natural generalization to $2 + 1$ dimensions is

$$iu_t = \nabla^2 u + a|u|^2 u . \tag{5.28}$$

We claim this equation cannot be solved by IST, because (5.28) has a similarity solution in the form

$$u(x, y, t) = R(\sqrt{x^2 + y^2}; \lambda) \exp[i\lambda t] , \tag{5.29}$$

and the ODE for $R(r)$ is not of P-type. Thus the nonlinear Schrödinger equation is solvable in $1+1$ dimensions, but not in $2+1$ dimensions, or in $3+1$ dimensions. On the same grounds, we claim that the equation for water waves in deep water,

$$iu_t + u_{xx} - u_{yy} + \sigma|u|^2 u = 0 \,, \tag{5.30}$$

cannot be solved by IST.

Example. If the Painlevé-property is as closely tied to complete integrability as we have claimed, it ought to identify values of the parameters for which the Lorenz model (5.1) is completely integrable. Thus we may ask whether the Lorenz model is ever of P-type. The answer is that there are exactly four choices of (σ, r, b) for which (5.1) is of P-type.

i) $\sigma = 0$. In this case the equations are effectively linear and, therefore, predictable. Certainly the solutions exhibit no chaotic behavior.

ii) $\sigma = \frac{1}{2}$, $b = 1$, $r = 0$. The equations have two exact integrals

$$y^2 + z^2 = A^2 \exp[-2t] \,, \tag{5.31}$$

$$x^2 - z = B \exp[-t] \,, \tag{5.32}$$

after which the third integration may be obtained by quadrature, or the solution may be expressed in terms of elliptic functions.

iii) $\sigma = 1$, $b = 2$, $r = \frac{1}{9}$. A first integral is

$$x^2 - 2z = C \exp[-2t] \,. \tag{5.33}$$

After an involved change of variables, the resulting second-order equation becomes P_{II}. Again, the problem is predictable.

iv) $\sigma = \frac{1}{3}$, $b = 0$, r arbitrary. (The analysis of this case is due to A. RAMANI.) We may write $y = 3\dot{x} + x$ from (5.1a), and replace (5.1) with a third-order equation for x. It has a first integral

$$\ddot{x}x - \dot{x}^2 + \frac{x^4}{4} = C \exp[-4/3t] \,. \tag{5.34}$$

With the substitution

$$T = \exp[-t/3] \,, \qquad x(t) = TW(T) \,, \tag{5.35}$$

(5.34) reduces to P_{III} with $\alpha = \beta = 0$. Thus the Lorenz model (5.1) has at least one integral, and reduces to a classically known equation of lower order, whenever the coefficients σ, r, b are chosen so that (5.1) is of P-type.

These isolated points of parameter space are embedded in larger regions in which the equations have first integrals, although they may not be completely integrable.

v) If $b = 1$, $r = 0$, then (5.31) obtains for any σ. The existence of this integral precludes ergodic trajectories.

vi) If $b = 2\sigma$, then for any (r, σ)

$$x^2 - 2\sigma z = C \exp[-2\sigma t]. \tag{5.36}$$

Again, ergodic trajectories are impossible.

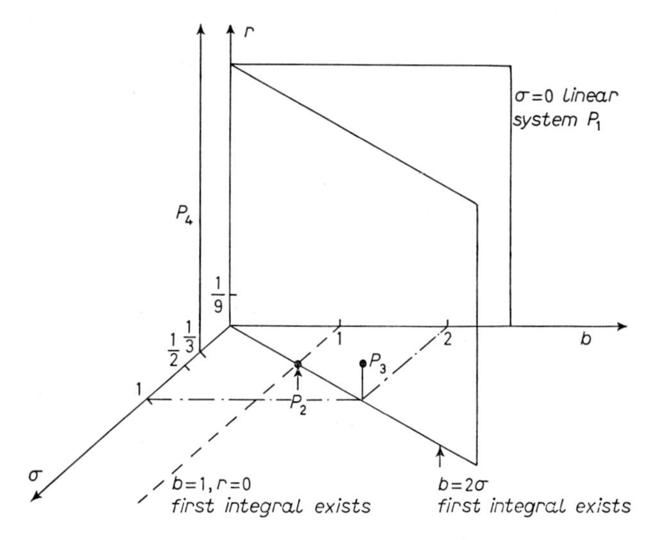

Fig. 8. – Map of parameter space for Lorentz model (5.1), showing where equations have Painlevé-property (P_1, ..., P_4) and where they admit exact integrals.

These different possibilities are shown in a map of parameter space in fig. 8. The point here is that, although the Lorenz model may be chaotic for some values of the parameters, it is completely integrable for others, and looking for the Painlevé-property provides an effective means of identifying points of predictable behavior. Of course, this notion is not restricted to the Lorenz model.

* * *

This work was supported by the U.S. Office of Naval Research and by the U.S. Army Research Office.

Note added in proofs.

We may let (σ, r or b) become infinite in (5.1), if we rescale the equations appropriately in this limit. We find in this way three additional limits in which (5.1) admits a first integral. These are: *a*) $r \to \infty$ with b and σ finite, *b*) $\sigma \to \infty$ with b and r finite, *c*) $r \to \infty$, $\sigma \to \infty$ with b and σ/r finite. In the first of these limits, ROBBINS [50] gave the complete solutions of (5.1) in terms of elliptic functions.

REFERENCES

General Bibliography

By the time the Proceedings of this school appear in print, a number of books on the theory of equations with solitons also will have appeared in print, including

G. LAMB: *Elements of Soliton Theory* (New York, N. Y., 1980).
V. E. ZAKHAROV, S. V. MANAKOV, S. P. NOVIKOV and L. P. PITAYEVSKY: *Theory of Solitons and the Method of the Inverse Problem* (in Russian) (Moscow, 1980).
M. J. ABLOWITZ and H. SEGUR: *Solitons and the Inverse Scattering Transform* (SIAM, 1981).

Most of the material in these lectures was drawn from this last reference.

[1] H. LAMB: *Hydrodynamics* (New York, N. Y., 1932).
[2] J. J. STOKER: *Water Waves* (New York, N. Y., 1957).
[3] J. V. WEHAUSEN and E. V. LAITONE: *Surface waves*, in *Handbuch der Physik*, Vol. **9** (Berlin, 1960).
[4] D. J. KORTEWEG and G. DEVRIES: *Philos. Mag.*, **39**, 422 (1895).
[5] J. L. HAMMACK and H. SEGUR: *J. Fluid Mech.*, **65**, 289 (1974).
[6] J. L. HAMMACK and H. SEGUR: *J. Fluid Mech.*, **84**, 337 (1978).
[7] T. B. BENJAMIN: *J. Fluid Mech.*, **25**, 241 (1966).
[8] H. ONO: *J. Phys. Soc. Jpn.*, **39**, 1082 (1975).
[9] V. E. ZAKHAROV: *Sov. J. Appl. Mech. Tech. Phys.*, **4**, 190 (1968).
[10] H. HASIMOTO and H. ONO: *J. Phys. Soc. Jpn.*, **33**, 805 (1972).
[11] H. C. YUEN and B. M. LAKE: *Phys. Fluids*, **18**, 956 (1975).
[12] J. D. GIBBON, I. N. JAMES and I. M. MOROZ: *Proc. R. Soc. London Ser. A*, **367**, 219 (1979).
[13] B. B. KADOMTSEV and V. I. PETVIASHVILI: *Sov. Phys. Dokl.*, **15**, 539 (1970).
[14] Lord RAYLEIGH (J. W. STRUTT): *Philos. Mag.*, **1**, 257 (1876).
[15] M. J. ABLOWITZ, D. J. KAUP, A. C. NEWELL and H. SEGUR: *Stud. Appl. Math.*, **53**, 249 (1974).
[16] G. B. WHITHAM: *Linear and Nonlinear Waves* (New York, N. Y., 1974).
[17] P. DEIFT and E. TRUBOWITZ: *Commun. Pure Appl. Math.*, **32**, 121 (1979).
[18] I. M. GEL'FAND and B. M. LEVITAN: *Izv. Akad. Nauk SSSR Ser. Mat.*, **15**, 309 (1951) (English translation: *Am. Math. Soc. Transl.*, Ser. 2, Vol. **1**, p. 253 (1955)).
[19] C. S. GARDNER, J. M. GREENE, M. D. KRUSKAL and R. M. MIURA: *Phys. Rev. Lett.*, **19**, 1095 (1967); also *Commun. Pure Appl. Math.*, **27**, 97 (1974).

[20] F. CALOGERO and A. DEGASPERIS: *Nuovo Cimento B*, **32**, 201 (1976).
[21] V. E. ZAKHAROV and A. B. SHABAT: *Sov. Phys. JETP*, **34**, 62 (1972).
[22] V. E. ZAKHAROV and L. D. FADDEEV: *Funct. Anal. Appl.*, **5**, 280 (1971).
[23] V. I. ARNOLD: *Mathematical Methods in Classical Mechanics* (New York, N. Y., 1978).
[24] H. GOLDSTEIN: *Classical Mechanics* (Reading, Mass., 1950).
[25] N. I. MUSKHELISVILI: *Singular Integral Equations* (Groningen, 1953).
[26] V. E. ZAKHAROV and S. V. MANAKOV: *Sov. Phys. Rev.*, **1**, 133 (1979).
[27] T. B. BENJAMIN and J. E. FEIR: *J. Fluid Mech.*, **27**, 417 (1967).
[28] M. J. ABLOWITZ and H. SEGUR: *J. Fluid Mech.*, **92**, 691 (1979).
[29] V. E. ZAKHAROV and S. V. MANAKOV: *Sov. Phys. JETP*, **44**, 106 (1976).
[30] H. SEGUR and M. J. ABLOWITZ: *J. Math. Phys. (N.Y.)*, **17**, 710 (1976).
[31] H. SEGUR: *J. Math. Phys. (N.Y.)*, **17**, 714 (1976).
[32] N. J. ZABUSKY and M. D. KRUSKAL: *Phys. Rev. Lett.*, **15**, 240 (1965).
[33] H. P. MCKEAN and E. TRUBOWITZ: *Commun. Pure Appl. Math.*, **29**, 143 (1976).
[34] B. A. DUBROVIN and S. P. NOVIKOV: *Sov. Phys. JETP*, **40**, 1058 (1975).
[35] P. F. BYRD and M. D. FRIEDMAN: *Handbook of Elliptic Integrals* (New York, N. Y., 1971).
[36] E. FERMI, J. PASTA and S. ULAM: *Studies of Nonlinear Problems*, Los Alamos Report LA 1940 (1955); reprinted in *The Collected Papers of Enrico Fermi* (Chicago, Ill., 1965).
[37] R. M. MIURA, C. S. GARDNER and M. D. KRUSKAL: *J. Math. Phys. (N. Y.)*, **9**, 1204 (1968).
[38] W. MAGNUS and S. WINKLER: *Hill's Equation* (New York, N. Y., 1979).
[39] C. L. SIEGEL: *Topics in Complex Function Theory*, Vol. II (New York, N. Y., 1971).
[40] N. J. ZABUSKY: *J. Phys. Soc. Jpn.*, **26**, supplement 196 (1969).
[41] E. N. LORENZ: *J. Atmos. Sci.*, **20**, 130 (1963).
[42] M. J. ABLOWITZ, A. RAMANI and H. SEGUR: *Lett. Nuovo Cimento*, **23**, 333 (1978).
[43] M. J. ABLOWITZ, A. RAMANI and H. SEGUR: *J. Math. Phys. (N. Y.)*, **21**, 715 (1980).
[44] M. J. ABLOWITZ, A. RAMANI and H. SEGUR: *J. Math. Phys. (N. Y.)*, **21**, 1006 (1980).
[45] F. L. INCE: *Ordinary Differential Equations* (New York, N. Y., 1956).
[46] V. E. ZAKHAROV: *Sov. Phys. JETP*, **38**, 108 (1974).
[47] M. WADATI: *J. Phys. Soc. Jpn.*, **32**, 1681 (1972).
[48] M. J. ABLOWITZ, D. J. KAUP, A. C. NEWELL and H. SEGUR: *Phys. Rev. Lett.*, **30**, 1262 (1973).
[49] J. B. MCLEOD and P. J. OLVER: *The connection between completely integrable, partial differential equations and ordinary differential equations of Painlevé-type*, preprint.
[50] K. A. ROBBINS: *SIAM J. Appl. Math.*, **36**, 457 (1979).

Small-Scale Ocean Waves.

J. L. HAMMACK

Department of Civil Engineering, University of California - Berkeley, Cal.

1. – Introduction.

Presented below are three lectures on various aspects and types of ocean waves. In all cases we are concerned with small-scale waves for which gravitation provides the dominant restoring force. The relevant length and time scales of these waves are small compared to the Kelvin-Rossby radius of deformation and inertial period, respectively, and Coriolis effects may be neglected. In addition, the small length scales (relative to Earth's radius) permit us to ignore the curvature of the ocean surface and adopt a plane-Earth approximation. Various other approximations will be adopted as appropriate in order to obtain model equations which are tractable analytically. (Fortunately, these tractable models appear to remain relevant for geophysical phenomena!) In particular, we are interested in weakly nonlinear systems in which the nonlinear effects manifest slowly; both nonlinear self-interacting and resonant-triad systems are considered.

An outline of these lectures is as follows. In sect. **2** we examine two nonlinear model equations for the evolution of gravity waves: the Korteweg-deVries (KdV) equation for long barotropic and baroclinic waves and the nonlinear Schrödinger (NLS) equation for short barotropic waves. After a brief review of the asymptotic (large time) solution of these equations by inverse-scattering theory, experimental data are presented which demonstrate the reality of soliton predictions—at least on laboratory scales. In sect. **3** we exploit the exact solution of the KdV equation and its linear approximations in order to derive rather precise criteria for modelling the evolution of long-wave initial data on geophysical or laboratory scales. The modelling criteria of sect. **2** are then applied to a typical oceanic tsunami in order to choose relevant model equations for propagating the (barotropic) wave from its generation region to specific target sites. Finally, in sect. **4** we examine the nonlinear excitation of « edge » waves near shore by linear wave trains incident from deep water. Both theoretical and experimental results are presented which document the excitation of two progressive edge wave modes through a nonlinear resonant-triad interaction with the wave reflected after normal incidence from offshore.

Much of the material in sect. **2** and **3** has appeared previously in the literature. Hence, many of the details concerning experimental equipment, procedures and analysis is omitted; a thorough discussion may be found in the cited references. The material presented in sect. **4** on edge waves is recent and not presently available in the literature. For clarity, a more detailed discussion of this material is presented. We also note here that the notation between sect. **2**, **3** and **4** is not necessarily consistent.

2. – Water wave solitons.

2·1. *Long gravity waves.* – Consider two fluid layers with uniform densities $\varrho_1 \leqslant \varrho_2$ resting in a gravitationally stable configuration (see fig. 1) on a horizontal and impermeable bed of infinite lateral extent. The upper layer possesses a free surface S_f along which pressure is constant; surface energy effects on S_f and the fluid-fluid interface S_i are negligible. Required are the two-dimensional, inviscid, irrotational motions which are bounded everywhere and evolve

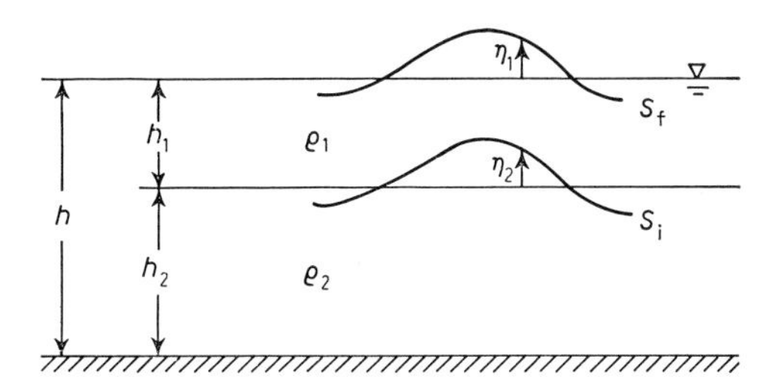

Fig. 1. – Definition sketch of the fluid domain.

from given initial distribution of velocities, free-surface deformation and interfacial deformation—the classical water wave problem. In particular, we are interested in the barotropic displacement η_1 of S_f from its static-equilibrium position and the baroclinic displacement η_2 of S_i from its equilibrium position. To derive the KdV equation as the appropriate model for both displacements from the governing equations, the following assumptions are necessary. First, characteristic wave lengths k^{-1} (an inverse wave number) must be long relative to the total fluid depth, *i.e.* $(kh)^2 \ll 1$, so that dispersive effects are weak. Second, characteristic wave amplitudes a are small relative to the total depth, *i.e.* $a/h \ll 1$, so that nonlinearity is weak. Third, both weak effects of dispersion and nonlinearity are approximately equal, *i.e.* $\varepsilon \equiv a/h \sim (kh)^2$. For simplicity, we will further assume at the outset that density differences in our two-layer ocean are small (as in its geophysical counterpart) so that $\Delta \equiv (\varrho_2 - \varrho_1)/\varrho_2 \ll 1$. Expanding the dependent parameters in the governing equations in a power series in ε and introducing the multiple time scales $t_0 = t$, $t_1 = \varepsilon t, \ldots$, one finds at leading order that the evolution equations for the initial wave are hyperbolic (nondispersive) and linear. At this order, which corresponds to the fast time scale t_0, an initial disturbance decomposes into four modes consisting of left- and right-running, barotropic and baroclinic modes. The phase speeds C_1 and C_2 of these modes are

(1a) barotropic (surface) waves: $C_1^2 = gh$;

(1b) baroclinic (interfacial) waves: $C_2^2 = g' h_1 h_2/h$;

where we have invoked the Boussinesq limit $\Delta \to 0$ with $g\Delta \equiv g' \ll g$ remaining finite. All wave modes propagate with permanent form and do not interact with each other or themselves; the baroclinic modes propagate much slower, $O(\varepsilon^{\frac{1}{2}})$, than the barotropic modes.

At the next order (ε^2) weak nonlinear effects and dispersion occur. Each wave mode experiences a self-interaction on the slow time scale $t_1 = \varepsilon t$, but no interactions between modes occur due to their rapid separation by the phase speed differences of (1). The self-interaction of the right-running barotropic mode is governed by a dimensional equation of the form

$$\eta_{1_t} + C_1 \eta_{1_x} + \tfrac{3}{2} C_1 \eta_1 \eta_{1_x} + \tfrac{1}{6} C_1 h^2 \eta_{1_{xxx}} = 0 . \tag{2}$$

A more convenient choice of nondimensional variables for describing these waves is

$$\begin{cases} \chi = (x - C_1 t)/h , & \tau = \frac{1}{6}(g/h)^{\frac{1}{2}} t , \\ f(\chi, \tau) = \frac{3}{2}\eta(x, t)/h . \end{cases} \tag{3}$$

In terms of these variables, (2) reduces to the KdV equation with the common form

$$f_\tau + 6 f f_\chi + f_{\chi\chi\chi} = 0 . \tag{4}$$

In a similar manner, the dimensional equation governing the self-interaction of the right-running baroclinic waves is

$$\eta_{2_t} + C_2 \eta_{2_x} + \frac{3}{2} C_2 \left(\frac{1}{h_2} - \frac{1}{h_1}\right) \eta_2 \eta_{2_x} + \frac{1}{6} C_2 h_1 h_2 \eta_{2_{xxx}} = 0 . \tag{5}$$

By introducing the normalized variables

$$(6)\qquad \begin{cases} \chi = (x - C_2 t)/(h_1 h_2)^{\frac{1}{2}}\,, \\[2mm] \tau = \dfrac{1}{6}\,(g'/h)^{\frac{1}{2}} t\,, \\[2mm] f = \dfrac{3}{2}\left(\dfrac{1}{h_2} - \dfrac{1}{h_1}\right)\eta\,, \end{cases}$$

eq. (5) reduces to the KdV equation given by (4).

The asymptotic solution of the KdV equation for arbitrary initial data $f(\chi, 0) \equiv f_0$ by inverse-scattering theory has been described in detail by SEGUR [1] *inter alios*. Here we briefly list features of the asymptotic solution to be illustrated in the laboratory experiments.

i) An initial disturbance evolves into a finite number of permanent waves (solitons) ordered by their amplitude. When the solitons are well separated, the local shape of each is given by

$$(7)\qquad f = \alpha\,\mathrm{sech}^2\left\{(\alpha/2)^{\frac{1}{2}}(\chi - \chi_0 - 2\alpha\tau)\right\},$$

where α and χ_0 are constants. The rank-ordered solitons are followed by a dispersive train (radiation) of oscillatory waves.

ii) The number N of solitons evolving from initial data of finite extent, say $f_0 = 0$ for $\chi < \chi_1$ and $\chi > \chi_2$, is equivalent to the number of zeros of φ for $\chi > \chi_1$, where φ satisfies

$$(8)\qquad \begin{cases} \dfrac{d^2\varphi}{d\chi^2} + f_0(\chi)\varphi = 0\,, \\[2mm] \varphi(\chi_1) = 1\,, \qquad \dfrac{d\varphi}{d\chi}(\chi_1) = 0\,. \end{cases}$$

iii) When the net volume V (or mass) in the initial wave is finite and positive, *i.e.*

$$(9)\qquad V = \int_{-\infty}^{\infty} f_0(\chi)\,d\chi > 0\,,$$

at least one soliton emerges.

iv) When $f_0 \leqslant 0$ everywhere, no solitons emerge and the asymptotic solution consists (only) of the radiation components.

v) Two other important classes of data are those for which $V < 0$ with $f_0 \nleqslant 0$ for all χ and those for which $V = 0$. No general statements regarding the asymptotic solution for these cases is provided theoretically. Experiments demonstrate that the evolution of solitons depends on the detailed structure of the initial data.

2˙1.1. Experimental procedures. In order to illustrate the applicability of the KdV equation as a model for long water waves, a series of experiments is conducted in a laboratory wave tank 31.6 m long, 61 cm deep and 39.4 cm wide. For studying the barotropic motions in a system with small density differences, the density stratification plays no role; hence, in these experiments a uniform-density fluid of depth h is used. A detailed description of these experiments is given by HAMMACK and SEGUR [2]. In order to study the solution of baroclinic waves, the tank is stratified with fresh water and brine. Details of these experiments can be found in [3] as well as [4].

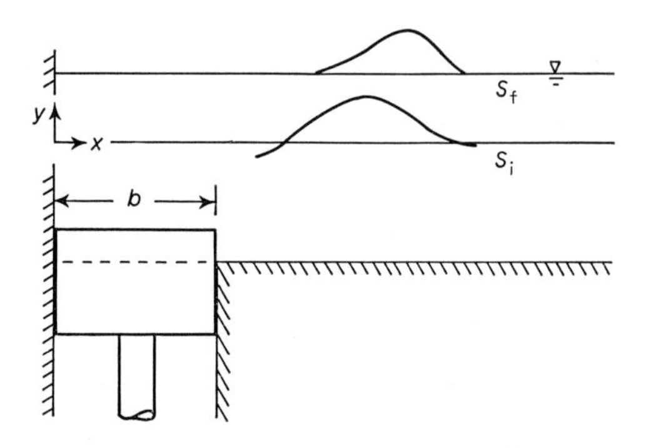

Fig. 2. – Schematic drawing of the wave generator.

In the long-wave experiments reported here, the wavemaker consists of a rectangular piston located in the tank bottom adjacent to an upstream endwall. The piston spans the tank width and has a length b in the direction of wave motion (see fig. 2); lengths of $b = 30.5$ cm and 61 cm are used. The vertical motion of the piston is controlled by an electro-hydraulic servo-system and completely user specified. In a typical experiment, the piston is moved for a finite time interval. Differences between the initial and final position of the piston permit the net volume V in the generated wave train to be calculated. Various initial data are generated by varying the time-displacement history of the piston.

Both surface and internal wave amplitudes are measured at fixed locations (stations) along the tank. We note here that differences between the temporal variation in wave amplitude at a fixed spatial position and the spatial variation of wave amplitude at a corresponding fixed time are small, $O(\varepsilon)$, and neglected in all calculations.

2'1.2. Results for barotropic wave evolution.

2'1.2.1. Initial data with $V > 0$. – Figures 3 and 4 illustrate the evolution of two barotropic waves with a net positive volume. Normalized wave amplitudes are presented at four succeeding stations along the tank in a co-ordinate system which moves with the linear (nondispersive) speed C_1. Note that the leading portion of the wave system appears at the left in these figures. Shifts of the waves to the left (right) at succeeding stations indicate phase speeds greater (less) than C_1 in this co-ordinate system.

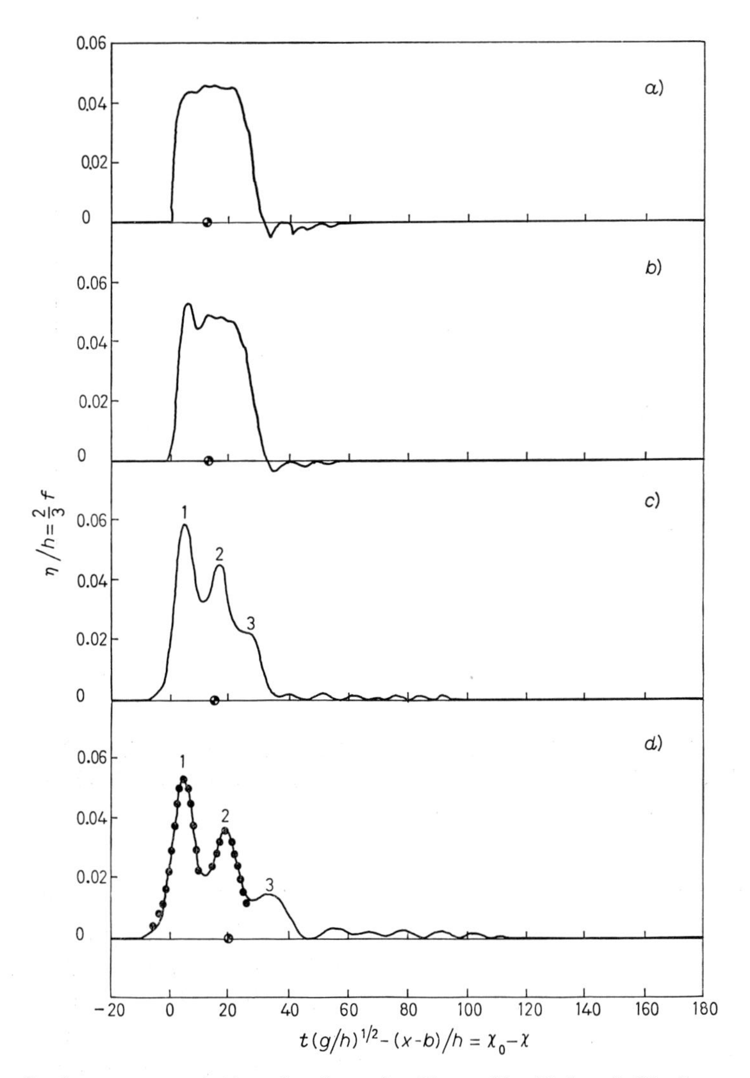

Fig. 3. – Surface wave evolution: $h=5$ cm, $b=61$ cm, $V=30.5$ cm², $N=3$; —— measured profiles, • soliton profiles computed using (7); *a*) $(x-b)/h=0$, *b*) $(x-b)/h=20$, *c*) $(x-b)/h=120$, *d*) $(x-b)/h=400$.

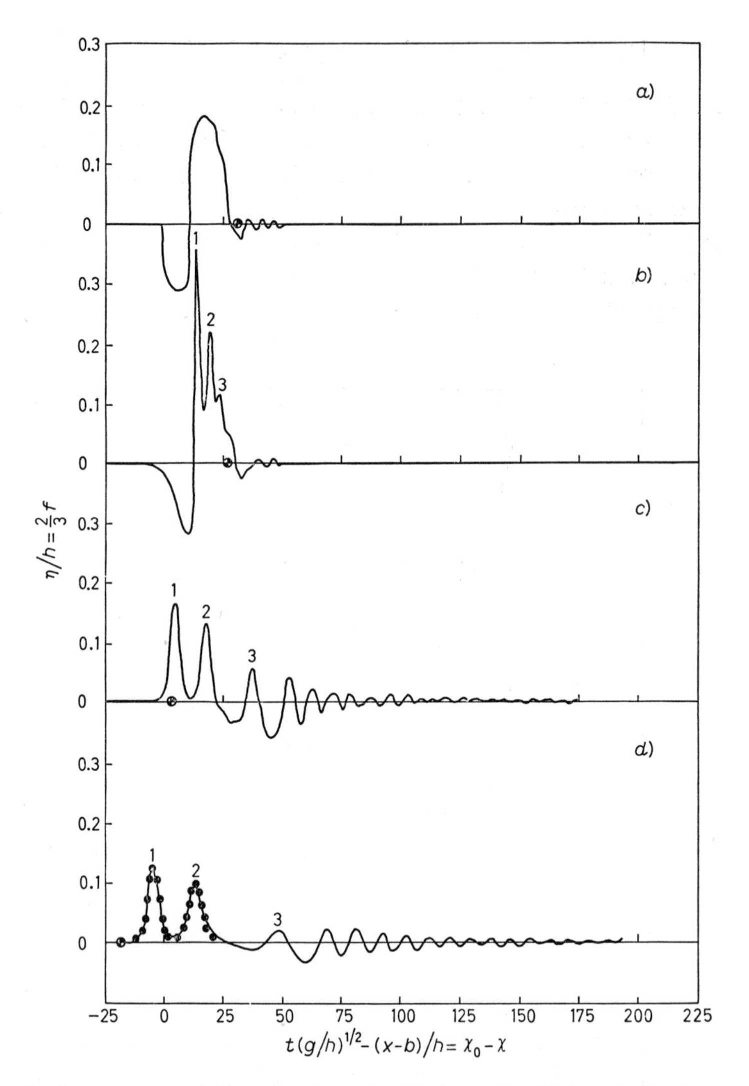

Fig. 4. – Surface wave evolution: $h=5$ cm, $b=30.5$ cm, $V=30.5$ cm^2, $N=3$; ——— measured profiles, • soliton profiles computed using (7); *a*) $(x-b)/h=0$, *b*) $(x-b)/h=20$, *c*) $(x-b)/h=200$, *d*) $(x-b)/h=400$.

The initial wave at $(x-b)/h=0$ in fig. 3 is rectangular and positive and appears to sort itself into three positive waves identified by separate crests (local maxima) during propagation. These three waves are rank-ordered by amplitude and are followed by a weak train of dispersive waves whose speed is much less than C_1. In order to examine the local shape of the leading two waves, theoretical soliton profiles defined by (7) have been superposed on the measured data at the last measurement station (using the measured wave amplitudes for specifying the parameter α). These profiles are not extended into regions where the two waves are still interacting strongy with adjacent wave structure. Clearly, the measured waves appear to be locally KdV solitons. Further evidence that the three lead waves at the last measurement station are solitons is provided by numerically integrating (8) using the wave profile at $(x-b)/h=0$ as the initial data f_0. Computations indicate that $N=3$ solitons should evolve in agreement with the observed pattern of evolution.

The evolution of a more complicated initial wave with $V>0$ is shown in fig. 4, where a leading negative wave is followed by a (larger) positive wave. After only twenty depths of propagation, the positive wave has separated into

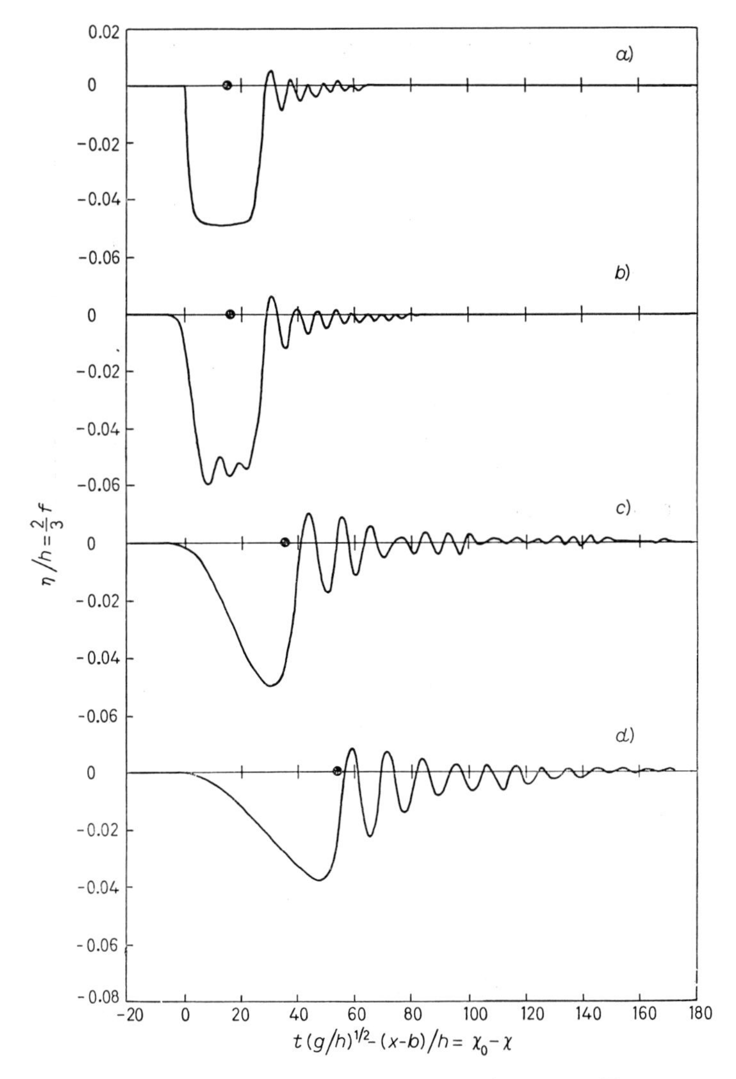

Fig. 5. – Surface wave evolution: $h=5$ cm, $b=61$ cm, $V=30.5$ cm^2, $N=0$; $a)$ $(x-b)/h=0$, $b)$ $(x-b)/h=20$, $c)$ $(x-b)/h=120$, $d)$ $(x-b)/h=400$.

three separate crests, while the negative wave remains essentially unchanged. During subsequent propagation the three labelled crests of the positive wave appear to retain their integrity as they progress through the leading negative wave and emerge at the front of the wave train. At the last measurement station, labelled waves 1 and 2 clearly resemble KdV solitons. The third wave is still interacting with the once-leading negative wave at the last station and cannot be unmistakably identified as a soliton. However, computations with (8) using the wave at $(x - b)/h = 0$ as f_0 yield $N = 3$, strongly suggesting the third wave is indeed a soliton. We note that other experiments also indicate that

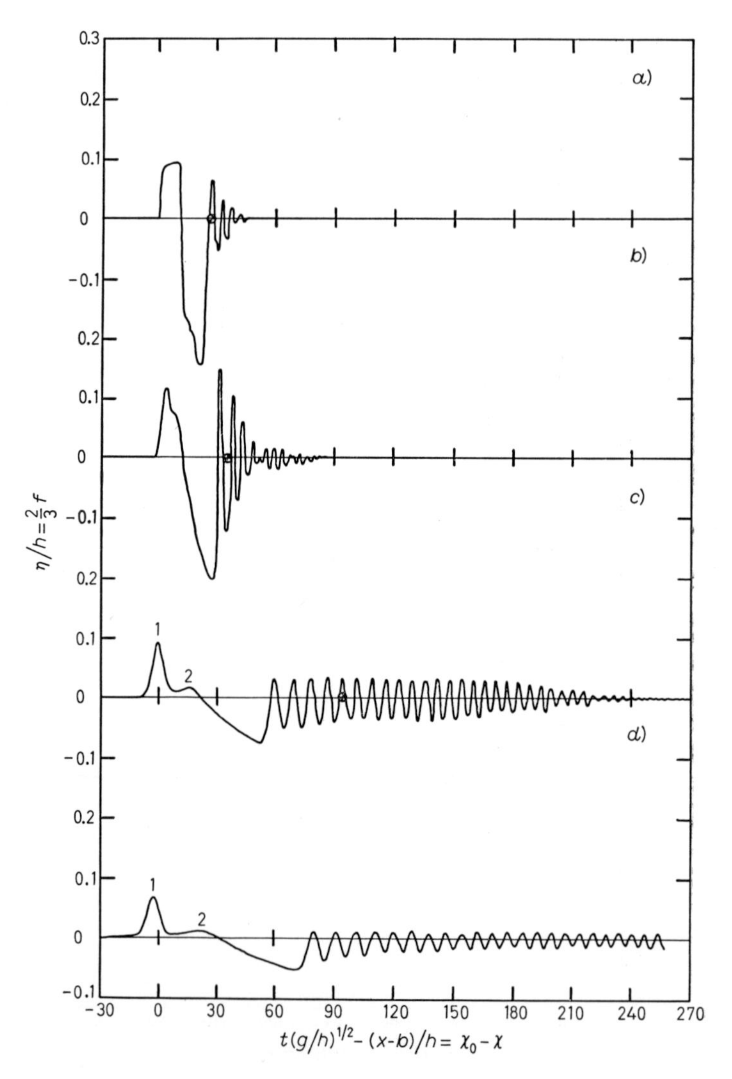

Fig. 6. – Surface wave evolution: $h = 5$ cm, $b = 30.5$ cm, $V = -30.5$ cm², $N = 1$; *a*) $(x - b)/h = 0$, *b*) $(x - b)/h = 20$, *c*) $(x - b)/h = 200$, *d*) $(x - b)/h = 400$.

solitons evolve from the positive waves in the initial wave and can be identified long before asymptotic conditions are achieved.

2'1.2.2. Initial waves with $V < 0$. – Results for the evolution of two barotropic wave systems with $V < 0$ are shown in fig. 5 and 6. The initial wave in fig. 5 is the negative counterpart of the experiment shown in fig. 3, where three solitons appeared to evolve. Over similar distances of propagation, no solitons appear to evolve in fig. 5—just as expected for initial data with $f_0 \leqslant 0$

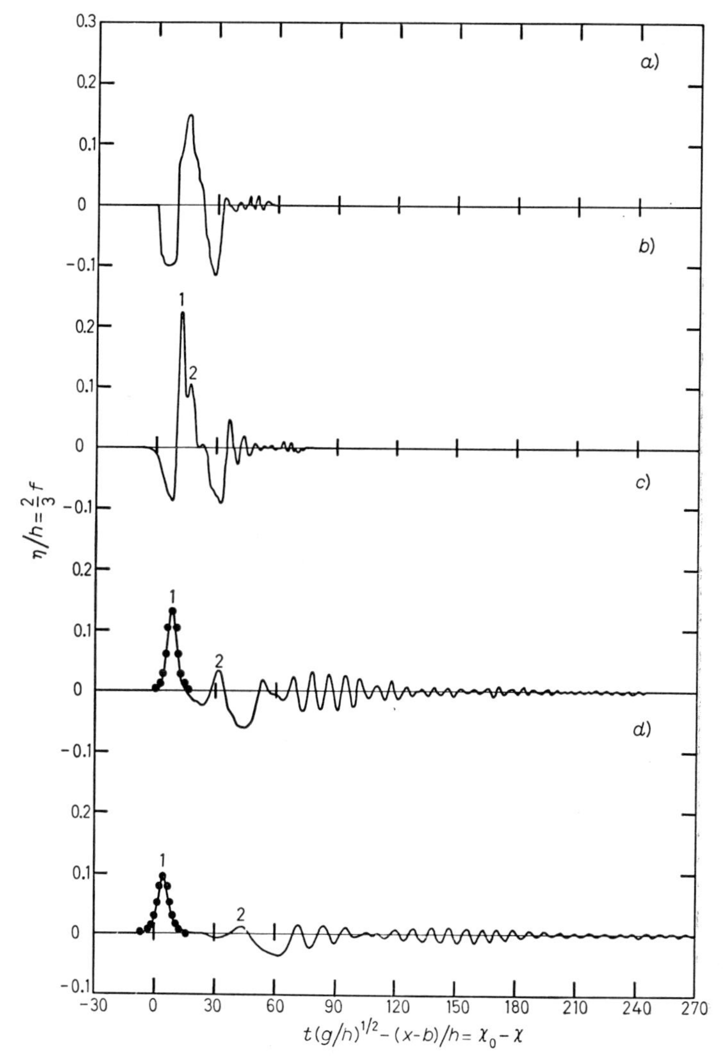

Fig. 7. – Surface wave evolution: $h = 5$ cm, $b = 30.5$ cm, $V = 0$, $N = 2$; ——— measured profiles, • soliton profiles computed using (7); *a*) $(x-b)/h = 0$, *b*) $(x-b)/h = 20$, *c*) $(x-b)/h = 200$, *d*) $(x-b)/h = 400$.

for all χ. Instead, a negative wave evolves whose frontal slope decreases and lengthens with time. This lead wave is followed by a train of strongly dispersive waves with phase speeds much less than C_1. In fact, the wave structure of fig. 5 represents the radiation solution of the KdV equation as shown with quantitative tests by HAMMACK and SEGUR [5].

Further evidence for the evolution of complicated waves with $V < 0$ is shown in fig. 6. In this case the lead positive wave evolves into one or possibly two waves, while the negative wave evolves in a manner similar to that observed in fig. 5. Computations using the initial wave in (8) indicate that one soliton should develop asymptotically. Hence, the second labelled wave is expected to eventually disappear, as it appears to be doing.

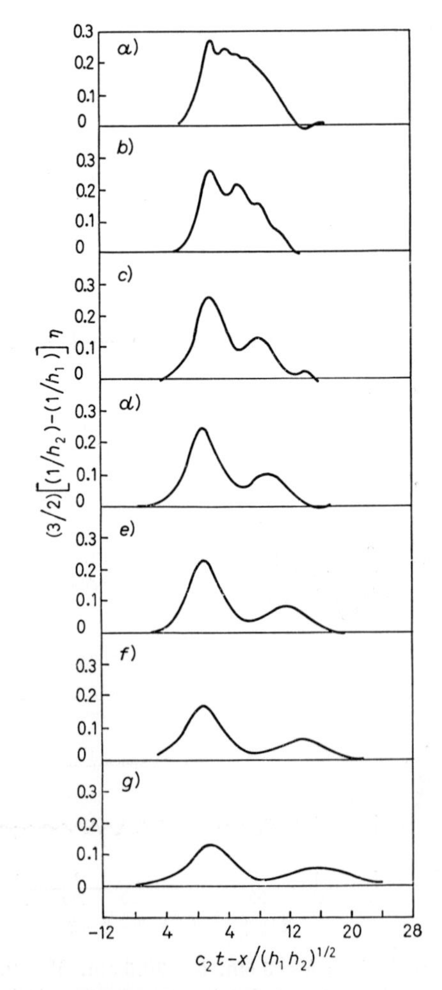

Fig. 8. – Internal-wave evolution: $h_1 = 45$ cm, $h_2 = 5$ cm, $b = 61$ cm, $\Delta = 0.05$; *a*) $x/\sqrt{h_1 h_2} = 1.7$, *b*) $x/\sqrt{h_1 h_2} = 6.7$, *c*) $x/\sqrt{h_1 h_2} = 33.3$, *d*) $x/\sqrt{h_1 h_2} = 60.0$, *e*) $x/\sqrt{h_1 h_2} = 100$, *f*) $x/\sqrt{h_1 h_2} = 151$, *g*) $x/\sqrt{h_1 h_2} = 191$.

2'1.2.3. Initial waves with $V = 0$. – Figure 7 illustrates the evolution of initial data with $V = 0$ consisting of a positive wave preceded and followed by negative waves. Applications of (8) to the initial wave suggest that two solitons should evolve. The large positive wave in the initial data quickly separates into two crests which appear to migrate through the lead negative wave during subsequent propagation. At the last station, labelled wave 1 has progressed to the front and clearly has the shape of a soliton everywhere. The second labelled wave still appears to be interacting with the frontal slope of the once-leading negative wave. A trailing train of dispersive waves similar to that of fig. 5 also evolves.

2'1.3. Results for baroclinic wave evolution. The evolution of baroclinic long waves at the interface of a stratified fluid with $h_1 = 45$ cm, $h_2 = 5$ cm and $\Delta = 0.05$ is shown in fig. 8. Wave amplitudes recorded at seven stations along the tank in a co-ordinate system that moves with the linear speed C_2 are presented. The initial wave is positive, and, according to calculations by (8), should evolve two solitons. The observed evolution is in agreement with this prediction. (A strongly damped train of oscillatory waves has been omitted in fig. 8.) A more quantitative comparison of the lead-wave profile with the theoretical shape (7) at the last four stations of measurements is shown in fig. 9. The agreement with the measured data in fig. 9 is excellent. Further results for baroclinic wave evolution are presented in [4].

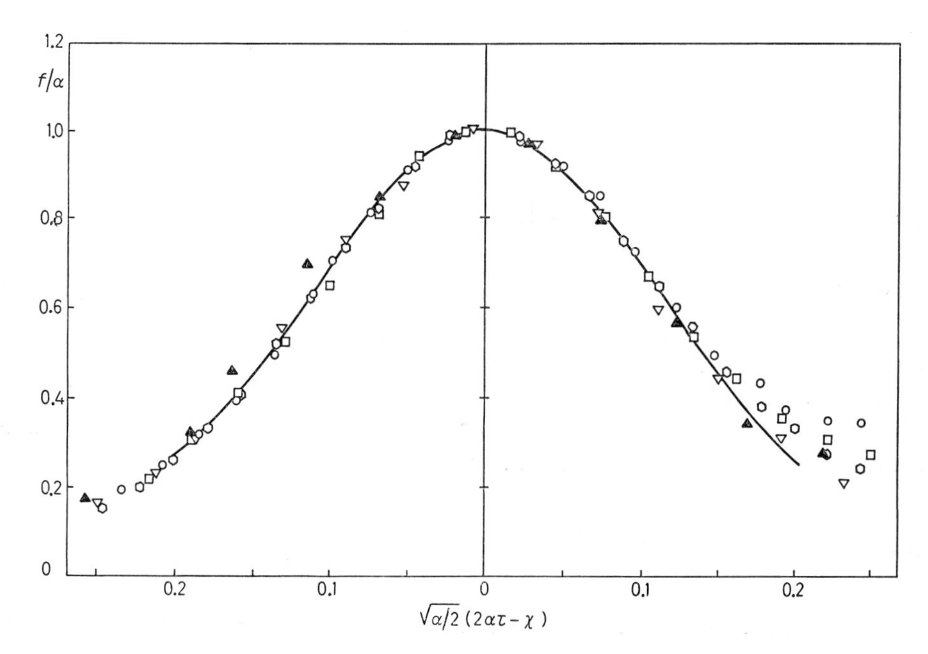

Fig. 9. – Comparison of leading wave profiles with theoretical (internal) KdV soliton: ○ $\chi = 33.3$, □ $\chi = 60$, ⬡ $\chi = 100$, ▲ $\chi = 151$, △ $\chi = 191$.

2'2. *Short gravity waves.* – We now turn our attention to the evolution of short gravity waves, *i.e.* waves whose characteristic length k^{-1} is comparable to or greater than the local ocean depth; only barotropic modes are considered. In fact, our interest in solitons requires $kh > 1.36$, which we adopt as the definition of the short-wave regime herein. In order to derive the nonlinear Schrödinger (NLS) equation as an evolution model, we consider a wave train with a dominant and identifiable (mean) wave number k_0. As for long waves, we require weak nonlinearity, which is now characterized by the wave steepness $\varepsilon \equiv ak_0 \ll 1$. The wave system is permitted to have weak modulations such that the variation in wave number δk is small, *i.e.* $\delta k/k_0 \ll 1$. In other words, we are concerned with « narrow-band » wave systems that are weakly nonlinear. Thirdly, we postulate a balance of both small effects so that $\varepsilon = ak_0 \sim \sim \delta k/k_0$. Multiple scale analysis of the governing equations again yields a hierarchy of problems at different orders of ε. At lowest order we recover the linear dispersive waves of Stokes [6] with an amplitude a that is constant. At the next time scale, $t_1 = \varepsilon t$, the wave amplitude is modulated, and we find that the amplitude modulations propagate with the linear group speed C_g. Continuing to the third order with a time scale $t_2 = \varepsilon^2 t$, one finds that the complex amplitude modulation a must satisfy the NLS equation. If we define nondimensional co-ordinates as

$$\left\{\begin{aligned} \chi &= \varepsilon k_0(\chi - C_g t)\,, \\ \tau &= \varepsilon^2 (gk_0)^{\frac{1}{2}} t\,, \\ A &= k_0^2 (gk_0)^{-\frac{1}{2}} a\,, \end{aligned}\right. \tag{10}$$

the NLS equation takes the form

$$iA_\tau + \lambda A_{\chi\chi} + \nu |A|^2 A = 0\,, \tag{11}$$

where λ and ν are known functions of the water depth h, gravitation g and carrier wave number k_0 (or frequency ω_0). Details of the derivation for finite depth can be found in [7] *inter alia.*

Like the KdV equation, the NLS equation can be solved exactly for arbitrary initial data by inverse-scattering theory. When $k_0 h > 1.36$, envelope solitons can occur; the one-soliton solution of (11) is

$$A = \alpha \left|\frac{2\lambda}{\nu}\right|^{\frac{1}{2}} \operatorname{sech}\{\alpha\chi\} \exp[i\lambda\alpha^2\tau]\,, \tag{12}$$

where α is an arbitrary constant related to the envelope amplitude. In general, initial data of finite extent will evolve a finite number of envelope solitons, rank-ordered by the group velocities of their dominant carrier waves and embedded in a dispersive train (radiation) of oscillatory waves which decays in amplitude with time. It is important to note that the speeds of both the solitons and the radiation components are not impacted by nonlinearity at this order. Hence, unlike their long-wave counterparts of the KdV equation, these solitons and radiation components do not separate with time. This suggests that, in order to observe clearly the evolution of envelope solitons even in a contrived laboratory experiment, one must design the initial data such that the wave content of the radiation spectrum at the dominant frequency of the soliton carrier waves is small. Alternatively, one must observe evolution until the inviscid decay of the radiation (by frequency dispersion) combined with viscous decay has progressed sufficiently.

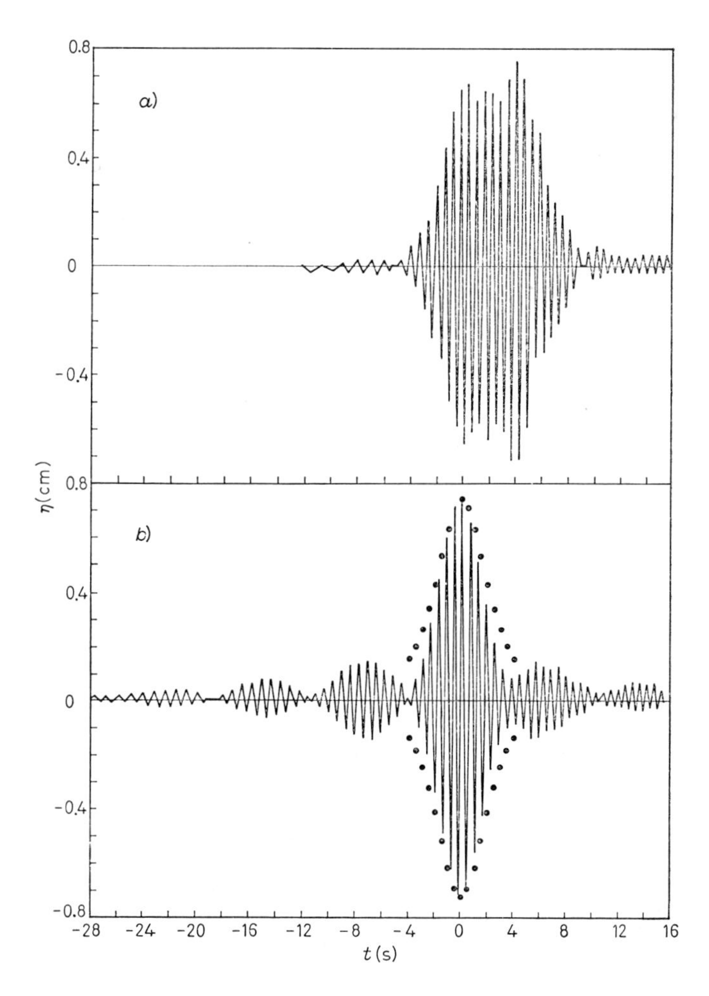

Fig. 10. – Evolution of a narrow-banded wave packet: —— measured profiles, • soliton profile computed using (12); *a*) $x = 0$, *b*) $x = 39$ m.

Although a quantitative comparison of theory and experiment for the NLS equation analogous to that of the KdV equation has not been performed, qualitative tests by YUEN *et al.* [8] provide evidence for its applicability to narrow-band systems. We present in fig. 10 an (unpublished) experiment which illustrates the evolution of an initial wave packet whose envelope amplitude and carrier wave frequency are modulated; the mean wave frequency is $\omega_0^{-1} = 0.6$ s. The experiment is conducted in a tank approximately 50 m long, 2.5 m wide, and a water depth of $h = 1$ m. After 45 m of propagation, the initial packet evolves into a collection of wave groups which are ordered by the group speed of their dominant carrier wave. Energy is concentrated in the group with carrier waves at the dominant period $\omega^{-1} \simeq 0.6$ s of the initial data. The envelope of this group clearly has the shape of the soliton profile (12) shown superposed on the measured wave. The envelopes for the other wave groups do not agree with (12), and apparently these groups represent the radiation components in the initial data.

Although testing of the NLS equation is not complete, there is growing evidence (see [9]) that the dynamics of short water waves is much more complicated than that of long waves, and more complicated models are required. The restriction of the NLS equation to narrow-band systems and its (probable) instability to transverse perturbations appear especially severe for many geophysical applications.

3. – Modelling criteria for long water waves.

Even with all the assumptions implicit in the classical water wave problem introduced in sect. **2**, the general equations remain intractable analytically. Hence, we resort to further approximations such as those required to yield the KdV and NLS equations. Generally, approximations are formalized by perturbation expansions in terms of a small parameter(s). In application of these approximate models, questions naturally arise as to when the inequalities used in ordering the physics are actually satisfied, *i.e.* how small is small? For evolution models, the relevant question distills into « during what time interval does a particular approximation correctly represent the general solution? ». Closely related to this topic are questions of similitude between laboratory models and their geophysical-scale counterparts. The answers to all of these questions involve determination of nondimensional parameters which characterize the phenomenon in question. In practice these parameters are often deduced through the ingenious use of dimensional analysis, by examination of the governing equations, or using the most preferred but least available method, examination of exact solutions of the general and approximate model equations. Here we examine the KdV equation and exploit exact solutions of its various approximations in order to develop rather precise criteria for modelling the propagation of long water waves. These criteria are then applied to the problem of tsunami propagation across ocean basins—the topic which precipitated the analysis. Further details are presented by HAMMACK and SEGUR [10].

3·1. *Analysis*. – In order to obtain definitive results, we adopt the following point of view. If the initial wave f_0 is sufficiently smooth and localized, and if $\varepsilon_1 = a/h$ and $\varepsilon_2 = (kh)^2$ based on the initial data are both small, then the KdV equation (4) will be the approximate model eventually. (In fact, we know that (4) is appropriate when $\varepsilon_1 = O(\varepsilon_2)$.) If ε_1 and ε_2 are not the same order of magnitude initially, then simpler forms of the KdV equation may be applicable for some time interval, *e.g.*,

(13) $$f_\tau = 0$$ (linear, nondispersive model),

(14) $$f_\tau + f_{\chi\chi\chi} = 0$$ (linear, dispersive model),

(15) $$f_\tau + ff_\chi = 0$$ (nonlinear, nondispersive model).

URSELL [11] showed that the nondimensional parameter

(16) $$U = \varepsilon_1/\varepsilon_2 = ak^{-2}/h^3,$$

which we shall call the Ursell number, is an important indicator to determine which of these models is appropriate. One should use (14), (4), or (15), depending on whether $U \ll 1$, $U = O(1)$, or $U \gg 1$, respectively. URSELL also provided evidence that U is time dependent and will tend to an order-one limit, so that all waves in this category eventually propagate according to (4) as postulated previously. Although there is general agreement with Ursell's results, there is disagreement on how to interpret « order unity » and how to define the relevant length scale k^{-1} for the evolving waves. To make his criteria more precise, we examine here initial data for which $\varepsilon_1 \ll \varepsilon_2 \ll 1$ initially, so that the linearized models (13) and (14) are the relevant approximations of (4). The dimensional length scale k_0^{-1} for the initial data is defined as its overall length, and the dimensional amplitude scale is a_0. Further, we will focus on criteria for modelling the leading wave only. Results for the trailing wave structure and the nonlinear approximation (15) are given in [10].

3·1.1. Linear dispersive theory. Since nonlinearity is small for the postulated initial data, we assume a small parameter $\mu \ll 1$ exists and seek a formal series solution of the KdV equation (4) in the form

(17) $$f = \mu f_1 + \mu^2 f_2 + O(\mu^3).$$

(The required definition of μ will come out of the analysis.) Substituting (17) into (4) yields a hierarchy of problems:

$$(f_1)_\tau + (f_1)_{\chi\chi\chi} = 0\,, \qquad \mu f_1(\chi, 0) = f_0\,, \tag{18}$$

$$(f_2)_\tau + (f_2)_{\chi\chi\chi} = -\,6f_1(f_1)_\chi\,, \quad f_2(\chi, 0) = 0\,, \quad \text{etc.} \tag{19}$$

The solution of (18), which is equivalent to (14), is well known:

$$\mu f_1 = \frac{1}{2\pi}\int_{-\infty}^{\infty} \hat{f}_0(\varkappa)\exp[i(\varkappa\chi + \varkappa^3\tau)]\,d\varkappa\,, \tag{20}$$

where

$$\hat{f}_0(\varkappa) = \int_{-\infty}^{\infty} f_0(\chi)\exp[-\,i\varkappa\chi]\,d\chi \tag{21}$$

is the Fourier transform of the initial data. For practical reasons, we are most interested in the asymptotic form ($\tau \to \infty$) of the solutions for both the linear and nonlinear models. As $\tau \to \infty$ with $|\chi|/\tau \to 0$ in order to remain at the wave front, the asymptotic form of (20) is

$$\mu f_1(\chi, \tau) = \hat{f}_0(0)(3\tau)^{-\frac{1}{3}}\,\mathrm{Ai}\,(\xi) - i\hat{f}_0'(0)(3\tau)^{-\frac{2}{3}}\,\mathrm{Ai}'\,(\xi) - \\ - \tfrac{1}{2}\hat{f}_0''(0)(3\tau)^{-1}\,\mathrm{Ai}''\,(\xi) + O[(3\tau)^{-\frac{4}{3}}]\,, \tag{22}$$

where $\xi = \chi/(3\tau)^{\frac{1}{3}}$ and Ai (ξ) is the Airy function. The coefficients in (22) have simple interpretations:

$$\hat{f}(0) = \int_{-\infty}^{\infty} f_0\,d\chi = V\,, \tag{23a}$$

$$-\,i\hat{f}'(0) = \int_{-\infty}^{\infty} \chi f_0\,d\chi = \beta_1 U_0\,, \tag{23b}$$

$$-\frac{1}{2}\hat{f}''(0) = \int_{-\infty}^{\infty} \chi^2 f_0\,d\chi = \beta_2\frac{U_0^2}{V}\,, \tag{23c}$$

where V is the nondimensional volume of the initial wave, U_0 is an Ursell number for the initial wave, and β_1, β_2 are constants that depend on the details of the initial wave. Thus (22) becomes

$$\mu f_1(\chi, \tau) = V(3\tau)^{-\frac{1}{3}}\left\{\mathrm{Ai}\,(\xi) + \beta_1\frac{U_0}{V}(3\tau)^{-\frac{1}{3}}\,\mathrm{Ai}'(\xi) + \right. \\ \left. + \beta_2\left(\frac{U_0}{V}\right)^2(3\tau)^{-\frac{2}{3}}\,\mathrm{Ai}''(\xi) + \ldots\right\}. \tag{24}$$

From (24) it follows that the time required for this representation to become asymptotic (second term smaller than first, etc.) is at least

$$(3\tau)^{\frac{1}{3}} \gg U_0/|V| . \tag{25}$$

A particular solution of (19) is

$$\mu^2 f_{2p}(\chi, t) = -\left\{ \int_{-\infty}^{\chi} \mu f_1(z, \tau)\, dz \right\}^2 \tag{26}$$

and, if we define

$$\theta(\chi) = \frac{1}{V^2}\left\{ \int_{-\infty}^{\chi} f_0(z)\, dz \right\}^2 , \tag{27}$$

the homogeneous solution of (19) is

$$\mu^2 f_{2h}(\chi, \tau) = V^2 \int_{-\infty}^{\chi} \mathrm{Ai}\,(z)\, dz + \frac{V^2}{2\pi} \int_{-\infty}^{\infty} \{\hat{\theta}(\varkappa) - (i\varkappa)^{-1}\} \exp\,[i(\varkappa\chi + \varkappa^3\tau)]\, d\varkappa . \tag{28}$$

Thus, as $\tau \to \infty$, $|\chi|/\tau \to 0$,

$$\mu^2 f_2(\chi, \tau) = V^2 \left\{ -\left[\int_{-\infty}^{\xi} \mathrm{Ai}\,(z)\, dz \right]^2 + \int_{-\infty}^{\xi} \mathrm{Ai}\,(z)\, dz + \ldots \right\} . \tag{29}$$

If we compare (24) and (29), the appropriate definition of μ is found to be simply

$$\mu = V . \tag{30}$$

In other words, linear dispersive theory along with its large-time asymptotics requires the dimensionless wave volume V to be small. (Note that V is independent of time.) Since the solution of (29) remains $O(1)$ as $\tau \to \infty$, the series (17) cannot remain asymptotic after

$$(3\tau)^{\frac{1}{3}} \sim |V|^{-1} . \tag{31}$$

Thus, if the initial wave has a dimensionless volume V (assumed $\neq 0$) and Ursell number U_0 based on its initial (given) dimensions, asymptotic linear dispersive theory is valid in an interval no longer than

$$U_0/|V| \ll (3\tau)^{\frac{1}{3}} \ll 1/|V| . \tag{32}$$

3·1.2. Nondispersive linear theory. Since our initial wave is postulated to be linear, then $U_0 \ll 1$ and necessarily $|V| \lll 1$, as well. The derivation of the KdV equation outlined in sect. **2** indicated that linear nondispersive theory (13) occurs on the first time scale of evolution. Hence we may write

$$3\tau \ll \varepsilon_2 = (k_0 h)^2 = (U_0/V)^2 \tag{33}$$

for (13) to be applicable.

3·2. *Summary and application of criteria to tsunami propagation.* – In summary, we have examined the evolution of long-wave initial data which is parameterized by an initial Ursell number $U_0 \ll 1$ and volume (or mass) $V \lll 1$. We may model the evolution of these initial data using linear nondispersive theory ($f_\tau = 0$) during a time interval

$$0 \leqslant 3\tau \ll \tau_1 = \left(\frac{U_0}{V}\right)^2 . \tag{34}$$

The next relevant model is linear dispersive theory ($f_\tau + f_{\chi\chi\chi} = 0$) which is valid during

$$\tau_1 \ll 3\tau \ll \tau_2 = \left(\frac{U_0}{V}\right)^3 \tag{35}$$

with its asymptotics becoming valid during

$$\tau_2 \ll 3\tau \ll \tau_3 = \frac{1}{V^2} . \tag{36}$$

Subsequently, we must use the KdV equation ($f_\tau + 6ff_\chi + f_{\chi\chi\chi} = 0$) for

$$3\tau \gg \tau_3 . \tag{37}$$

(The question of the time scale for applicability of KdV asymptotics is discussed by HAMMACK and SEGUR [10].) As an example of the application of these results, consider the dimensional scales adopted by CARRIER [12] for major tsunamis which impact entire ocean basins: $a_0 \simeq 10$ ft, $h \simeq 1.5 \cdot 10^4$ ft, $k_0^{-1} = 2 \cdot 10^5$ ft. Then $U_0 \sim 0.1$ and $V \sim 0.01$ and linear nondispersive theory is valid for times corresponding to propagation distances (using $C_1^2 = gh$) of

$$0 < x \ll 600 \text{ miles} .$$

Linear dispersive theory is appropriate for

$$600 \text{ miles} \ll x \ll 6000 \text{ miles} ,$$

indicating that dispersion may affect this wave over much of typical ocean trajectories. Even so, linear asymptotics do not apply until $x \gg 6000$ miles, which exceeds the length of realistic trajectories. Hence, linear dispersive asymptotics and the KdV equation are not required for describing the lead wave of the tsunami discussed here. (The KdV equation may be used, but it is unnecessarily complicated.) Of course, other factors may be required to accurately model long-term tsunami propagation (such as the variable bathymetry along trajectories and three-dimensional spreading of wave energy); the intent of the analysis here is only to develop insight into the relative importance of dispersion and nonlinearity.

4. – Excitation of standing edge waves on beaches.

4·1. *Introduction.* – In recent years considerable attention has been focused on the occurrence of ocean waves which become trapped and capable of concentrating energy in localized regions (wave guides). Wave trapping can occur whenever gradients exist in a parameter which affects the wave's phase speed, *e.g.* Coriolis parameter, Brunt-Väisälä frequency, current speeds and water depth. STOKES [13] provided the first theoretical evidence of trapping for surface (gravity) waves near the shoreline of a plane sloping beach. STOKES found a normal-mode solution for the (barotropic) departure η of the water surface from its static-equilibrium position (see fig. 11*a*)) of the form

$$\eta_m(x, y, t) = \alpha \sin\beta \exp[-k_m y \cos\beta] \cos k_m x \sin\omega_m t, \tag{38}$$

where $-\operatorname{tg}\beta$ is the beach slope, x points in the longshore direction and y points offshore. The (linear) dispersion relation for the various longshore mode numbers $m = 1, 2, \ldots$ is

$$\omega_m^2 = g k_m \sin\beta, \tag{39}$$

where $k_m = m\pi/b$ is fixed by the beach width b. Note that the Stokes' mode of (38) is periodic in the longshore direction with crests pointing offshore which decay in amplitude with an e-folding distance $y_e = (k_m \cos\beta)^{-1}$. The crest amplitude is maximum at the shoreline with a magnitude $\alpha \sin\beta$, where α is referred to as the run-up amplitude. (Of course, run-up phenomena cannot be represented by linear solutions; however, the reality of run-up is acknowledged since α is the up-beach length of the horizontal projection of the vertical shoreline amplitude.) URSELL [14] demonstrated that the Stokes' mode is only the lowest ($j = 0$) of a discrete set $j = 0, 1 \ldots, J$ of trapped modes, where J is the greatest integer satisfying

$$(2J + 1)\beta \leqslant \frac{\pi}{2}. \tag{40}$$

These higher discrete modes are algebraically complicated, but retain longshore periodic behavior. Their crest amplitude is maximum at the shoreline, but oscillates in the offshore direction with j nodes while decaying exponentially (see fig. 11*b*)). To complete the set of normal modes, HANSON [15] showed that there exists a continuous spectrum of waves with wave number magnitude k satisfying $\omega^2 > gk$. These modes are even more complicated to describe algebraically, but resemble simple deep-water wave trains far offshore. An example of one of these modes with crests parallel to the shoreline is shown in fig. 11*b*). We emphasize here that the continuous spectrum modes are not trapped like the discrete modes.

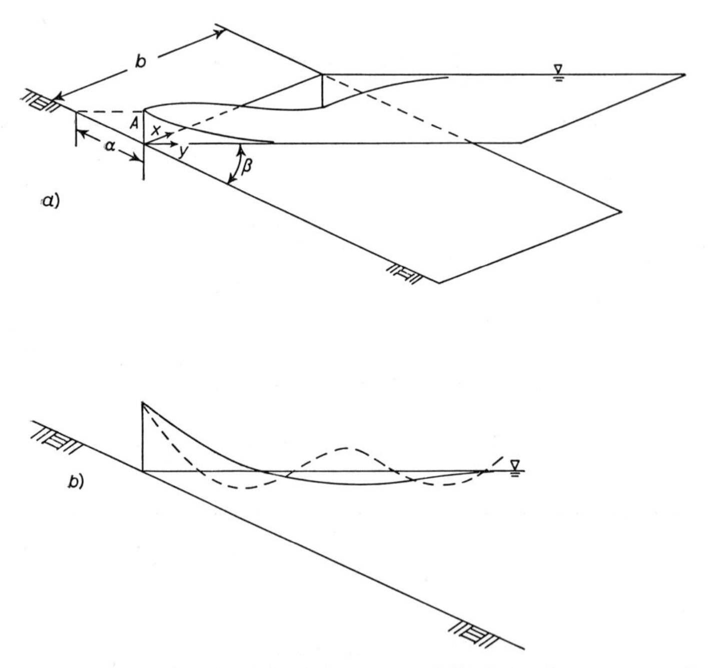

Fig. 11. – *a*) Definition sketch of fluid domain and Stoke's edge wave mode. *b*) Sample offshore profiles of higher edge wave modes: ——— discrete spectrum (trapped) mode, — — — continuous spectrum (untrapped) mode.

Oceanographic interest in edge waves appears to have originated in Isaacs *et al.* [16] suggestion that these waves might be responsible for « surf beat » phenomena. Since then, edge waves have been indicted as a potential mechanism for numerous features of coastal dynamics such as beach cusps [17], crescentic bars [18] and rip currents [19]. The lack of direct observational evidence of edge wave modes on beaches (*i.e.* wave crests pointing offshore!) naturally raises questions as to how and if these modes can be excited. The most important mechanism for extensive generation opportunities of edge waves was provided by GUZA and DAVIS [20], who demonstrated theoretically that Stokes' modes could be excited by simple wave trains normally incident onto a perfectly reflecting beach from deep water. Basically, the nonlinear coupling between the incident/reflected wave of frequency W and background edge wave « noise » of subharmonic frequency $\omega = W/2$ leads to a resonant interaction and growth of the edge wave noise. Eventually other nonlinear processes develop to limit growth; however, at steady state the edge wave amplitude exceeds that of the incident/reflected wave. The entire evolution of the edge wave has been investigated theoretically by GUZA and BOWEN [21] (hereafter referred to as G-B), MINZONI and WHITHAM [22] (hereafter referred to as M-W) and ROCKLIFF [23].

Herein we present quantitative experimental measurements on edge wave excitation by normally incident wave trains. Some of the predictions of G-B and M-W are tested and necessary modifications for imperfectly reflecting beaches are presented. An outline of the presentation is as follows. In subsect. **4**'2 we review the theoretical results with emphasis on the predictions which can be explicitly tested by the experiments. In subsect. **4**'3 a brief description of the experimental facilities and data analysis techniques is presented. A comparison of measured and predicted data appears in subsect. **4**'4, followed by a summary of the major results in subsect. **4**'5. More details of each aspect of this note may be found in [24].

4'2. *Review of the theories.* – The first description of both edge wave excitation and subsequent evolution was developed by GUZA and BOWEN [21] using the shallow-water equations; hence, their results are limited to small beach angles ($\beta \ll 1$). With clever but intuitive reasoning, G-B isolate several processes, analyze each separately, and combine linearly to yield a complete evolution model. (More recently, ROCKLIFF [23] has reproduced some of the G-B results relying more formally on the governing shallow-water equations.) MINZONI and WHITHAM [22] use the full equations and provide the most formal (mathematically) description of the edge wave evolution process; their results remain valid for arbitrary beach slopes $\beta \leqslant \pi/2$. The greater formality of the M-W formulation permits more justifiable ordering of different processes. Hence, the outline below of edge wave evolution is based primarily on their formalism. Following the qualitative description, a quantitative listing of relevant parameters predicted by both M-W and G-B will be presented.

The classical water wave problem for the inviscid, irrotational, barotropic motions of an incompressible ocean in the wedge-shaped region of fig. 11a) may be formulated in terms of a complex velocity potential φ. Initially, we have a linear wave train normally incident onto the beach from deep water; the incident-wave amplitude far offshore is a_i. We also take the incident wave to be perfectly reflected from the beach. (The reflection coefficient is $R = a_r/a_i = 1$, where a_r is reflected-wave amplitude.) Hence, the forcing for the onshore edge wave noise is a standing-wave mode which is a member of the continuous spectrum of normal modes discussed in subsect. 4·1. To simplify (somewhat) the standing-wave description, it is further assumed that the beach slopes are a member of the denumerably infinite set $\beta = \pi/2N$ with $N = 1, 2, \ldots$. The complex potential for the standing wave may be written as $\varphi_s \sim \exp[iWt]$, and we note that $|\varphi_s| \sim \varepsilon \ll 1$. In addition to the primary standing wave, we assume three-dimensional perturbations by Stokes edge wave with complex potential $\varphi_e \sim \exp[i\omega t]$, where $|\varphi_e| \ll \varepsilon \ll 1$ initially. On a very short time scale (stage 1), nonlinear interactions between the primary wave and the edge wave noise are negligible. However, on a longer time scale (stage 2), quadratic interactions arise as a consequence of the nonlinear boundary condition at the free surface and become significant. If we assume the edge wave noise that satisfies the resonance condition $\omega = W/2$ to dominate, quadratic interactions $\varphi_s \varphi_e^*$ (where * denotes complex conjugate) give rise to terms of the form $\exp[iWt/2]$ and contribute to the growth of the subharmonic edge wave noise. This interaction between the linear standing wave and linear edge wave noise produces exponential growth of the form $a \sim \exp[\gamma t]$, where a is the (real) edge wave amplitude at any crest location (*i.e.* $a = a(x, y, t)$, and we take $a(x, 0, t) \equiv A(x, t)$ as the shoreline amplitude) and γ is the initial growth rate. The edge wave noise grows until a later time scale (stage 3) is reached where its finite amplitude leads to the development of processes which limit further growth. First, quadratic self-interactions of the edge wave with the form $\varphi_e \varphi_e \sim \exp[iWt]$ contribute to the offshore standing wave; this is termed radiation by G-B and corresponds to the fact that nonlinear (even second-order) standing edge waves leak energy at frequency 2ω to deep water. Since $|\varphi_s| \sim \varepsilon$ for the disturbed standing wave in stage 3, radiation feedback suggests that a steady state may be reached when $|\varphi_e| \sim \varepsilon^{\frac{1}{2}}$, *i.e.* the edge wave becomes larger locally than the forcing wave. Second, cubic self-interaction terms of the form $\varphi_e \varphi_e \varphi_e^* \sim \exp[iWt/2]$ contribute further to edge wave nonlinearity by modifying its dispersion relation. Hence, the system is retuned and the resonant frequency is shifted. If the forcing was perfectly resonant initially, it will now be off resonance and an effective reduction of the on-resonance growth rate γ will occur. (Of course, if the initial forcing is slightly off resonance, a similar modification in γ is required.) When combined, all of these processes lead to an evolution for (inviscid) edge waves of the form

$$\frac{\mathrm{d}A}{\mathrm{d}t} = (\gamma^2 - \varkappa^2)^{\frac{1}{2}} A - \mu A^3 , \tag{41}$$

where γ is the initial on-resonance growth rate, $\varkappa$ is a measure of the reduction in forcing efficiency due to nonlinear retuning, and μ is a feeback coefficient due to both radiation and retuning. (Note that in (41) we tacitly assume that the edge wave phase remains constant through stage 3. Theoretically, this assumption is invalid; however, the experimental data support its applicability.)

At the risk of misquoting G-B and M-W and possibly introducing numerical errors, we now list quantitative expressions distilled from their studies for the parameters appearing in (41). The initial (on-resonance) growth rate predicted for the edge waves is

$$\text{(42}a\text{)} \qquad \text{G-B:} \quad \gamma = \frac{0.0424 a_i W^3}{g\sqrt{\beta}\beta^2}\,;$$

$$\text{(42}b\text{)} \qquad \text{M-W:} \quad \gamma = \frac{0.0426 a_i W^3}{g\sqrt{\beta}\,\mathrm{tg}^2\beta}\,;$$

where the M-W result is accurate to within 3% for $\beta \leqslant \pi/4$. These predictions are in agreement in the shallow-water limit ($\beta \to 0$). Modification of the initial growth rate by nonlinear retuning in steady state may be conveniently written to the same order of small parameter $\varepsilon_0 = \alpha_0 k$ (where α_0 is the steady-state run-up amplitude of the edge wave) as

$$\text{(43)} \qquad \varkappa = \omega^{(2)} - \omega^{(0)}$$

with $\omega^{(2)}$ and $\omega^{(0)}$ representing the nonlinear and linear natural frequencies of the edge wave, respectively. The shallow-water limits for $\omega^{(2)}$ given explicitly by G-B and inferred from the presentation of M-W are

$$\text{(44}a\text{)} \qquad \text{G-B:} \quad \omega^{(2)} = \omega^{(0)}[1 + 0.055\,\varepsilon_0^2]\,;$$

$$\text{(44}b\text{)} \qquad \text{M-W:} \quad \omega^{(2)} = \omega^{(0)}[1 + 0.012\,\varepsilon_0^2]\,;$$

$\omega^{(0)} = (gk\beta)^{\frac{1}{2}}$ in both cases. The different results of (44a) and (44b) do not appear reconciliable. Predictions for the feedback coefficient μ in the shallow-water limit also differ at the same order of approximation according to

$$\text{(45}a\text{)} \qquad \text{G-B:} \quad \mu = 0.001\,795\,\frac{W^5}{g^2\beta^4}\,;$$

$$\text{(45}b\text{)} \qquad \text{M-W:} \quad \mu = 0.002\,323\,\frac{W^5}{g^2\beta^4}\,.$$

Finally, we note that the steady-state amplitude of the edge wave at the shoreline is found from (41) to be

$$\text{(46}a\text{)} \qquad A_0 = \left[\frac{(\gamma^2 - \varkappa^2)^{\frac{1}{2}}}{\mu}\right]^{\frac{1}{2}}.$$

4'3. *Experimental facilities and procedures.* – A set of five experiments is conducted to study the excitation of standing edge waves on a beach by waves normally incident from offshore. Experiments are performed in a laboratory basin 1.83 m wide, 4.0 m long and 30 cm deep as sketched in fig. 12. The tank is equipped with a mechanical wave generator and a smooth beach whose slope can be varied. An actual beach width of $b = 1.80$ m was used in the experiments by incorporating internal sidewalls over the beach section as indicated in fig. 12. This enables the offshore tank section of uniform depth to be surrounded by energy-absorbing material to minimize unwanted re-

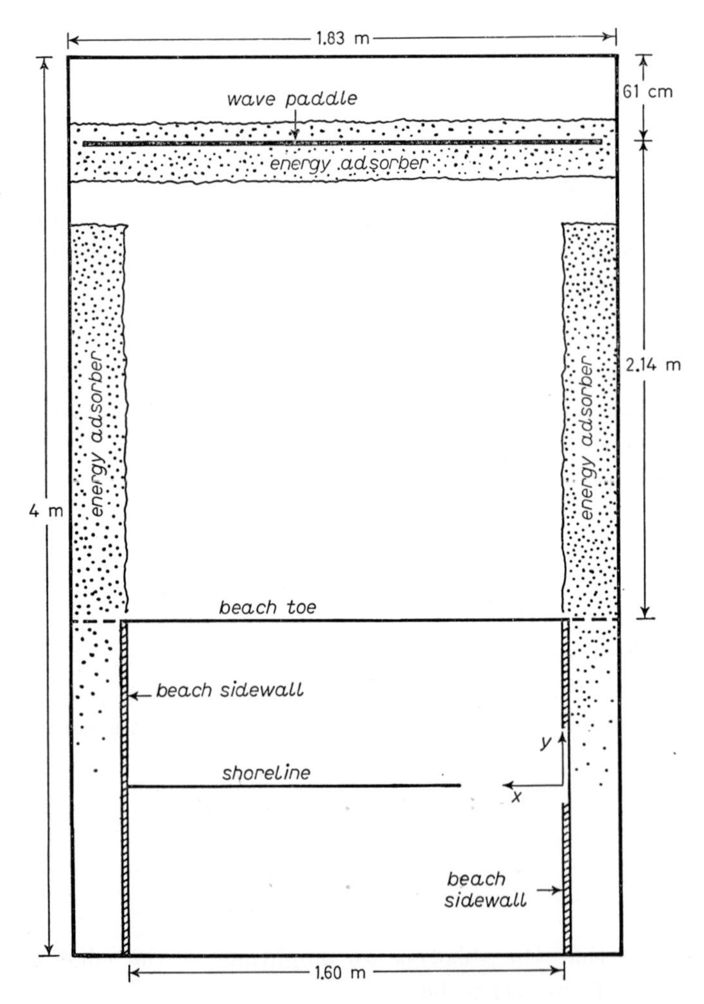

Fig. 12. – Schematic drawing of wave basin—plan view.

flections and simulate open-ocean conditions. Extensive precautions were taken to ensure that the shoreline and wavemaker were parallel and to prevent beach « pumping » during actual experiments.

An offshore water depth of $h = 25.4$ cm was used in all five experiments while the beach slope was varied from $\beta = 15°$ ($N = 6$) to $\beta = 22.5°$ ($N = 4$);

one experiment, run 4, was conducted at $\beta = 20°$, which does not correspond to a slope of $\pi/2N$ for any integer N. It is immediately apparent that the offshore uniform-depth region does not conform to the (unrealistic) mathematical model where the water depth increases linearly offshore. However, the offshore distance over the sloping beach section always exceeded three e-folding distances ($> 95\%$ decay) for the edge wave modes excited (assuming Stokes' modes). Based on [25], it is not expected that the edge wave dispersion relation is affected by the uniform depth offshore in any of the experiments.

In all experiments wave amplitudes are measured at three locations; one gage onshore over the sloping beach and two gages offshore in the uniform-depth region. The onshore gage is always positioned at a longshore location corresponding to an edge wave antinode. The reality of edge waves very near the shoreline requires that they deviate from the profile predicted by (38); hence, the onshore gage is located approximately one e-folding distance away from the shoreline and always outside of the surf zone.

The procedures adopted for each of the five experimental runs are as follows. First, the frequency ω of the desired edge wave is calculated based on the beach width $b = 1.6$ m; the offshore wavemaker is then adjusted to a frequency $W = 2\omega$. The incident and reflected wave amplitudes (a_i and a_r) of the partial standing-wave systems are then determined from simultaneous measurements at the offshore gages in the absence of edge wave excitation. Edge wave excitation is suppressed by inserting a thin plate perpendicular to the shoreline which penetrates the surf zone; the plate introduces boundary conditions which destroy the resonance condition necessary for rapid edge wave growth. The thin plate is then removed permitting the evolution of the edge wave. Once the edge wave attains a steady state, the wavemaker is stopped and the (unforced) decay of the edge wave over the sloping beach is measured. Hence, each experiment produces a time series of the water surface elevation onshore and offshore containing stage 1, stage 2, stage 3, and additional stage 4 where edge wave forcing is terminated and viscous damping forces dominate.

4·4. *Comparison of experiment and theory.* – Table I summarizes the measured data for the partial standing waves generated by the wavemaker/beach system in the absence of edge waves. The incident and reflected wave am-

TABLE I. – *Beach and offshore standing-wave properties.*

Run	β	W (rad/s)	a_i (cm)	a_r (cm)	R
1	15°	9.83	0.82	0.11	0.134
2	15°	7.60	0.62	0.31	0.500
3	18°	9.82	0.87	0.30	0.345
4	20°	10.65	0.90	0.32	0.356
5	22.5°	10.86	0.63	0.29	0.460

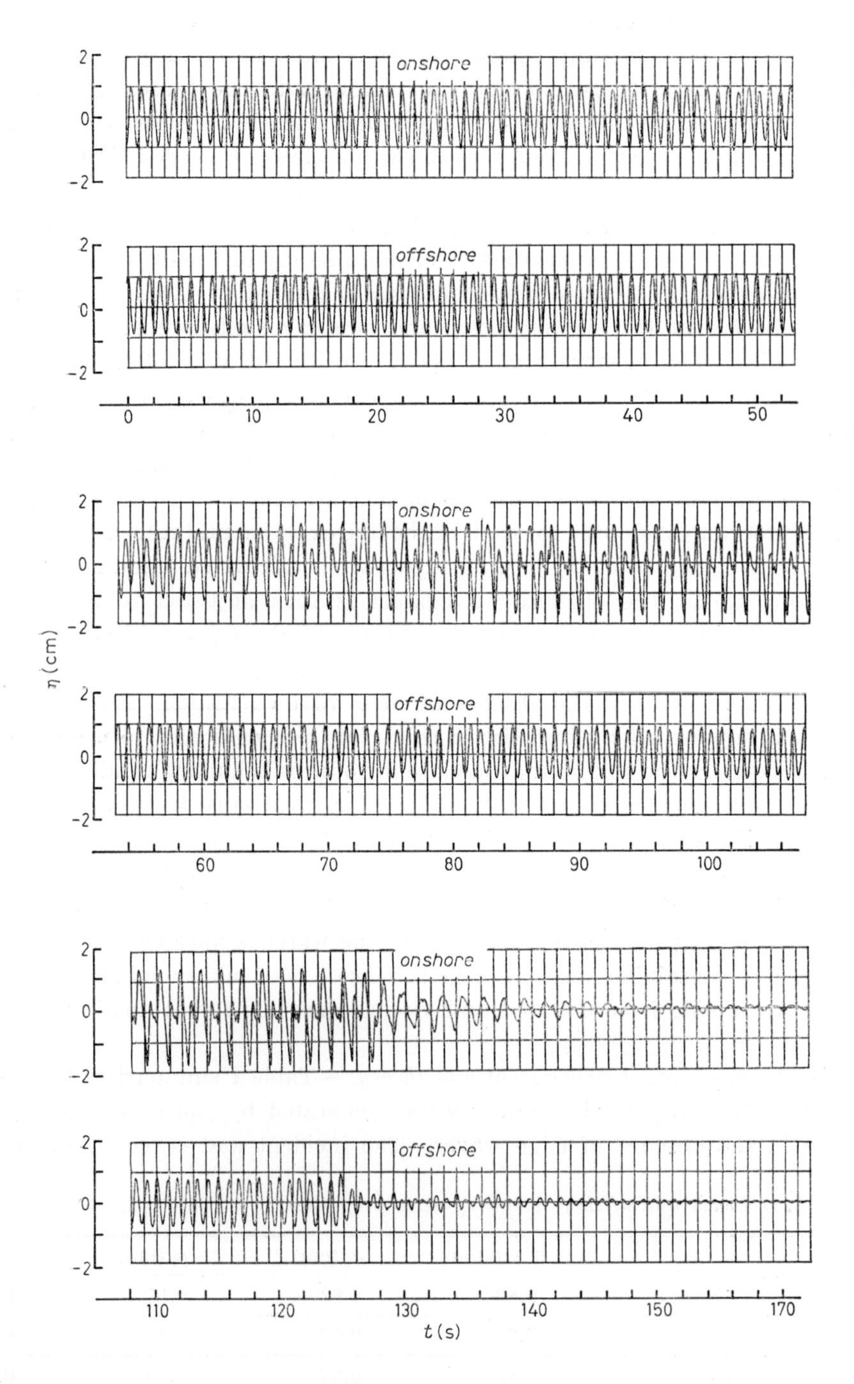

Fig. 13. – Time series of water surface elevations onshore and offshore.

plitudes are presented along with the respective reflection coefficients which range from $R = 0.134$ to $R = 0.500$. (Recall that theoretical results assume perfect reflection with $R = 1$.) It should be noted that wave breaking near the shoreline was observed in each experiment and was especially strong in run 1.

A typical time series (run 2) taken onshore and offshore during edge wave excitation, evolution and eventual damping is shown in fig. 13. The onshore record clearly shows the effect of edge wave growth as it alternates between constructive and destructive interference with successive crests of the incident/reflected wave system. (This behavior is a direct consequence of subharmonic excitation with $W = 2\omega$.) A periodogram computed using the fast Fourier transform of the onshore wave record for run 2 is shown in fig. 14. The first

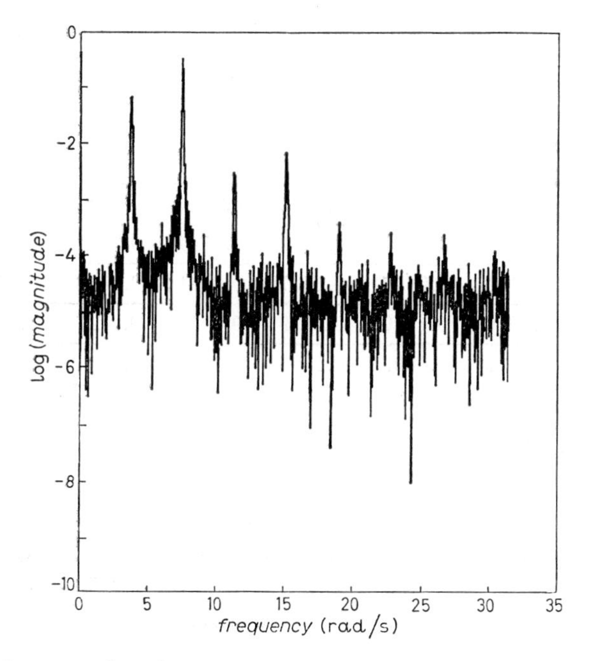

Fig. 14. – Periodogram of onshore time series.

peak in the spectrum corresponds to the excited edge wave mode, while the second peak corresponds to the standing wave generated from offshore. The equally spaced peaks at higher frequencies represent the superharmonics of both the edge wave and standing wave. The centering of wave content about well-separated and narrow bands with identifiable dominant frequencies (ω, 2ω, 3ω, ...) such as that exhibited in fig. 14 permits the use of complex demodulation techniques to view the real-time evolution of both the amplitude and phase of each dominant wave component (see [26]). By means of the periodogram, an initial guess, say ω_g, for the dominant frequency in a band of interest is determined. The time series is then multiplied by $\exp[-i\omega_g t]$ to shift the desired wave content to zero frequency. This demodulated signal is then low-pass filtered in time to yield the instantaneous phase and amplitude of the signal component with frequency ω_g. If the estimated frequency ω_g is in error, a linear change in instantaneous phase with time will be observed; the slope of the linear change represents the error in ω_g. In this manner the dominant frequency of each narrow band in the periodogram may be determined very accurately as well as its instantaneous amplitude and phase. (More details on the application of complex demodulation technique including a discussion of the low-pass filter properties may be found in [24].) The edge wave frequencies in each experiment were determined using this technique and are summarized in table II. Figures 15 and 16 show the instantaneous amplitude

TABLE II. – *Measured and predicted properties of excited edge waves.*

Run	m	ω (rad/s)	γ_m (s^{-1})	δ (s^{-1})	γ'_m (s^{-1})	γ (s^{-1})	γR (s^{-1})	a_0 measured (cm)	a_0 G-B (cm)	a_0 M-W (cm)
1	5	4.93	0.0295	0.1086	0.1381	0.9652	0.1293	0.28	0.55	0.49
2	3	3.80	0.0880	0.0771	0.1651	0.3144	0.1572	0.75	1.52	1.36
3	4	4.83	0.1120	0.0677	0.1797	0.6051	0.2088	0.67	1.01	0.93
4	4	5.29	0.0910	0.1289	0.2199	0.6031	0.2147	1.38	1.14	1.26
5	4	5.42	0.1000	0.0693	0.1693	0.3264	0.1501	0.90	1.07	1.05

and phase, respectively, of the edge wave harmonic at the onshore gage in run 2. Both parameters oscillate rather wildly during the initial time $t < 25$ s; this behavior is a characteristic result when the signal-to-noise ratio is small. (Note that during this period this instantaneous amplitude of the edge wave is less than 0.07 cm!) The edge wave amplitude in fig. 15 then begins to grow in an exponential manner, while the edge wave phase in fig. 16 becomes constant. As time continues to increase, edge wave growth slows and a steady state is achieved for $t > 70$ s; the gage site amplitude of the edge wave at steady state is $a_0 = 0.754$ cm. The measured growth rates γ_m and steady-state amplitudes a_0 are shown in table II. The inferred shoreline amplitude A_0 based on the measured a_0 at gage site is easily calculated; the ratio of A_0/a_i ranges from 3 to 6 in all experiments in agreement with the theoretical prediction that steady-state amplitudes of the edge wave should exceed those of the offshore standing wave. It is important to note that measured growth rates are significantly influenced by viscous and turbulent damping forces which must be clarified before a legitimate comparison with inviscid and irrotational theoretical models. The viscous damping rate is easily measured in the laboratory model simply by turning off the wavemaker and monitoring the damping

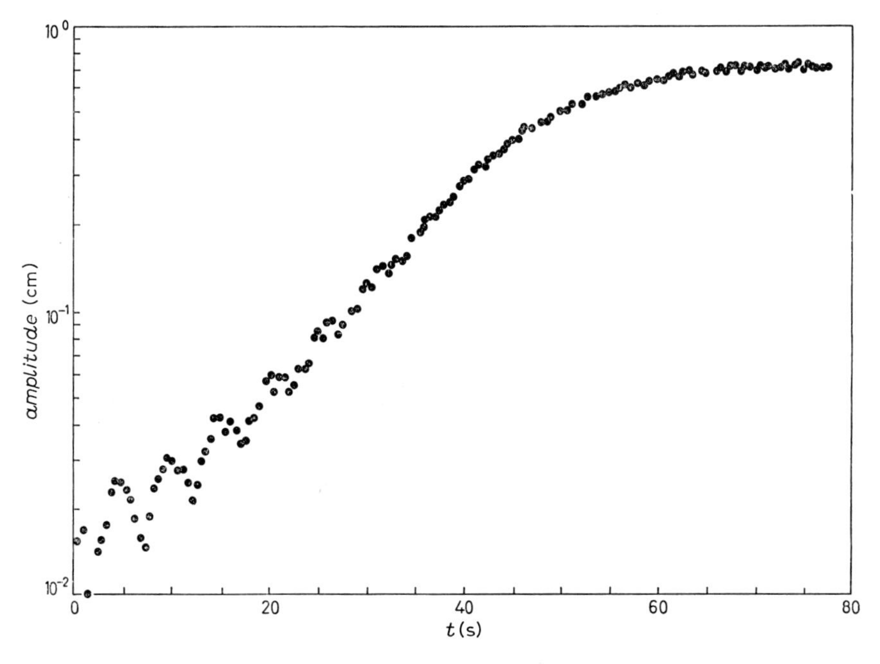

Fig. 15. – Amplitude evolution of subharmonic edge wave for run 2.

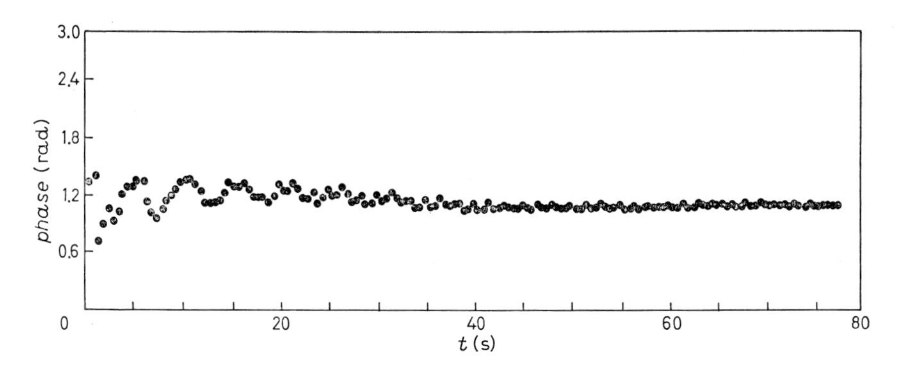

Fig. 16. – Phase evolution of subharmonic edge wave in run 2.

of the edge wave harmonic with time. Typical results are shown in fig. 17, which corresponds to run 2. As expected, exponential decay occurs when the edge wave becomes sufficiently small (and linear theory is applicable). The measured decay rates δ in the exponential stage for each experiment are presented in table II. As noted by GUZA and BOWEN [27], the damping of edge waves in the presence of breaking incident waves is likely to be dominated by turbulent exchange mechanisms rather than (laminar) viscosity. Viscous effects are enhanced by laboratory model scales and should play an important role; however, based on the results to follow, turbulent exchange mechanisms also

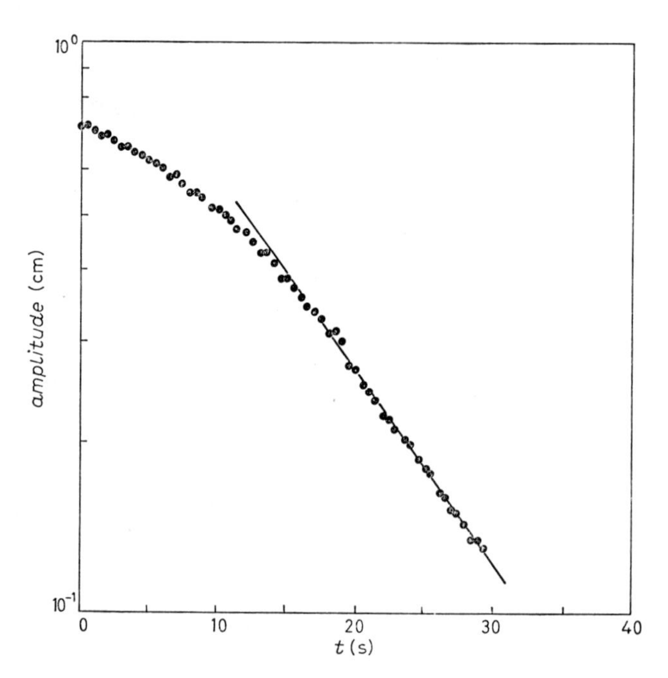

Fig. 17. – Viscous damping of subharmonic edge wave in run 2.

appear to be important at laboratory scales. Ignoring the effects on laboratory damping of edge waves due to incident-wave breaking, a measured « inviscid » estimate of the initial growth rate may be calculated by $\gamma'_m = \gamma_m + \delta$. Results for γ'_m are tabulated in table II along with the theoretical predictions for the growth rate according to (42a). It is evident that large discrepancies still exit; however, it should also be remembered that the theoretical predictions assume a perfectly reflecting beach. To examine the potential effect of the imperfect reflection in the experiments we have plotted the ratio of γ'_m/γ *vs.* the reflection coefficient R in fig. 18. Note that excellent correlation exists (data lie on 45° line) suggesting that the growth rates should be calculated using the reflected-wave amplitude a_r instead of the incident-wave amplitude a_i, *i.e.* we should replace γ by γR. Theoretical estimates of γR are listed in table II. It is important to emphasize that this behavior of the measured data clearly demonstrates that it is the reflected-wave component of the offshore standing wave which drives the edge waves. It further supports the observation [17] that edge waves do not occur on dissipative beaches which produce little reflection.

In order to compute the theoretical amplitude of the edge waves in steady state, the results above suggest that we should modify the theoretical growth rate in (46) to the form

$$A_0 = \left[\frac{R(\gamma^2 - \varkappa^2)^{\frac{1}{2}}}{\mu}\right]^{\frac{1}{2}}. \tag{46b}$$

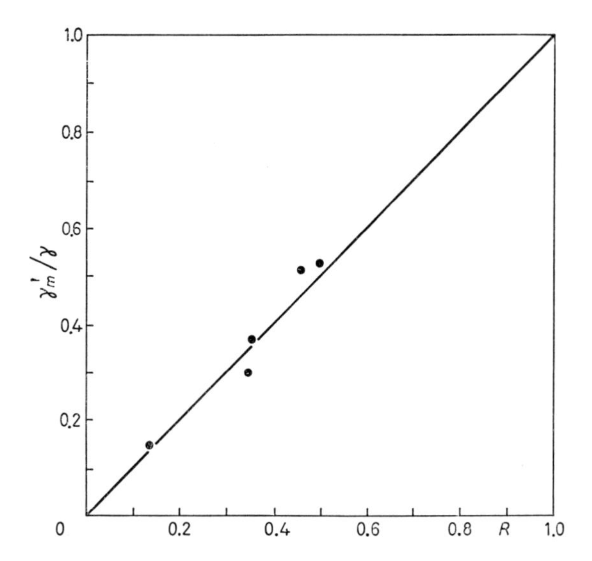

Fig. 18. – Correlation of measured and theoretical « inviscid » growth rates with beach reflection coefficient.

The theoretical amplitude at a specific gage site then becomes

$$a_0 = A_0 \cos k_m x \exp[-k_m y \cos \beta]; \tag{47}$$

results of computations for the predictions of both G-B and M-W are shown in table II. Except for run 4, theoretical results exceed measured data by factors of 1.18 to 1.91. This behavior pattern is expected, since the turbulent damping of the edge waves in the presence of breaking incident waves has been neglected. Also consistent with this hypothesis is the fact that agreement between experiment and prediction is best in run 5, where the least wave breaking was observed. Run 4 is anomalous as the measured amplitude exceeds theoretical prediction. (Interestingly, absolute agreement between measured and predicted data is best for run 4!) There is also the following evidence that run 4 is unusual relative to the other experiments. Note from table I that the offshore standing wave changes little between runs 3 and 4, while the steady-state amplitude of the edge wave in table II doubles! In other words, a major difference in dynamical response occurs even though the forcing remains approximately constant. The only significant difference between runs 3 and 4 is the slightly greater beach slope (2°) in run 4 which no longer corresponds to one of the set $\pi/2N$ required by the theoretical models. Based on this limited datum, it does appear that the predictions for steady-state response on beach with slopes satisfying $\beta = \pi/2N$ cannot be extrapolated to intermediate slopes. (However, extrapolation for the initial growth rates does appear permissible with the modifications due to partial reflection described earlier.)

4'5. *Conclusions.* – Based on the experiments described herein, the following major conclusions may be stated regarding edge wave excitation on beaches by normally incident waves from offshore.

i) The theoretical growth rates calculated for perfectly reflecting beaches may be generalized to beaches with imperfect reflection simply by reducing the growth rates in direct proportion to the reflection coefficient of the beach. This result suggests that the reflected wave is indeed the driving force for edge wave response and absolutely necessary in order to excite these modes.

ii) Growth rates modified for partial-reflection effects may be extrapolated to beaches with slopes not satisfying $\beta = \pi/2N$.

iii) No change in the edge wave phase occurs once the exponential growth stage is encountered. Hence, the evolution equation for the *real* amplitude of the edge wave is applicable.

iv) There is limited evidence that the steady-state amplitudes of edge waves on beaches with $\beta \neq \pi/2N$ are significantly larger than those for nearby beach slopes with $\beta = \pi/2N$.

v) Damping effects on edge waves are significantly influenced by turbulent exchange mechanisms resulting from breaking incident waves, even on the laboratory scale.

* * *

Financial support for much of the work reported in these lectures was provided by the Office of Naval Research, Fluid Dynamics Division, and the National Science Foundation, Engineering Division. The Center for Studies of Nonlinear Dynamics at the La Jolla Institute also provided support for the preparation of the lecture notes. The experiments reported in sect. **4** were performed by N. K. Lin, at the University of California. The author gratefully acknowledges the support and assistance from each of these sources.

REFERENCES

[1] H. Segur: *J. Fluid Mech.*, **59**, 721 (1973).
[2] J. Hammack and H. Segur: *J. Fluid Mech.*, **65**, 289 (1974).
[3] J. Hammack: *J. Phys. Oceanogr.*, **10**, 1455 (1980).
[4] J. Hammack and H. Segur: *J. Fluid. Mech.*, subjudice (1981).
[5] J. Hammack and H. Segur: *J. Fluid Mech.*, **84**, 337 (1978).
[6] G. G. Stokes: *Trans. Cambridge Philos. Soc.*, **8**, 441 (1849).
[7] H. Hasimoto and H. Ono: *J. Phys. Soc. Jpn.*, **33**, 805 (1972).
[8] H. Yuen and B. Lake: *Phys. Fluids*, **18**, 956 (1975).
[9] H. Yuen and B. Lake: *Annu. Rev. Fluid Mech.*, **12**, 303 (1980).

[10] J. Hammack and H. Segur: *J. Fluid Mech.*, **84**, 359 (1978).
[11] F. Ursell: *Proc. Cambridge Philos. Soc.*, **49**, 685 (1953).
[12] G. Carrier: *Lect. Appl. Math. Am. Math. Soc.*, **14**, 157 (1971).
[13] G. G. Stokes: *Brit. Ass. Adv. Sci. Rep.*, part 1, p. 1 (1846).
[14] F. Ursell: *Proc. R. Soc. London Ser. A*, **214**, 79 (1952).
[15] E. T. Hanson: *Proc. R. Soc. London Ser. A*, **111**, 49 (1926).
[16] J. D. Isaacs, E. A. Williams and C. Ekart: *EOS*, **32**, 37 (1951).
[17] R. T. Guza and D. L. Inman: *J. Geophys. Res.*, **80**, 2997 (1975).
[18] A. J. Bowen and D. L. Inman: *J. Geophys. Res.*, **76**, 8662 (1971).
[19] A. J. Bowen and D. L. Inman: *J. Geophys. Res.*, **74**, 5479 (1969).
[20] R. T. Guza and R. E. Davis: *J. Geophys. Res.*, **79**, 1285 (1974).
[21] R. T. Guza and A. J. Bowen: *J. Mar. Res.*, **34**, 269 (1976).
[22] A. A. Minzoni and G. B. Whitham: *J. Fluid Mech.*, **79**, 273 (1977).
[23] R. Rockliff: *Math. Proc. Cambridge Philos. Soc.*, **83**, 463 (1978).
[24] N. K. Lin: Thesis, Department of Civil Engineering, University of California, Berkeley, Cal. (1981).
[25] A. A. Minzoni: *J. Fluid Mech.*, **74**, 369 (1976).
[26] P. Bloomfield: *Fourier Analysis of Time Series: An Introduction* (New York. N. Y., 1976), p. 118.
[27] R. T. Guza and A. J. Bowen: *Proceedings of the XV Coastal Engineering Conference* (1976), p. 560.

Internal Solitons in the Andaman Sea (*).

A. R. OSBORNE (**)

Exxon Production Research Company - Houston, Tex. 77001

T. L. BURCH

EG&G Environmental Consultants - Waltham, Mass. 02154

1. – Introduction.

Internal waves occur within subsurface layers of marine waters which are stratified because of temperature and salinity variations. Disturbances created within the ocean give rise to these waves which represent a significant mechanism for the transport of momentum and energy within the ocean. Historically, oceanographers have given careful attention to internal waves and their side effects, for they can significantly influence oceanic-current measurements, undersea navigation, antisubmarine warfare operations and even the feeding habits of marine animals.

In late 1975 and early 1976 internal-wave currents as high as 1.8 m/s were observed during a four-month measurement program and during subsequent drilling operations conducted by Exxon in (600÷1100) m of water in the Andaman Sea, offshore Thailand [1]. As a result of this study, Exxon concluded that knowledge of internal-wave behavior would be necessary for the design of future deepwater offshore production facilities.

Herein we present an analysis of internal-wave data obtained during an ensuing four-day measurement program conducted in the southern Andaman Sea in October, 1976. The program was designed to simultaneously measure internal waves and the associated response of the drillship Discoverer 534, which was in the process of setting a new world record for deepwater drilling at that time in more than 1030 m of water. The problem of drillship response to these waves has been previously reported in [1, 2], and will not be discussed here. In this paper: 1) We discuss a theoretical framework for inter-

(*) This paper first appeared in *Science*, Vol. **208**, 2 May 1980.

(**) Present address: Istituto di Cosmogeofisica, 10133 Torino, Italia.

pretation of internal-soliton data. 2) We present a preliminary analysis of the Andaman Sea data in light of this framework and show that the data corroborate the occurrence in nature of solitons, *i.e.* solitary waves which by definition retain their shape and speed after a collision with each other. 3) We discuss the interaction of internal solitons with surface waves in light of our observations of an unusual surface wave phenomenon previously referred to as the « tide rip ».

In order to understand the behavior of internal solitary waves and solitons, it is helpful to first understand surface solitary waves, *i.e.* waves which travel on the surface of the water rather than beneath it. To this end we first present some relevant historical evidence for surface solitary waves and we further discuss evidence for interactions between internal solitary waves and surface waves.

2. – Historical setting.

The study of wave phenomena in physics is presently undergoing dramatic changes as a result of the discovery of solitons, *i.e.* localized nonlinear waves which exhibit a number of unusual properties in strong contrast to linear waves which have formed the main conceptual framework for many physical theories involving wave phenomena. Since the discovery of the soliton [3], the study of the solitary wave in several fields of endeavor has received new impetus. Many differential equations have been derived which describe a variety of nonlinear systems with dispersion and which admit solitary-wave solutions. Historically, solitary waves were thought to self-destruct upon collision with each other. Subsequent to the discovery of the soliton, emphasis has been placed on determining whether certain solitary waves behave as solitons. A wide spectrum of mathematical and numerical techniques have been devised to test whether solitary-wave solutions of a particular system can survive a collision with each other, *i.e.* whether they retain their shape and speed after the collision and, therefore, deserve the designation soliton. Examples of physical systems which have been shown to have soliton solutions are water waves, ion-acoustic waves in a plasma, magnetohydrodynamic waves in a plasma, pressure waves in liquid-gas bubble mixtures, propagation of sound waves through a crystal lattice and phonon packets in low-temperature nonlinear crystals [4].

Apparently, the first documented observation of the solitary wave was made by John Scott Russell [5, p. 319] in the last century. In his own words he recounted the following:

« I was observing the motion of a boat which was rapidly drawn along a narrow channel by a pair of horses, when the boat suddenly stopped—not so the mass of water in the channel which it had put in motion; it accumulated round the prow of the vessel in a state of violent agitation, then, suddenly leaving it behind, rolled forward with great velocity, assuming the form of a large solitary elevation, a rounded, smooth and well-defined heap of water, which continued its course along the channel apparently without change of form or diminution of speed. I followed it on horseback, and overtook it still rolling on at a rate of some eight or nine miles an hour, preserving its figure some thirty feet long and a foot to a foot and a half in height. Its height gradually diminished and after a chase of one or two miles I lost it in the windings of the channel. Such, in the month of August, 1834, was my first chance interview with that singular and beautiful phenomenon ... ».

RUSSELL went on to make rather precise, well-controlled observations of solitary waves. A model differential equation to describe solitary-wave behavior was not developed until 1895 when KORTEWEG and DEVRIES [6] approximated the Navier-Stokes equations for small, but finite-amplitude, waves in a shallow channel to yield the KdV equation (an excellent modern discussion of this topic is given by MIURA [7]). We shall describe this equation in the following section.

Another historical precursor which we shall find relevant to our data interpretation and to the interaction of surface and internal waves is given in [8], which is a description by HORSBURGH (made sometime before 1861) of a « tide rip ».

« In the entrance of the Malacca Straits, near the Nicobar and Acheen Islands, and between them and Junkseylon, there are often very strong ripplings, particularly in the southwest monsoon; these are alarming to persons unacquainted, for the broken water makes a great noise when the ship is passing through the ripplings in the night. In most places ripplings are thought to be produced by strong currents, but here they are frequently seen when there is no perceptible current ... so as to produce an error in the course and distance sailed, yet the surface of the water is impelled forward by some undiscovered cause. The ripplings are seen in calm weather approaching from a distance, and in the night their noise is heard a considerable time before they come near. They beat against the sides of a ship with great violence, and pass on, the spray sometimes coming on deck; and a small boat could not always resist the turbulence of these remarkable ripplings. »

Because the above observation occurred in relatively deep water, the phenomena described cannot be the result of a classical tide rip. Tide rips are due to strong tidal currents flowing over bars and through inlets which oppose the propagation of surface waves, creating regions of short, choppy, breaking waves. For this reason, we shall simply refer to these strange waves first observed by HORSBURGH as « rips » throughout this paper.

PERRY and SCHIMKE [9] made observations of rips north of Sumatra from the U.S. Coast and Geodetic Survey ship Pioneer in 1964. They described

« ... distinct zones of whitecaps ranging from 200 to 800 m in width and stretching from horizon to horizon (approximately 30 km) in a north-south direction ... in the Andaman Sea north of Sumatra. At least five of these zones, with a spacing of about 3200 m between each zone, were observed. The observed zones or bands of choppy water had short, steep, randomly oriented waves with heights of about 0.3 to 0.6 m. Each band stood out distinctly in a otherwise undisturbed sea. A 4 m/s NNW wind and a surface water temperature of 29 °C were observed, but neither changed significantly as the ship crossed the bands of choppy water. »

Using a mechanical bathythermograph, PERRY and SCHIMKE obtained several profiles of water temperature between the surface and 250 m depth during the rip passages and were able to associate the rips with internal waves as large as 80 m. PHILLIPS [10] has reviewed Perry and Schimke's rip observations in light of interactions between long internal waves and surface waves.

LAFOND also observed similar surface phenomena in the Indian Ocean [11] and the *Marine Observer* [9] reports these waves in the Bay of Bengal and adjacent waters and variously describes them as « current rips », « tide rips », « lines of demarcation », and « disturbed » and « rippled water ». We remark on our own observations of these waves and to their relationship to internal solitons in a later section.

3. – The soliton.

The observations of solitary waves by RUSSELL and the subsequent discovery by KORTEWEG and DEVRIES of a theoretical description in terms of a nonlinear differential equation represented the extent of physical understanding of solitary waves at the beginning of this century. For nearly 70 years thereafter, the solitary wave was considered to be a relatively unimportant curiosity in the field of nonlinear wave theory. It was generally thought that the collision of two solitary waves would result in a strong nonlinear interaction and ultimately end in the destruction of the original waves (see [4] for a penetrating review of this topic). Thus the stage was set for the remarkable results of a computer experiment conducted by ZABUSKY and KRUSKAL [3], who simulated the KdV equation for the collision of two solitary waves. The results of the simulation were surprising: the waves retained their shapes and propagation velocities after the collision! Because of the somewhat « elementary particle-like » behavior of these waves ZABUSKY and KRUSKAL coined the word « soliton » to describe them.

We first briefly recount the solitary wave theory of Korteweg and deVries. The KdV equation for the propagation of surface solitary waves is given by

$$\eta_t + c_0\eta_x + \alpha\eta\eta_x + \gamma\eta_{xxx} = 0\,, \tag{1}$$

$$\begin{cases} c_0 = \sqrt{gh}\,, \\ \alpha = 3c_0/2h\,, \\ \gamma = c_0 h^2/6\,. \end{cases} \tag{2}$$

Here $\eta(x, t)$ is the amplitude of the solitary wave as a function of horizontal displacement x down the channel and time t, h is the water depth, g is the acceleration of gravity, and c_0 is the phase speed of the associated linear wave. The subscripts in eq. (1) refer to partial derivatives with respect to x or t. We shall see later that not only surface solitary waves but also internal solitary waves may be described by the KdV equation. The key to understanding this equation lies in the competition between the nonlinear term $(\alpha\eta\eta_x)$ and the dispersive term $(\gamma\eta_{xxx})$. Under certain conditions, these terms balance and the result is a stable configuration called a solitary wave which has the following analytical form (and is a special solution to eq. (1)):

$$\eta(x, t) = \eta_0 \operatorname{sech}^2[(x - ct)/L]\,, \tag{3}$$

where η_0 is the maximum amplitude of the solitary wave, $c = c_0(1 + \eta_0/2h)$ is the phase speed of the wave and $L = \sqrt{4h^3/3\eta_0}$ is its « characteristic » length. A surface solitary wave is shown in fig. 1 and clearly may be described as a « bump on the water », in strong contrast to the appearance of the linear « sine » wave.

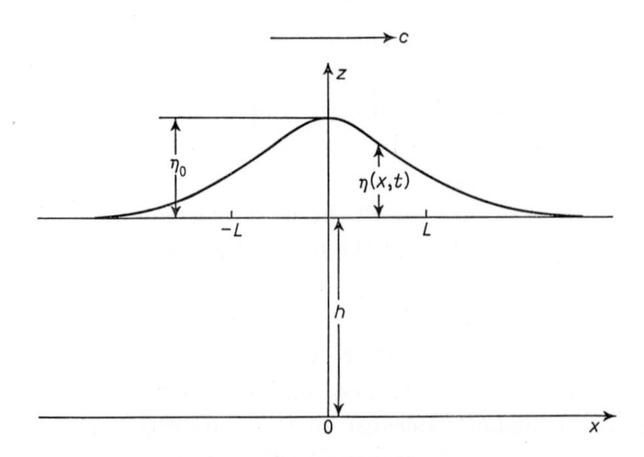

Fig. 1. – A surface solitary wave with amplitude η_0 moves to the right with phase speed c in water of depth h.

In 1967, GARDNER *et al.* [12] published an important paper which presents an analytical solution to the initial-value problem governed by the KdV evolution equation and which analytically predicts the behavior of the soliton. Given an arbitrary initial wave form $\eta(x, 0)$, how does it evolve thereafter in time according to the KdV equation? The character of the Gardner *et al.* solution is illustrated in fig. 2. A « sufficiently localized » initial wave form

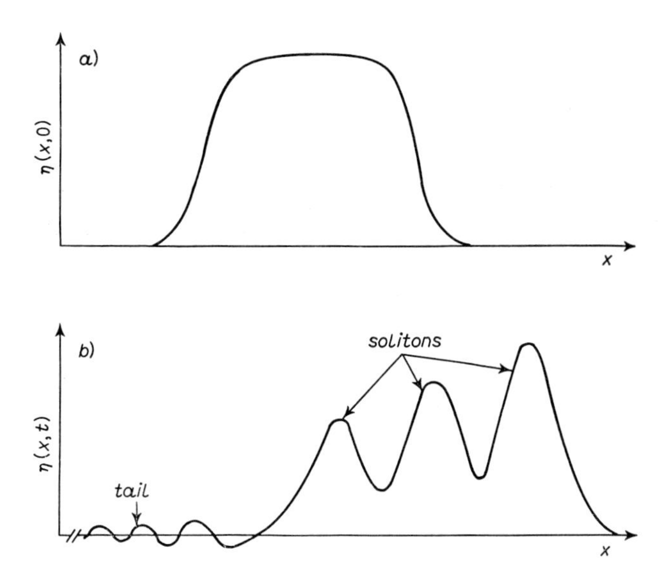

Fig. 2. – A « sufficiently localized » initial wave form (a)) evolves into solitons and a dispersive linear wave train or « tail » in the far field (b)).

$\eta(x, 0)$ will evolve into one or more solitons and a dispersive « tail » or wave trains as $t \to \infty$. The integral conditions which specify the requirement for a sufficiently localized initial wave may be found in [12]-[14]. With the normalizations for amplitude, horizontal co-ordinate and time.

$$\left\{ \begin{aligned} u &= 3\eta/2h \,, \\ r &= (x - c_0 t)/h \,, \\ \tau &= c_0 t/6h \,, \end{aligned} \right. \tag{4}$$

the KdV equation may be written in the following particularly convenient form:

$$u_\tau + 6uu_r + u_{rrr} = 0 \,. \tag{5}$$

The essential results of Gardner *et al.* may then be easily summarized as a set of « rules » which govern the behavior of long, nonlinear waves (solitons) described by the KdV equation for a constant-depth, constant-breadth channel (see [14] for a detailed mathematical description). We emphasize here the importance of these rules in the interpretation of soliton signals in experimental data:

1) The soliton amplitudes must be positive definite, $u(r, \tau) \geqslant 0$, and are given by the dimensionless solution

$$u(r, \tau) = 2A \operatorname{sech}^2[\sqrt{A}(r - 4A\tau)] , \tag{6}$$

where the amplitude $2A$ is found from eq. (8) below.

2) The dimensional soliton phase speed exceeds the phase speed of the associated linear wave c_0 by an amount proportional to its amplitude

$$c = c_0(1 + 2A/3) . \tag{7}$$

3) Since larger solitons travel faster, they evolve into groups which are rank-ordered, the largest leading the rest (note fig. 2).

4) No acceptable initial wave form can evolve into two solitons of the same phase speed, *i.e.* two identical solitons cannot be created by a single initial wave form.

5) Interacting solitons experience at most a phase shift, advancing the faster and retarding the slower. Their phase speeds and shapes remain unaltered after a collision with each other.

6) If the area under the initial wave form is positive (and « sufficiently local »), then at least one soliton emerges.

7) If the initial wave form is everywhere negative, no soliton emerges, only the « tail » or train of dispersive linear waves is formed.

8) The nondimensional amplitude of the largest soliton (assuming the initial wave form is small) is

$$A = \tfrac{1}{2}\left(\int_{-\infty}^{\infty} u(r, 0)\, \mathrm{d}r\right)^2 \geqslant 0 . \tag{8}$$

9) The number of solitons N which emerge may be found from the Schrödinger equation where the « potential well » is defined by the dimensionless initial wave form $u(r, 0)$:

$$\mathrm{d}^2\varphi/\mathrm{d}r^2 + u(r, 0)\,\varphi = 0 , \qquad \varphi(r_0) = 1 , \qquad \mathrm{d}\varphi(r_0)/\mathrm{d}r = 0 , \tag{9}$$

where r_0 is to the left of the localized function $u(r, 0)$. Then the number of zeros of φ is the number of solitons which evolve from the initial wave form.

10) The amplitudes A for each member of a soliton packet may be written in renormalized form for $u(r, 0) = u_0 \operatorname{sech}^2[\sqrt{A}(r - 4A\tau)]$, u_0 arbitrary (see [15]),

$$\tilde{A}_n = [(N - n)/(N - 1)]^2, \tag{10}$$

where n is the number of a particular soliton in a packet, *i.e.* $n = 1$ corresponds to the first and largest soliton, while $n = 2$ is the next largest, etc. Thus the renormalized amplitude (denoted by the « tilde ») of the first soliton is unity and each succeeding soliton is smaller according to eq. (10).

11) Since each of the solitons evolves with independent amplitude and phase speed, the separation distances between solitons in a packet are a direct measure of the propagation distance back to the source. Thus a means is available for estimating the location of the initial wave form, *i.e.* the distance to the source of the solitons.

4. – Internal solitons.

The characteristic behavior of internal solitons depends on the ratio of water depth to the soliton scale length $\lambda = 2L$. The three regimes which may be encountered are *a*) very deep (where $h/\lambda \gg 1$), *b*) very shallow (where $h/\lambda \ll 1$) and *c*) intermediate depths ($h/\lambda \sim 1$). The deep-water case is described by the Benjamin-Ono equation (B-O) [16, 17], the shallow-water case is governed by the KdV equation, and the intermediate-water-depth case has been investigated by JOSEPH [18], KUBOTA, KO and DOBBS [19] and CHEN and LEE [20]. For our purposes we are not concerned with the direct effect of the various subtleties occurring in the deep and intermediate-depth cases, primarily because of the incomplete nature of the theories and because of the present lack of understanding as to how waves traveling from deep to shallow water make the transition from the B-O regime to the KdV regime [20]. Because our measurement program was conducted in water about half a scale length deep ($h/\lambda \sim 0.4 \div 0.6$), we feel that the data may be adequately described, to a reasonable order of approximation, by the theoretically well-understood KdV equation; thus a detailed investigation of this equation for a two-layer fluid is warranted. We will allude to the possible effects of intermediate and deep water depths in our comparison to data.

In a two-layer fluid, the dimensional form of the KdV equation is given by eq. (1), where now $\eta(x, t)$ is the interface displacement between the two fluids as shown in fig. 3 [21]. The upper layer is assumed to have depth h_1 and

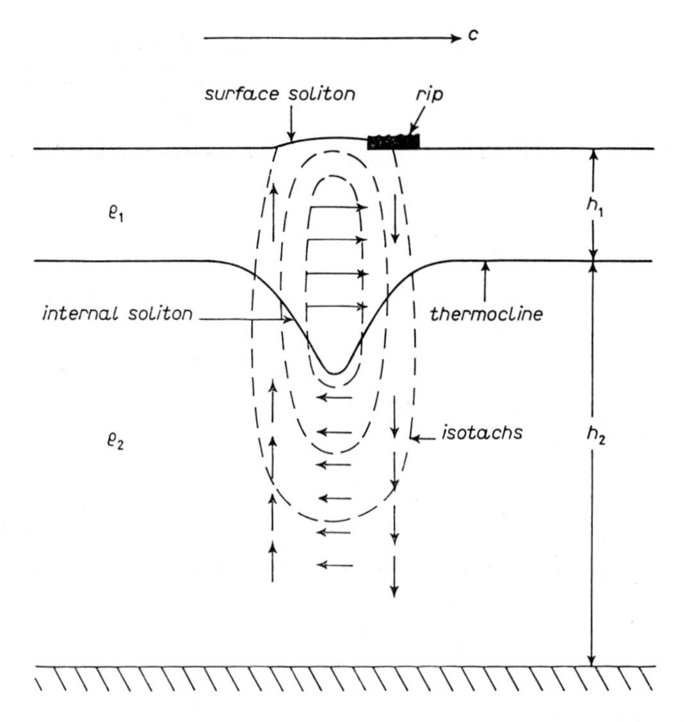

Fig. 3. – An internal soliton in a two-layer, finite-depth fluid is a wave of depression when $h_1 < h_2$. The dashed lines correspond to lines of constant water particle speed, « isotachs », and the arrows indicate the magnitude and direction of the water particles. A small surface soliton (amplitude $\sim(\varrho_2 - \varrho_1)\eta_0$) accompanies the internal soliton [22]. The approximate location of the surface rips, as observed in our Andaman Sea measurements, is also shown.

density ϱ_1, while in the lower layer the respective quantities are h_2 and ϱ_2. For the constant coefficients of eq. (1), we have approximately

$$c_0 \simeq [g(\Delta\varrho/\varrho)\, h_1/(1+r)]^{\frac{1}{2}}, \tag{11}$$

$$\alpha \simeq -(3c_0/2)[(1-r)/h_1], \tag{12}$$

$$\gamma \simeq c_0 h_1 h_2/6 . \tag{13}$$

The approximate expressions shown in eqs. (11)-(13) were made assuming $\varrho \simeq \varrho_1 \simeq \varrho_2$, which is true in the ocean, where the small density differences are due primarily to temperature and salinity variations. Here $\Delta\varrho = \varrho_2 - \varrho_1$ and $r = h_1/h_2$.

The internal-soliton solution to eq. (1) is

$$\eta(x, t) = -\eta_0 \operatorname{sech}^2[(x-ct)/L], \tag{14}$$

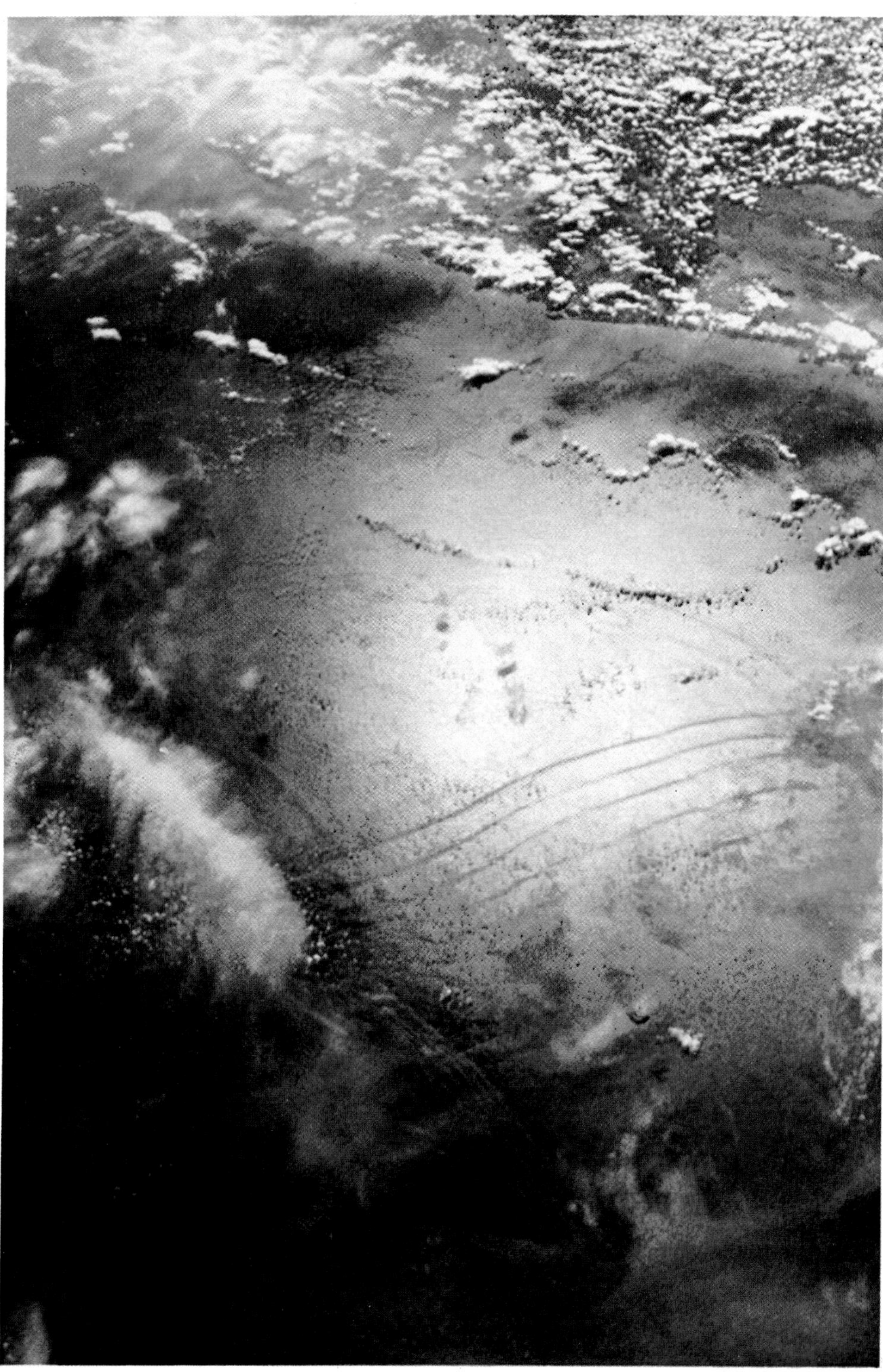

Apollo-Soyuz photograph of the Andaman Sea surface; north is to the left. The west coast of Thailand is in the upper part of the picture. The light puffy regions are clouds, while the long striations (~ 100 km long) in the lower center are caused by the presence of large internal waves propagating towards Thailand. Such striations are visible from orbit because the internal waves modify the reflectivity properties and amplitudes of ocean surface waves.

where we have assumed the upper layer to be thinner than the lower layer. This results in a downward displacement of the fluid interface, as can be noted by the negative sign in front of η_0 in eq. (14). The soliton phase speed is

$$c = c_0[1 - \eta_0 \alpha / 3c_0] \tag{15}$$

and the scale length is

$$L = [-12\gamma/\eta_0 \alpha]^{\frac{1}{2}}. \tag{16}$$

The horizontal water particle velocities of the soliton in the upper layer are given by

$$u_1(x, t) = (c_0 \eta_0 / h_1) \operatorname{sech}^2[(x - ct)/L] \tag{17}$$

and, in the lower layer, we have

$$u_2(x, t) = -(c_0 \eta_0 / h_2) \operatorname{sech}^2[(x - ct)/L]. \tag{18}$$

Note that the amplitudes of the horizontal velocities in both layers do not decay with depth, but that the lower-layer velocities move in the opposite direction to those in the upper layer. The vertical velocities do decay with depth; however, we do not present these equations, since they are not relevant to our data analysis.

Because we are concerned with the case in which $h_1 < h_2$, the internal soliton is a wave of depression, as pointed out in regard to eq. (14). Thus we shall be concerned with modifying statements 1), 6), 7) of our summary of the Gardner *et al.* solution in order to address internal-soliton behavior. For internal solitons when $h_1 < h_2$: Negative (downwardly displaced) internal solitons may emerge from an initial wave form only when its area is negative (an average downward displacement of the thermocline). No solitons emerge if the initial wave form is everywhere positive (an upward displacement of the thermocline).

According to PHILLIPS [23] the total internal-wave energy per unit crest length is approximately

$$E_{\mathrm{T}} = (\varrho_2 - \varrho_1)\, g \overline{\eta^2}. \tag{19}$$

Using eq. (16), we find

$$\overline{\eta^2} = \int_{-L}^{L} \eta^2(x, t)\, \mathrm{d}x = \tfrac{4}{3}\eta_0^2 L. \tag{20}$$

Hence an estimate of the total energy per unit crest length is

$$E_{\mathrm{T}} = \tfrac{4}{3}(\varrho_2 - \varrho_1)\, g\eta_0^2 L\,. \tag{21}$$

The above results have been derived assuming equipartition between kinetic and potential energy, which is exact for linear waves, and, as can be readily shown, is a good approximation for solitary waves.

When a single solitary wave propagates from deep water into shallow water, say from the deep ocean onto the continental shelf, the initial solitary wave may then evolve into one or more rank-ordered solitons, a process called soliton fission. This process is equivalent to assuming that the solitary wave may be placed directly onto the shelf and then allowed to evolve according to the Gardner *et al.* solution. Using the form of the deep-ocean solitary wave developed by JOSEPH [18], DJORDJEVIC and REDEKOPP [21] derived a fission law appropriate for determining the number of KdV internal solitons N appearing on the shelf in a two-layer fluid:

$$N \leqslant 1 + [(32/3)(h_1/h_{2s})(1 - h_1/h_{2s})/(\eta_0/h_1)]^{\frac{1}{2}} \ln[(6/\pi)(\eta_0/h_1)(h_{2\infty}/h_1)]\,. \tag{22}$$

Here h_{2s} is the lower-layer thickness on the shelf and $h_{2\infty}$ is the lower-layer thickness in deep water.

Thus rank-ordered internal solitons may occur from two sources: 1) evolution from the initial wave form over constant water depth, or 2) fission from an initially stable solitary wave which moves over decreasing water depth into shallow water (say, onto the continental shelf).

5. – The Andaman Sea internal-wave data.

In 1975 Exxon began planning drilling operations in the southern Andaman Sea, offshore Thailand, in (600÷1000) m of water. Because there was little information available about ocean currents in the area, a four-month measurement program was conducted at two locations to determine the severity of the local current environment. Our goal was to assess the potential impact on the design of the drilling riser to be used in this environment. Our preliminary measurement program established the existence of large internal waves in the area, with currents as high as 1.8 m/s. However, the time resolution coded into the instruments used in the program was rather long (which, however, allowed us to measure for an extended period of time) and we were, therefore, unable to determine the detailed structure of the waves.

As a result of these preliminary measurements, we elected to conduct a program with improved time resolution to allow us to examine the internal waves in detail while simultaneously measuring the response of a drillship to the wave kinematics. In October, 1976, Exxon Production Research Company contracted EG&G Environmental Consultants to conduct an internal-wave measurement program in the Southern Andaman Sea in 1093 m of water, 7 km west of the drillship Discoverer 534 at 6° 53′ N latitude and 97° 04′ W longitude (see fig. 4). Current meters (EG&G model CT/3) were placed on a taut subsurface mooring at approximate depths below the ocean surface of 53 m, 87 m, 116 m, 164 m and 254 m. These electromagnetic, digitally recording instruments measure water temperature in addition to vector components of horizontal current velocity. Additionally, Savonius rotor current meters (EG&G model 102) were located at depths beneath the surface of 121 m, 437m, 635 m, 895 m and 1001 m.

All instruments were set to record nearly continuously, and, as a result, the total recording capacity was limited to only four days. Because of results obtained from our previous measurement program in this area, we knew that the internal waves arrived in packets of about five or six waves. The packets were spaced about 12 h 26 min apart and occurred most noticeably near the twice monthly spring tides for periods up to one week. Thus there was some confidence that a mooring could be deployed at a time when the internal waves were active. In addition to the moored instrumentation, XBT (expendable bathythermograph) casts were made during periods of internal-wave activity. These devices consist of a weighted thermistor which falls at a prescribed rate through the water column and sends temperature signals to a recording device via a small-diameter wire. CTD (conductivity, temperature and depth) casts had previously allowed us to describe the *T-S* (temperature, salinity) curve for the region, and thus allowed computation of the spatial and temporal variation of the water density directly from the XBT measurements. Shipboard observations were also used to establish air and surface water temperature, wind speed, cloud cover, ship position and surface wave height.

Prior to the inception of the measurement program contacts with NASA (National Aeronautics and Space Administration) and NOAA (National Oceanic and Atmospheric Administration) provided for the use of the LANDSAT and NIMBUS satellites, respectively, for simultaneous observation of the sea surface during our *in situ* measurements [24]. The motivation for this effort was guided by the results of Apel *et al.* [25], who showed that internal waves may be occasionally recorded on satellite images as a consequence of the fact that the reflectivity properties of surface waves can be altered due to the presence of internal waves. We had also received reports of visual observations of surface rips from the drilling vessel and hoped that satellite photography would provide information about the spatial scale of the rips and ultimately of the internal waves. Unfortunately, we were not successful in obtaining satellite photographs during the four days of measurements, primarily due to local cloud cover.

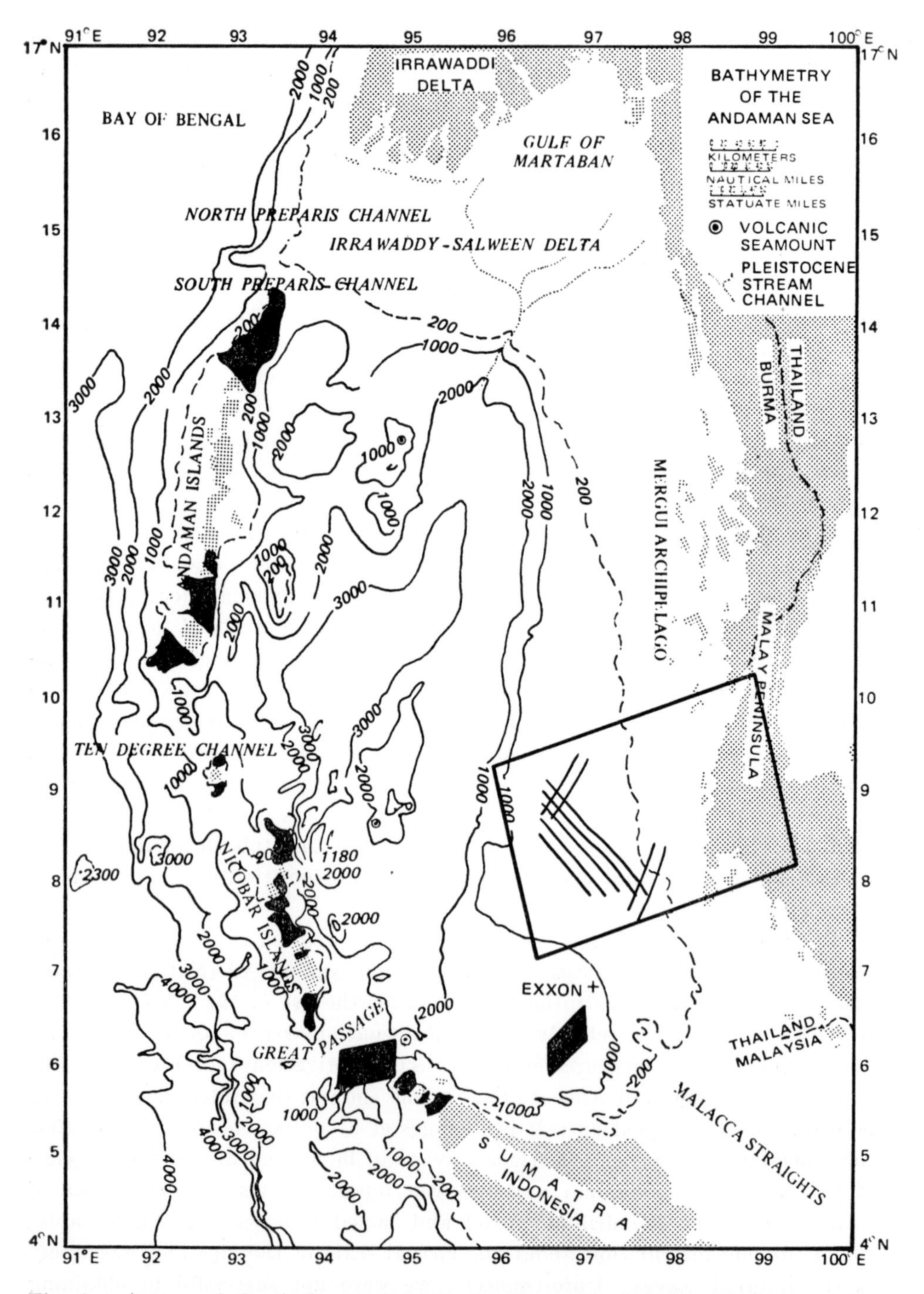

Fig. 4. – A map of the Andaman Sea showing the major bathymetric features, the location of the satellite photograph [24] on the color plate and the location of the observed internal waves (the box denotes the boundary of the photo). Potential source regions are shown as blackened areas near the Andaman Islands, the Nicobar Islands and northern Sumatra. The darkened parallelograms mark the location of the surface rip observations (« zones of whitecaps ») made by PERRY and SCHIMKE [9].

We have, however, found over 50 photographs of the Andaman Sea surface from LANDSAT files (5 of which show internal waves with crests as long as 150 km and wave lengths as great as 15 km) and APEL [26] has discussed a photograph of the Andaman Sea taken during the Apollo-Soyuz mission. This photograph (see color plate) shows three packets of internal waves, two moving from the NW and one from the SW. Figure 4 shows the boundary of this photograph projected onto a map of the Andaman Sea; also shown are the internal-wave traces. Apparently, based on this single photograph, there are at least two distinct sources for these internal waves. One source is likely somewhere in the Andaman Islands and the other is probably near the southernmost of the Nicobar Islands or near northern Sumatra. Since APEL [26] suggests a tidal origin and because we later present evidence for the tidal generation of our measured waves, we infer that shallow-water regions near these islands may be source candidates. We have, therefore, arbitrarily blackened areas inside the 200 m contour in fig. 4 to emphasize the possible location of potential source regions near these islands.

In order to document why we feel that the internal waves in the Andaman Sea may be so easily observed from satellite orbit, we include fig. 5, which shows a sequence of photographs taken on board the survey vessel during the passage of a rip band associated with an internal soliton. The sequence begins on October 27, 1976, 10:15 local time (GMT + 7 hours) when the local winds were calm. Observable in the distance was a long band of breaking waves stretching from one horizon to the other, consisting of waves of about 1.8 m in height, and approaching from due west at about 2.2 m/s in an eastwardly moving background sea state of 0.6 m (fig. 5*a*)). In fig. 5*b*) (10:16), 5*c*) (10:17) the rip continues to approach. In fig. 5*d*) (10:19) the motion of the survey vessel (29 m long) was quite pronounced in 1.8 m seas. This condition persisted for several minutes until about 10:25 when the trailing edge of the rip passed by (fig. 5*e*), *f*), *g*)) and the wave heights quickly dropped to less than 0.1 m. The surface of the Andaman Sea had the appearance of a mill pond. Several minutes later another rip approached and the entire process was thereafter repeated at approximately 40 min intervals during the next 4 h. We observed six distinct rips, each approximately (600÷1200) m wide. Every rip accompanied an internal soliton, as will be shown later in this section.

While there is no direct proof that the rip phenomenon is the same effect observable on the satellite photographs (*i.e.* we do not have sea surface observations occurring simultaneously with the satellite photographs), the simplest conclusion is that the rips are the explanation for the long parallel bands so readily observed in the satellite photographs.

Figure 6 shows the result of a XBT cast taken during a time when large internal waves were not present. The temperature structure during this time of year was found to be quite stable. A well-mixed surface layer extends to about 60 m, where the temperature then drops from 28 °C to 12 °C at a depth

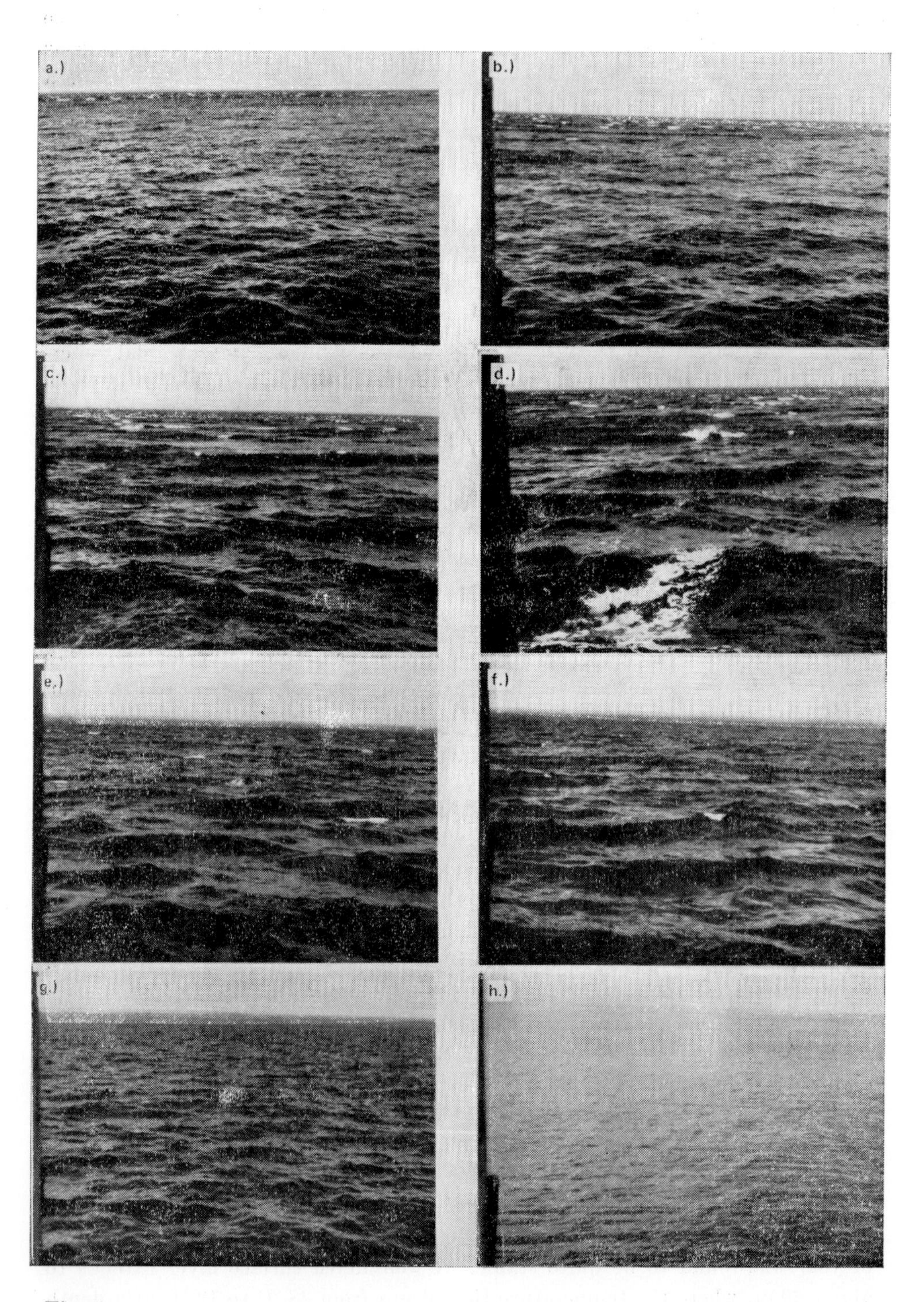

Fig. 5.

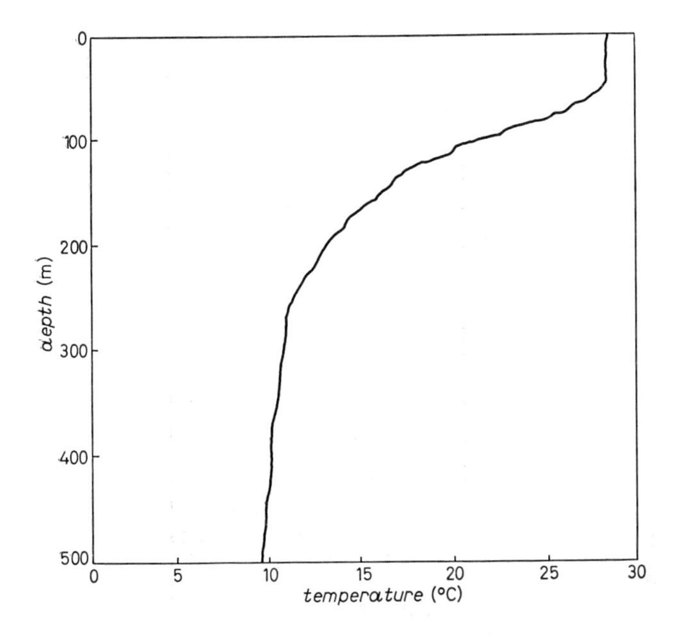

Fig. 6. – Typical quiescent temperature structure at the measurement site in the Andaman Sea.

of 200 m, and continues to drop to about 8 °C at 1000 m. This temperature structure represents a significant density stratification with depth.

Figure 7 displays the thermistor temperature measurements for the first internal-wave packet observed during our program. The temperature signals were obtained at several depths, as shown. Concentrating on the temperature variations which occurred at a depth of 164 m, we note a random, background, small-amplitude internal-wave field punctuated by large-amplitude temperature enhancements which are positive-definite, rank-ordered signals beginning at

---

Fig. 5. – Sequence of photographs of the Andaman Sea surface taken as a rip band approached from the west at a speed of 2.2 m/s and passed the survey vessel on October 27, 1976, at 10:15 local time. The air temperature was 30 °C and the winds were calm during the sequence. *a*) 10:15, the rip was seen in the distance, stretching from one horizon to the other, as a well-defined line of breaking waves. The background sea state preceding the rip was $\sim$0.6 m, and approached from the west. *b*) 10:16, the rip continued to approach in background waves $\sim$0.6 m. *c*) 10:17, the rip had just arrived at the vessel with wave heights $\sim$1.8 m. *d*) 10:19, the survey vessel was tossed about in the 1.8 m waves of the rip band. *e*) 10:22, the rearward edge of the rip was visible in 1.8 m waves. *f*) 10:23, the rearward edge of the rip receded as the waves dropped to 1.3 m. *g*) 10:25, the wave amplitudes dropped to 0.6 m. *h*) 10:32, the rip had completely passed as the waves dropped to ripples $\sim$0.1 m.

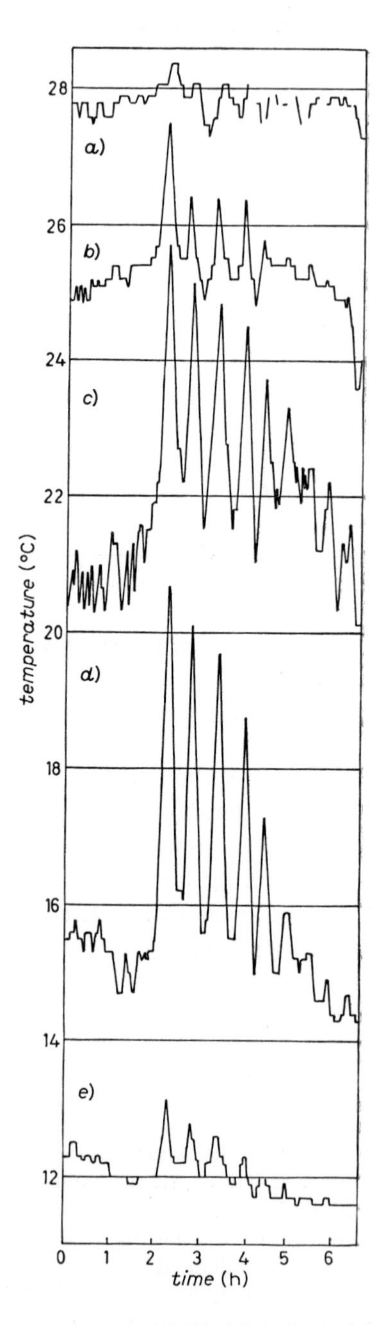

Fig. 7. – Temperature signals of internal solitons recorded at several depths on October 24, 1976, beginning at 19:30 local time: *a*) 53 m, *b*) 87 m, *c*) 116 m, *d*) 164 m, *e*) 254 m.

2 h 12 min into the record (which commenced on October 24, 1976, at 19:30). The wave amplitudes can be determined approximately by converting the temperature signal to amplitude units using the temperature gradient obtained from fig. 6. The first temperature enhancement in fig. 7 is 5.2 °C, corresponding to a 60 m internal wave. Each succeeding wave is smaller than its predecessor, revealing a pattern similar to that in fig. 2 and thus leading to a soliton interpretation of the data. The difference in the rank-ordering observed in the data and that in fig. 2 can be ascribed to the fact that the horizontal co-ordinate in the latter is space x, while in the former the co-ordinate is time t. In fig. 7, we note that time advances to the right, while the space co-ordinate increases to the left. Therefore, in terms of the space co-ordinate, the leftmost soliton of fig. 7 is the largest and leads the rest of the packet, which is consistent with fig. 2 and the Gardner *et al.* theory. The data obtained during the four-day measurement program show that passages of similar packets occurred, on the average, every 12 h 26 min, which is the semi-diurnal tidal period. This is clear evidence that the occurrence of the wave packets is linked in some way to the semi-diurnal tide.

The measured eastward components of water particle velocity, obtained simultaneously with the above temperature records, are shown in fig. 8 as a function of time and depth. Because the eastward component is within 5° of the propagation direction of the internal waves, these velocities are virtually aligned with the wave direction. First note that the records at the various depths have a nonzero mean due to the presence of an internal tide having maximum particle speeds of about 30 cm/s. The presence of the internal solitons is clearly evident, although they do not appear rank-ordered unless the tidal signal is removed. Note that the particle velocities for 164 m depth and above show an eastward enhancement in the flow, while for 437 m and below the solitons cause flow enhancements which are in a westerly direction.

In order to give a preliminary interpretation to these velocity data, we note that the horizontal velocities for a two-layer model are given by eqs. (17) and (18), where the negative sign before the amplitude in eq. (18) predicts flow reversal in the lower layer. Note also that the ratio of the maximum horizontal velocity in the upper layer to that in the lower layer is h_2/h_1. Using $h_1 = 230$ m (obtained from the Benny model [27], which takes into account the general stratification found at the measurement site) and $h_2 = 863$ m, we find $h_2/h_1 = 3.75$. From the data, the maximum particle velocities for the first wave (with amplitude ~ 60 m) at 87 m and 1001 m were 55 cm/s and 15 cm/s, respectively. Thus the measured ratio for this single event is 3.67, which agrees well with the expected theoretical value. For the same wave, we also find from eq. (11) that $c_0 = 2.14$ m/s. Hence the theory predicts that a maximum velocity occurs in the upper layer of $c_0\eta_0/h_1 = 55.8$ cm/s and in the lower layer $c_0\eta_0/h_2 = 14.9$ cm/s, both of which agree well with the data. We emphasize here not how accurate the two-layer model is (for we obtained improved

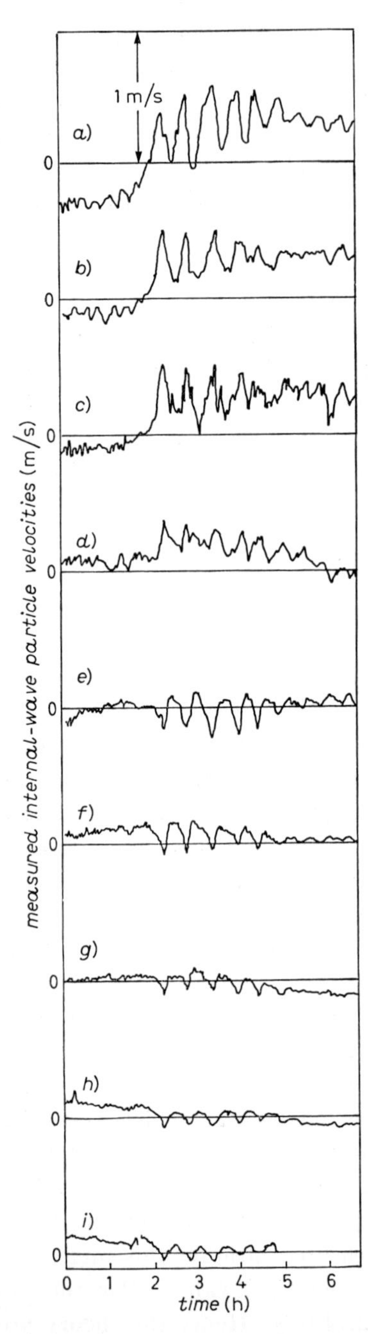

Fig. 8. – Eastward particle velocity signals of internal solitons superimposed on an internal tide recorded at several depths on October 24, 1976, beginning at 19:30 local time: *a*) CT/3, 87 m; *b*) CT/3, 116 m; *c*) M102, 121 m; *d*) CT/3, 164 m; *e*) CT/3, 254 m; *f*) M102, 437 m; *g*) M102, 635 m; *h*) M102, 895 m; *i*) M102, 1001 m.

agreement via the general stratification model, the details of which we discuss in a later publication), but the fact that *a*) there is no particle velocity decay below 600 m as evidenced by the measured velocities for the lower three current meters, and *b*) the ratio of measured particle speed between the upper and lower layers (u_1/u_2) is close to the ratio of the depths of the two layers (h_2/h_1). These observations are predicted by the KdV equation (but are not predicted by the deep-water B-O theory or any of the theories for intermediate depth) and we, therefore, feel justified in interpreting our results in light of the KdV equation.

In fig. 9 we display the results of several XBT casts (launched every 90 s) which were recorded on October 25, 1976, during the passage of a 60 m internal soliton. Shown as a function of depth are the isotherm contours which clearly indicate a wave of depression. The time axis increases to the right, while the wave advances to the left. The surface rip is seen to be on the leading edge of the soliton, a result which we observed consistently for all of the measured waves.

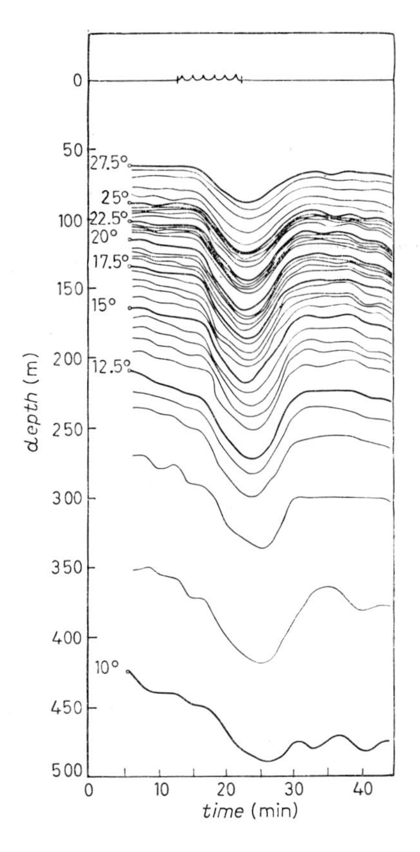

Fig. 9. – Isotherm contours of an internal soliton obtained from XBT casts on October 25, 1976, beginning at 8:40.

In fig. 10 we show the rank-ordering of the solitons measured at a single depth of 116 m. Six soliton packets were observed with an average of nearly 5 solitons per packet. The amplitude of the first soliton in each packet has been normalized to unity, so that decreasing amplitude with increasing soliton number indicates the appropriate rank-ordered behavior. The vertical bar

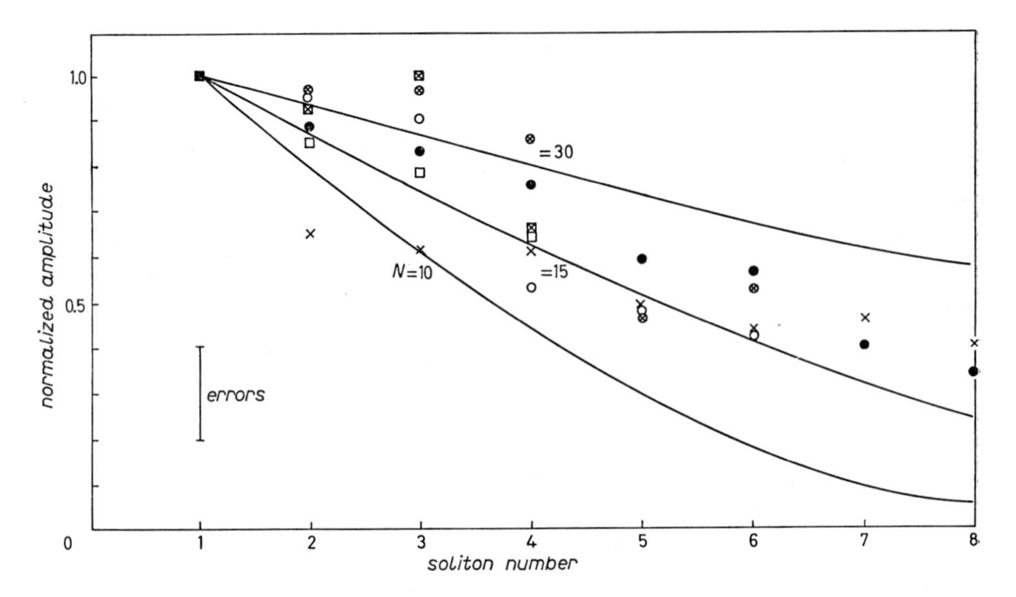

Fig. 10. – Normalized soliton amplitudes for 6 soliton packets obtained from temperatures measured 116 m below the surface. The error bar indicates the approximate uncertainty in amplitude due to the presence of the background random internal-wave field which is viewed as noise in the present experiment. The different symbols correspond to each of the six observed packets.

shown in the figure is representative of the magnitude of the random background internal-wave field which is viewed as « noise » in this experiment. The length of the error bar is based on temperature measurements taken during « quiet » time between the passage of the soliton packets. Thus the amplitude of a single wave is considered to be uncertain by an amount equal to the error bar length. Also shown in the figure are the curves for normalized amplitude which would result if the number of solitons to evolve were $N = 10$, 15 and 30, as computed from eq. (10). These curves are predicted by the Gardner *et al.* theory, which assumes that the solitons evolve in a constant-breadth, constant-depth channel, with no dissipation. Variations in these parameters could flatten the curves (*i.e.* decrease their slope) for the normalized amplitudes (because the influences of these effects are greater for the larger leading solitons than for the smaller trailing solitons). This could result in the conjecture that N must be greater than actually observed in order to get reasonable agreement with the data. In this paper we make no effort to correct for the effects of variable bathymetry, radial spreading or dissipation. We note, however, that the $N = 15$ curve gives reasonable agreement with the centroid of the amplitudes shown in fig. 10.

6. – Discussion.

After having briefly reviewed the data, we present our assessment of the generation, propagation and dissipation of internal solitons in the Andaman Sea. While the behavior of these waves must be inferred for locations other than our measurement site, we shall rely on our knowledge of internal-wave behavior based on the KdV equation. We shall also consider other physical effects which may occur and hence cause deviations from one-dimensional KdV behavior. We, therefore, feel that our overall assessment of this problem is qualitatively accurate and is also relevant to previous measurement programs which have detected internal solitary waves, *i.e.* [28] (amplitudes ~ 10 m) and [29] (amplitudes ~ 20 m) and to another experiment consisting of *in situ* measurements which are interpreted in terms of internal solitons, *i.e.* [30] (amplitudes ~ 20 m).

Because the measurements indicate a connection with the semi-diurnal tide (*i.e.* wave packets are detected every 12 h 26 min) and because the source region must lie to the west of the measurement site, the Nicobar Island chain and northern Sumatra represent likely candidates for potential source regions. In this light we may apply rule 11) of our summary of the Gardner *et al.* results and determine that the distance from our measurement site to the source is on the order of 300 km, which is consistent with the approximate location of the above-mentioned source regions (we have estimated the mean water depth as 1400 m and neglected radial spreading and energy dissipation). Figure 4 shows shallow-water areas near these locations where tidal flow over uneven bathymetry might result in an initial wave form which could ultimately evolve into a rank-ordered soliton packet. We note that the average spring tide at Galathea Bay, Great Nicobar Island (6° 47′ N, 93° 51′ E), is 1.4 m.

Lee and Beardsley [31] and Maxworthy [32], though not totally in agreement with each other, establish that a « sufficiently localized » wave form can develop from the effects of current flow over uneven bathymetry. The Maxworthy results indicate the likelihood of a lee wave forming to the west of a ridge during westward tidal flow out of the Andaman Sea. Upon reversal of the tidal flow, the lee wave ultimately propagates over the ridge and develops into the warm trough discussed by Lee and Beardsley. A plausible inference is that the « initial » wave form (*i.e.* the remnant of Maxworthy's lee wave) develops to the east of one or more of the shallow-water areas near the Nicobar Islands, and, further, that the initial wave form is a depression of the thermocline (a warm trough), as this is the only way, theoretically, that internal solitons can arise when the upper layer is thinner than the lower. An initial elevation (a cold dome) will evolve only into a dispersive wave train with no solitons. Because the scale of the sources is small compared to the areal extent of the waves on the satellite photographs, we view the initia wave form as a localized source that immediately begins to propagate, and evolves into solitons while undergoind radial spreading, encounteıing variable topography and slowly losing energy through dissipation.

The simple one-dimensional, constant-depth, constant-width channel model for the KdV equation may be modified to include dissipation effects [30], variable topography and channel width [33, 34], and radial spreading [35, 36]. Unfortunately, the Gardner *et al.* results no longer strictly apply and numerical models for these complex cases must be developed in order to obtain quantitative results. However, we may answer questions about the relative qualitative effects on certain predictions of the Gardner *et al.* theory. For example, radial spreading, dissipation and increasing water depth will reduce the amplitudes of larger leading solitons, in a relative sense, as compared to the smaller trailing solitons. These combined effects will result in a less rapid decay of soliton amplitude from the front of the packet to the back than that predicted by eq. (10) for the observed number of solitons. At the present time, however, there is no way to distinguish between the relative importance of radial speading, dissipation and variable topography on our data.

Another possible scenario for internal-soliton production in the Andaman Sea is through the fission process. Because of the extremely varied topography of the Andaman Basin, especially immediately to the east of the Andaman and Nicobar Islands, it is possible that the initial warm trough (developed in a potential shallow-water source region) might immediately propagate into water depths greater than, say, 3000 m and hence enter the deepwater Benjamin-Ono regime. Such a wave might then undergo fission when propagating up onto the eastern shelf of the Andaman Sea. If we use eq. (22) for $h_1 = 230$ m, $h_2 = 3000$ m and $\eta_0 = 70$ m, 6 or 7 solitons are predicted to evolve via the fission process. This is close to the average number of solitons which we actually observed in our program. Thus the fission mode is a possible means for soliton production in this area. Based only on our measurements, it is not possible to distinguish which of the two soliton production modes is more likely to exist in the Andaman Sea.

We feel that, in order to describe our own observations, we must infer the existence of a shallow-water source region near the southernmost Nicobar island or northern Sumatra where semi-diurnal tidal flow over a ridge could lead to an initial wave form (a warm trough) which would ultimately evolve into solitons via either of two methods. If the warm trough traverses relatively constant water depth, then the Gardner *et al.* solution seems most likely. However, if the warm trough first propagates into deep water and then onto the shelf, the fission process may be the preferred choice.

From the satellite photographs and our observations, an estimate of the energy in a typical internal soliton packet (using eq. (21)) is on the order of 10^{14} J (we assumed a packet of 6 waves, with leading crest length ~ 100 km). Thus an internal soliton colliding with the bathymetry near the continental shelf west of Thailand could dissipate power (per kilometer of crest length) at the rate of 2000 MW during the time $\tau = L/C \sim 10$ min. However, the duration of such events is rather short and the quiet period between them is long, so that there appears to be little likelihood of ever using this energy for practical purposes.

In regard to our observations of the surface rips, PHILLIPS [10] has pointed out the possibility that the « zones of whitecaps » observed by PERRY and SCHIMKE are due to the interaction of surface waves and long, large-amplitude internal waves. PHILLIPS develops arguments which lead to the conclusion that the internal-wave orbital velocities sweep the surface wave energy from regions of flow divergence and accumulate it in regions of flow convergence. We have established that the internal waves in the Andaman Sea have characteristics of long, large-amplitude internal solitons and that the rips always lie above the leading edge of the solitons, which is a region of flow convergence (see fig. 3). The resonance condition considered by PHILLIPS occurs when the internal-wave phase speed c equals the group velocity of the surface waves c_g, which requires that the surface wavelength be substantially less than the scale length of the internal waves. For the sequence of surface waves shown in fig. 5 the associated measured internal-wave speed was 2.2 m/s. Assuming the surface waves were in resonance with the internal soliton and further assuming that the surface background waves may be described by a Pierson-Moskowitz spectrum, we may estimate the zero-crossing period to be ~ 3 s and the significant wave height to be ~ 0.6 m. These results are consistent with our previously mentioned visual observations of the surface waves, and we conclude that the surface rips are accounted for by the Phillips theory.

In conclusion we may summarize the properties of the Andaman Sea data which are consistent with a soliton interpretation:

1) A random, small-amplitude, background internal-wave field is punctuated by occasional large-amplitude depressions of the thermocline which we interpret as internal solitary waves.

2) The internal solitary waves occur in packets of rank-ordered waves, the largest leading the rest. We infer that the waves are internal solitons and that they have evolved either from an initial wave form (over approximately constant water depth) or from the fission process (over variable water depth).

3) The packets occur every 12 h 26 min, clear evidence of a tidal origin for the internal solitons.

4) The likely source regions for the internal solitons lie in shallow water

near northern Sumatra or the southernmost of the Nicobar Islands where the average spring tidal range is $\sim$1.4 m.

5) The initial disturbance in the source region must lead to a depression of the thermocline, as this is the only way that internal solitons can be created when the upper layer is thinner than the lower.

6) The measured particle velocities at any given depth tend not to reverse in time, supporting the internal-soliton interpretation.

7) Based on satellite photography, the crest lengths of the internal solitons may be as long as 150 km and the separation distance between solitons in a packet may be as high as 15 km.

8) The internal solitons interact strongly with surface waves resulting in surface « rips » (short, choppy, breaking waves) which extend from horizon to horizon and are about 1 km wide.

We are presently involved in comparing the data to the general nonlinear stratification model of Benny [27], results which we will publish in the near future. A detailed comparison of this nonlinear analysis with conventional linear internal-wave theory is also forthcoming.

REFERENCES

[1] A. R. OSBORNE, T. L. BURCH and R. I. SCARLET: *J. Petr. Technol.*, **30**, 1497 (1978).
[2] R. A. WARRINER and D. H. SHUMWAY: *Annual Fall Technical Conference and Exhibition of the Society of Petroleum Engineers of* AIME, *Denver, Colo., 1977*, SPW6831.
[3] N. J. ZABUSKY and M. D. KRUSKAL: *Phys. Rev. Lett.*, **15**, 240 (1965).
[4] A. C. SCOTT, F. Y. F. CHU and D. W. McLAUGHLIN: *Proc. IEEE*, **61**, 1443 (1973).
[5] J. SCOTT RUSSELL: *Report on Waves* (London, 1844), p. 311.
[6] D. J. KORTEWEG and G. DEVRIES: *Philos. Mag.*, **39**, 422 (1895).
[7] R. M. MIURA: *SIAM Rev.*, **18**, 412 (1976).
[8] M. F. MAURY: *The Physical Geography of the Sea and its Meteorology* (New York, N. Y., 1861), p. 404.
[9] R. B. PERRY and G. R. SCHIMKE: *J. Geophys. Res.*, **70**, 2319 (1965).
[10] O. M. PHILLIPS: in *Annual Reviews of Fluid Mechanics*, Vol. **7**, edited by M. VAN DYKE and W. G. VINCENTI (Palo Alto, Cal., 1974), p. 93.
[11] E. C. LAFOND and C. BORRESWARA RAO: *Andhra University Memoirs in Oceanography*, Vol. **1**, Andhra University Series No. 49 (Waltair, 1954), p. 102.
[12] C. S. GARDNER, J. M. GREEN, M. D. KRUSKAL and R. M. MIURA: *Phys. Rev. Lett.*, **19**, 1095 (1967).
[13] H. SEGUR: *J. Fluid Mech.*, **59**, 721 (1973).
[14] J. L. HAMMACK and H. SEGUR: *J. Fluid Mech.*, **65**, 289 (1974).
[15] G. B. WHITHAM: *Linear and Nonlinear Waves* (New York, N. Y., 1964), p. 577.
[16] T. BROOKE BENJAMIN: *J. Fluid Mech.*, **25**, 241 (1966); **29**, 559 (1967).
[17] H. ONO: *J. Phys. Soc. Jpn.*, **39**, 1082 (1975).

[18] R. J. JOSEPH: *J. Phys. A: Math. Nucl. Gen.*, **10**, L225 (1977).
[19] T. KUBOTA, D. R. S. KO and L. D. DOBBS: *J. Hydronaut.*, **12**, 157 (1978).
[20] H. H. CHEN and Y. C. LEE: *Phys. Rev. Lett.*, **43**, 264 (1979).
[21] V. D. DJORDJEVIC and L. G. REDEKOPP: *J. Phys. Oceanogr.*, **8**, 1016 (1978).
[22] A. S. PETERS and J. J. STOKER: *Commun. Pure Appl. Math.*, **13**, 115 (1960).
[23] O. M. PHILLIPS: *Dynamics of the Upper Ocean* (London, 1969), p. 169.
[24] R. I. SCARLET played a major role in the design of the program. We are indebted to S. C. FREDEN (NASA Goddard Space Flight Center) and A. STRONG (NOAA, National Environmental Service Center (NESS)) for help in accessing the LANDSAT and NIMBUS satellites, respectively. M. N. GREER graciously pointed out ref. [8]. J. R. APEL kindly brought to our attention ref. [26] and the Apollo-Soyuz photograph of the Andaman Sea. R. L. GORDON provided valuable assistance in a previous phase of our analysis. Ka-Kit TUNG provided stimulating conversations.
[25] J. R. APEL, H. M. BYRNE, J. R. PRONI and R. L. CHARNELL: *J. Geophys. Res.*, **80**, 865 (1975); J. R. APEL, J. R. PRONI, H. M. BYRNE and R. L. SELLERS: *Geophys. Res. Lett.*, **2**, 128 (1975).
[26] J. R. APEL: *Apollo-Soyuz Test Project Summary Science Report*, Vol. **2**, NASA SP-412 (1978).
[27] D. J. BENNY: *J. Math. Phys. (N. Y.)*, **45**, 52 (1966).
[28] D. FARMER and J. D. SMITH: in *Hydrodynamics of Estuaries and Fjords*, edited by J. NIHOUL (New York, N. Y., 1978), p. 465.
[29] A. E. GARGETT: *Deep-Sea Res.*, **23**, 17 (1976).
[30] K. HUNKINS and M. FLIEGAL: *J. Geophys. Res.*, **78**, 539 (1973).
[31] C.-Y. LEE and R. C. BEARDSLEY: *J. Geophys. Res.*, **79**, 453 (1974).
[32] T. MAXWORTHY: *J. Geophys. Res.*, **84**, 338 (1979).
[33] R. S. JOHNSON: *J. Fluid Mech.*, **60**, 813 (1973).
[34] J. W. MILES: *J. Fluid Mech.*, **91**, 181 (1979).
[35] J. W. MILES: *J. Fluid Mech.*, **84**, 181 (1978).
[36] K. KO and H. H. KUEHL: *Phys. Fluids*, **22**, 1343 (1979).

PART III

MIXED-LAYER DYNAMICS

Mixed-Layer Physics.

P. P. NIILER

School of Oceanography, Oregon State University - Corvallis, Ore. 97330

1. – Observations of the ocean mixed layer and its equations of motion.

1·1. *Observations of variability in the upper ocean.* – Ocean circulation is maintained by the exchange of momentum, heat and water vapor between the ocean and the atmosphere. The momentum is transferred to the ocean through the action of atmospheric winds on its surface and heating of the ocean occurs primarily by the absorption of incident solar radiation. In the top few meters of the ocean the vertical convergence of these atmospheric fluxes has a few-thousand-kilometer spatial scale and a few-hour temporal scale of variability. The upper-ocean flow is variable both on the time and space scale of the atmospheric forcing, as well as on those scales which are governed by the internal, dynamical motions of water movement in ocean basins. The natural ocean scales can be a few hundred kilometers horizontally and a few years temporally and most typically are very different from natural atmospheric scales. The transfer of heat back to the atmosphere by long-wave radiation and evaporative and sensible cooling depends upon the ocean surface temperature and this temperature pattern is partly established by ocean circulation. Therefore, the ocean also imposes its own scales of variability on the atmosphere. Description and modeling of the physics of air-sea interaction and the upper ocean is complex, because the physical processes at work in both the air and water must be taken into account simultaneously. However, when the fluxes into and out of the ocean are directly measured, an independent description of the processes of the ocean response to atmospheric forcing is rendered. Here we describe how these specified atmospheric fluxes of momentum and heat are absorbed and distributed vertically in the upper ocean.

Half of the incident solar radiation on the ocean surface is composed of visible, short-wave (0.4 μm to 0.7 μm) irradiance and the other half is infra-red. Experimentally it is determined that clear ocean water absorbs 74 % of this flux in the top 1.5 m and the remaining portion is absorbed exponentially over

15 m [1, 2]. Figure 1 displays total downward flux as a function of water depth measured in the North Pacific at 34° N and 154° W. In this way, the top 15 m of the ocean surface is heated daily and, in the absence of ocean motion (and cooling), a 15 m thin and warm surface layer would develop over the ocean every summer day. In fig. 2 are two observations of the upper-ocean temperature (and salinity) distribution with depth on 21 and 24 August, 1977,

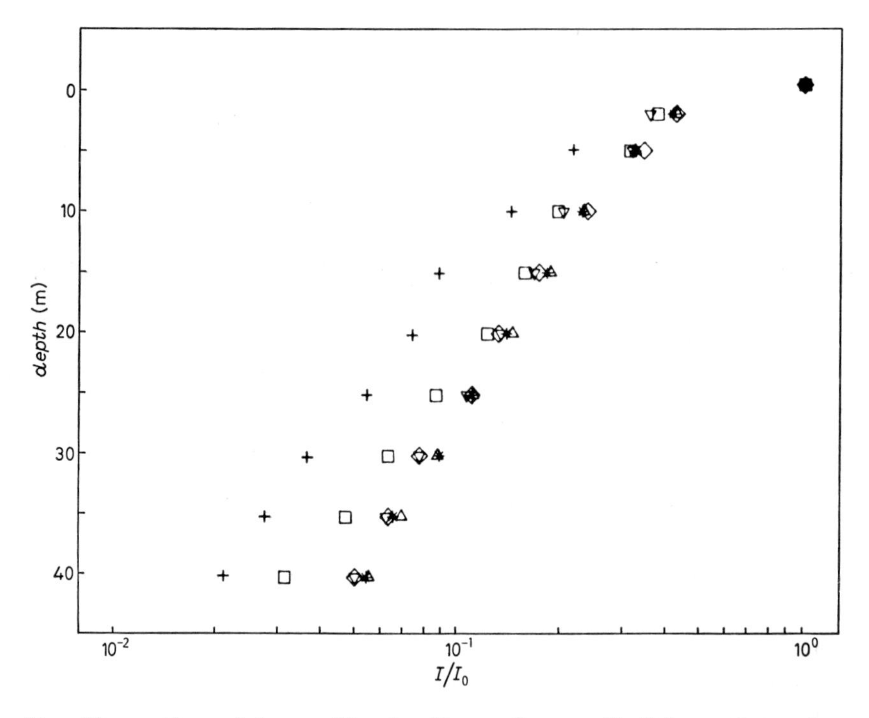

Fig. 1. – Observations of downwelling irradiance, I, normalized by surface values, I_0, as a function of depth at 34° N, 154° W. The conditions for each run are given by PAULSON and SIMPSON [2]: + run 1, * run 4, ◇ run 5, ▵ run 6, ▫ run 9, ▿ run 10.

in the North Pacific at 50° N and 145° W. Evident is a relatively homogeneous temperature layer of 30 m. Furthermore, when these profiles are compared with a profile in winter, it is also evident that the solar flux has « penetrated » to 70 m depth. The seasonal heating of the upper ocean cannot be accomplished by direct radiation absorption, nor by molecular diffusion. If the molecular conductivity of heat in the ocean is k, then in time t heat can diffuse to a depth $z = (kt)^{\frac{1}{2}}$, or with $k = 1.4 \cdot 10^{-3}\ \text{cm}^2/\text{s}$, $t = 3$ months $= 10^7$ s, $z \simeq 1.2$ m. Water motion in the upper ocean, therefore, must vigorously carry heat vertically and distribute it to greater depths than where it is absorbed and to where it can diffuse by molecular processes. Near the surface, water temperature at times appears to be very well « mixed vertically ». In the time

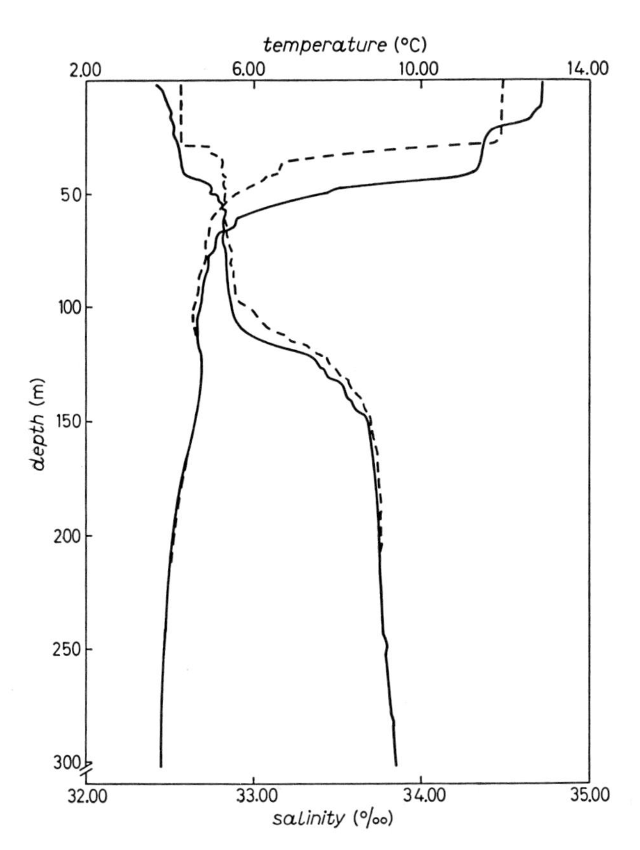

Fig. 2. – Profiles of temperature and salinity at 49° 34.9′ N, 145° 08.2′ W, 0845 GMT, 21 August 1977 (solid), 49° 38.2′ N, 145° 07.6′ W, 1320 GMT, 24 August 1977 (dashed). A strong storm passed over this area between the times these profiles were taken. Changes of profiles below 35 m are due to internal waves, above 35 m due to air-sea interaction (adapted from [3]).

interval between two temperatures traces in fig. 2, there was a strong ocean storm with 15 m/s wind, and the surface layer became well mixed during this storm.

The understanding of the physical process which causes an ocean « mixed layer » and how it evolves with time depends upon the complete description of the flow patterns near the ocean surface. Accurate observations of circulations which carry water parcels vertically and which are responsible for the mixing are very difficult to make because this circulation is very weak, very transient and velocity sensors cannot be kept stable due to the action of surface waves. Horizontal flow and some elements of very-small-scale motions (or the dissipation of ocean circulation to heat), on the other hand, can be measured. We find that the intensity of both increases in the upper ocean as atmospheric winds increase [4, 5]. Figure 3 displays the coherence and phase between

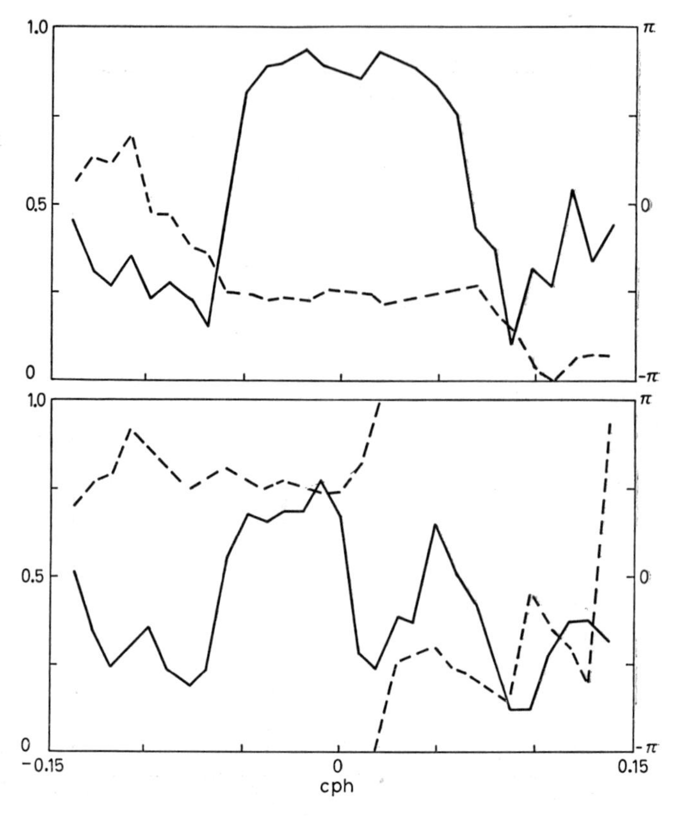

Fig. 3. – Coherence (solid curve, left scale) and phase (dashed, right scale) of wind stress and currents at 5 m (upper panel) and 42 m (lower panel) in MILE (adapted from [3]).

winds and horizontal currents at 5 m depth within the mixed layer, and at 42 m depth, below the mixed layer in the period 19 August-September 1977, at 50° N, 145° W [3]. The mixed-layer currents respond to the change in wind, while those below do not. Also note that there is not much coherence at periods less than a pendulum day and 15 min. These motions are due to deep-water internal waves which appear not to change with storms. Figure 4 is a graph of mechanical-energy dissipation *vs.* depth in the same location during a stormy and a quiet period [5]. The energy level of very-small-scale ocean motions, which dissipate to heat, also increases in the mixed layer during a storm. We observe that the momentum flux from the atmosphere is also transported vertically by upper-ocean circulation, rather than by molecular processes, and on time scales of a storm this transfer does not go deeper than the mixed layer. This transfer occurs by high-frequency motions. Therefore, as with the heat flux from radiation, momentum flux from wind stress is stored in the upper ocean. The modeling and description of how the heat and momentum which

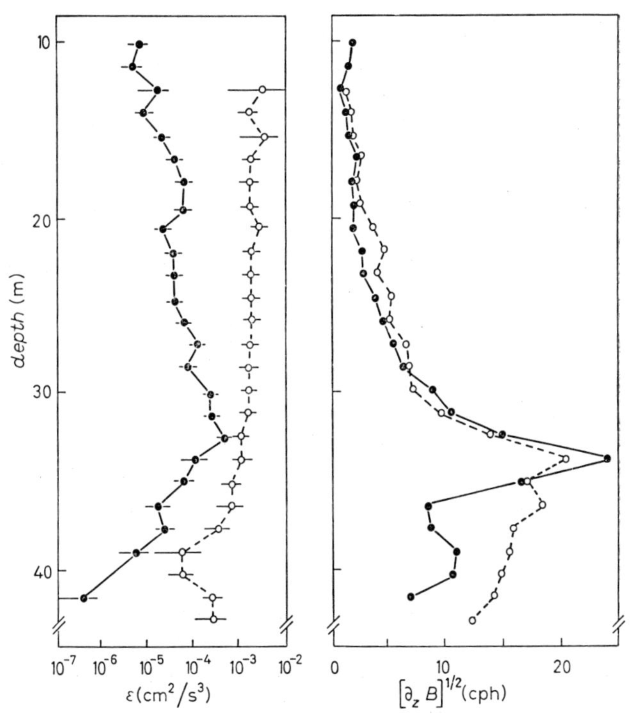

Fig. 4. – Mechanical dissipation rate, ε, and buoyancy frequency, $(\partial_z B)^{\frac{1}{2}}$, in low wind (solid circles) and high wind (open circles) speed in MILE (adapted from [5]).

are imparted to the ocean surface are vertically distributed in the upper ocean over a few days or weeks is the central question in this treatment of the physics of the ocean mixed layer.

In these lectures we first discuss the hydrodynamic theory and formulation of models of ocean mixed layers, next simple laboratory experiments and simple ocean situations are discussed and finally specific ocean observations are presented in context of the theory.

1·2. *Conservation equations for wind-driven motions, waves and turbulence.* – Consider time-dependent flow of water on a plane ocean which rotates with angular velocity $\Omega = \frac{1}{2}f$, where east, north and vertical direction are measured by x, y, z, time is measured by t and the respective velocity components are u, v, w. In vector notation, the Boussinesq form of the conservation equations for momentum, buoyancy and mass are

$$\partial_t \boldsymbol{v} + \boldsymbol{v}\cdot\nabla\boldsymbol{v} + f\hat{k}\times\boldsymbol{v} = -\nabla p + \hat{k}b + \nu\nabla^2\boldsymbol{v}\,, \tag{1}$$

$$\partial_t b + \boldsymbol{v}\cdot\nabla b = \alpha g k_T \nabla^2 T - \beta g k_s \nabla^2 s + \partial_z q + \Phi\,, \tag{2}$$

$$\nabla\cdot\boldsymbol{v} = 0\,. \tag{3}$$

In the above, the equation of state for the density is $\varrho = \varrho_0[1 - \alpha(T - T_0) + \beta(s - s_0)]$. The buoyancy of the water is $b \equiv \alpha g(T - T_0) - \beta g(s - s_0)$, where T is the thermodynamic temperature measured in °C and s is the salinity measured as the mass of dissolved salt in parts per thousand of sea water, ‰. In upper 200 m of subpolar oceans, the variations of density are represented by this equation to a few percent with $\alpha = 1.8 \cdot 10^{-4}/°C$, $\beta = 0.8 \cdot 10^{-4}/‰$ and $T_0 = 5$ °C, $s_0 = 34.0$.

The quantity $p = -gz + \not{p}/\varrho_0$ and $\not{p}$ is the dynamic pressure; for ocean application, the Coriolis parameter is $f = 2\Omega \sin \Theta_0$, where Ω is the rate of rotation of the Earth and Θ_0 is the latitude. Equation (3) is derived from adding αg times the conservation of heat $-\beta g$ times the conservation of salt. In this equation k_s and k_T are molecular conductivities of salt and heat, respectively, g is the gravitational acceleration and c is the specific heat of sea water at constant pressure. Due to the penetration of solar radiation into sea water, the water is heated locally at a rate $c\varrho_0 q/\alpha g$. Experimentally, a good representation for this function is

$$q = (\alpha g/c\varrho_0) I_0[R \exp[z/\xi_1] + (1 - R) \exp[z/\xi_2]], \tag{4}$$

where I_0 is the net incident solar radiation flux in W/m² on the ocean surface. In the North Pacific, $R \simeq 0.74$, $\xi_1 \simeq 1.4$ m, $\xi_2 = 14$ m [2]. Φ is the buoyancy produced by dissipation of mechanical energy to heat. It is small compared to any other term in eq. (3) and will not be considered further.

Elementary considerations of the spatial scales and amplitude of motion which cause vertical transport in the upper ocean or which are due to the known surface stresses excited by wind lead us to conclude that the motion is turbulent at least over that volume of water in which momentum transfer takes place. The Reynolds number of a 1 m size water parcel moving at 1 cm/s relative to another parcel is 10^4, and for upper-ocean situations these equations cannot be integrated analytically or on today's computers. Since individual water elements cannot be described, we make plausible hypotheses of how their aggregate or average motion might be described and use experiments to check these hypotheses. In what follows, only rather simple hypotheses are considered and these are examined in light of ocean experiments and descriptions.

To describe motion in our model of upper ocean, we consider it to be made of three components designated as

$$\begin{cases} \boldsymbol{v} = \boldsymbol{V}(z, t) + \tilde{\boldsymbol{v}}(x, y, z, t) + \boldsymbol{v}'(x, y, z, t), \\ b = B(z, t) + \tilde{b}(x, y, z, t) + b'(x, y, z, t). \end{cases} \tag{5}$$

The largest spatial-scale motions are described by $\boldsymbol{V}$ and these motions are forced directly by atmospheric fluxes. The x and y dependence of this motion is, for simplicity, not considered and atmospheric fluxes which cause these vary only in time. Next, $\tilde{\boldsymbol{v}}$ represents indigenous internal gravity waves in the upper ocean. Their linear dynamics and their energy level as a function of period and wavelength are reasonably well known [6, 7]. Experimentally we find that these propagate to the upper ocean from great depths where they are generated and they are most energetic in periods ranging from a pendulum day ($2\pi/f$) to twenty minutes. The internal wave field is stationary and homogeneous and, therefore, does not vary when averaged over large horizontal areas of the ocean. In our model the spatial average is taken over an area larger than the longest wave. Finally, $\boldsymbol{v}'$ is due to surface gravity waves and all other motions, or turbulence, which have a time scale shorter than a minute and spatial scales of 100 m to millimeters. The ensemble, or 10 min, temporal average of $\boldsymbol{v}'$ is smaller than $\tilde{\boldsymbol{v}}$. Both internal gravity waves and surface gravity waves are adiabatic and are easily identified because a small-amplitude motion of a specified period will have a predictable wavelength. Turbulent motions in the upper ocean, as in other high-Reynolds-number flows, are intermittent in time and localized in space. They are nonadiabatic because they account for the largest portion of mechanical-energy dissipation to heat. The equations of motion are nonlinear, therefore the different scales of motion, though not corrolated horizontally, or over many ensembles, interact.

We substitute (5) into eqs. (1)-(3) and average the resulting expression horizontally. If we denote the spatial average by an overbar, and because $\boldsymbol{V} = \bar{\boldsymbol{v}}$, the equations for the mean motion are

$$\partial_t \boldsymbol{V} + f\hat{k}\times\boldsymbol{V} = -\,\partial_z(\overline{\tilde{w}\tilde{\boldsymbol{v}}} + \overline{w'\boldsymbol{v}'})\,, \tag{6}$$

$$0 = -\,\partial_z \mathscr{P} + B\,, \tag{7}$$

$$\partial_t B = -\,\partial_z(\overline{\tilde{w}\tilde{b}} + \overline{w'b'}) + \partial_z q\,, \tag{8}$$

$$\partial_z W = 0\,. \tag{9}$$

These motions are of high Reynolds number, therefore effects of molecular processes are not considered. The influence of the internal wave and turbulent motion is now apparent in that these are responsible for the vertical transport and distribution of momentum and buoyancy in the upper ocean. Because turbulence is produced by both internal waves and mean motion, there is an interaction of all three components.

The equations for the internal wave field result from the ensemble average of eqs. (1)-(3) and using the balances in (6)-(9). Denote the ensemble average by $\langle\ \rangle$ and, since $\langle\boldsymbol{v}'\rangle = \langle b'\rangle = 0$ and $\langle\boldsymbol{V}\rangle = \boldsymbol{V}$, $\langle B\rangle = B$, $\langle\tilde{\boldsymbol{v}}\rangle = \tilde{\boldsymbol{v}}$, $\langle\tilde{b}\rangle = \tilde{b}$,

$$\begin{aligned}\partial_t\tilde{\boldsymbol{v}} + \tilde{\boldsymbol{v}}\cdot\nabla\tilde{\boldsymbol{v}} + f\hat{k}\times\tilde{\boldsymbol{v}} + \hat{k}\tilde{b} + \nabla\tilde{p} &= \\ = -\,[\boldsymbol{V}\cdot\nabla\tilde{\boldsymbol{v}} + \tilde{w}\partial_z\boldsymbol{V} - \partial_z\overline{\tilde{w}\tilde{\boldsymbol{v}}} &- \partial_z\overline{w'\boldsymbol{v}'} + \partial_z\langle w'\boldsymbol{v}'\rangle]\,,\end{aligned} \tag{10}$$

$$\partial_t \tilde{b} + \tilde{\boldsymbol{v}} \cdot \nabla \tilde{b} + \tilde{w} \partial_z B = -[\boldsymbol{V} \cdot \nabla \tilde{b} - \partial_z \overline{\tilde{w}\tilde{b}} - \partial_z \overline{w'b'} + \partial_z \langle w'b' \rangle], \tag{11}$$

$$\nabla \cdot \tilde{\boldsymbol{v}} = 0 . \tag{12}$$

The terms on the left-hand side of eqs. (10)-(12) govern the motion of internal waves. The terms on the right are due to the interaction produced by mean motion and turbulence acting on internal waves. We note that in general $\langle w'b' \rangle \neq \overline{w'b'}$, however $\overline{\langle w'b' \rangle} = \overline{w'b'}$. This becomes clear later when explicit models for turbulence stresses are used. Internal-wave theory is included in this presentation primarily to point out that very few calculations of the interaction of upper-ocean internal waves with the wind-produced changes in the mixed layer exist at this time [8, 9] and some very simple observations need to be explained. A typical seasonal thermocline in the middle of the ocean can theoretically support internal waves with periods of three to four minutes, however no internal waves with periods in excess of twenty minutes are measured [3]. This dearth of high-frequency thermocline waves surely must be due to the effects of either strong shear of mean flow ($\partial_z \boldsymbol{V}$), or intense turbulence ($\boldsymbol{v}'$) which occurs during storms. Provided the effects of $\boldsymbol{v}'$, b' are parameterized, eqs. (6)-(12) can be integrated numerically.

1·3. *Parameterization of turbulent transfer.* – The equations of $\boldsymbol{v}'$, b' are the full Navier-Stokes equations and these cannot be solved explicitly at the large Reynolds numbers of interest here. For a description of mixed layer and internal waves, however, rather deceptively few and simple moments of the nonadiabatic motions are required, *i.e.* $\langle w'b' \rangle$ and $\langle w'\boldsymbol{v}' \rangle$. A theory of, or experiments on the turbulent vertical transfer of buoyancy and momentum as a function of the state of the internal waves and mean conditions is required.

The simplest hypothesis is that turbulent transfer of momentum and buoyancy occurs just like molecular transfer,

$$-\langle w'\boldsymbol{v}' \rangle = k_{\text{M}} \partial_z (\boldsymbol{V} + \tilde{\boldsymbol{v}}), \tag{13}$$

$$-\langle w'b' \rangle = k_{\text{B}} \partial_z (B + \tilde{b}), \tag{14}$$

where k_{M} and k_{B} are now the constant turbulent viscosity and conductivity. In this way currents are accelerated with depth from the surface input of momentum and a seasonal thermocline is formed during the heating season. In such an ocean there is no well-defined mixed layer, and the wind-driven momentum and temperature vertical scale depths of frequency are independent and equal to $l_{\text{M}} = [k_{\text{M}}/(f \pm \omega)]^{\frac{1}{2}}$ and $l_{\text{B}} = (k_{\text{B}}/\omega)^{\frac{1}{2}}$, respectively. This structure is not born out by ocean observations or laboratory experiments.

In the laboratory it is found that vertical turbulent transfer is inhibited by vertical stratification and enhanced by vertical shear of horizontal currents. This observation can be parameterized by allowing k_{M} and k_{B} to be decreasing functions of increasing Richardson number, $\mathrm{Ri} = \partial_z(B + \tilde{b})/|\partial_z(\boldsymbol{V} + \tilde{\boldsymbol{v}})|^2$ (*e.g.*, [10]). However, only specific functions are consistent with simple considerations of energy balance of the turbulent fluid. In simple situations of the shearing of a stratified fluid, turbulence does not occur unless the local value of $\mathrm{Ri} \leqslant \frac{1}{4}$, which severely limits turbulent transfer above this value. Laboratory experiments [11] of rotating flows show that, if $\partial_z B = 0$, or $\mathrm{Ri} = 0$, the turbulent coefficient for momentum transfer is proportional to $u_*^2 f$, where the applied stress of the flat water surface is $\tau = \varrho_0 u_*^2$. In nonrotating stratified flow, without shear, it is proportional to $(u_*^4 \partial_z B)^{\frac{1}{2}}$. In the seasonal thermocline where mechanical energy is dissipated by an equilibrium internal-wave distribution at a rate equal to ε, it should be proportional to $\varepsilon/\partial_z B$ [12]. Because there is no *ad hoc* and consistent scheme for choosing a scaling or a consistent functional dependence of $\langle w' b' \rangle$ and $\langle w' \boldsymbol{v}' \rangle$ on the parameters over the time-dependent development of the mixed layer, theories of turbulence are used which are at least not inconsistent with laboratory data in unstratified, fully developed turbulent flows.

To develop an energetically consistent theory for the turbulent fluxes $\langle w'b' \rangle$ and $\langle w'\boldsymbol{v}' \rangle$, equations for these quantities are needed. In the form which is convenient for the specific problems that follow, these are formed from the second moments of the equations for $\boldsymbol{v}'$ and b', *viz.*

$$\tfrac{1}{2}\partial_t\langle \boldsymbol{v}' \cdot \boldsymbol{v}' \rangle = - \langle w'\boldsymbol{v}' \rangle \cdot \partial_z(\boldsymbol{V} + \tilde{\boldsymbol{v}}) + \langle w'b' \rangle - \partial_z \langle w'(\boldsymbol{v}' \cdot \boldsymbol{v}'/2 + p') \rangle - \varepsilon, \tag{15}$$

$$\tfrac{1}{2}\partial_t\langle b'^2 \rangle = - \langle w'b' \rangle \partial_z(B + \tilde{b}) - \partial_z \left\langle \frac{w'\, b'\, b'}{2} \right\rangle - \varphi\,, \tag{16}$$

$$\partial_t \langle w'\boldsymbol{v}' \rangle = - \langle w'w' \rangle \partial_z(\boldsymbol{v} + \tilde{\boldsymbol{v}}) - \partial_z \langle w'w'\boldsymbol{v}' \rangle + \\ + \langle b'\boldsymbol{v}' \rangle - \langle \partial_z p' \boldsymbol{v}' + w' \nabla p' \rangle - \tfrac{2}{3}\hat{k}\varepsilon, \tag{17}$$

$$\partial_t \langle b'\boldsymbol{v}' \rangle = - \langle w'\boldsymbol{v}' \rangle \cdot \partial_z(B + \tilde{b}) - \langle w'b' \rangle \partial_z(\boldsymbol{V} + \tilde{\boldsymbol{v}}) + \\ + \hat{k}\langle b'^2 \rangle - \partial_z \langle w b' \boldsymbol{v}' \rangle - \langle b' \nabla p' \rangle . \tag{18}$$

In the above, the horizontal convergence of the turbulent fluxes and Coriolis force are neglected in favor of the vertical convergence.

Equation (15) is the turbulent-mechanical-energy equation in which turbulent-mechanical-energy dissipation rate to heat is expressed as ε. Equation (16) is the turbulent-buoyancy production equation, and φ represents the dissipation rate of turbulent buoyancy. These are equations for the evolution of the second-order turbulent moments. To solve these, specific physical assumptions are now made about the unknown third-order moments in terms of lower-order moments, the role of pressure and velocity correlations and the dissipation of mechanical energy and buoyancy. In the next section, these hypotheses are developed further.

2. – Models of vertical transport of heat and momentum.

2·1. *Isotropy, mixing length and local-equilibrium hypothesis.* – In very simple laboratory situations, it is experimentally found that, when a turbulence-generating source and a mean buoyancy gradient are locally absent, anisotropic turbulence quickly decays to an isotropic state in which $\langle u'^2 \rangle = \langle v'^2 \rangle = \langle w'^2 \rangle$ and $\langle w'\boldsymbol{v}' \rangle = 0$ [13]. As expressed by the last term on the right-hand side of eq. (17), this process is produced locally by the pressure gradient and velocity correlation terms, and it is most simply modeled as [14]

$$\langle \boldsymbol{v}'\partial_z p' + w'\nabla p' \rangle = \frac{q}{3l_1}\left[\langle w'\boldsymbol{v}' \rangle - \frac{\hat{k}}{3}q^2\right] - C_1 q^2 \partial_z(\boldsymbol{V} + \tilde{\boldsymbol{v}}) \,, \tag{19a}$$

$$\langle b'\nabla p' \rangle = \frac{q}{3l_2}[\langle b'\boldsymbol{v}' \rangle] \,. \tag{19b}$$

Therefore, a decay to isotropy and a decay of the momentum flux and buoyancy flux from an initial or boundary source occur on a time scale $q/3l_1$ and $q/3l_2$, respectively. The time scale for this process is based on the time a turbulent « eddy » of energy $q^2 = \frac{1}{2}\langle \boldsymbol{v}'\boldsymbol{v}' \rangle$ transverses or « mixes » through a length l_1 (or l_2). Therefore, l_1 and l_2 are termed the *turbulent mixing lengths.*

Secondly, the third-order moments are considered to represent the self-diffusion of turbulence and these are modeled according to the diffusion laws [15]

$$-\langle w'[(\boldsymbol{v}'\cdot\boldsymbol{v}')/2 + p'] \rangle = q\lambda_1 \partial_z q^2 \,, \tag{20}$$

$$-\langle w'(w'\boldsymbol{v}') \rangle = q\lambda_1 \partial_z \langle w'\boldsymbol{v}' \rangle \,, \tag{21}$$

$$-\langle w'(b'^2) \rangle = q\lambda_2 \partial_z \langle b'^2 \rangle \,, \tag{22}$$

$$-\langle w'(b'\boldsymbol{v}') \rangle = q\lambda_3 \partial_z \langle b'\boldsymbol{v}' \rangle \,. \tag{23}$$

Note the analogy of expressions (20)-(23) with expressions (13) and (14), provided the diffusion coefficients are proportional to $q\lambda$. The diffusion assumption here is made for the third-order moments, rather than the second-order moments; λ_1, λ_2, λ_3 are additional mixing lengths.

Finally, Kolomogoroff [16] local-isotropy hypothesis is used to express the dissipation rates as

$$\varepsilon = q^3/\Lambda_1 \,, \tag{24}$$

$$\varphi = q\langle b'^2 \rangle/\Lambda_2 \,. \tag{25}$$

Again, Λ_1 and Λ_2 are the spatial scales in which a distant turbulence source « diffuses » or decays to heat and q/Λ_1 and q/Λ_2 are the comparable times after which an initial distribution of turbulence cannot be detected.

Equations (15)-(25) are now a closed set of partial differential equations for the turbulent moments and these contain seven unknown mixing lengths and a constant C_1. Our ignorance of the role of the self-diffusion or decay of turbulence is expressed in the belief that there are well-determined (perhaps experimentally) length or space scales over which turbulent elements lose their identity. Though not physically appealing for all situations, some simple phenomena, as turbulent stresses near ocean bottom, can be quite well understood and summarized both conceptually and energetically with rather simple prescription of mixing lengths [17]. Whether upper-ocean mixing can be represented adequately by these hypotheses is presently a topic of fascinating research, because turbulent structures on some limited scales can now be measured in the mixed layer and the seasonal thermocline [5, 18].

To simplify the problem, we further assume that all mixing lengths are proportional to each other,

$$l_1, l_2, \lambda_1, \lambda_2, \lambda_3, \Lambda_1, \Lambda_2 = (A_1, A_2, D, D, D, B_1, B_2)l, \tag{26}$$

and l is now the basic quantity to be specified with $A_1, \ldots, B_2$ being « universal » constants. Their numerical values can be judiciously chosen from nearly neutral laboratory experiments.

Adjacent to a solid boundary, as sea ice or ocean bottom,

$$ql \sim w_* z/\varkappa, \tag{27}$$

where w_* is an appropriately chosen friction velocity and $\varkappa$ is von Karman's constant. This prescription cannot be justified near the wavy, free, ocean surface which is not rigid, nevertheless very recent observations of turbulent dissipation in the mixed layer indicate this prescription is also appropriate during strong storms [19]. Therefore, prescriptions that l is proportional to an integral, or differential, scale of the distribution of $q(z)$ within the upper ocean might be

$$ql = \frac{w_* z}{\varkappa}\Bigg/\left(1 + \frac{w_* z}{\varkappa l_\infty}\right), \qquad l_\infty = a\int_{-\infty}^{0} |z| q(z)\,\mathrm{d}z \Bigg/ \int_{-\infty}^{0} q(z)\,\mathrm{d}z. \tag{28}$$

A prescription where $l = l_\infty$ has interesting consequences, for, in the absence of $\boldsymbol{V}$, $\tilde{\boldsymbol{v}}$ and b, eq. (15) is

$$\partial_t q^2 = D\partial_z ql\partial_z q^2 - q^3/B_1 l. \tag{29}$$

There is no intrinsic or natural space or time scale in eqs. (28) and (29) (recall that D and B_1 are dimensionless constants). Therefore, in these models, turbulent energy will diffuse on the space scale established by its initial spatial distribution and decay on the time scale which depends upon its initial strength $q(z, 0)$ and initial distribution $l(z, 0)$ according to $q(z, 0)/l(z, 0)$. If, initially, $q(z, 0) = 0$, and a boundary flux of turbulent energy is specified, as might be produced by gravity waves, $ql\partial_z q^2|_{z=0} = mu_*^3(t)$ [20], these equations cannot be integrated. In other words, in this integral scale model, turbulence cannot diffuse into a region that is not turbulent already and some care must be exercised how an initial mixing length is chosen. In the ocean application, this length scale evolves to the length scale which produces turbulence in the mixed layer. During storm conditions the principal source of turbulence is the vertical shear of horizontal currents in and at the base of the mixed layer.

A review of some of the general properties of these equations and the choice of the « universal » constants is given by YAMADA [21] under conditions of a stable atmosphere boundary layer, which most closely resembles the oceanic storm conditions. He shows that the turbulent vertical fluxes of momentum and buoyancy during boundary layer development can be expressed as

$$-\langle w' \boldsymbol{v}' \rangle = S_{\mathrm{M}} ql\partial_z \boldsymbol{V}, \qquad -\langle b' w' \rangle = S_{\mathrm{B}} ql\partial_z B, \tag{30}$$

and S_{M} and S_{B} (as well as ql) are rather simple, decreasing functions of increasing Richardson number. In the equilibrium situation, where ∂_t and $\lambda_1, \lambda_2, \lambda_3 = 0$, the equations for the turbulent fluxes are algebraic and analytic expression can be obtained for both S_{M} and S_{B}. These also are decreasing functions of increasing Richardson number for $\mathrm{Ri} < R_{\mathrm{c}} = 0.23$ (for the values of « universal » constants chosen by YAMADA). In the water column where $\mathrm{Ri} > R_{\mathrm{c}}$ vertical transfer of both momentum and buoyancy ceases. When turbulent « self-diffusion » is included in the model ($\lambda_1, \lambda_2, \lambda_3 \neq 0$), turbulent energy can be transferred to a more stable region, however at a much reduced rate. Also, the ratio of $S_{\mathrm{B}}/S_{\mathrm{M}}$ is a very slowly decreasing function of increasing Ri, and for all practical purposes equal to $1.2 \div 1.3$. Thus the turbulent-buoyancy diffusion coefficient is a bit larger than the momentum diffusion coefficient. These results are independent of the particular functional form of l. Figure 5 displays the equilibrium, stability functions S_{M} and S_{B} along with those proposed by MUNK and ANDERSON [10].

A simple way to see the dependence of the turbulent diffusion coefficients on Ri (and other parameters) is to consider the turbulent-mechanical-energy equation (15) and eqs. (29) and (30), which in the *equilibrium* form reduce to

$$S_{\mathrm{M}} ql|\partial_z \boldsymbol{V}|^2 - S_{\mathrm{B}} ql\partial_z B = q^3/B_1 l. \tag{31}$$

Multiply (31) by l^2 and divide by q and, since $S_{\mathrm{B}}/S_{\mathrm{M}}$ is approximately a constant, R_0^{-1}, there obtains a rather simple expressions for the turbulent-transfer coef-

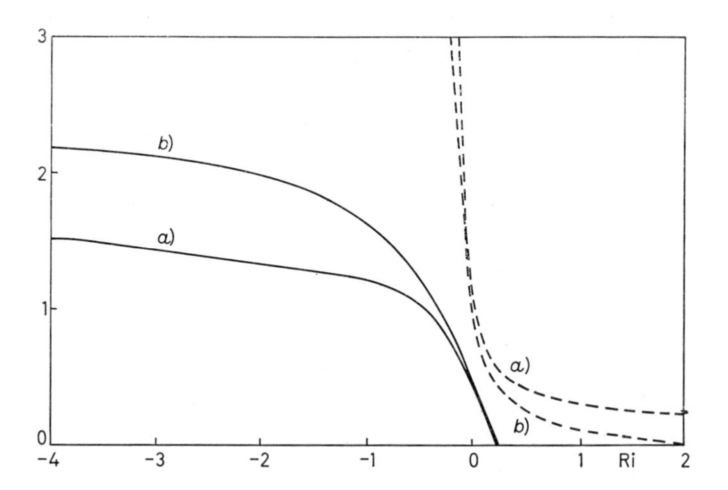

Fig. 5. – The equilibrium stability function for vertical turbulent transport of *a*) momentum, S_M, and *b*) buoyancy, S_B, as function of Richardson number Ri. The solid curves are proposed by the closure scheme of Mellor and Yamada [14] and dashed curves are those proposed by MUNK and ANDERSON [10] (adapted from [22]).

ficients for momentum and buoyancy, respectively,

$$k_M = S_M^{\frac{3}{2}} l^2 |\partial_z V| \{1 - \mathrm{Ri}/R_0\}^{\frac{1}{2}} B_1^{\frac{1}{2}}, \tag{32}$$

$$k_B = \frac{1}{R_0} S_M^{\frac{3}{2}} l^2 |\partial_z V| \{1 - \mathrm{Ri}/R_0\}^{\frac{1}{2}} B_1^{\frac{1}{2}}. \tag{33}$$

Therefore, in this scheme, the turbulence is limited by the functional dependence of S_M (or S_B) on Ri, because it vanishes before $\mathrm{Ri} = R_0$.

If we assume that S_M and S_B are constant, in which case Rotta's hypothesis is not needed, and in which case $k_M \sim ql$, $k_B \sim ql/R_0$, or a diffusion hypothesis is made of the second-order moments (*e.g.*, [23]), then there still is an effective damping of turbulence (q vanishes) for $\mathrm{Ri} \geqslant R_0$. In the atmospheric measurements R_0 is determined experimentally to be between 1.1 and 0.6. As we shall see later, the vertical extent of oceanic mixing under storms depends upon one fourth power of R_c (or R_0) and the mixed-layer response is rather insensitive to the precise value of this quantity. However, the character of the ocean turbulence which is predicted by these models is very different depending upon whether S_M or S_B are a function of Ri.

In summary, if we use the hypotheses of the existence of a predominant scale of energy-containing eddies, and that these eddies are all dissipated, a rapid return to local isotropy and a condition of near local equilibrium yields a model of vertical transport of momentum and buoyancy in which this transfer is severely damped by stratification. Vertical mixing can occur only if the local value of horizontal shear in the upper ocean exceeds a minimum critical value. The self-diffusion of turbulent energy enhances vertical transfer into the stable water mass below the mixed layer, but at a greatly reduced rate from what occurs in the region of supercritical shear.

2·2. *The entrainment interface and the bulk dissipation hypotheses.* – It is observed that during and after storms the mixed-layer buoyancy is very uniform with depth and there is a large buoyancy gradient at its base. Also, during summer storms, when the ocean is heated, as the mixed layer deepens, the surface temperature becomes colder, indicating that cold or heavy water is *entrained* at the mixed-layer base and is mixed throughout the upper water column. In this process, the potential energy of the mixed-layer water mass is raised, because heavy water is lifted up from below and net work is accomplished on the ocean by the storm. Figure 6 displays the ocean conditions during a storm on the west Florida continental shelf on 10 Febraury, 1973 (26° N, 84° W) [24]. This figure also shows that there is a dramatic increase

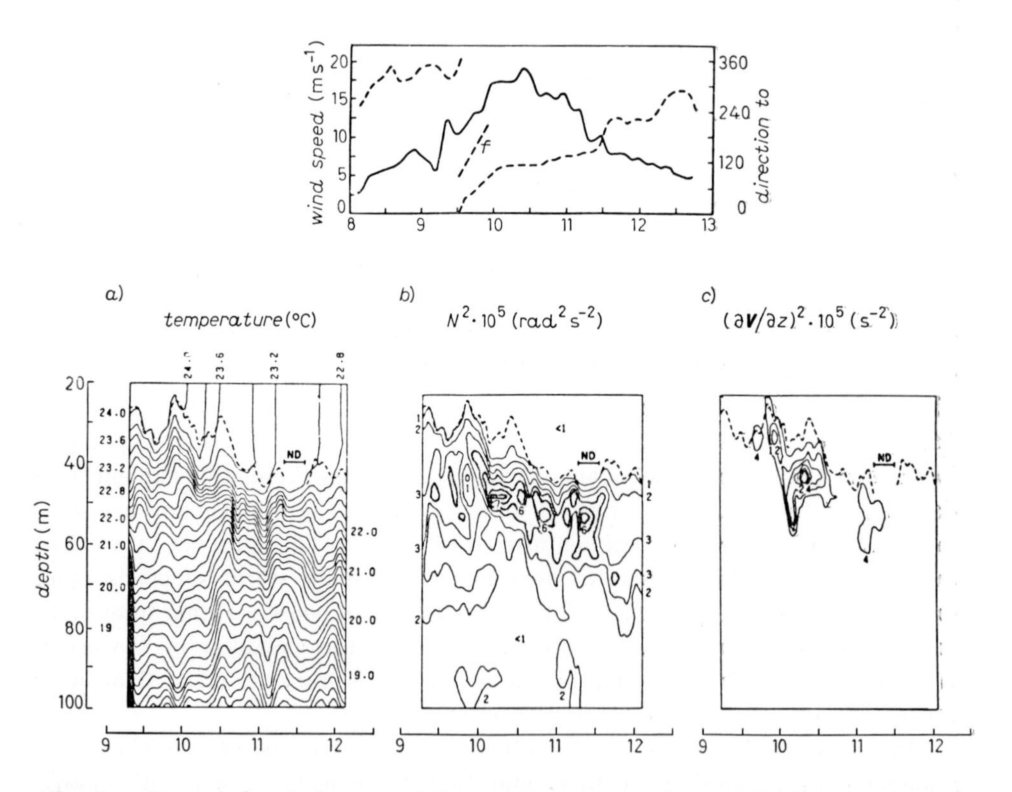

Fig. 6. – Storm-induced changes of the upper ocean on the west Florida continental self (26° N, 84° W). In the center panel $N^2 \equiv \partial_z B$. Note the increase of stability and increase of $|\partial_z \boldsymbol{V}|^2$ at the base of the mixed layer (dashed line in lower panels) during the storm (adapted from [24]). Upper panel: ——— speed, – – – direction.

of the vertical shear of horizontal currents at the mixed-layer base directly after the storm. We can use these observations during an active mixing time to model the vertical structures of mean conditions and, using conservation equations for mean buoyancy and momentum, infer what are the consistent vertical structures of the turbulent fluxes of buoyancy and momentum.

Consider eq. (8) in the case where B is independent of z in a mixed layer of depth h. It thus follows that $-\overline{w'b'}+q$ must be a linear function of z (where again for simplicity set $\overline{\tilde{w}\tilde{b}}=0$),

$$(34)\qquad -\overline{w'b'}+q=-\overline{w'b'}|_0+ \\ +q(0)+(z/h)[-\overline{w'b'}|_0+q(0)+\overline{w'b'}|_{-h}-q(-h)]\,.$$

Also, if in the stormy period the horizontal velocity field is independent of z, or at least its predominant shear is concentrated at the mixed-layer base (as seen in fig. 6), then $\overline{w'\boldsymbol{v}'}$ must also be a linear function of depth,

$$(35)\qquad -\overline{w'\boldsymbol{v}'}=-\overline{w'\boldsymbol{v}'}|_0+(z/h)[-\overline{w'\boldsymbol{v}'}|_0+\overline{w'\boldsymbol{v}'}|_{-h}]\,.$$

In the above expressions, the surface flux of buoyancy is $-\overline{w'b'}|_0+q_0=\dot{B}_0$, and $-\varrho_0\overline{w'\boldsymbol{v}'}=\boldsymbol{\tau}_0$ is the surface stress (for a discussion of these fluxes in terms of atmospheric and oceanic conditions, see [22]); $-\overline{w'b'}|_{-h}$ and $-\varrho_0\overline{w'\boldsymbol{v}'}|_{-h}$ are termed the entrainment fluxes of buoyancy and momentum, respectively, and these result from the incorporation of nonturbulent, heavy and slowly moving water mass (or a mass moving in different direction) into the turbulent, lighter and more rapidly moving mixed layer as the mixed layer deepens. Therefore, it is a simple matter to show that, if the thickness between the turbulent and nonstratified column and the nonturbulent and stratified column is δ and $\delta/h\ll 1$, then [25]

$$(36)\qquad -w'b'|_{-h}=(B-B_+)\,\partial_t h+\overline{w'b'}_+\,,$$

$$(37)\qquad -\varrho_0\overline{w'\boldsymbol{v}'}|_{-h}=\varrho_0(\boldsymbol{V}-\boldsymbol{V}_+)\,\partial_t h+\boldsymbol{\tau}_+\,,$$

where + designates the conditions directly below the turbulent interface. To study the effects of ocean storms, we can set $-\overline{w'b'_+}=\boldsymbol{\tau}_+=\boldsymbol{V}_+\simeq 0$, because we expect the turbulence levels there to be greatly reduced below the mixed layer and the wind-driven flow there is small. With the incorporation of expressions (34)-(37) into eqs. (6) and (8), the momentum and buoyancy conservation equations for the mixed layer are

$$(38)\qquad \partial_t h\boldsymbol{V}+f\hat{k}\times h\boldsymbol{V}=\boldsymbol{\tau}_0/\varrho_0\,,$$

$$(39)\qquad h\partial_t B+(B-B_+)\,\partial_t h=\dot{B}_0-q(-h)\,.$$

Equation (15) is now used to express an energetically consistent balance between the production of turbulence and its destruction by stratification (or lifting of cold water into the mixed layer or vertical mixing the surface warm layers) or its destruction by dissipation. Integrate (15) vertically from below the mixed layer to the surface, and there results the expression

$$(40)\qquad \int_{-h}^{0}\partial_t q^2\,\mathrm{d}z + \overline{w'(p'+\boldsymbol{v}'\cdot\boldsymbol{v}'/2)}|_0 + \int_{-h}^{0}\overline{w'\boldsymbol{v}'}\cdot\partial_z\boldsymbol{V}\,\mathrm{d}z = \int_{-h}^{0}\overline{w'b'}\,\mathrm{d}z - \int_{-h}^{0}\varepsilon\,\mathrm{d}z\,.$$

For open-ocean conditions, where there is a free surface, a net flux of turbulent energy occurs from the gravity wave and atmospheric-pressure perturbation interactions into the mixed layer [20]

$$(41)\qquad -\overline{w'(p'+\tfrac{1}{2}\boldsymbol{v}'\cdot\boldsymbol{v}')}|_0 = mu_*^3\,.$$

This flux is unique to the open-ocean surface and it does not occur in atmospheric boundary layers, on the ocean bottom or under sea ice.

Secondly, turbulent energy is produced at the base of the mixed layer by the interaction of the entrainment stress with the shear of the mean flow and, using expression (37), with $\boldsymbol{\tau}_+ = \boldsymbol{V}_+ = 0$, we obtain

$$(42)\qquad -\int_{-h}^{0}\overline{w'\boldsymbol{v}'}\cdot\partial_z\boldsymbol{V}\,\mathrm{d}z = \tfrac{1}{2}\boldsymbol{v}\cdot\boldsymbol{v}\,\partial_t h\,.$$

Finally, a sink (or source) of turbulent energy results from the mixing of heated water (or the convection of cooled water) from the surface or lifting of cold water from the mixed-layer base. This process is expressed in the first term on the right-hand side of (39) and with expressions (34) and (36) it is

$$(43)\qquad -\int_{-h}^{0}\overline{w'b'}\,\mathrm{d}z = \tfrac{1}{2}h\dot{B}_0 + \tfrac{1}{2}h(B-B_+)\,\partial_t h - \int_{-h}^{0}q\,\mathrm{d}z + \tfrac{1}{2}hq(-h)\,.$$

So far the principal assumptions have been that a very turbulent mixed layer occurs under a storm and the interface between this layer and the non-turbulent and stratified layer is thin. Expressions (41)-(43) are true for all vertical-transport models which produce well-mixed buoyancy and momentum layers. The essential element of the bulk models, as developed here, is that sufficient turbulent energy exists to render the upper ocean homogeneous and specification of how turbulence is dissipated (recall Kolmogoroff's hypothesis for laboratory situations). The simplest approach which is appropriate for the time scales associated with ocean storms is to assume that, of each *source*, a fraction is dissipated, or

$$\int_{-h}^{0} \varepsilon \, dz = m_* u_*^3 + \frac{m_s}{2} \boldsymbol{V}\cdot\boldsymbol{V} \partial_t h + \frac{m_c}{4} h(|\dot{B}_0| - \dot{B}_0) . \tag{44}$$

In expression (44) the terms are the fraction m_* which the surface production is dissipated, the fraction m_s which the shear source is dissipated and, m_c, the fraction which the surface convective ($\dot{B}_0 < 0$) energy source is dissipated (for a discussion of the effects of dissipation for seasonal time scales see [26]). Substitute expressions (41)-(44) into (40), and, with the neglect of the small term $\int_{-h}^{0} q^2 \, dz$, there results the equation for prediction of the mixed-layer depth

$$\begin{aligned} \frac{1}{2}[(B - B_+)h + (1 - m_s)\boldsymbol{V}\cdot\boldsymbol{V}]\partial_t h = \\ = (m - m_*)u_*^3 - \frac{h}{2}\left[\dot{B}_0 - \frac{2}{h}\int_{-h}^{0} q \, dz + q(-h)\right] - \frac{1}{4} m_c h[|\dot{B}_0| - B_0] . \end{aligned} \tag{45}$$

Equations (38), (39) and (45) now describe the evolution of $\boldsymbol{V}$, B and h during periods of intense air-sea interaction. If $\partial_t h < 0$, the evolution of $h(t)$ is predicted from the solution of the right-hand side of (45) set equal to zero.

In summary, we can use the ocean observation that during periods of intense storms the transition zone at the mixed-layer base is a thin entrainment interface and within the mixed layer the buoyancy and velocity fields are uniform with depth to construct a model of the vertical distribution of turbulent fluxes in the mixed layer. The entrainment fluxes of momentum and buoyancy are not known *a priori*, however we see that these can produce and destroy turbulent energy. As in continuous models, the turbulent-energy balance in the mixed layer is used to constrain these fluxes to be energetically consistent. The dissipation constants, m_*, m_s, m_c, are evaluated from oceanic (or atmospheric) field data. This in principle can be done from different temporal segments of a single ocean data series, of « mean » $B(z, t)$ and $\boldsymbol{V}(z, t)$, because in ocean conditions these represent processes on different time and vertical space scales. We shall discuss a number of these attempts in sect. **3**. A much more satisfactory and fundamentally important ocean data set, however, would be a description of the actual turbulent structures (or Reynolds stresses) and their statistics in the mixed layer and the entrainment interface, because then the fundamental assumptions of various models can be tested in greater detail (for example are the mixed layer « eddies » anisotropic, etc.). Though a difficult problem, the direct measurement of turbulent structures in the ocean mixed layer are now being made and it is a much needed, fruitful and exciting reasearch field.

2·3. *Models of the formation of the mixed layer during storms.* – The simplest model of the formation of an ocean mixed layer under a storm is to consider that initially the ocean is quiescent and linearly stratified and a unidirectional and subsequently time-independent wind stress is suddenly applied on the ocean surface. There are no surface buoyancy fluxes ($\dot{B}_0$, $q = 0$), surface wave energy fluxes ($m = 0$) or internal waves. Although idealized, this problem is extremely revealing in understanding more complex situations. Numerical studies have been made of this problem by MELLOR and DURBIN [27], MARTIN [28], WARN-VARNAS and PIACSEK [29] and KUNDU [30] both with equilibrium and nonequilibrium turbulence (self-diffusion) closure models for turbulent fluxes. A variety of prescriptions for mixing lengths are used. The results of these numerical experiments are summarized below.

i) After the onset of the wind, the mixed layer evolves in two rather distinct stages. There is a rapid formation and deepening of the mixed layer in the first quarter inertial period, which is then followed by a very slow erosion. In the models which exhibit a sharp reduction of turbulence above a critical value of Ri, the rapid erosion of the stratified column is arrested at a depth h_* below which turbulence ceases,

$$h_* = au_*(R_c)^{\frac{1}{4}}/f^{\frac{1}{2}}(\partial_z B(0))^{\frac{1}{4}}, \tag{46}$$

where a is a constant near unity, $\partial_z B(0)$ is the initial constant buoyancy gradient and $\varrho_0 u_*^2$ is the applied surface stress. The depth h_* is independent of the specific prescription of the mixing lengths or the equilibrium assumption. At the end of the rapid deepening period, the shear production of turbulent energy very nearly (to 95%) balances the dissipation at each level. During the deepening period, however, the shear production is strongest at the mixed-layer base where local $\text{Ri} < R_c$, and there it is in considerable excess of local production (on the mixed-layer average, the excess is 25%). This excess is used to rapidly mix the entrained fluid through the deepening interface and the « mixed layer ». At the end of the initial deepening period, the horizontal velocity field has reached a peak amplitude and inertial motions are established.

ii) During the process of the rapid mixing, two rather distinct horizontal flow patterns are established. Strong, time-variable currents of inertial period continue unabated, and these are relatively independent of depth within the mixed layer. Because they have no shear, these motions produce no turbulence within the mixed layer. In addition, there is a quasi-steady mean drift which closely resembles a steady, nonstratified, turbulent Ekman spiral with the prescribed constant wind stress on the surface and *zero bottom stress* above the mixed-layer base. The slow erosion process and the vertical structure of the quasi-steady Ekman spiral, of course, depend upon the mixing length prescription and the specific internal balances of turbulent production, diffusion and dissipation. The most general closure models predict that turbulence which is associated with the Ekman spiral is anisotropic, such that, if the wind is in the x-direction, then $\overline{u'^2} > \overline{w'^2} \simeq \overline{v'^2}$. After the first quarter inertial period, the buoyancy flux $\overline{w'b'}$ is a linear function of depth with a maximum value at the mixed-layer bottom, while dissipation and shear production profiles have considerable curvature, with a minimum value near the mixed-layer bottom. After a short time, $\int_{-h}^{0} \varepsilon\, dz/u_*^3 \to$ const, which is model dependent. These simple quantitative predictions can presumably be checked by observations.

The identical problem can be treated analytically within the framework of the bulk model which was developed in subsect. **2**·2. Under the conditions described above, the solutions to eqs. (38) and (39) are, respectively,

$$hV = \frac{u_*^2}{f}[\sin ft, \cos ft - 1] \tag{47}$$

and

$$h(B - B_+) = \tfrac{1}{2} h^2 \partial_z B(0) . \tag{48}$$

Substitute eqs. (47) and (48) into (45), which results in

$$\tfrac{1}{2}\partial_t h \left[\frac{h^2}{2}\partial_z B(0) - \frac{2(1-m_s)}{h^2}\frac{u_*^4}{f^2}(1-\cos ft)\right] = (m - m_*)u_*^3 . \tag{49}$$

DESZOEKE and RHINES [31] show that under oceanic conditions $(\text{of } u_*, f, \partial_z B(0))$ there are two different regimes of mixed-layer deepening after the onset of a storm. Within the first quarter inertial period, a good approximate solution to (49) is obtained by balancing the first two terms of the left-hand side and

$$h \simeq \sqrt{2}\,(1 - m_s)^{\frac{1}{4}}[u_*/f^{\frac{1}{2}}(\partial_z B(0))^{\frac{1}{4}}](1 - \cos ft)^{\frac{1}{4}} , \tag{50}$$

which expression reaches a maximum value, h_*, a quarter inertial period after the onset of the storm,

$$h_* = \sqrt{2}\,(1 - m_s)^{\frac{1}{4}} u_*/f^{\frac{1}{2}}(\partial_z B(0))^{\frac{1}{4}} . \tag{51}$$

During this period, the bulk Richardson number as defined by $R_B \equiv \equiv h(B - B_+)/\boldsymbol{V}\cdot\boldsymbol{V} = R_{BC}$ is a constant, equal to $1 - m_s$, and the maximum depth is proportional to $R_{BC}^{\frac{1}{4}}$. The physics of this process is identical to the shear mixing process described in continuous models. KUNDU [30] shows that during this process in continuous closure models the bulk Richardson number is approximately equal to twice the critical local Richardson number (above which local turbulence is damped).

After the first quarter inertial period, the balance of terms in (49) is shifted to the term in the left-hand side and the term on the right-hand side. The solution is for $t > \pi/2f$

$$\frac{1}{6}h^3 \simeq \frac{1}{6}h_*^3 + u_*^3\frac{m - m_*}{\partial_z B(0)}\left(t - \frac{\pi}{2f}\right), \tag{52}$$

and the subsequent deepening is a rather slow erosion process due to the surface wave energy flux. In closure models, this flux can be transported only through the self-diffusion of turbulence. In its absence, the deepening of mixed layers would not occur beyond the first few hours after a mid-latitude storm.

In summary, under ocean storms, two principal energy sources are parameterized in vertical mixing: the shear of the upper-ocean inertial currents and gravity wave flux from a breaking, confused sea. When these are parameterized in energetically consistent models, we discover that these processes act on different time scales and produce mixed layers of very different depths. If the mixed layer at the onset of a storm is deeper than h_*, the shear of inertial currents cannot produce further entrainment very effectively, while the vertical diffusion of turbulence can slowly erode the upper ocean to much greater depths. We shall now examine ocean observations for evidence of these processes in operation.

3. – Mixed-layer experiments and their simulations with models.

3'1. *Model simulations of observed upper-ocean changes during storms.* – To test the hypotheses which are made about turbulent transfer, properly chosen ocean observations of the mean buoyancy and velocity can be used to compare with model simulations of the same quantities. A more desirable comparison is between the measured and predicted turbulent fluxes (or second-order moments), but very few of these measurements have been made, especially under storm conditions. Most open-ocean observations of temperature and horizontal velocity during storm conditions are made from moored instruments at a fixed location. When the most rapid sampling is used, these contain time series of only a portion of the environmental spectrum. We measure the sum of the somewhat stationary internal waves, the storm-produced events and variability due to the energetic subinertial frequency mesoscale of the upper ocean. It is not possible to design a filter in time (or frequency space) which removes the internal waves, especially of tidal period, and at the same time retains the signal of the rapid changes associated with storms. Spatial filtering, though effective for isolating storm effects, requires many observational points, or spatial averaging, at moderate time intervals, however such observations at depth are now not available. Therefore, to remove the internal waves, various temporal averages are used. To remove or identify the advective effects of the lower-frequency mesoscale, which are not modeled here, we assume that the horizontal convergence of mass, buoyancy and momentum in the mixed layer which is due to the low-frequency motions is equal to what it is below the mixed layer. The mixed-layer depth determination is especially sensitive to vertical displacements by internal waves, and the buoyancy and momentum budgets are affected both by internal waves and low-frequency mesoscale. The sea surface temperature, or « bucket » temperature, on the other hand, does not reflect the vertical motions of linear internal waves or mesoscale and, when horizontal advective effects due to the mesoscale are removed (or are not present), its changes reflect primarily the effects of air-sea interaction. In each of the three examples of ocean observations which are discussed below, very different methods of measurements are used and somewhat different methods of data analysis are needed to isolate the effecst of air-sea interaction from other naturally occurring phenomena.

The first careful study of the effect of storms on the upper ocean is by DENMAN and MIYAKE [32]. During a period from 12 June to 25 June, 1970, they deployed a large number of XBT's (approximately every half hour) from the Canadian weather ship of station P at approximately 50° W and 145° W. At the same time, direct measurements of the incident solar radiation and surface wind at the ship's mast was carried out. With these data, estimates of three hourly averages of bucket temperature surface stress and buoyancy flux were made and the evolution of the subsurface temperature was obtained on a six-hourly basis. DENMAN and MIYAKE then show that over a two-week period they can compute the bucket temperature quite well from the model which is based on eqs. (4), (38), (39) and (45), where $R = 0$, $1 - m_s = 0$, and $m - m_* \simeq 1.0$ (for the specific value of 10 m drag coefficient used to estimate u_*^3). Simply, in this calculation, the mixing depth for heat is accurately modeled with two parameters, $m - m_*$ and ξ_2. The XBT data are quite sparse and internal waves are large, therefore the modeling of temporal and depth-dependent structure of the temperature field cannot be verified as well, however it is apparent that the excursion of the mixed layer under storms is modeled well. The maximum depth to which the mixed layer evolves during a storm is governed by the parameter $m - m_*$. The minimum value to which it retreats in calm periods is thus governed by the choice ξ_2. No correction of these data or model was needed for advective motion. Their simulation is presented in fig. 7.

PRICE *et al.* [24] present an analysis of changes of ocean conditions under three storms on the west Florida continental shelf at 26° N, 84° W. Both horizontal velocity and temperature are measured by lowering and raising sensors through the water column approximately every hour. Therefore, the time series of temperature is very much like the XBT samples obtained by DENMAN and MIYAKE, but vertical profiles of horizontal currents are also obtained. The surface buoyancy flux and wind stress are estimated from bulk

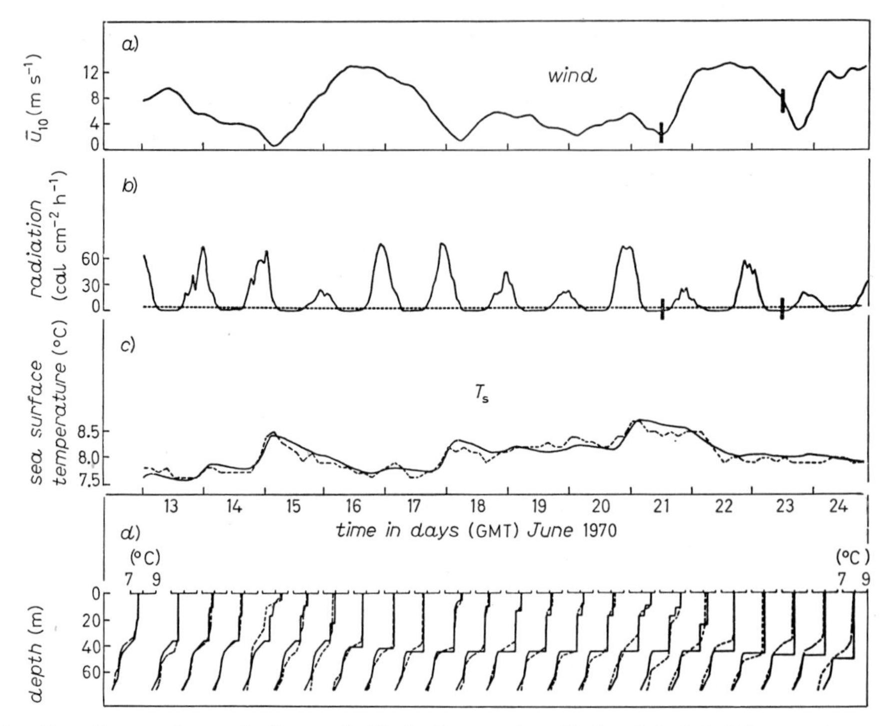

Fig. 7. – Comparison of observed (dashed curve) and simulated (solid curve) upper-ocean changes at ocean weather station PAPA (50° N, 145° W) with model coefficients $m - m_* = 1.02$, $\xi_2 = 3.3$ m (panels *c*) and *d*)). The surface friction velocity is computed from $u_* = (\varrho_a C_{10}/\varrho_w)^{\frac{1}{2}} \bar{U}_{10}$, C_{10} is the 10 m drag coefficient on panel *a*), and panel *b*) presents the surface heat flux, from which the surface buoyancy is computed as described in sect. **1** (adapted from [32]).

air-sea interaction formula and ship observations of bulk parameters. PRICE *et al.* show that the heat, or buoyancy, content of the water column at the measurement site is also affected by vertical and horizontal advection. However, when a correction is made for these advective fluxes into the base of the mixed layer, the mixed-layer depth and buoyancy are quite adequately computed again from eqs. (4), (38), (39) and (45) with the parameter $m - m_* = 0$ and $1 - m_s = 0.6$. They then suggest that upper-ocean mixing is entirely produced by shear of upper-ocean mean currents and detailed ocean current prediction is essential for computation of mixing in upper ocean as is the case in equilibrium, closure models. This interpretation is based on a least-square comparison of the model computed and observed values of $(B - B_+)\, \partial_t h$, or the entrainment buoyancy flux. While in a model this parameter is sensitive to the assumptions of which physical turbulent-energy source (or parameter) is operational, in the data it is also the most sensitive to the manner in which internal waves are filtered from the vertical motion of the mixed-layer depth, and is subject directly to the vagaries of how heat advection is parameterized. Nevertheless, the potential importance of the energy source from vertical shear of mean currents was established. Figure 8 is their simulation of observations during June 8 to June 14, 1972, when advection effects were minimal. This shows that either the process of « wave mixing » or « shear mixing » can supply sufficient energy to mix the water column during storm conditions. A comprehensive list of the most recent simulations, not discussed in detail here, is found in [33].

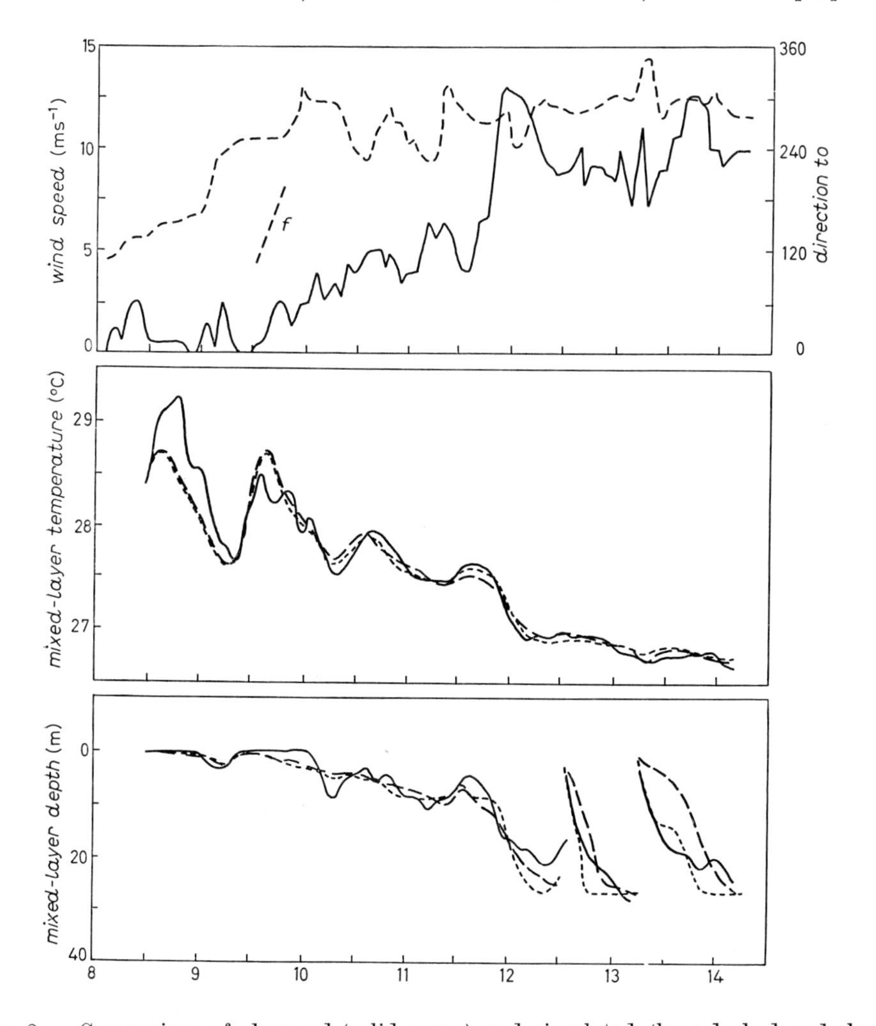

Fig. 8. – Comparison of observed (solid curve) and simulated (long-dashed and short-dashed curve) upper-ocean changes on the west Florida continental shelf (26° N, 84° W) in June 1972. The long-dashed curve is for $1-m_s=0.6$, $m-m_*=0$, and the short-dashed curve is for $1-m_s=0$, $m-m_*=0.9$. (For the method of computing stresses and buoyancy flux see [24].) Upper panel: ——— speed, — — — direction.

3·2. *The mixed-layer experiment* MILE. – In the period of August 19-September 5, 1977, two suites of current meters were deployed on taught surface moorings at 50° N, 145° W, the vicinity of ocean weather station PAPA. The purpose of the experiment was to measure the simultaneous evolution of the velocity and temperature fields in the upper ocean during early fall storms. The instruments, which were 3 m apart vertically through the mixed layer and seasonal thermocline, sampled the ocean approximately every two minutes, for a period of 19 days, so the internal wave field was adequately resolved in time and vertical direction. Air-sea fluxes of momentum and buoyancy were based on bulk estimates of hourly averaged parameters measured on shipboard. A detailed description of the mooring and relative performance of the instruments is given by HALPERN *et al.* [34] and the principal observations of the physical variability of the upper ocean is found in detail in [3]. Here we discuss the principal results of MILE in the context of the physical processes which are parameterized in the equations of the evolution of the mean flow (eqs. (4), (38), (39), (45)).

During the MILE period there was a net flux of buoyancy from the atmosphere to the ocean and this flux was vertically stirred, over a depth of 32 m. Two thermoclines were present in the upper ocean: a transient thermocline above 32 m, which appeared and disappeared with the vagaries of daytime heating, nighttime cooling and occasional wind stirring, and, below that, there was a seasonal thermocline, in which the thermal variability was principally related to the semi-diurnal internal tide and a few longer-period mesoscale features. To remove the internal waves from the signal, DAVIS *et al.* use the overbar of subsect. 1·2 as a 12.5 hour average, and assume that, because the mixed layer did not go deeper than 32 m, the turbulent fluxes, $\overline{w'b'}$ and $\overline{w'\boldsymbol{v}'}$, are much larger above 32 m than below 32 m. However, because variability was observed below the level which turbulent fluxes penetrate, there must be a horizontal gradient $\partial_x P$, $\partial_y P$ to achieve a momentum balance and a $W\partial_z B$ to achieve a buoyancy balance (the low-frequency, horizontal advection of buoyancy is small). These fields are not expected to vanish in the mixed layer and they contribute to the momentum and buoyancy budgets there as well. DAVIS *et al.* now assume that above 32 m $\partial_x P$ and $\partial_y P$ are independent of depth and $W(z, t)$ is a linear function of depth, with the 32 m values computed from the values of $-[\partial_t \boldsymbol{V} + f\hat{k}\times \boldsymbol{V}]$ and $-\partial_t B/\partial_z B$, respectively. Salinity variations were small and heat and buoyancy balances are synonymous. With these assumptions, it is shown that, to about 30 %, a budget of momentum and heat (or buoyancy) within the layer above 32 m is achieved. As expressed in the conservation equations, these are

$$\int_{-32}^{0}\{\partial_t(\boldsymbol{V}-\boldsymbol{V}_{32})+f\hat{k}\times(\boldsymbol{V}-\boldsymbol{V}_{32})\}\,\mathrm{d}z=\boldsymbol{\tau}_0/\varrho_0\,, \tag{53}$$

$$\int_{-32}^{0}\{\partial_t T+W\partial_z T\}=\dot{Q}_0/c\varrho\,, \tag{54}$$

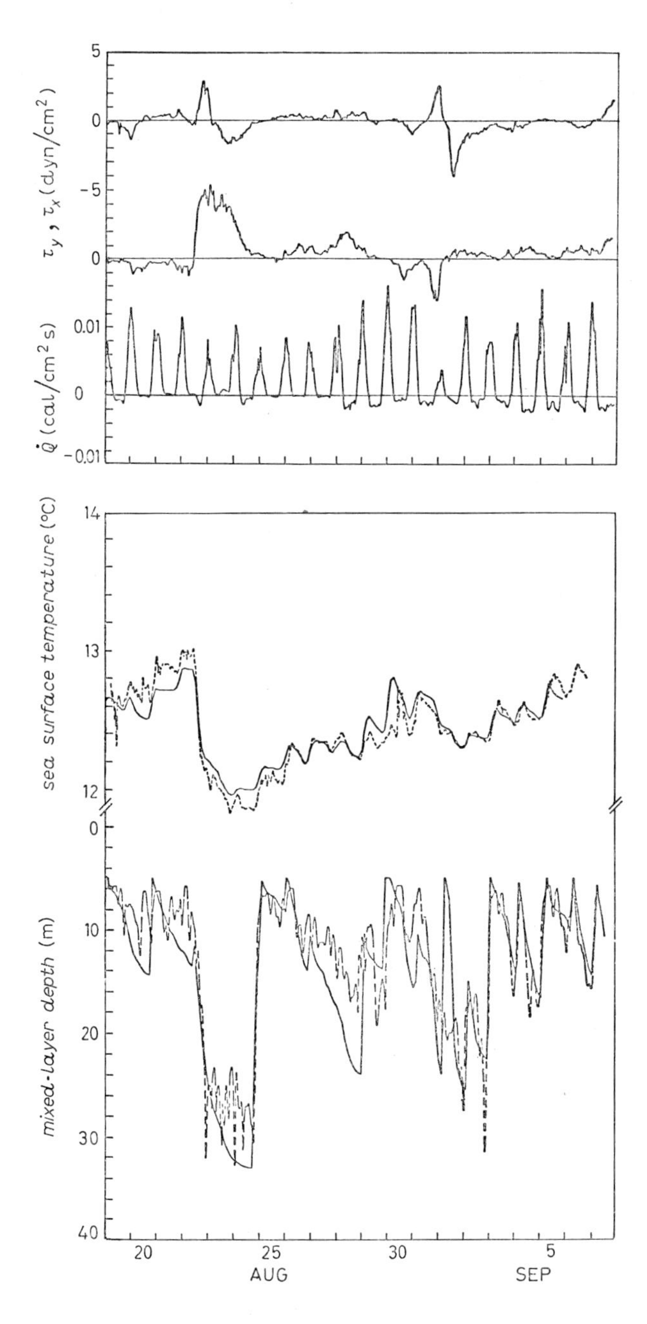

Fig. 9. – Comparison of observed (dashed) and simulated (solid) upper-ocean changes at ocean weather station PAPA (50° W, 145° W) in August-September, 1977. The simulation is for $1 - m_s = 0.49$, $m - m_* = 0.39$. (For the method of computing stresses and buoyancy flux, see [3].)

where $\boldsymbol{\tau}_0$ is the surface wind stress and $\dot{Q}_0$ is the surface heat flux and subscript 32 refers to a quantity evaluated at 32 m depth.

Therefore, when making a comparison of the MILE data set with a one-dimensional model computation, it is important either to correct the data for three-dimensional phenomena which are represented in the pressure gradients and vertical heat advection, or to parameterize these effects in a model.

The MILE data on both temperature and velocity are now used in a comparison with model integrations. The mixed-layer model equations are essentially (38), (39) and (45). Below the mixed layer, $T(z, t)$ is taken directly from observation and the velocity field is integrated with a turbulent stress law expressed by $-\overline{w'\boldsymbol{v}'} = l_U^2|\partial_z \boldsymbol{V}|\partial_z \boldsymbol{V}$, with $l_U = 8.2$ cm. The stress between the mixed-layer bottom and the top of the seasonal termocline is $\boldsymbol{\tau}_+ = \varrho C_U|\boldsymbol{V} - \boldsymbol{V}_+|(\boldsymbol{V} - \boldsymbol{V}_+)$, with $C_U = 5\cdot 10^{-4}$. Initially, $\boldsymbol{V}(z, 0) = 0$ and the initial temperature profile is taken as the first 12.5 h average of August 19, 1977. Surface values are given for $\boldsymbol{\tau}_0$ (fig. 9*a*)) and $\dot{Q}_0$ (fig. 9*b*)). As was done by PRICE *et al.* [24], a large number of integrations were made for different values of $m - m_*$ and $1 - m_s$. (The values of ξ_1, ξ_2 and R are considered already determined by other experimental data and the parameter m_c plays no role because during MILE there was no intense cooling). As was pointed out by DESZOEKE and RHINES [31], the parameter $1 - m_s$ governs the initial rate of a shallow layer deepening directly after the storm and $m - m_*$ governs the maximum depth to which the mixed layer penetrates over a few days of stormy period. DAVIS *et al.* [3] find that a good replication of the mixed-layer temperature and mixed-layer depth is achieved with $1 - m_s \simeq 0.48$ and $m - m_* = 0.39$, although acceptable solutions for the conditions produced by the August 22 storm can be achieved with $1 - m_s = 0$, $m - m_* = 0.69$ or $1 - m_s = 0.67$, $m_s = 0.13$. The equivalent value of $m - m_*$ in Denman and Miyake's [32] study is 0.53, when a proper conversion of their u_*^3 to MILE ship-mast heigth and drag coefficient is made. Figure 9 presents the best MILE simulation for mixed-layer temperature and depth.

The comparison between the observed and model currents is made by computing the model spectra of $\boldsymbol{G} = \partial_t \boldsymbol{V} + f\hat{k}\times\boldsymbol{V}$, and comparing this to the wind-coherent part of the same quantity where $\boldsymbol{V}$ has been replaced by $\boldsymbol{V} - \boldsymbol{V}_{32}$ in the observations. Figure 10 displays the model and observed values of the complex transfer function

$$\mu(z, \omega) = \hat{G}(z, \omega)/\hat{\tau}_0(\omega)\,, \tag{55}$$

where $\hat{G}(z, \omega)$ is the Fourier transform of the vector, $i(\omega + f)(U + iV)$ is the Fourier transform of $\tau_0^{(x)} + i\tau_0^{(y)}$. Within the error bounds of determining the observed values of μ there is good agreement between observation and theory. DAVIS *et al.* [3] note that it is not surprising that simple slab or bulk models of the upper ocean can produce good replication of the observations, because these are simply an expression of the budgets of upper-ocean momentum and heat in which the mixing depth is predicted. In bulk models, with two parameters, one can independently adjust both the rate of mixed-layer deepening (and sea surface temperature cooling) and the maximum depth to which the mixing occurs after strong storms. The physical process of shear production of turbulence through wind generation of inertial currents produces the initial rapid response, and the more languid erosion of the main thermocline is produced by a parameterized « turbulence diffusion ». Bulk models are, thus, the limit of the continuous models, with turbulence limitation above a critical Ri, where the mixing length is always comparable to the mixed-layer depth.

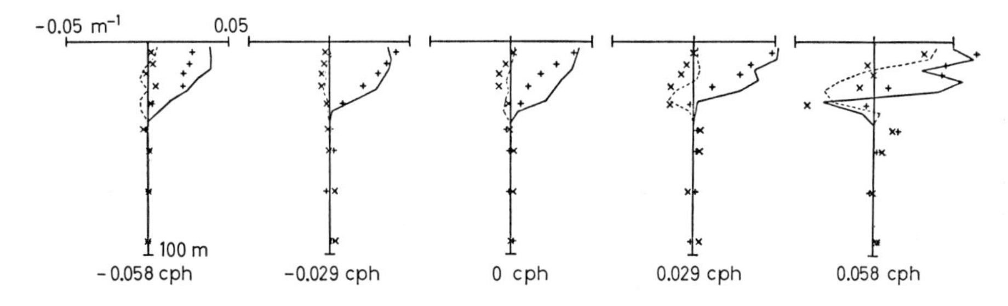

Fig. 10. – Comparison of observed (+, real part; ×, imaginary part) with simulated (solid curve: real part; dashed curve: imaginary part) complex transfer function between the momentum flux convergence in the upper ocean and the surface momentum flux. Model parameters are same as in fig. 9. (For the method of computation and error analysis, see [3].)

The unresolved central questions in the physics of the ocean mixed layer are more observational than theoretical. However difficult to make, direct measurements of the turbulent fluxes are needed, as is a description of the physical processes which produces this flux. Vertical profiles of the low-frequency currents can now be made in the mixed layer with 10 % accuracy in moderate weather and many long-time series of these are required in different conditions of stratification (MILE was only 19 days long). In MILE we see that « one-dimensional » vertical mixing can represent the upper-ocean air-sea interaction even in the presence of three-dimensional advective field of heat and momentum. STEVENSON [26] shows that this should theoretically also be the case for seasonal changes on top of a mid-ocean eddy field as is found in the central North Atlantic. However, we know that the processes represented in the formalism of sect. **1** and **2** cannot represent the accumulated effects of vertical or horizontal mixing by « interleaving » water masses. As is suggested by the somewhat formal decomposition of the fields in sect. **1**, the effect of horizontal shear of internal waves, especially at the mixed-layer base where large density gradients occur, can produce enhanced mixing during storms. Observation can tell us where and when in the water column is internal wave breaking the most frequent. As in other areas of physical oceanography, mixed-layer physics is largely an observational science.

* * *

The support of the Office of Naval Research is gratefully acknowledged for the compilation of this manuscript, under N000014-79-C-0004 with Oregon State University.

REFERENCES

[1] A. Ivanoff: *Oceanic absorption of solar energy*, in *Modeling and Prediction of the upper Layers of the Ocean*, edited by E. B. Kraus (London, 1977).

[2] C. A. Paulson and J. J. Simpson: *J. Phys. Oceanogr.*, **7**, 952 (1977).

[3] R. Davis, R. deSzoeke, D. Halpern and P. Niiler: *Variability and dynamics of the upper ocean during MILE*, *Deep-Sea Res.*, in press (1981)

[4] R T. Pollard and R. C. Millard jr.: *Deep-Sea Res.*, **17**, 812 (1970).

[5] T. M. Dillon and D. R. Caldwell: *J. Geophys. Res.*, **85**, 1910 (1980).

[6] R. Käse and A. R. Clarke: *Deep-Sea Res.*, **28**, 815 (1978).

[7] T. J. Spoering: *Towed observations of internal waves in the upper ocean*, Oregon State University, School of Oceanography, Ref. 79-10 (1979).

[8] M. E. Stern: *J. Mar. Res.*, **35**, 479 (1977).

[9] E. Mollo-Christensen: *J. Phys. Oceanogr.*, **7**, 684 (1977).

[10] W. H. Munk and E. R. Anderson: *J. Mar. Res.*, **7**, 276 (1948).

[11] D. R. Caldwell and C. van Atta: *J. Fluid Mech.*, **44**, 79 (1970).

[12] T. Osborn: *J. Phys. Oceanogr.*, **10**, 85 (1980).

[13] J. C. Rotta: *Z. Phys.*, **129**, 547 (1951); **131**, 51 (1951).

[14] G. L. Mellor and T. Yamada: *J. Atmos. Sci.*, **31**, 1791 (1974).

[15] G. L. Mellor: *J. Atmos. Sci.*, **30**, 1061 (1973).

[16] A. N. Kolmogoroff: *Izv. Akad. Nauk SSSR*, **30**, 301 (1941) (translation by S. K. Friedlander and L. Topper: in *Turbulence, Classic Papers in Statistical Theory* (New York, N. Y., 1961)).

[17] D. R. Caldwell and T. M. Chriss: *Science*, **205**, 1131 (1979).

[18] N. S. Oakey and T. A. Elliott: *Dissipation in the mixed layer near Emerald Basin*, in *Marine Turbulence*, edited by J. Nihoul (Amsterdam, 1980).

[19] T. M. Dillon: personal communication (1980).

[20] J. Richman and C. Garrett: *J. Phys. Oceanogr.*, **7**, 876 (1978).

[21] T. Yamada: *J. Atmos. Sci.*, **32**, 926 (1976).

[22] E. B. Kraus, Editor: *Modeling and Prediction of the Upper Layers of the Ocean* (London, 1977).

[23] B. G. Vager and S. S. Zilitinkevich: *A theoretical model of the diurnal variations of the meteorological fields*, *Meteorol. Gidrol*, **7** (1968).

[24] J. F. Price, C. N. K. Mooers and J. C. Van Leer: *J. Phys. Oceanogr.*, **8**, 582 (1978).

[25] P. P. Niiler: *J. Mar. Res.*, **33**, 405 (1975).

[26] J. W. Stevenson: *J. Phys. Oceanogr.*, **9**, 57 (1979).

[27] G. L. MELLOR and P. A. DURBIN: *J. Phys. Oceanogr.*, **6**, 718 (1975).
[28] P. J. MARTIN: *A comparison of three diffusion models of the upper mixed layer of the ocean*, NRL-GRD/OTEC, technical report under ERDA Contract E(49-26)1005 (1978).
[29] A. C. WARN-VARNAS and S. PIACSEK: *Geophys. Astrophys. Fluid Dyn.*, **13**, 224 (1979).
[30] P. K. KUNDU: *J. Phys. Oceanogr.*, **10**, 220 (1980).
[31] R. A. DESZOEKE and P. B. RHINES: *J. Mar. Res.*, **34**, 111 (1976).
[32] K. DENMAN and M. MIYAKE: *J. Phys. Oceanogr.*, **3**, 185 (1973).
[33] R. W. GARWOOD: *Rev. Geophys. Space Phys.*, **17**, 1507 (1979).
[34] D. HALPERN, R. WELLER, M. BRISCOE, R. DAVIS and J. MCCULLOUGH: *The* JASIN *and* MILE *current meter intercomparison test, J. Geophys. Res.*, in press (1981).

PART IV

MODELS OF OCEAN SURFACE WAVES

Summary of Probability Laws for Wave Properties.

L. E. BORGMAN

University of Wyoming - Laramie, Wy.

1. – Introduction.

The multivariate normal probability density for linear wave properties forms the initial starting point for deriving wave property probability formula. The wave property is defined in terms of the random variables occurring within the wave and the probability laws are developed through various functional transformations. M. S. LONGUET-HIGGINS has been a pioneer in much of this work.

2. – The Rayleigh distribution for wave heights.

The Rayleigh distribution is intimately related to the bivariate normal probability law for the two wave properties $\eta(t)$ and $\dot{\eta}(t) = \mathrm{d}\eta/\mathrm{d}t$. The spatial variables, x and y, are set equal to zero and suppressed in the notation, since both quantities are referenced to the same locality. The origin of the co-ordinate system might as well be placed at that same locality.

The probabilistic models for $\eta(t)$ and $\dot{\eta}(t)$ are [1, 2]

$$\left\{\begin{aligned} \eta(t) &= 2\int_0^{\infty}\int_0^{2\pi}\sqrt{S(f,\theta)\,\mathrm{d}f\,\mathrm{d}\theta}\,\cos(2\pi f t - \Phi)\,,\\ \dot{\eta}(t) &= -2\int_0^{\infty}\int_0^{2\pi}2\pi f\sqrt{S(f,\theta)\,\mathrm{d}f\,\mathrm{d}\theta}\,\sin(2\pi f t - \Phi)\,. \end{aligned}\right. \tag{1}$$

The covariance matrix

$$\operatorname{cov}(\eta, \dot{\eta}) = \begin{pmatrix} m_0 & 0 \\ 0 & m_2 \end{pmatrix}, \tag{2}$$

where

$$(3)\quad \begin{cases} m_0 = 2\int\limits_0^\infty \int\limits_0^2 S(f,\theta)\,d\theta\,df\,, \\ m_2 = 2\int\limits_0^\infty \int\limits_0^2 (2\pi f)^2 S(f,\theta)\,d\theta\,df\,. \end{cases}$$

Consequently the normal density for $(\eta, \dot{\eta})$ is

$$(4)\quad f_{\eta,\dot{\eta}}(u,v) = \exp\left[-(u,v)\begin{pmatrix} m_0 & 0 \\ 0 & m_2\end{pmatrix}^{-1}\begin{pmatrix} u \\ v\end{pmatrix}^{-1}\right]\Big/ 2\pi\sqrt{\begin{vmatrix} m_0 & 0 \\ 0 & m_2\end{vmatrix}} = \\ = \exp\left[-u^2/2m_0 - v^2/2m_2\right].$$

Let the envelope function $E(t)$ be defined as

$$(5)\quad E(t) = \sqrt{\eta^2 + m_0\dot{\eta}^2/m_2}\,.$$

This function will equal $|\eta|$ wherever $\dot{\eta} = 0$, which is at the crest and trough of the wave trace. In between, it will vary somewhat smoothly between those values.

For a cosine-shape wave with a narrow-band spectral density

$$(6)\quad S(f) = \int\limits_0^{2\pi} S(f,\theta)\,d\theta = [m_0/2]\,\delta(f - f_0)$$

(where $\delta(f)$ is the Dirac delta), the envelope $E(t)$ will precisely equal the wave amplitude throughout the wave cycle. For this case, from (3)

$$(7)\quad m_2 = 2\int\limits_0^\infty (2\pi f)^2[m_0/2]\,\delta(f - f_0)\,df = (2\pi f_0)^2 m_0\,,$$

$$(8)\quad E(t) = \sqrt{\eta^2 + (\dot{\eta}/2\pi f_0)^2}\,.$$

If

$$(9)\quad \eta = \alpha\cos(2\pi f_0 t + \Phi)$$

is substituted into (8), the result is

$$(10)\quad E(t) = \alpha\,,$$

as was claimed above.

The probability law for $E(t)$ may be derived as follows. First the variables are transformed through the equations

$$\eta = \alpha \cos \psi , \qquad \alpha = \sqrt{\eta^2 + (m_0 \dot{\eta}/m_2)^2} = E(t) ,$$

or

$$\dot{\eta} = \sqrt{m_2/m_0}\alpha \sin \psi , \qquad \psi = \operatorname{arctg}\left[(\dot{\eta}/\eta)\sqrt{m_0/m_2}\right] , \tag{11}$$

and then ψ is integrated out to give the marginal density for α. The Jacobian of the transformation is

$$J = \begin{vmatrix} \cos \psi & -\alpha \sin \psi \\ \sqrt{m_2/m_0} \sin \psi & \sqrt{m_2/m_0}\, \alpha \cos \psi \end{vmatrix} = \alpha \sqrt{m_2/m_0} . \tag{12}$$

Consequently the probability density for (E, Ψ) is from (4)

$$f_{(E,\Psi)}(\alpha, \psi) = \alpha \exp[-\alpha^2/2m_0]/2\pi m_0 , \qquad a > 0,\ 0 \leqslant \psi \leqslant 2\pi . \tag{13}$$

Hence the marginal density for E is

$$f_E(\alpha) = \int_0^{2\pi} f_{E,\Psi}(\alpha, \psi)\, d\psi = \alpha \exp[-\alpha^2/2m_0]/m_0 , \qquad E > 0 . \tag{14}$$

This is the Rayleigh probability density for the wave envelope (*i.e.* wave amplitude), where m_0 is the variance of the sea surface.

The Rayleigh density is often expressed in other terms. For example, let $H = 2\alpha$ be the wave height. The Rayleigh density under this transformation becomes

$$f_H(h) = h \exp[-h^2/8m_0]/4m_0 , \qquad h > 0 . \tag{15}$$

The distribution function for wave height is

$$F_H(h) = P[H \leqslant h] = \int_0^h [x \exp[-x^2/8m_0]/4m_0]\, dx = 1 - \exp[-h^2/8m_0] , \qquad h > 0 . \tag{16}$$

3. – The significant wave height.

What is the value of $h = h_{\frac{2}{3}}$ such that $P[H \leqslant h_{\frac{2}{3}}] = \frac{2}{3}$? From (16) this would be

$$1 - \exp[-h_{\frac{2}{3}}^2/8m_0] = \tfrac{2}{3} ,$$

or

$$h_{\frac{2}{3}} = \sqrt{8m_0 \ln 3}\,. \tag{17}$$

The expected value of H given $H > h_{\frac{2}{3}}$ is defined as the significant wave height H_S:

$$H_S = E[H|H > h_{\frac{2}{3}}] = \int_{h_{\frac{2}{3}}}^{\infty} [h[f_H(h)]/P(H > h_{\frac{2}{3}})]\,dh = \tag{18}$$

$$= 3\int_{8m_0 \ln 3}^{\infty} [h^2 \exp[-h^2/8m_0]/4m_0]\,dh\,.$$

After integration by parts and a rearrangement in terms of the standard normal

$$H_S = [\sqrt{8\ln 3} + 3\sqrt{8\pi}(1 - \Phi\sqrt{2\ln 3})]\sqrt{m_0} = 4.004\sqrt{m_0}\,. \tag{19}$$

Since the constant is so close to an integer, the relation is usually rounded off and expressed as

$$H_S \approx 4\sqrt{m_0} = 4\sqrt{2\int_0^{\infty}\int_0^{2} S(f,\theta)\,d\theta\,df} = 4 \text{ std dev } (\eta)\,. \tag{20}$$

Also the Rayleigh density is often parameterized by substituting $H_S/4$ for m_0 as

$$f_H(h) = [4h/H_S^2] \exp[-2h^2/H_S^2]\,, \qquad h > 0\,. \tag{21}$$

One other wave height measure is sometimes used. This is the root-mean-square wave height defined by

$$H_{\text{r.m.s.}} = \int_0^{\infty} h^2[4h/H_S^2] \exp[-2h^2/H_S^2]\,dh = H_S^2/2 \tag{22}$$

after integration by parts. In summary (except for a minor numerical approximation introduced)

$$H_S = 4\sqrt{m_0} = \sqrt{2}\,H_{\text{r.m.s.}}\,. \tag{23}$$

It should be emphasized that the above relationships are based on the Rayleigh distribution and its underlying assumptions. For real wave trains, the use of (20) to convert from the wave spectral density to a significant wave height is still open to controversy. Some authors argue that it is a fair approximation, while others dispute it. The accuracy probably depends on the sea conditions present and their statistical peculiarities.

The basic paper on the use of the Rayleigh distribution for sea waves is by LONGUET-HIGGINS [3].

4. – A distribution for crest elevation.

CARTWRIGHT and LONGUET-HIGGINS [4] applied a derivation by RICE [5] to obtain a family of distributions for the sea surface maxima. The development is as follows. Let $f_{\eta,\dot\eta,\ddot\eta}(x, y, z)$ be the joint-probability density function for the water level elevation at time t at a given location and for the first and second time derivatives at that same location.

A Taylor's series expansion for $\dot\eta(t)$ gives to first order that

$$\dot\eta(t) = \dot\eta(t_0) + \ddot\eta(t_0)(t - t_0)\,. \tag{24}$$

If the above relation holds exactly, an extremum of $\eta(t)$ ($i.e.$ a point where $\dot\eta(t) = 0$) will occur in the interval $(t_0 - \Delta t, t_0 + \Delta t)$, if and only if

$$|\dot\eta(t_0)| \leqslant |\ddot\eta(t_0)|\Delta t\,. \tag{25}$$

If the extremum is a maximum, then $\ddot\eta(t)$ will be negative. Under the assumption that (25) holds and $\ddot\eta(t)$ at the extremum is negative, let M be the value of $\eta(t)$ at the extremum. Then, to a fair approximation, if Δt is very small,

$$F_{\max}(\alpha) = P[\eta \leqslant \alpha \,\big|\, |\dot\eta(t_0)| \leqslant |\ddot\eta(t_0)|\Delta t,\ \ddot\eta(t) < 0] \approx \tag{26}$$

$$\approx P[\eta(t_0) \leqslant \alpha \,\big|\, |\dot\eta(t_n)| \leqslant |\ddot\eta(t_0)|\Delta t,\ \ddot\eta(t_0) < 0]\,.$$

The approximation replaces conditions on the sea surface at time t, which is basically unknown, with related conditions at t_0. Consequently

$$F_{\max}(\alpha) = P[\eta \leqslant \alpha,\ |\dot\eta| \leqslant |\ddot\eta|\Delta t,\ \ddot\eta < 0]/P[|\dot\eta| \leqslant |\ddot\eta|\Delta t,\ \ddot\eta < 0] = \tag{27}$$

$$= \int_{-\infty}^{\alpha}\int_{-\infty}^{0}\int_{-z\Delta t}^{z\Delta t} f_{\eta,\dot\eta,\ddot\eta}(x, y, z)\,\mathrm{d}y\,\mathrm{d}z\,\mathrm{d}x \Big/ \int_{-\infty}^{0}\int_{-z\Delta t}^{z\Delta t} f_{\dot\eta,\ddot\eta}(y, z)\,\mathrm{d}z\,\mathrm{d}y\,.$$

Since Δt is quite small,

$$\int_{-z\Delta t}^{z\Delta t} g(z)\,\mathrm{d}z \approx g(0)\,2\,\Delta t \tag{28}$$

by the mean-value theorem of calculus. If this theorem is applied to (27),

$$F_{\max}(\alpha)=\int_{-\infty}^{\alpha}\int_{-\infty}^{0} f_{\eta,\dot{\eta},\ddot{\eta}}(x,0,z)\,|z|\,2\,\Delta t\,\mathrm{d}z\,\mathrm{d}x \Big/ \int_{-\infty}^{0} f_{\dot{\eta},\ddot{\eta}}(0,z)\,|z|\,2\,\Delta t\,\mathrm{d}z\,. \tag{29}$$

The $2\,\Delta t$ can be divided out of both numerator and denominator and the resulting expression differentiated with respect to α to obtain a probability density for the maximum

$$f_{\max}(\alpha)=[\mathrm{d}/\mathrm{d}\alpha]F_{\max}(\alpha)=\int_{-\infty}^{0} f_{\eta,\dot{\eta},\ddot{\eta}}(\alpha,0,z)\,|z|\,\mathrm{d}z \Big/ \int_{-\infty}^{0} f_{\dot{\eta},\ddot{\eta}}(0,z)\,|z|\,\mathrm{d}z\,. \tag{30}$$

The probabilistic model for η and $\dot{\eta}$ was given earlier (1). The model of $\ddot{\eta}$ would similarly be

$$\ddot{\eta}(t)=-2\int_0^{\infty}\int_0^{2\pi}\sqrt{S(f,\theta)\,\mathrm{d}\theta\,\mathrm{d}f}\,(2\pi f)^2\cos(2\pi f t-\Phi)\,. \tag{31}$$

Consequently

$$E(\eta)=E(\dot{\eta})=E(\ddot{\eta})=0\,, \tag{32}$$

$$\operatorname{cov}(\eta,\dot{\eta},\ddot{\eta})=\begin{pmatrix} m_0 & 0 & -m_2\\ 0 & m_2 & 0\\ -m_2 & 0 & m_4\end{pmatrix}, \tag{33}$$

where m_0 and m_2 are defined by (3) and

$$m_4=2\int_0^{\infty}\int_0^{2\pi}(2\pi f)^4 S(f,\theta)\,\mathrm{d}\theta\,\mathrm{d}f\,. \tag{34}$$

If these results are substituted into (30),

$$f_{\max}(\alpha)=\left[-\int_{-\infty}^{0}\left\{-\frac{1}{2}(\alpha,0,z)\begin{pmatrix} m_0 & 0 & -m_2\\ 0 & m_2 & 0\\ -m_2 & 0 & m_4\end{pmatrix}^{-1}\begin{pmatrix}\alpha\\0\\z\end{pmatrix}\Big/ (2\pi)^{\frac{3}{2}}\sqrt{\begin{vmatrix} m_0 & 0 & -m_2\\ 0 & m_2 & 0\\ -m_2 & 0 & m_4\end{vmatrix}}\right\} z\,\mathrm{d}z\right]\Big/ \left[-\int_{-\infty}^{0}\left\{\exp\left[-\frac{1}{2}(0,z)\begin{pmatrix} m_2 & 0\\ 0 & m_4\end{pmatrix}^{-1}\begin{pmatrix}0\\z\end{pmatrix}\right]\Big/ 2\pi\sqrt{\begin{vmatrix} m_2 & 0\\ 0 & m_4\end{vmatrix}}\right\} z\,\mathrm{d}z\right]. \tag{35}$$

Since

$$\begin{pmatrix} m_0 & 0 & -m_2 \\ 0 & m_2 & 0 \\ -m_2 & 0 & m_4 \end{pmatrix}^{-1} = \begin{pmatrix} m_2 m_4 & 0 & m_2^2 \\ 0 & m_0 m_4 - m_2^2 & 0 \\ m_2^2 & 0 & m_0 m_2 \end{pmatrix} \Big/ m_2(m_0 m_4 - m_2^2), \tag{36}$$

it follows that

$$f_{\max}(\alpha) = \Big[\int_{-\infty}^{0} \exp\,[-\tfrac{1}{2}[m_2 m_4 \alpha^2 + m_0 m_2 z^2 + 2m_2^2 \alpha z]/m_2(m_0 m_4 - m_2^2)]\, z\, \mathrm{d}z \Big/ \tag{37}$$

$$\Big/ (2\pi)^{\frac{3}{2}} \sqrt{m_2(m_0 m_4 - m_2^2)}\Big] \Big/ \int_{-\infty}^{0} [\exp\,[-(z^2/2m_4)]/2\pi \sqrt{m_2 m_4}]\, z\, \mathrm{d}z\,.$$

Completing the square on z in the numerator exponent and evaluating the integral in the denominator gives

$$f_{\max}(\alpha) = \{-\exp\,[-\alpha^2/2m_0]/\sqrt{2\pi m_4(m_0 m_4 - m_2^2)}\}\cdot \tag{38}$$

$$\cdot \int_{-\infty}^{0} \exp\,[-(m_0 z + m_2 \alpha)^2/2m_0(m_0 m_4 - m_2^2)]\, z\, \mathrm{d}z\,.$$

Now change the variable of integration to

$$u = [m_0 z + m_2 \alpha]/\sqrt{m_0(m_0 m_4 - m_2^2)}\,. \tag{39}$$

The expression becomes, after some algebra,

$$f_{\max}(\alpha) = [1/\sqrt{2\pi m_0}]\Big\{\sqrt{(m_0 m_4 - m_2^2)/m_0 m_4}\, \exp\,[-(\alpha^2/2)[m_4/(m_0 m_4 - m_2^2)]] + \tag{40}$$

$$+ [m_2/m_0\sqrt{m_4}]\sqrt{2\pi}\,\alpha \exp\,[-\alpha^2/2m_0]\,\Phi[m_2\alpha/\sqrt{m_0(m_0 m_4 - m_2^2)}]\Big\}.$$

If $m_0 m_4 - m_2^2 = 0$, as it does for waves with a narrow-band spectral density, the probability density reduces to

$$f_{\max}(\alpha) = [\alpha/m_0] \exp\,[-\alpha^2/2m_0], \tag{41}$$

which is the Rayleigh density. If $m_2/\sqrt{m_0 m_4} = 0$, then

$$f_{\max}(\alpha) = \exp\,[-\alpha^2/2m_0]/\sqrt{2\pi m_0}\,. \tag{42}$$

This is a normal probability density with a zero mean and a variance equal to m_0. Thus the density derived (sometimes called a Rice density) is a mixture of a normal density and a Rayleigh density.

5. – A probability law for wave periods and amplitudes.

A joint-probability density for wave periods and amplitudes may be developed by a procedure outlined by LONGUET-HIGGINS [6]. Let f_0 be defined as

$$f_0 = 2\int_0^\infty \int_0^{2\pi} f S(f, \theta)\, d\theta\, df \Big/ 2\int_0^\infty \int_0^{2\pi} S(f, \theta)\, d\theta\, df \,. \tag{43}$$

The statistical model for the water level elevations about mean water level may be expressed as

$$\begin{aligned}\eta(t) &= 2\int_0^\infty \int_0^{2\pi} \sqrt{S(f, \theta)\, d\theta\, df}\, \cos\left[2\pi(f - f_0)t - \Phi + 2\pi f_0 t\right] = \\ &= \left\{2\int_0^\infty \int_0^{2\pi} \sqrt{S(f, \theta)\, d\theta\, df}\, \cos\left[2\pi(f - f_0)t - \Phi\right]\right\} \cos 2\pi f_0 t - \\ &\qquad - \left\{2\int_0^\infty \int_0^{2\pi} \sqrt{S(f, \theta)\, d\theta\, df}\, \sin\left[2\pi(f - f_0)t - \Phi\right]\right\} \sin 2\pi f_0 t .\end{aligned} \tag{44}$$

Thus, if two new random variables ξ_1 and ξ_2 are defined by

$$\left\{\begin{aligned}\xi_1 &= 2\int_0^\infty \int_0^{2\pi} \sqrt{S(f, \theta)\, d\theta\, df}\, \cos\left[2\pi(f - f_0)t - \Phi\right] , \\ \xi_2 &= 2\int_0^\infty \int_0^{2\pi} \sqrt{S(f, \theta)\, d\theta\, df}\, \sin\left[2\pi(f - f_0)t - \Phi\right] ,\end{aligned}\right. \tag{45}$$

then

$$\eta(t) = \xi_1 \cos 2\pi f_0 t - \xi_2 \sin 2\pi f_0 t \,. \tag{46}$$

The derivation will also involve the time derivatives of ξ_1 and ξ_2:

$$\left\{\begin{aligned}\dot{\xi}_1 &= -2\int_0^\infty \int_0^{2\pi} 2\pi(f - f_0) \sqrt{S(f, \theta)\, d\theta\, df}\, \sin\left[2\pi(f - f_0)t - \Phi\right] , \\ \dot{\xi}_2 &= 2\int_0^\infty \int_0^{2\pi} 2\pi(f - f_0) \sqrt{S(f, \theta)\, d\theta\, df}\, \cos\left[2\pi(f - f_0)t - \Phi\right] .\end{aligned}\right. \tag{47}$$

Before proceeding to the method of bringing in wave period, it is useful to develop the joint density for $(\xi_1, \xi_2, \dot{\xi}_1, \dot{\xi}_2)$. These wave properties will have a multivariate normal density with mean zero and a covariance matrix

$$\operatorname{cov}(\xi_1, \xi_2, \dot{\xi}_1, \dot{\xi}_2) = \begin{pmatrix} \mu_0 & 0 & 0 & 0 \\ 0 & \mu_0 & 0 & 0 \\ 0 & 0 & \mu_2 & 0 \\ 0 & 0 & 0 & \mu_2 \end{pmatrix}, \tag{48}$$

where

$$\begin{cases} \mu_0 = 2\int_0^\infty \int_0^{2\pi} S(f, \theta)\, d\theta\, df\,, \\ \mu_2 = 2\int_0^\infty \int_0^{2\pi} 4\pi^2 (f - f_0)^2 S(f, \theta)\, d\theta\, df\,. \end{cases} \tag{49}$$

Some of the zeros in (48) are due to the fact that

$$2\int_0^\infty \int_0^{2\pi} (f - f_0) S(f, \theta)\, d\theta\, df = 0 \tag{50}$$

by virtue of the definition of f_0. Consequently, the joint density for $(\xi_1, \xi_2, \dot{\xi}_1, \dot{\xi}_2)$ will be

$$f_{\xi_1,\xi_2,\dot{\xi}_1,\dot{\xi}_2}(u, v, w, x) = \exp[-\tfrac{1}{2}[(u^2 + v^2/\mu_0) + (w^2 + x^2/\mu_2)]/(2\pi)^2 \mu_0 \mu_2\,. \tag{51}$$

Consider the transformation

$$\begin{cases} \xi_1 = R \cos \Psi\,, \\ \xi_2 = R \sin \Psi\,, \\ \dot{\xi}_1 = \dot{R} \cos \Psi - R\dot{\Psi} \sin \Psi\,, \\ \dot{\xi}_2 = \dot{R} \sin \Psi + R\dot{\Psi} \cos \Psi\,. \end{cases} \tag{52}$$

The Jacobian of the transformation is R^2 and, therefore, the joint density for $(R, \Psi, \dot{R}, \dot{\Psi})$ is

$$f_{R,\Psi,\dot{R},\dot{\Psi}}(\varrho, \psi, \dot{\varrho}, \dot{\psi}) = \varrho^2 \exp[-\varrho^2/2\mu_0] \exp[-(\dot{\varrho}^2 + \varrho^2\dot{\psi}^2)/2\mu_2]/(2\pi)^2 \mu_0 \mu_2\,, \tag{53}$$

because

$$u^2 + v^2 = \varrho^2\,, \qquad w^2 + x^2 = \dot{\varrho}^2 + \varrho^2 \dot{\psi}^2\,. \tag{54}$$

With these technical details out of the way, attention can be returned again to the period. From (47)

$$\eta(t) = R\cos\Psi\cos 2\pi f_0 t - R\sin\Psi\sin 2\pi f_0 t = R\cos(\Psi + 2\pi f_0 t)\,. \tag{55}$$

Let

$$\chi = \Psi + 2\pi f_0 t\,, \tag{56}$$

a Taylor series expansion of χ to the first order is

$$\chi(t) = \chi(t_0) + \dot{\chi}(t_0)(t - t_0)\,. \tag{57}$$

When the argument of the cosine in (55) increases by 2π, the time will have increased through one wave period. To first-order accuracy

$$2\pi = \chi(t_0 + T) - \chi(t_0) = \dot{\chi}(t_0)\,T\,, \tag{58}$$

or

$$T \approx 2\pi/\dot{\chi} = 2\pi/(\dot{\Psi} + 2\pi f_0)\,. \tag{59}$$

By (55), R is the wave amplitude. Hence the joint density R and $T = 2\pi/(\dot{\Psi} + 2\pi f_0)$ will be the density for amplitude and period, at least approximately.

The joint density for R and $\dot{\Psi}$ is obtained by integrating out Ψ and $\dot{R}$:

$$\begin{aligned} f_{R,\dot{\Psi}}(\varrho, \dot{\Psi}) &= \{\varrho^2 \exp[-\varrho^2/2\mu_0]\exp[-\varrho^2\dot{\psi}/2\mu_2]/\sqrt{2\pi}\mu_0\sqrt{\mu_2}\}\cdot \\ &\quad\cdot\int_{-\infty}^{\infty}[\exp[-\dot{\varrho}^2/2\mu_2]/\sqrt{2\pi\mu_2}]\,\mathrm{d}\dot{\varrho}\int_0^{2\pi}\mathrm{d}\psi/2\pi = \\ &= \varrho^2\exp[-\varrho^2/2\mu_0]\exp[-\varrho^2\dot{\psi}/2\mu_2]/\sqrt{2\pi}\,\mu_0\sqrt{\mu_2}\,. \end{aligned} \tag{60}$$

The separate densities for R and $\dot{\Psi}$ are

$$\begin{aligned} f_R(\varrho) &= \{\varrho\exp[-\varrho^2/2\mu_0]/\mu_0\}\int_{-\infty}^{\infty}[\exp[-\varrho^2\dot{\psi}^2/2\mu_2]/\sqrt{2\pi\mu_2/\varrho^2}]\,\mathrm{d}\dot{\psi} = \\ &= \varrho\exp[-\varrho^2/2\mu_0]/\mu_0 \qquad \text{(Rayleigh density)}\,, \end{aligned} \tag{61}$$

$$f_{\dot{\Psi}}(\dot{\psi}) = \int_0^{\infty}[\varrho^2\exp[-(\varrho^2/2\mu_0)[1 + (\mu_0/\mu_2)\psi^2]/\sqrt{2\pi}\,\mu_0\sqrt{\mu_2}]\,\mathrm{d}\varrho\,. \tag{62}$$

The integral in (62) may be expressed in terms of the new variable

$$u = \varrho\sqrt{[1 + (\mu_0/\mu_2)\,\dot{\psi}^2]/\mu_0} \tag{63}$$

as

$$(64)\qquad f_{\dot{\Psi}}(\dot{\psi}) = [1/\mu_0\sqrt{2\pi\mu_2}]\int_0^\infty [u^2\mu_0/(1+(\mu_0/\mu_2)\,\dot{\psi}^2)]\cdot$$

$$\cdot[\exp[-u^2/2]\sqrt{\mu_0}/\sqrt{1+(\mu_0/\mu_2)\,\dot{\psi}^2}]\,du = \{\sqrt{\mu_0/\mu_2}/[1+(\mu_0/\mu_2)\,\dot{\psi}^2]^{\frac{3}{2}}\}\cdot$$

$$\cdot\int_0^\infty [u^2\exp[-u^2/2]/\sqrt{2\pi}]\,du = \sqrt{\mu_0/\mu_2}/2\{1+(\mu_0/\mu_2)\,\dot{\psi}^2\}^{\frac{3}{2}}\,.$$

The density for T (see (59)) can be developed from (64) by the transformation

$$(65)\qquad \tau = 2\pi/(\dot{\psi}+2\pi f_0)\,,$$

or

$$(66)\qquad \dot{\psi} = 2\pi/\tau - 2\pi f_0\,,\qquad \partial\dot{\psi}/\partial\tau = -2\pi/\tau^2\,.$$

Consequently

$$(67)\qquad f_T(\tau) = [\sqrt{\mu_0/\mu_2}/2\{1+(\mu_0/\mu_2)(2\pi/\tau-2\pi f_0)^2\}^{\frac{3}{2}}](2\pi/\tau^2)\,.$$

LONGUET-HIGGINS chooses to make an additional approximation to (59)

$$(68)\qquad T \approx [1/f_0]/[1+\dot{\Psi}/2\pi f_0] \approx (1/f_0)[1-\dot{\Psi}/2\pi f_0]\,.$$

The assumption here is that the rate of change of Ψ (*i.e.* $\dot{\Psi}$) is small relative to f_0, so that terms like $(\dot{\Psi}/2\pi f_0)^k$ for $k \geqslant 2$ can be ignored.

The density for T as defined by (68) may be derived from (64) by the transformation

$$(69)\qquad \tau = [1-\dot{\psi}/2\pi f_0]/f_0\,.$$

Let

$$(70)\qquad \bar{\tau} = 1/f_0\,.$$

Then

$$(71)\qquad \dot{\psi} = [2\pi(\bar{\tau}-\tau)]/\bar{\tau}^2\,,\qquad \partial\dot{\psi}/\partial\tau = -2\pi/\bar{\tau}^2\,.$$

Hence

$$(72)\qquad f_T(\tau) = [\mu_2\bar{\tau}^4/4\pi^2\mu_0]/[2[\mu_2\bar{\tau}^4/4\pi^2\mu_0+(\tau-\bar{\tau})^2]^{\frac{3}{2}}\,.$$

The joint density for R and T may be determined from (60) by the transformation

$$(73)\qquad R = R\,,\qquad \dot{\Psi} = 2\pi(\bar{\tau}-\tau)/\bar{\tau}^2\,.$$

The Jacobian of the transformation is $-2\pi/\bar{\tau}^2$ and so

$$(74)\qquad f_{R,T}(\varrho,\tau)=\{\sqrt{2\pi}\,\varrho^2/\mu_0\sqrt{\mu_2}\,\bar{\tau}^2\}\exp\left[-(\varrho^2/2\mu_0)[1+(\mu_0/\mu_2)(4\pi^2/\bar{\tau}^4)(\tau-\bar{\tau})^2]\right].$$

This is the joint density of the wave amplitude and period under the assumption that the approximation in (68) is satisfactory.

One further density is of some interest. This is the probability density for period, given the amplitude has some value, say ϱ_0. From (61)

$$(75)\qquad f_{T|R=\varrho_0}(\tau)=f_{R,T}(\varrho_0,\tau)/F_R(\varrho_0)=\{\sqrt{2\pi}\,\varrho_0^2/\mu_0\sqrt{\mu_2}\,\bar{\tau}^2\}\cdot$$

$$\cdot\exp\left[-[\varrho_0^2/2\mu_0][1+(\mu_0/\mu_2)(4\pi^2/\bar{\tau}^4)(\tau-\bar{\tau})^2]\right]/[\varrho_0/\mu_0]\exp\left[-\varrho_0/2\mu_0\right]=$$

$$=\exp\left[-\tfrac{1}{2}[(\tau-\bar{\tau})/(\sqrt{\mu_2}\,\bar{\tau}^2/2\pi\varrho_0)]^2\right]/\sqrt{2\pi}(\sqrt{\mu_2}\,\bar{\tau}^2/2\pi\varrho_0),$$

which is a normal probability density with mean $\bar{\tau}$ and standard deviation $\sqrt{\mu_2}\,\bar{\tau}^2/2\pi\varrho_0$.

An alternative height-period distribution has been developed by members of the C.N.E.X.O. group in France [7, 8]. For a cosine wave the period may be defined as

$$(76)\qquad T=2\pi\sqrt{-\eta/\ddot{\eta}},$$

where $\ddot{\eta}$ denotes the second time derivative of η. Thus one can start with the trivariate normal density for η, $\dot{\eta}$ and $\ddot{\eta}$ (see eqs. (1), (31)-(33)) and derive the joint density of α (see eq. (27)) and of T. If one uses the same conditions as eq. (25) ff. for restriction to near-crest values,

$$(77)\qquad P[\eta\leqslant u,\ \ddot{\eta}\leqslant w;\ \text{given } |\dot{\eta}|\leqslant|\ddot{\eta}|\Delta t,\ \ddot{\eta}<0]\approx$$

$$\approx\int_{-\infty}^{u}\int_{-\infty}^{w}f_{\eta,\dot{\eta},\ddot{\eta}}(u',0,w')|w'|\,dw'\,du'\Big/\int_{-\infty}^{0}f_{\dot{\eta},\ddot{\eta}}(0,w')|w'|\,dw'$$

for $w\leqslant 0$ and $-\infty\leqslant u\leqslant\infty$. The conditional density for η and $\ddot{\eta}$, given the specified « near-crest » conditions, is then

$$(78)\qquad f_{\eta,\ddot{\eta}|\text{near-crest}}(u,w)=wf_{\eta,\dot{\eta},\ddot{\eta}}(u,0,w)\Big/\int_{-\infty}^{0}w'f_{\eta,\ddot{\eta}}(0,w')\,dw'.$$

Under the conditions of being near the crest, the new variables

$$(79)\qquad H=2\eta,\qquad T=2\pi\sqrt{-\eta/\ddot{\eta}}$$

will then give a joint density of height and period in the same sense as eq. (76) approximates the period for irregular waves. The resulting density, after making the transformation in eq. (78), is

$$f_{H,T}(h,\tau)=\{4\pi^4/\sqrt{2\pi m_4\Delta}\}[h^2/\tau^5]\cdot \cdot\exp\left[-[h^2/8\Delta\tau^4](m_4\tau^4-8m_2\pi^2\tau^2+16m_0\pi^4)\right], \tag{80}$$

where

$$\Delta=m_0m_4-m_2^2. \tag{81}$$

Several workers have compared the Longuet-Higgins and the C.N.E.X.O. height-period densities with data [9-11]. The C.N.E.X.O. formula demonstrates many characteristics of small wave heights and periods, while the Longuet-Higgins formula works better for larger waves.

One of the difficulties in both densities is that the spectral moment m_4 is infinite for most spectral-density models. Thus in data analysis the value for m_4 obtained from spectral estimates varies with the choice of the Nyquist frequency. There are various ways around this dilemma which are based on other properties of the data [4].

One procedure was developed by CHEN and BORGMAN [10, 11] for waves greater than the significant wave height. This procedure (see eq. (75)) is based on the standard deviation, σ_{HS}, for wave period, given the wave height is equal to the significant wave height. From the Longuet-Higgins theory,

$$\sigma_{HS}^2=m_2T_0^4/\pi^2H_S^2, \tag{82}$$

where H_S denotes the significant wave height. The conditional probability law for period, given wave height equals h, is then normally distributed with some mean period T_0 and a variance given by $\sigma_{HS}H_S^2/h^2$. Data analysis of hurricanes Carla and Camille in the Gulf of Mexico lead to the conclusions [10] that the expected value of period for waves greater than H_S is approximately constant with value

$$T_0=0.85T_p, \tag{83}$$

where T_p is the reciprocal of the modal frequency of the wave spectral density. Furthermore, the standard deviation of the wave period for wave heights greater than H_S is well approximated by

$$(\text{standard deviation of period, given } H=h,\ h\geqslant H_S)=0.15T_pH_S/h. \tag{84}$$

Thus the last two equations together with the normality assumption for period given wave height and the Rayleigh distribution for wave height provide an engineering solution for waves larger than the significant wave height.

REFERENCES

[1] L. E. BORGMAN: *Statistical models for ocean waves and wave forces*, in *Advances in Hydroscience*, Vol. **8**, edited by T. VAN CHOW (New York, N. Y., 1972).

[2] L. E. BORGMAN: *Directional spectra from wave sensors*, in *Ocean Wave Climate, Marine Science Series*, Vol. **8** (New York, N. Y., 1979), p. 269.

[3] M. S. LONGET-HIGGINS: *J. Marine Res.*, **11**, 245 (1952).

[4] D. E. CARTWRIGHT and M. S. LONGUET-HIGGINS: *Hindcasting the directional spectra of hurricane generated waves*, OTC paper 2332 (Houston, Tex., 1956).

[5] S. O. RICE: *Am. J. Math.*, **61**, 409 (1959).

[6] M. S. LONGUET-HIGGINS: *J. Geophys. Res.*, **80**, 2688 (1975).

[7] A. CAVANIE, M. ARHAN and R. EZRATY: *A Statistical Relationship between Individual Height and Periods of Storm Waves*, BOSS 76 (Trondheim, 1976).

[8] R. EZRATY, M. LAURENT and M. ARHAN: *Comparison with Observations at Sea of Period and Height Dependent Sea State Parameters from a Theoretical Model*, OTC 2744 (1977), p. 149.

[9] Y. GODA: *The observed joint distribution of periods and heights of sea waves*, in *Proceedings of the XVI Conference on Coastal Engineering* (Hamburg, 1978).

[10] E. CHEN: *Statistical distributions of wave heights and periods for hurricane waves*, Ph. D. Thesis, University of Wyoming (1979).

[11] E. CHEN, L. E. BORGMAN and E. YFANTIS: *Height and period distribution for hurricane waves*, in *Proceedings of the Civil Engineering in Oceans*, Vol. **4** (San Francisco, Calif., 1979), p. 321.

Techniques for Computer Simulation of Ocean Waves.

L. E. BORGMAN

University of Wyoming - Laramie, Wy.

1. – Introduction.

Oceanographers and design engineers are becoming more and more familiar with the concept of a directional wave spectrum (*i.e.* a function of direction and frequency which gives the contribution to the overall variance of the sea surface, of waves traveling in that direction and having that frequency). It is fairly standard now to hindcast directional spectra from meteorological data [1]. Thus it is expected that compilations of climatological assemblages of directional spectra will soon be available to designers. In addition, a variety, of methods for estimating the directional spectrum from various types of ocean wave sensors have been developed [2] and used in an operational mode [3, 4].

If it is assumed that these hindcast and measured directional spectral densities will eventually become widely available to engineers, the next natural question is « How can directional spectra best be used in the design process? » Frequency spectra have been used directly in fatigue studies and in vibrational analysis of linearized structural models for a number of years. It is a natural and obvious extension to introduce directional spectra in these situations. As climatological assemblages of directional spectra become available, they can be used to treat operational scheduling problems. The directional spectrum can be used to develop the probability laws for wave height, period and direction, which in turn can be used to interpret scheduling interruptions and other such problems.

A real problem lies in the best way to use directional spectral information in extremal-event design. The spectrum is an expression of « average » conditions relative to the second-moment behavior of the sea surface elevations, while the extremal event depends primarily on the right-hand tail of the probability law for wave heights and may depend on higher-order moments than second order. Despite the differences, there is some evidence that linear wave theory coupled with the directional wave spectrum provides a good first approximation to extremal events.

The standard design procedure of using unidirectional, single-frequency, nonlinear (Stokes) waves of permanent form also differs significantly from the short-crested, transient, directional, confused seas which actually occur in most storms. Thus the design engineer faces the question of which is better—the deterministic, nonlinear approach or the linear, directional, statistical approach. Eventually, there will probably be an acceptable theory lying somewhere between these two special cases.

However, suppose the engineer chooses to use the linear, directional, statistical wave theory. How does he then introduce it into his design? One approach is by simulation on a digital computer. In this method, the computer generates « random » numbers and uses them to produce synthetic wave records consistent with the directional wave spectrum and with linear wave theory.

The preferred data base for studying an engineering problem would, of course, be a number of actual measurements of environmental forces taken during major storms at the site of interest. However, in most cases this is unavailable. Simulated data provide an artificial substitute for such records. The simulations will only reproduce the particular features which are built into it by the programmer. The common choice for ocean waves is to assume that the wave properties jointly follow a multivariate normal probability law with zero mean and a covariance matrix consistent with the spectra and cross-spectra determined by the ocean wave directional spectra. An alternative assumption is that the water level elevations follow the form

$$B_{jm}^2 = S(f_m, \theta_j)\,\Delta f_m \Delta\theta_j\,, \tag{1}$$

$$\eta(x, y, t) = 2 \sum_{m=1}^{M} \sum_{j=1}^{J} B_{jm} \cos\,(k_m x \cos\theta_j + k_m y \sin\theta_j - 2\pi f_m t + \Phi_{jm})\,, \tag{2}$$

where $S(f, \theta)$ is the directional spectrum, f_m and θ_j are the central frequency and angle in the (m, j) cell of the frequency-angle plane, Δf_m and $\Delta\theta_j$ are the dimensions of the cell, x and y are horizontal rectangular co-ordinates of position, t is time, and k_m is the wave number associated with f_m. Also, for each j and m, Φ_{jm} is an independent random variable which is uniformly distributed in the interval $(0, 2\pi)$ radians. The other wave properties are determined from analogy with linear wave theory. The same set of Φ_{jm} are used for all simultaneously occurring wave properties.

The normality assumption and the random-phase assumption lead to the same behavior if J is relatively large. Under these circumstances, a central-limit theorem leads to normality of the wave properties. If J is small, a modification may be introduced to force the asymptotic normality by inserting a « replication » index, r, to give

$$\eta(x, y, t) = 2 \sum_{m=1}^{M} \sum_{j=1}^{J} \sum_{r=1}^{R} [B_{jm}/\sqrt{R}]\cdot \cos\,(k_{jmr} x \cos\theta_{jmr} + k_{jmr} y \sin\theta_{jmr} - 2\pi f_{jmr} t + \Phi_{jmr})\,. \tag{3}$$

Here f_{jmr} and θ_{jmr} (for $r = 1, 2, ..., R$) are randomly selected from the (m, j) cell and Φ_{jmr} are all independent and uniform in $(0, 2\pi)$. If R is four or larger, this will behave approximately as a normal, even if J is one.

All simulation schemes on digital computers are based on the introduction of random numbers somehow into the computations. Some recommended procedures for generating these numbers are outlined below.

2. – Pseudorandom numbers.

Number sequences which have many of the properties of a set of independent random numbers uniformily distributed in the interval $(0, 1)$ may be conveniently developed using modulo arithmetic. Such sequences are not truly random, because they are produced by a fixed and deterministic generation scheme. However, the numbers strongly resemble a random sequence, and are, therefore, called pseudorandom numbers.

Modulo arithmetic [5] replaces any number by its remainder after division by a fixed base number. Thus

$$12 \bmod 5 = 2 \tag{4}$$

specifies that when 12 is divided by 5 the remainder is 2. The essential property for the present application may be stated abstractly as follows. The product of two numbers, x and y, when reduced modulo B gives a number z

$$xy \bmod B = z \tag{5}$$

with the property that $0 \leqslant z \leqslant B$. Furthermore, if x and y are not small compared with B, the number z occupies an apparently accidental position in the interval $(0, B)$. Consequently, $U = z/B$ is a more or less accidental number in $(0, 1)$.

In practice, x is assigned a fixed integer value which is not small relative to B and is prime. The integer y is called the seed number and is arbitrarily chosen to start the computation off. Equation (5) then gives a value z and a uniform number U. This z integer is used as the seed number for the next generation and a second z is produced. The procedure cycles until enough uniform numbers are produced. Some choices of x and B are given by ZELEN and SEVERO [6].

A pseudorandom number, U, which is uniform in $(0, 1)$ can easily be changed to be uniform in (a, b) by the formula

$$U^* = a + (b - a)\, U\,. \tag{6}$$

Pseudorandom numbers which are uniformly distributed in the interval (0, 1) can be converted to normal pseudorandom numbers by several techniques (*op. cit.*, p. 952). Two schemes are particularly useful. Let U_1 and U_2 be two independent uniform (0, 1) random variables. Then

$$Z_1 = \sqrt{-2 \ln U_1} \cos 2\pi U_2 \tag{7}$$

and

$$Z_2 = \sqrt{-2 \ln U_1} \sin 2\pi U_2 \tag{8}$$

will be independent normal random deviates with mean zero and variance one. The second recommended scheme is an « inverse » method. Let U be a uniform (0, 1) random variable. Then if, by the same reference (*op. cit.*, eq. (26, 2.23)),

$$\begin{cases} \varrho = \begin{cases} U\,, & \text{if } U \leqslant 0.5\,, \\ 1-U\,, & \text{if } U \geqslant 0.5\,, \end{cases} \\ t = \sqrt{-2 \ln \varrho}\,, \\ Z = t - (c_0 + c_1 t + c_2 t^2)/(1 + d_1 t + d_2 t^2 + d_3 t^3)\,, \end{cases} \tag{9}$$

$$\begin{cases} c_0 = 2.515\,517\,, & d_1 = 1.432\,788\,, \\ c_1 = 0.802\,853\,, & d_2 = 0.189\,269\,, \\ c_2 = 0.010\,328\,, & d_3 = 0.001\,308\,. \end{cases} \tag{10}$$

The quantity Z will be a normally distributed random variable with inaccuracy bounded by $4.5 \cdot 10^{-4}$. The relative speed on the computer for the two schemes depends on the rapidity with which the computer algorithm involved can calculate logarithms and trigonometric functions, as compared with polynomials.

A column vector of correlated normal pseudorandom numbers $\mathbf{X}$, having a covariance matrix C and a mean vector $\boldsymbol{\mu}$, may be generated by matrix operation on a vector $\boldsymbol{Z}$ of independent normal random variables having zero mean and unit variance by a method given by Scheuer and Stoller [7,8]. In this procedure

$$\mathbf{X} = \boldsymbol{\mu} + A\boldsymbol{Z}\,, \tag{11}$$

where A is a triangular matrix whose element, A_{ij}, in the i-th row and j-th column is given by the recursive formulae

$$
(12) \quad \begin{cases} a_{i1} = c_{i1}/c_{11}\,, & 1 \leqslant i \leqslant \text{number of rows}, \\ a_{ij} = \left[c_{ij} - \sum_{k=1}^{j-1} a_{ik}a_{jk}\right]/a_{jj}\,, & 1 < j < i \leqslant \text{number of rows}, \\ a_{ii} = \left[c_{ii} - \sum_{k=1}^{i-1} a_{ik}^2\right]^{\frac{1}{2}}, & 1 < i \leqslant \text{number of rows}, \end{cases}
$$

where c_{ij} is the i-th row and j-th column of C.

Problems occasionally arise with this method in the square root of $c_{ii} - \sum a_{ik}^2$ due to computer round-off. Theoretically, if C is a legitimate covariance matrix and so is positive definite, the quantity under the square root cannot be negative. However, if there are many components in $\boldsymbol{X}$ and the computer does not carry enough significant figures, negative values can occur due to round-off. This may also arise if the components of $\boldsymbol{X}$ are highly correlated. One remedy may be to use double or triple precision in the calculations.

An alternative procedure based on eigenvectors is useful if the components of $\boldsymbol{X}$ are highly correlated. Let B be a matrix whose columns are the eigenvectors of the covariance matrix C. Let L be the diagonal matrix whose main diagonal elements are the eigenvalues and whose off-diagonal elements are zero. Then the eigenvector-eigenvalue relation to C can be expressed as

$$
(13) \qquad CB = BL\,.
$$

The matrix B is orthogonal if the eigenvalues are distinct, or may be made orthogonal, in the generalized eigenvector sense, if there are eigenvalues with identical values. Define

$$
(14) \qquad \boldsymbol{X} = \boldsymbol{\mu} + BL^{\frac{1}{2}}\boldsymbol{Z}\,,
$$

where $\boldsymbol{Z}$ is the vector of independent normal pseudorandom numbers (mean zero, variance one) used previously and $L^{\frac{1}{2}}$ is the diagonal matrix with elements which are the square root of the eigenvalues. Then $\boldsymbol{X}$ has mean $\boldsymbol{\mu}$ and covariance matrix C.

There are several advantages to this procedure. For one thing, standard computer routines are available to calculate B and L. For another, if the components of $\boldsymbol{X}$ are highly correlated, only a few of the eigenvalues will be significantly different from zero. Suppose $(\lambda_j,\, j > \nu)$ are negligible compared with the earlier eigenvalues. The vector L and the matrix B are partitioned into the first ν elements and the remaining elements:

$$L = \begin{pmatrix} L_1 & 0 \\ 0 & L_2 \end{pmatrix}, \tag{15}$$

$$B = (B_1, B_2) . \tag{16}$$

That is, L_1 is a $\nu \times \nu$ diagonal matrix with main diagonal elements $(\lambda_1, \lambda_2, \ldots, \lambda_\nu)$ and zeros elsewhere. Similarly B_1 is a $N \times \nu$ matrix consisting of the first ν columns of B. Then

$$\boldsymbol{X} = \boldsymbol{\mu} + (B_1, B_2) \begin{pmatrix} L_1^{\frac{1}{2}} & 0 \\ 0 & L_2^{\frac{1}{2}} \end{pmatrix} \begin{pmatrix} \boldsymbol{Z}_1 \\ \boldsymbol{Z}_2 \end{pmatrix} = \boldsymbol{\mu} + B_1 L_1^{\frac{1}{2}} \boldsymbol{Z}_1 + \boldsymbol{E}, \tag{17}$$

where

$$\boldsymbol{E} = B_2 L_2^{\frac{1}{2}} \boldsymbol{Z}_2 . \tag{18}$$

The elements of $\boldsymbol{E}$ have mean zero and a sum of variances which is less than $(\sum \lambda_j, j > \nu)$. Thus, if the deleted eigenvalues are truly negligible, $\boldsymbol{E}$ is a negligible random vector.

The primary advantage of the above is that a N-component correlated random normal vector $\boldsymbol{X}$ is generated from a ν-component vector of independent standard normals. If $\nu \ll N$, this can be a substantial savings. In either of the above methods the matrices, A or B, depended only on the covariance matrix and so can be computed one time only. Thereafter, the matrix can be used to generate as many simulated sets as needed.

3. – Filters.

Some simulation procedures are based on filtered Gaussian (*i.e.* normal) white noise. The underlying structure can be developed either with respect to a continuous stochastic process (as was done in the referenced paper) or with respect to a discrete random sequence. In light of the convenience of the fast Fourier transform in calculations on digital computers, the filter theory will be developed here in the discrete formulation. The stochastic process X_n $(0 \leqslant n < N)$ will be thought of as a sampled version (sampling increment Δt) of a process continuous in time.

The finite or discrete Fourier transform (FFT) of a sequence $\{w_n, 0 \leqslant n < N\}$ is the sequence $\{W_m, 0 \leqslant m < N\}$ defined by

$$W_m = \Delta t \sum_{n=0}^{N-1} w_n \exp[-i2m\pi n/N], \tag{19}$$

where $i = \sqrt{-1}$ and e is the base for natural logarithms. If $\Delta f = 1/N\,\Delta t$, it can be shown that

$$w_n = \Delta f \sum_{m=0}^{N-1} W_n \exp\left[i2\pi mn/N\right]. \tag{20}$$

An integral linear filter with weight function $w(t)$ is in equation form

$$X(t) = \int_{-\infty}^{\infty} w(\tau)\, Z(t-\tau)\, \mathrm{d}\tau . \tag{21}$$

This same operation, in discrete format, is

$$X_n = \Delta t \sum_{j=0}^{N-1} w_j\, Z_{n-j} \tag{22}$$

with the convention that

$$w_j = \begin{cases} w(j\,\Delta t) & \text{for } 0 \leqslant j \leqslant N/2, \\ w(\{j-N\}\,\Delta t) & \text{for } N/2 < j < N. \end{cases} \tag{23}$$

That is, the values of $w(t)$ for negative argument are shifted to the interval $(N\,\Delta t/2,\ N\,\Delta t)$. Alternatively, this can be thought of as repeating $w(t)$, $|t| \leqslant N\,\Delta t/2$, periodically with period $N\,\Delta t$, and then selecting the interval of representation as $(0,\ N\,\Delta t)$. All discrete sequences, such as w_n, are thought of as being periodically repeated with period N in the various formulae. The choices of N and Δt should be selected so that $w(t)$ is essentially zero for $|t| > N\,\Delta t/2$.

Let $\tilde{X}_m$ be the FFT of X_n and $\tilde{Z}_m$ be the FFT of Z_n. It can be shown that

$$\tilde{X}_m = W_m \tilde{Z}_m \tag{24}$$

if eq. (22) holds.

Define $\{X_n^{(l)};\ 1 \leqslant l \leqslant L,\ 0 \leqslant n < N\}$ as a set of L random, real-valued, time sequences, each of length N. The discrete cross-covariance function between $X_n^{(l)}$ and $X_n^{(l')}$, with lag k, is defined as

$$C_{ll'}(k) = \mathrm{E}\left[\{X_n^{(l)} - \mathrm{E}(X_n^{(l)})\}\{X_{n+k}^{(l')} - \mathrm{E}(X_{n+k}^{(l')})\}\right], \tag{25}$$

where $\mathrm{E}(\cdot)$ is the expectation operator from mathematical statistics. If $l = l'$, this is just the ordinary covariance function. The discrete cross-spectrum is then

$$S_{ll'}(m) = \sum_{k=0}^{N-1} C_{ll'}(k) \exp\left[-\,i2\pi km/N\right] \Delta t \tag{26}$$

(*i.e.* the FFT of $C_{ll'}(k)$). This function $S_{ll'}(m)$ is, in general, complex-valued.

A procedure for developing interrelated, correlated sequences of normal variates from normal sequences $\{Z_n^{(l)}; 1 \leqslant l \leqslant L, 0 \leqslant n < N\}$ which are independent with mean zero and variance one is as follows:

$$X_n^{(l)} = \Delta t \sum_{u=1}^{l} \sum_{j=0}^{N-1} w_j^{(l,u)} Z_{n-j}^{(u)} + \mu_n^{(l)} . \tag{27}$$

The filter functions are selected to force all spectra and cross-spectra to have previously assigned values. The system of equations which must hold for this to be true is, for $1 \leqslant l' \leqslant l \leqslant L$,

$$S_{ll'}(m) = \Delta t \sum_{v=1}^{l'} \overline{W_m^{(l,v)}} W_m^{(l',v)} . \tag{28}$$

(The overbar denotes the complex conjugate.)

Expressed in expanded form this is

$$\begin{cases} S_{11}(m) = \Delta t |W_m^{(1,1)}|^2 , \\ S_{21}(m) = \Delta t \, \overline{W_m^{(2,1)}} W_m^{(1,1)} , \\ S_{22}(m) = \Delta t [|W_m^{(2,1)}|^2 + |W_m^{(2,2)}|^2] , \\ S_{31}(m) = \Delta t \, \overline{W_m^{(3,1)}} W_m^{(1,1)} , \\ S_{32}(m) = \Delta t [\overline{W_m^{(3,1)}} W_m^{(2,1)} + \overline{W_m^{(3,2)}} W_m^{(2,2)}] , \\ S_{33}(m) = \Delta t [|W_m^{(3,1)}|^2 + |W_m^{(3,2)}|^2 + |W_m^{(3,3)}|^2] , \end{cases} \tag{29}$$

etc.

Any set of filters with frequency domain filter functions $W_m^{(l',l)}$ which satisfy these equations will produce a simulation. The solution is not unique, but, with some arbitrary choices, a set of filter functions which work may be generated recursively. For example, let $W_m^{(1,1)}$ have zero phase shift. Then $W_m^{(1,1)} = \sqrt{S_{11}(m)/\Delta t}$. Consequently, $W_m^{(2,1)}$ may be solved from the second equation. Then $W_m^{(2,2)}$ may be determined from the third equation, provided the zero-phase-shift version is selected, etc. Actually, any arbitrary phase shifts could be introduced where they are needed as zero is just a convenient choice.

4. – Simulation concepts.

Any simulation is artificial. The programmer builds in the wave properties he wishes to preserve. The procedures discussed in the following are methods which produce Gaussian simulations having the desired mean and spectral relationships. Within that constraint, there are a number of methods which can be used. The most straightforward procedure for simulating ocean wave properties is to develop the simulated values for each time, one after the other. This will be called time domain simulation. It has the advantage of directness and ease of understanding, but it may be costly in terms of computer time. A much faster procedure, but more indirect, is frequency domain simulation. In this type of procedure, the discrete Fourier transform of the time series is simulated first. Then the fast Fourier transform is used to revert to the time domain. Frequency domain procedures exploit the speed of the fast Fourier transform and the fact that the Fourier coefficients of correlated time sequences are uncorrelated for frequencies less than the Nyquist frequency.

A variety of procedures may be used to produce the simulation either in time or in frequency. Four basic techniques in each domain will be discussed in the subsequent sections. Although the specifics of application are usually quite different in time and frequency domains, the same category names may be used in both domains. These are 1) matrix multiplication of independent normal random variables with the matrix determined recursively from the covariance matrix, 2) matrix multiplication of independent normal random variables with the matrix determined by eigenvector techniques, 3) filtered white noise and 4) wave property construction with the introduction of random phases and the use of replication to force normality through a central-limit theorem. The main emphasis will be on the frequency domain techniques because of their great time savings in digital-computer processing.

5. – Time domain simulation by matrix multiplication.

This procedure will be illustrated by an example. Suppose one wishes to simulate the water level elevations at two locations (x_1, y_1) and (x_2, y_2). Let $\eta^{(1)}$ and $\eta^{(2)}$ be the elevations above mean water level at the two locations and define η_n as the combined sequence obtained by listing the time series for $\eta^{(2)}$ after the time series for $\eta^{(1)}$. The simulation problem then may be restated as the production of the sequence $\{\eta_n, 0 \leqslant n < 2N\}$ of multinormal random variables. The covariance matrix C with elements

$$c_{ij} = \mathrm{E}[\eta_i \eta_j], \qquad 0 \leqslant i < 2N, \; 0 \leqslant j < 2N, \tag{30}$$

will be a $2N$ by $2N$ matrix. The covariance function between $\eta(x, y, t)$ and $\eta(x', y', t+\tau)$ may be written in terms of the directional spectral density as

$$\begin{aligned} C_{\eta\eta'}(\tau) &= \mathrm{E}[\eta(x, y, t)\eta(x', y', t+\tau)] = \\ &= 2\int_0^\infty \int_0^{2\pi} S(f, \theta) \cos[k(x'-x)\cos\theta + k(y'-y)\sin\theta - 2\pi f t]\,\mathrm{d}\theta\,\mathrm{d}f\,. \end{aligned} \tag{31}$$

The elements c_{ij} may be computed from this formula.

The time domain simulation by matrix multiplication consists of the following steps:

1) The $2N$ by $2N$ covariance matrix is computed.

2) The multiplication matrix is determined, either by recursion (eq. (11) ff.) or by eigenvector-eigenvalue computations (eq. (14) ff.).

3) A sequence of $2N$ pseudorandom independent normal (mean zero, variance one) numbers are generated.

4) The matrix is multiplied by the vector of random numbers to produce a simulation of η.

The first half of $\{\eta\}$ is the simulated time sequence at location (x_1, y_1), while the second half is the simulation for location (x_2, y_2).

Clearly, there are computer memory problems if the covariance matrix is too large. Thus, if $N = 200$, the covariance matrix is a 400×400 matrix. The matrix alone would require 160 000 storage locations, although reduction to about half of this can be arranged from the fact that a covariance matrix is symmetric. The multiplication matrix will be triangular and will need another 80 000 locations. With some computational ingenuity, the multiplication matrix can be stored into the same location originally occupied by the covariance matrix. Consequently, an overall simulation of two 200 time step simulations will require about 100 000 memory locations of storage. (This adds 10 000 for program and system and 10 000 for other arrays involved in the computations.) Most computers begin having indigestion if the arrays get much bigger than this.

Symmetries and regularities in the covariance matrix can be used to reduce the space required for the covariance matrix. Also the multiplication matrix can be read in from peripheral storage (tape or disk pack) row by row as needed in the multiplication. Computer time then becomes a problem because of the read operations required. In any case, the outlook is not very optimistic for using the matrix multiplication in time domain for long sequences. The covariance matrix for most wave properties is such that the eigenvalue approach does not greatly improve the outlook. However, the stationarity usually assumed for the time series leads to the covariance function being only a function of time lag. This behavior forces a regularity on the covariance matrix and reduces the whole concept of matrix multiplication to an equivalence with the filtering of white noise.

6. – Time domain simulation by filtered white noise.

The cross-spectral matrix for the two time series $\{\eta_n^{(1)}\}$ and $\{\eta_n^{(2)}\}$ defined in the last section can be used to derive filters by eqs. (28), (29) if the time series are stationary. The cross-spectral matrix for $(\eta^{(1)}, \eta^{(2)})$ is

$$\text{cross-spectral matrix} = \begin{bmatrix} S_{11}(m) & S_{12}(m) \\ S_{21}(m) & S_{22}(m) \end{bmatrix}, \tag{32}$$

where, with $f_m = m\,\Delta f = m/N\,\Delta t$,

$$S_{11}(m) = S_{22}(m) = \int_0^{2\pi} S(f_m, \theta)\, d\theta\,, \tag{33}$$

$$\begin{aligned} S_{12}(m) = \overline{S_{21}(m)} = &\int_0^{2\pi} S(f_m, \theta) \cos\,[k(x_2 - x_1)\cos\theta + k(y_2 - y_1)\sin\theta]\, d\theta\, - \\ &- i\int_0^{2\pi} S(f_m, \theta) \sin\,[k(x_2 - x_1)\cos\theta + k(y_2 - y_1)\sin\theta]\, d\theta\,. \end{aligned} \tag{34}$$

The time domain simulation by filtering white noise proceeds by the following steps:

1) The spectra are computed from eqs. (33) and (34).

2) These are used to calculate the frequency domain filter functions $W^{(1,1)}(m)$, $W^{(2,1)}(m)$ and $W^{(2,2)}(m)$.

3) The frequency domain filter functions are inverted with the fast Fourier transform to evaluate the time domain filter functions $w_n^{(1,1)}$, $w_n^{(2,1)}$ and $w_n^{(2,2)}$.

4) Two sequences of independent, standard normal (mean zero, variance one) pseudorandom numbers $\{z_n^{(1)}, 0 \leqslant n < N\}$ and $\{z_n^{(2)}, 0 \leqslant n < N\}$ are generated. Then by eq. (27) with $\mu_n^{(1)} = 0$, $\mu_n^{(2)} = 0$,

$$\begin{cases} \eta_n^{(1)} = \Delta t \sum\limits_{j=0}^{N-1} w_j^{(1,1)} z_{n-j}^{(1)}\,, \\ \eta_n^{(2)} = \Delta t \sum\limits_{j=0}^{N-1} w_j^{(2,1)} z_{n-j}^{(1)} + \Delta t \sum\limits_{j=0}^{N-1} w_j^{(2,2)} z_{n-j}^{(2)}\,. \end{cases} \tag{35}$$

Once the sequences $\{w_j^{(1,1)}\}$, $\{w_j^{(2,1)}\}$ and $\{w_j^{(2,2)}\}$ are computed, they can be used over and over with different sets of $\{z_n^{(l)}\}$ to produce as many simulations as needed.

It should be pointed out that eqs. (35) can be computed much more efficiently in the frequency domain. Consequently, the time domain form of this procedure is really only appropriate if, for some reason, the simulation must proceed in real time.

7. – Time domain simulation with random-phase structure.

The obvious and direct procedure for time domain simulation of wave properties is expressed in eq. (3) and is discussed in the associated text. The time domain simulation proceeds through the following steps. 1) The JMR random phases, Φ_{jmr}, are generated by developing JMR uniform, independent, pseudorandom numbers and multiplying each by 2π. 2) Let the (j, m)-th cell have midpoints (f_m, θ_j) and widths Δf_m and $\Delta\theta_j$. The JMR values, f_{jmr}, are developed by choosing JMR independent, uniform, pseudorandom numbers U'_{jmr} and computing the formula

$$f_{jmr} = f_m + (U'_{jmr} - 0.5)\Delta f_m . \tag{36}$$

Similarly, the directions, θ_{jmr}, are computed from the JMR independent, uniform, pseudorandom numbers U''_{jmr} from the equation

$$\theta_{jmr} = \theta_j + (U''_{jmr} - 0.5)\Delta\theta_j . \tag{37}$$

The set of wave numbers, k_{jmr}, is related to the frequencies by the usual dispersion relation. 3) Once the sets Φ_{jmr}, f_{jmr}, θ_{jmr} and k_{jmr} are developed, the wave properties may be computed from eq. (3) or from analogous formulae from linear theory for other wave properties. The same realizations for the random sets are used for any wave properties simultaneously occurring.

If there are many different wave properties being developed for various locations within the wave and if the simulation must proceed in real time, there are reasonable advantages to the random-phase simulation. However, it will be shown later that the analogous frequency domain simulation is much more efficient relative to the expenditure of computer time.

8. – Frequency domain probability structure.

Suppose that $\boldsymbol{X}$ is any multivariate normal vector, $\boldsymbol{M}$ is any matrix of deterministic scalars, and $\boldsymbol{c}$ is any vector of constants. Then $\boldsymbol{MX} + \boldsymbol{c}$ will also follow a multivariate normal probability law [9]. This relationship can be applied to the finite Fourier transform of a Gaussian process. Let U_m and

V_m be defined by

$$A_m = U_m - iV_m, \tag{38}$$

$$U_m = \Delta t \sum_{n=0}^{N-1} X_n \cos(2\pi mn/N), \tag{39}$$

$$V_m = \Delta t \sum_{n=0}^{N-1} X_n \sin(2\pi mn/N). \tag{40}$$

Hence the vector $\boldsymbol{A}$, whose transpose is defined as

$$\tilde{\boldsymbol{A}} = (U_0, U_1, V_1, U_2, V_2, \ldots, U_{N/2}), \tag{41}$$

will be multivariate normal if $\{X_n\}$ is a multivariate normal vector. Both V_0 and $V_{N/2}$ were deleted from $\tilde{\boldsymbol{X}}$ because they are identically zero if $\{A_m\}$ is a real-valued sequence. Also, for $0 < m < N/2$,

$$U_{N-m} = U_m, \tag{42}$$

$$V_{N-m} = -V_m, \tag{43}$$

if $\{X_n\}$ is real-valued. Consequently the complete sequence $\{A_m, 0 \leqslant m < n\}$ is determined if $\boldsymbol{A}$ is known.

The whole procedure can be reversed. The vector $\boldsymbol{A}$ can be generated as a multivariate normal vector. The components of $\boldsymbol{A}$ can be used to specify $\{A_m, 0 \leqslant m < N\}$. This sequence is then inverted with a discrete Fourier transform to obtain $\mathbf{X}$. At first glance, this would appear to be more time-consuming than time domain simulation. In actual fact, the variance-covariance structure of the vector $\boldsymbol{A}$ is so simple, at least for stationary sequences, that the frequency domain method is much faster in most applications.

Consider now two real-valued stationary, periodic, Gaussian time series $\{X_n^{(l)}, l = 1, 2\}$ with transforms $\{A_m^{(l)}\}$, covariance function $C_{ll}(k)$, spectral density $S_{ll}(m)$ and zero mean. Let

$$C_{12}(k) = \mathrm{E}[X_n^{(1)} X_{n+k}^{(2)}], \tag{44}$$

then

$$S_{12}(m) = \Delta t \sum_{n=0}^{N-1} C_{12}(k) \exp[-i2\pi km/N], \tag{45}$$

$$= c_{12}(m) - iq_{12}(m) \tag{46}$$

is the cross-spectral density of $\{X_n^{(1)}\}$ with $\{X_n^{(2)}\}$. The sequence $c_{12}(m)$ is the co-spectral density and $q_{12}(m)$ is the quad-spectral density of $\{X_n^{(1)}\}$ with $\{X_n^{(2)}\}$. Let $U_m^{(l)}$ and $V_m^{(l)}$ be defined by analogy with the earlier definitions. The four values $U_m^{(1)}$, $V_m^{(1)}$, $U_m^{(2)}$, $V_m^{(2)}$ for $0 \leqslant m \leqslant N/2$ have mean zero and are uncorrelated with the corresponding four values for other $0 \leqslant m \leqslant N/2$. For $0 < m < N/2$, these coefficients follow a multivariate normal probability law with covariance matrix [10]

$$\text{(47)} \qquad \operatorname{cov} = \begin{bmatrix} U_m^{(1)} \\ V_m^{(1)} \\ U_m^{(2)} \\ V_m^{(2)} \end{bmatrix} = \{N\Delta t/2\} \begin{bmatrix} S_{11}(m) & 0 & c_{12}(m) & q_{12}(m) \\ 0 & S_{11}(m) & -q_{12}(m) & c_{12}(m) \\ c_{12}(m) & -q_{12}(m) & S_{22}(m) & 0 \\ q_{12}(m) & c_{12}(m) & 0 & S_{22}(m) \end{bmatrix}.$$

The covariance matrix for $m = 0$ or $m = N/2$ is different, because the imaginary parts of the coefficients are identically zero. The covariance matrix for $m = 0$ or $N/2$ is

$$\text{(48)} \qquad \operatorname{cov} = \begin{bmatrix} U_m^{(1)} \\ U_m^{(2)} \end{bmatrix} = N\,\Delta t \begin{bmatrix} S_{11}(0) & c_{12}(0) \\ c_{12}(0) & S_{22}(0) \end{bmatrix}.$$

To apply the results to ocean wave simulation one needs only to introduce the appropriate cross-spectral values as derived from linear wave theory. Any linear wave property can be written as the linear superposition (one for each frequency) of formulae of the form

$$\text{(49)} \qquad P_l(x, y, z, t) = aQ_l(z, d)\,G_l(\theta)\,H_l(kx\cos\theta + ky\sin\theta - 2\pi ft + \Phi)\,.$$

In this equation, a is the wave amplitude at the frequency f, $Q_l(z, d)$ is an attenuation function, z is the vertical co-ordinate (positive upward, zero at mean water level), d is water depth, k is wave number, t is time, and Φ is the phase. The variables x and y are horizontal co-ordinates such that (x, y, z) for a right-handed Cartesian co-ordinate system and θ is the direction of wave travel as measured counterclockwise from the positive x-direction. The function $G_l(\theta)$ is $\cos\theta$ for x-components of directional properties like velocities, $\sin\theta$ for y-components and 1.0 for nondirectional properties like water level elevation or pressure. The function $H_l(y)$ is $\cos y$ for properties in phase with the wave profile and $\sin y$ for properties out of phase with the wave profile.

The spectral relations for two properties $P_l(x_1, y_1, z_1, t)$ and $P_{l'}(x_2, y_2, z_2, t)$ are

$$\text{(50)} \qquad S_{ll}(m) = (\Delta f)^{-1} \int_{(m-0.5)\Delta f}^{(m+0.5)\Delta f} Q(f, z_1, d) \int_0^{2\pi} S(f, \theta)\, G_l^2(\theta)\, d\theta\, df\,,$$

$$\text{(51)} \qquad c_{ll'}(m) = (\Delta f)^{-1} \int_{(m-0.5)\Delta f}^{(m+0.5)\Delta f} Q_l(f, z_1, d)\,Q_{l'}(f, z_2, d) \int_0^{2\pi} S(f, \theta)\, G_l(\theta)\, G_{l'}(\theta)\,\cdot$$
$$\cdot H_{ll'}^*(k(x_2 - x_1)\cos\theta + k(y_2 - y_1)\sin\theta)\, d\theta\, df\,,$$

$$q_{ll'}(m) = (\Delta f)^{-1} \int_{(m-0.5)\Delta f}^{(m+0.5)\Delta f} Q_l(f, z_1, d) Q_{l'}(f, z_2, d) \int_0^{2\pi} S(f, \theta) G_l(\theta) G_{l'}(\theta) \cdot \cdot H^{**}_{ll'}(k(x_2 - x_1)\cos\theta + k(y_2 - y_1)\sin\theta)\, d\theta\, df . \tag{52}$$

In these equations

$$\begin{bmatrix} H^*_{ll'}(y) \\ H^{**}_{ll'}(y) \end{bmatrix} = \begin{cases} \begin{Bmatrix} \cos y \\ \sin y \end{Bmatrix}, & \text{if } H_l \text{ and } H_{l'} \text{ are both cosines or are both sines;} \\ \begin{Bmatrix} \sin y \\ -\cos y \end{Bmatrix}, & \text{if } H_l \text{ is a cosine and } H_{l'} \text{ is a sine;} \\ \begin{Bmatrix} -\sin y \\ \cos y \end{Bmatrix}, & \text{if } H_l \text{ is a sine and } H_{l'} \text{ is a cosine.} \end{cases} \tag{53}$$

The covariance matrix for $m = N/2$ is planned to have elements equal to zero by choosing Δt small enough so that the spectral densities are essentially zero for frequencies near the Nyquist frequency ($m = N/2$). The covariance matrix for $m = 0$ must be treated more carefully. This is, of course, related to the *DC* component or average value of the processes. Although typical spectral models such as the Pierson-Moskowitz formula predict $S(f, \theta) = 0$ at $f = 0$, $S_{ll}(0)$ will not ordinarily be zero. This follows from eqs. (50)-(52), where the quantities are not necessarily zero, but are the averages of $S(f, \theta)$ in the vicinity of zero frequency.

9. – Frequency domain simulation by matrix multiplication.

The basic techniques for generating pseudorandom, correlated, normal variables as given in eqs. (11)-(18) may be applied to frequency domain simulation. The Fourier coefficients $U^{(l)}_m$ and $V^{(l)}_m$ for the l-th property ($1 \leqslant l \leqslant L$) are generated by matrix multiplication of $2L$ independent normal (mean zero, variance one) random variables for each $0 < m < N/2$. This operation can be done independently for each m index because of the lack of correlation between the Fourier coefficients for frequencies less than the Nyquist frequency. A new set of $2L$ independent normal random numbers must be generated for each m. The values for N and Δt are selected so that the energy at $m = N/2$ is negligible and $U^{(l)}_{N/2} = 0$ for all time series being simulated. The generation of the Fourier coefficients for $m = 0$ requires only L independent normal random numbers since $V^{(l)}_0 = 0$ always. The Fourier coefficients for $N/2 < m < N$ are determined from the constraint that the time series being simulated must be real-valued. This constraint was previously stated in eqs. (42), (43).

If L is large, the eigenvector approach may be the most efficient procedure to use, since the properties being simulated are often highly correlated. The Fourier coefficients for the L time series are often all linear combinations of just a few eigenvectors. In either matrix multiplication method, the matrix for each m is completely determined by the covariance matrix for the $(U_m^{(l)}, V_m^{(l)})$. Once these matrices are determined, they can be used again and again in repeated simulation by generating new independent random numbers each time. Also the matrices are fairly continuous with respect to m as a variable. Thus the matrices can be determined only for a selected subset of the m-values, and the matrices for intermediate m-values obtained by interpolation.

Once the Fourier coefficients for $0 \leqslant m < N$ are all specified for each wave properly, the time series values are rapidly obtained by the fast Fourier transform:

$$X_n^{(l)} = \Delta f \sum_{m=0}^{N-1} [U_m^{(l)} - iV_m^{(l)}] \exp[i2\pi mn/N] . \tag{54}$$

10. – Frequency domain simulation by filtered white noise.

The filter theory outlined in eqs. (19)-(29) can be used to generate frequency domain simulations. The values for $W_m^{(l,u)}$ as derived from eq. (29) are applied to the frequency version of eq. (27) with $\mu_n^{(l)} = 0$. Let

$$\tilde{X}_m^{(l)} = \sum_{u=1}^{l} W_m^{(l,u)} \tilde{Z}_m^{(u)} \tag{55}$$

for $0 \leqslant m < N/2$, where $\tilde{Z}_m^{(u)}$ are the Fourier coefficients for Gaussian white noise. If $Z_n^{(u)}$ is normal with mean zero and variance one for each n, it follows that the real and imaginary parts of $\tilde{Z}_m^{(u)}$ will each be normal with mean zero and variance equal to $N(\Delta t)^2/2$ for each u and for $0 < m < N/2$. The imaginary parts of $\tilde{Z}_0^{(u)}$ and $\tilde{Z}_{N/2}^{(u)}$ will be zero and the real parts will be normal with mean zero and variance $N(\Delta t)^2$. As usual for the Fourier coefficients of a real-valued time series,

$$\tilde{Z}_m^{(u)} = \overline{\tilde{Z}_{N-m}^{(u)}} . \tag{56}$$

Thus the Fourier coefficients for $N/2 < m < N$ are completely determined by those for $0 < m < N/2$.

The simulated time series may be obtained by the following steps. Equation (55) is computed for $0 \leqslant m \leqslant N/2$. Then the terms of $\tilde{X}_m^{(l)}$ for $N/2 < m < N$ are obtained from the complex conjugate of $\tilde{X}_{N-m}^{(l)}$. Finally, the complete series $\{X_m^{(l)}, 0 \leqslant m < N\}$ are reverted to time domain by the fast Fourier transform.

11. – Frequency domain simulation by random-phase structure.

A straightforward extension of eq. (2) to the general wave property expressed in eq. (49) gives

$$P_l(x, y, z, t) = 2 \sum_{m=1}^{M} \sum_{j=1}^{J} \sum_{r=1}^{R} (B_{jm}^{(l)}/\sqrt{R})\, Q_l(f_m, z, d)\, G_l(\theta_{jmr}) H_l(\beta)\,, \tag{57}$$

where

$$\beta = k_m x \cos\theta_{jmr} + k_m y \sin\theta_{jmr} - 2\pi f_m t_n + \Phi_{jmr}\,, \tag{58}$$

$$B_{jm}^{(l)} = \left\{ \int_{f_m-\Delta f/2}^{f_m+\Delta f/2} \int_{\theta_j-\Delta\theta_j/2}^{\theta_j+\Delta\theta_j/2} S(f, \theta)\, \mathrm{d}\theta\, \mathrm{d}f \right\}^{\frac{1}{2}} \tag{59}$$

and, as usual, $t_n = n\,\Delta t$, $f_m = m\,\Delta f = m/N\,\Delta t$. The limits of frequency integration in eq. (59) are often modified for $m = 1$ to the interval $(0, f_1 + \Delta f/2)$ in order to include all the sea surface variance.

The formula in eq. (57) can be expressed as a discrete Fourier sum. The series length, N, is chosen so that $M < N/2$. Let

$$P_{jmr} = k_m x \cos\theta_{jmr} + k_m y \sin\theta_{jmr}\,. \tag{60}$$

Then β can be decomposed into

$$\beta = P_{jmr} - 2\pi\, mn/N + \Phi_{jmr} \tag{61}$$

and the complex-variable definitions for trigonometric functions introduced to give

$$P_l(x, y, z, t_n) = \sum_{m=0}^{N-1} A_m^{(l)} \exp[i2\pi mn/N]\,. \tag{62}$$

The coefficient $A_m^{(l)}$ is defined for $0 < m < N/2$ by

$$A_m^{(l)} = \sum_{j=1}^{J} \sum_{r=1}^{R} (B_{jm}/\sqrt{R})\, Q_l(k_m, z, d)\, G_l(\theta_{jmr}) \begin{pmatrix} 1 \\ i \end{pmatrix} \exp[-i(P_{jmr} + \Phi_{jmr})]\,, \tag{63}$$

where the first component of the vector $(1, i)$ is used if H_l is a cosine, while the second component is used if H_l is a sine. The values of $A_0^{(l)}$ and of $\{A_m^{(l)}, M < m \leqslant N/2\}$ are set to zero and the relation

$$A_m^{(l)} = \overline{A_{N-m}^{(l)}} \tag{64}$$

used to define the coefficients for $N/2 < m < N$.

The simulation proceeds as follows. The pseudorandom set of values of Φ_{jmr} and θ_{jmr} are first generated. These are used to develop the $\{A_m^{(l)}; 1 \leqslant l \leqslant L, 0 \leqslant m < N\}$ and then the individual time series are obtained from the fast Fourier transform as stated in eq. (62).

There is a minor difficulty in the above procedure. Although the theoretical mean of a wave property is zero, any short time interval from an actual record will have a sample mean which differs slightly from zero, due to the accidental randomness of that record. The above simulation procedure forces the simulated sample functions to have a zero mean exactly. If the slight variation of mean level in the sample functions is important, other modifications should be introduced. One procedure would be to use the matrix method just for $A_0^{(l)}$. Ordinarily the slight variation in sample mean for the various properties is not significant in the application and the adjustment for variation in sample mean is not made. The resulting simulations have been constrained to have sample mean zero. The whole concept of constrained simulation, and a related concept of conditional simulation, will be discussed later.

12. – Computer requirements for frequency domain simulations.

Frequency domain simulations save both memory and time as compared with time domain operations. Consider, for example, the discussion after eq. (31). The two water level elevations, η_1 and η_2, require only a 4×4 covariance matrix for the real and imaginary parts of the Fourier coefficients. (This actually only involves four numbers—the two spectra and the co- and quad-spectra.) Of course, it requires this at every frequency with significant wave energy. However, usually a hundred such 4×4 matrices will span the energy-containing frequencies and there is ordinarily sufficient continuity, so that the covariance matrix for intermediate frequencies can be obtained by interpolation.

Computer time is also substantially reduced. Consider, for example, time and frequency domain simulation using the random-phase structure. In eq. (3), MJR trigonometric functions must be computed for each of N times. Thus the product $MNJR$ is a measure of computer time requirements for that simulation. Now consider the method outlined by eqs. (62) and (63). There JR terms must be computed for each of M frequencies. Thus the computer time required is on the order of MJR. If N is large, this is quite a difference.

13. – Constrained and conditional simulations.

A constrained simulation is a simulation in which the resulting artificial sample functions are adjusted to have specified sample properties. The simplest example would be a sequence of standard normal independent pseudorandom numbers $\{Z_0, Z_1, Z_2, ..., Z_{N-1}\}$ which are adjusted by subtracting the sample average and dividing by the sample standard deviation. The resulting sequence will exactly have mean zero and variance one. The original sequence $\{Z_n\}$ was a sample from a population with theoretical mean zero, variance one, but its sample values will deviate slightly from these theoretical values due to the randomness and finite extent of the sample.

Correlated simulated sequences may be constrained to have other specified sample properties. For example, a specified sample covariance function or sample spectral density can be forced onto the simulated sequence. It is not always clear if the resulting sequence has the same distribution as the original one. In fact, in general it will not. The original example above of a sequence of independent normal random numbers will have a weak correlation between successive values after the sample mean is subtracted from each term. Nevertheless, it is occasionally useful to work with simulated time series which have been constrained.

What type of questions may be answered with constrained time series simulations? Generally such questions are related to behaviors which pivotally are concerned with a given sample property. Consider the following two questions:

1) What is the vibrational behavior of an oil-drilling structure which is experiencing ocean waves for one hour whose theoretical spectral density is specified?

2) What is the vibrational history of the same structure when it experiences waves for one hour whose sample spectral density is a specified function?

In a simulation for the first question, the actual sample spectra for the one hour of data will not equal the theoretical function. In fact, it may differ quite a bit from the theoretical value which holds for the population. A simulation for the second question will force the spectra for the one-hour record to equal the specified function. One source of variation, the sample fluctuation, will have been removed. Several simulations will all have exactly the same sample spectral density.

There are many unresolved theoretical questions connected with constrained simulations. The foregoing is intended only to be a brief introduction to the problem. However, it is common engineering practice to introduce constraints on simulations, at least for some types of studies.

Conditional simulations are less theoretically questionable. In a conditional simulation, one or more of the simulated values are assigned numerical values, and the rest are obtained from distribution theory and pseudorandom numbers. For example, suppose $\{X_0, X_1, X_2, ..., X_{N-1}\}$ to be a multivariate Gaussian sequence and it is known, *a priori*, that X_0 and X_{N-1} are both zero.

The sequence $\{X_1, X_2, \ldots, X_{N-2}\}$ can be simulated conditionally given $X_0 = = X_{N-1} = 0$. If the original sequence was highly positively correlated, the simulated X_1 will not differ appreciably from $X_0 = 0$. That is, the correlation will be preserved between the given and the simulated values. Techniques of conditional simulation have been used in geological problems [11]. The concept appears to be very promising for coastal engineering applications.

The concept can be extended to several simultaneous time series. One or more of the time series can have specified values and the remaining time series can be simulated conditionally. Thus the sea surface elevation time series could be set equal to a measured wave record, and the intercorrelated bottom pressure which was occurring simultaneously would be simulated conditionally. This latter type of conditional simulation (one or more time series specified, the remaining ones simulated) will be given the primary attention in the following.

Conditional simulations may be generated either by matrix multiplication or filtered white noise. (At the present time, it is not clear how the random-phase procedures could be used.) Only matrix multiplication procedures will be presented here. Two basic theorems for the matrix multiplication technique are as follows.

Let $\boldsymbol{W}$ be a normal random (column) vector with n components, which has mean vector $\boldsymbol{\mu}$ and covariance matrix C. Let $\boldsymbol{W}$ be partitioned into two vectors $\boldsymbol{W}_1$ and $\boldsymbol{W}_2$ with n_1 and n_2 components, respectively. The vector $\boldsymbol{\mu}$ and the matrix C are similarly partitioned. Thus

$$n = n_1 + n_2 , \tag{65}$$

$$\boldsymbol{W} = \begin{pmatrix} \boldsymbol{W}_1 \\ \boldsymbol{W}_2 \end{pmatrix}, \tag{66}$$

$$\boldsymbol{\mu} = \begin{pmatrix} \boldsymbol{\mu}_1 \\ \boldsymbol{\mu}_2 \end{pmatrix}, \tag{67}$$

$$C = \begin{pmatrix} C_{11} & C_{12} \\ C_{12}^{\mathrm{T}} & C_{22} \end{pmatrix}, \tag{68}$$

where the superscript T denotes the matrix transpose.

Theorem A. The conditional probability law for $\boldsymbol{W}_2$, given $\boldsymbol{W}_1 = \boldsymbol{w}_1$, is multivariate normal with conditional mean of

$$\boldsymbol{W}_2 = \boldsymbol{\mu}_2 + C_{12}^{\mathrm{T}} C_{11}^{-1}(\boldsymbol{w}_1 - \boldsymbol{\mu}_1) , \tag{69}$$

conditional covariance matrix

$$\boldsymbol{W}_2 = C_{22} - C_{12}^{\mathrm{T}} C_{11}^{-1} C_{12} . \tag{70}$$

Proof. See ref. [9], pp. 27-29.

Theorem B. Let $\boldsymbol{W}$ be an unconditional simulation of the random vector. That is, $\boldsymbol{W}$ follows a multivariate normal probability law with mean $\boldsymbol{\mu}$ and covariance matrix C. The vector $\widetilde{\boldsymbol{W}}_2$ defined by

$$\widetilde{\boldsymbol{W}}_2 = C_{12}^{\mathrm{T}} C_{11}^{-1}(\boldsymbol{w}_1 - \boldsymbol{W}_1) + \boldsymbol{W}_2 \tag{71}$$

will be a conditional simulation of $\boldsymbol{W}_2$, given $\boldsymbol{W}_1 = \boldsymbol{w}_1$. The mean vector and covariance matrix for $\boldsymbol{W}_2$ are the same as the conditional mean and covariance relations specified in theorem A and $\widetilde{\boldsymbol{W}}_2$ is a multivariate normal random vector.

Proof. Since every linear combination of multivariate normals is also multivariate normal [9], $\widetilde{\boldsymbol{W}}_2$ is a multivariate normal vector. Also

$$\mathrm{E}[\widetilde{\boldsymbol{W}}_2] = C_{12}^{\mathrm{T}} C_{11}^{-1}(\boldsymbol{w}_1 - \boldsymbol{\mu}_1) + \boldsymbol{\mu}_2 , \tag{72}$$

$$\begin{aligned} \operatorname{cov}(\widetilde{\boldsymbol{W}}_2) &= \mathrm{E}[\{\widetilde{\boldsymbol{W}}_2 - \mathrm{E}(\widetilde{\boldsymbol{W}}_2)\}\{\widetilde{\boldsymbol{W}}_2 - \mathrm{E}(\widetilde{\boldsymbol{W}}_2)\}^{\mathrm{T}}] = \\ &= \mathrm{E}[\{(\boldsymbol{W}_2 - \boldsymbol{\mu}_2) - C_{12}^{\mathrm{T}} C_{11}^{-1}(\boldsymbol{W}_1 - \boldsymbol{\mu}_1)\}\{(\boldsymbol{W}_2 - \boldsymbol{\mu}_2) - C_{12}^{\mathrm{T}} C_{11}^{-1}(\boldsymbol{W}_1 - \boldsymbol{\mu}_1)\}^{\mathrm{T}}] = \\ &= C_{22} - C_{12}^{\mathrm{T}} C_{11}^{-1} C_{12} . \qquad \textit{q.e.d.} \end{aligned} \tag{73}$$

For a time domain simulation by matrix multiplication, $\boldsymbol{W}_2'$ would contain the given values and $\boldsymbol{W}_1'$ would be simulated as a normal with the mean and covariance matrix listed in the theorem. Either of the previous techniques (triangular matrix or eigenvector) could be used to generate $\boldsymbol{W}_1'$. A mean-zero version of $\boldsymbol{W}_1'$ would be developed first, and then the appropriate mean vector as listed in the theorem would be added on.

It is usually easier to do the conditional simulation in frequency domain by matrix multiplication. The unknown and the given time series each have Fourier coefficients which are uncorrelated from frequency to frequency ($0 < m < N/2$), and have the covariance matrix for each m as previously discussed. The Fourier coefficient for the given time series can be easily developed by a fast-Fourier-transform computation. The coefficients for the unknown (*i.e.* to be simulated conditionally) time series are normally distributed with mean vector and covariance matrix as specified by the theorem. The simulation of these unknown coefficients, for each m, can be done by matrix multiplication. At each frequency, only a $2L$-component vector is involved if L is the number of unknown time series. This is a much smaller operation than the corresponding time domain simulations.

14. – Nonlinear wave simulation.

Historically, a linear wave theory is a solution to the wave equation based on a linearization of Bernoulli's equation to provide approximate boundary conditions assumed to hold at mean water level. By contrast, a nonlinear theory is a formulation in which the boundary conditions are not linearized and which are taken to hold on the actual free surface of the water.

When these definitions are transferred to a statistical wave theory, some adjustments of meaning result. The linear, directional, statistical wave theory is based on superposition of linear waves each with its own amplitude, frequency, direction of travel and independent random phase which is uniform in the interval $(0, 2)$. Asymptotically, as the number of wavelets increases indefinitely, the theory implies that the sea surface and associated kinematic quantities form a Gaussian field (*i.e.* follow a multivariate normal probability law). Thus nonlinearity for the directional, statistical, wave theory is interpreted as non-Gaussian probabilistic behavior.

In preparing a catalog of nonlinear wave behavior, one needs to keep clearly in mind whether the nonlinear behavior represents departures from the consequences of the linearized boundary conditions or whether the nonlinear behavior is a deviation from multivariate normal properties. From a statistical wave theory context, the second type of nonlinear behavior is generally implied. Hopefully, the modified nonlinear statistical theory will lead to kinematics which better satisfy boundary conditions.

A brief summary of nonlinear behavior in ocean wave follows.

1) Wave crests are more « peaked » and wave troughs are « flatter » than Gaussian theory would predict. The sea surface elevations above mean water level are skewed toward positive values. The wave kinematics also reflect this same « crest peakedness ».

2) Large waves may be grouped together in time of occurrence more than would be predicted from Gaussian theory.

3) Higher-order moments and their frequency domain counterparts (*i.e.* the bispectrum, trispectrum, etc.) may be different from the Gaussian properties.

4) The relation between wave number and frequency may be different from that predicted by linear wave theory, and in fact may be a probabilistic relation.

5) Waves may not travel through each other as predicted by linear superposition. There may be wave-wave interactions and transfer of energy from one frequency to another.

In connection with item 3) above, it should be noted that, for a Gaussian process, the spectral density completely determines all higher-order moments and the probabilistic behavior. If the process is non-Gaussian, the spectral density does not determine the probabilities. It characterizes only the second-order behavior. Two non-Gaussian processes with the same spectral density can have substantially different statistical behavior.

These various types of nonlinear behavior are very relevant to the decisions involved in choosing among the alternatives for computer simulation of statistical, nonlinear waves. Which properties need to be introduced into the simulation? Which can be left out? A substantial number of subjective judgements is required.

Just as there is no unique meaning to the phrase « a nonlinear wave theory », so there is no perfect way to simulate « nonlinear » wave fields. The various options available will be outlined under three headings: sea surface simulation procedures, simultaneous simulation of multiple wave properties and conditional simulation of associated wave properties.

15. – Sea surface simulation procedures.

The sea surface is a function of (x, y, t). Although in principle one could simulate $\eta(x, y, t)$ simultaneously over an (x, y, t) grid in 3-D space, it is not very efficient to do so. Here the emphasis will be on producing $\eta(x_0, y_0, t)$ as a function of t, with x_0, y_0 held fixed. This reduces the problem to a one-dimensional time series, or perhaps several time series if various space points are desired.

It is assumed in the following that $\eta_0(t) = \eta(x_0, y_0, t)$ is a mean-zero, stationary stochastic process with covariance and trivariance, respectively,

$$C(\tau) = \mathrm{E}[\eta_0(t)\eta_0(t+\tau)]\,, \tag{74}$$

$$T(\tau_1, \tau_2) = \mathrm{E}[\eta_0(t)\eta_0(t+\tau_1)\eta_0(t+\tau_2)]\,, \tag{75}$$

where $\mathrm{E}[\cdot]$ denotes the statistical expectation. The spectrum and bispectrum for η_0 will be

$$S(f) = \int_{-\infty}^{\infty} C(\tau) \exp[-i2\pi f\tau]\,\mathrm{d}\tau\,, \tag{76}$$

$$B(f_1, f_2) = \int_{-\infty}^{\infty} T(\tau_1, \tau_2) \exp[-i2\pi(f_1\tau_1 + f_2\tau_2)]\,\mathrm{d}\tau_1\,\mathrm{d}\tau_2\,. \tag{77}$$

A computer simulation of $\eta_0(t)$ which preserves the spectrum and bispectrum can be based on a quadratic filter applied to band-limited Gaussian white noise. Band-limited white noise, as used here, refers to a mean-zero stochastic process, $Z(t)$, having spectral density

$$S(F) = \begin{cases} 1/2B\,, & |f| \leqslant B\,, \\ 0\,, & |f| > B\,. \end{cases} \tag{78}$$

The quadratic nonlinear filter is mathematically represented by

$$\eta_0(t) = \int_{-\infty}^{\infty} w_1(\tau) Z(t-\tau)\,d\tau + \int_{-\infty}^{\infty}\int_{-\infty}^{\infty} w_2(\tau_1, \tau_2) Z(t-\tau_1) Z(t-\tau_2)\,d\tau_1\,d\tau_2\,. \tag{79}$$

It can be shown that the trivariance function and the bispectrum may be used to derive a reasonable $w_2(\tau_1, \tau_2)$ function [12]. The covariance function and spectrum then determine $w_1(\tau)$. Thus the procedure produces a sea surface time history with the correct spectrum and bispectrum.

A somewhat simpler procedure to introduce skewness into the sea surface (this is what a nonzero bispectrum affects) is to produce the sea surface by any linear-wave-theory procedure and then to make a zero-memory, Gaussian-to-gamma transformation. That is, if

$$\Phi(x; \mu, \sigma^2) = \int_{-\infty}^{x} \exp[-(u-\mu)^2/2\sigma^2]\,du/\sqrt{2\pi\sigma}\,, \tag{80}$$

$$G(y; c, \alpha, \beta) = \begin{cases} \int_c^y (v-c)^{\alpha-1} \exp[-(v-c)/\beta]\,dv\,, & \text{if } y > c\,, \\ 0\,, & \text{if } y \leqslant c\,. \end{cases} \tag{81}$$

The Gaussian-to-gamma transformation, as applied to a Gaussian sea surface profile, $w(t)$, having mean zero and variance σ^2, is

$$\eta_0(t) = G^{-1}\big(\Phi(w(t); 0, \sigma^2); c, \alpha, \beta\big)\,. \tag{82}$$

The nice feature of the gamma distribution function is that, if it is large enough while $\alpha\beta^2$ is held fixed, the probability law will asymptotically become normal. The mean, variance and skewness of the gamma specified previously are

$$\begin{cases} \text{mean} = c + \alpha\beta\,, \\ \text{variance} = \alpha\beta^2\,, \\ \text{skewness} = 2/\sqrt{\alpha}\,. \end{cases} \tag{83}$$

Thus, if

$$\left\{\begin{aligned} \alpha &= \text{skewness}/2\,,\\ \beta &= 2\sqrt{\text{variance}/\text{skewness}}\,,\\ c &= -\alpha\beta = -\,\text{skewness}\sqrt{\text{variance}}/2\,, \end{aligned}\right. \tag{84}$$

then the Gaussian-to-gamma transformation acting on a Gaussian sea surface profile will produce a mean-zero sea surface profile with the assigned (positive) skewness and the imposed variance.

The previous procedure will not introduce non-Gaussian wave grouping if that is the type of nonlinearity desired. Such behavior can be introduced by various amplitude modulation techniques. Let $Z(f)$ and $A(t)$ be two independent, mean-zero, random time series where the covariance function for $Z(t)$ dies to zero with increasing lag much quicker than $A(t)$. Then

$$\eta_0(t) = A^2(t)Z(t) \tag{85}$$

will behave as a sea surface which has large waves whenever $A^2(t)$ is large and small waves when $A^2(t)$ is small. The mean value of $\eta_0(t)$ will be zero and the variance will be $\mathrm{E}(A^4(t))\,\mathrm{E}(Z^2(t))$. The standard deviation of $\eta_0(t)$ (a measure of the average wave height) can be controlled through $\mathrm{E}(A^4(t))$ and $\mathrm{E}(Z^2(t))$ in the simulation.

In some simulation procedures, $Z(t)$ is given a special form which makes it more « wavelike »:

$$Z(t) = \cos\left(2\pi f_0 t - \Phi(t)\right)\,, \tag{86}$$

where $\Phi(t)$ is random time series which varies much more slowly than the period $T_0 = 1/f_0$. One advantage of this choice is that $Z(t)$ has unit amplitude and $A(t)$ becomes the envelope random function.

16. – Simultaneous simulation of multiple wave properties.

Rather than simulating a sea surface and then developing related kinematic properties by some method, it is appealing to directly model the complete wave system. However, this can only be done by making additional assumptions.

One method would be by assuming that nonlinear waves may be produced by superposition of Stokes waves

$$\eta(x, y, t) = \sum_{j=1}^{J}\sum_{n=1}^{N} a_{jn}\cos\left[n(k_j x\cos\theta_j + k_j y\sin\theta_j - 2\pi f_j t + \Phi_j)\right]\,, \tag{87}$$

where k_j and f_j have the usual Stokes interrelation, as do $(a_{j1}, a_{j2}, a_{j3}, \ldots, a_{jN})$. It is difficult to relate these amplitudes and phases to the spectrum, bispectrum, or other time series characteristics. However, the other kinematic properties are relatively easy to express in formula for a given set of $\{a_{jn}\}$ and $\{f_j\}$ by analogy to Stokes wave theory.

A somewhat similar procedure can be developed by simple departures from linear wave theory

$$\eta(x, y, t) = 2\int_0^\infty \int_0^{2\pi} \sqrt{S(f, \theta)\, d\theta\, df} \cos(kx\cos\theta + ky\sin\theta - 2\pi ft + \Phi). \tag{88}$$

Usually the Φ random values for each (f, θ) are assumed to be independent and uniformly distributed in $(0, 2\pi)$. A nonlinearity can be introduced by making Φ not independent from (f, θ) to other (f, θ) pairs. In simulations, correlated phases in (f, θ) space can be generated from a correlated Gaussian function on (f, θ) by setting

$$\Phi = 2\pi P(Z(f, \theta); 0, 1), \tag{89}$$

where the $P(z; 0, 1)$ is the Gaussian distribution function previously defined. As an additional randomness, k could be taken as random centered about the theoretical value. The exact form for the interdependence in $Z(f, \theta)$ and $k(f, \theta)$ would have to be studied further from actual wave data. However, once the appropriate nonstationary randomness for these stochastic fields is established, other kinematic properties could easily be simulated by analogy with linear theory.

The method of multiple quadratic filters could be extended from the similar treatment for sea surface simulations. Let $P_j(t)$ be the wave property at location (x_j, y_j, z_j). Suppose J wave properties $p_1(t), p_2(t), \ldots, p_J(t)$ are desired. Let $Z_1(t), Z_2(t), \ldots, Z_J(t)$ be independent, band-limited, Gaussian white noise and set

$$p_j(t) = \sum_{m=1}^{j} \int_{-\infty}^{\infty} w'_{jm}(\tau) Z_m(t-\tau)\, d\tau + \int_{-\infty}^{\infty}\int_{-\infty}^{\infty} w''_{jm}(\tau_1, \tau_2) Z_m(t-\tau_1) Z_m(t-\tau_2)\, d\tau_1 d\tau_2. \tag{90}$$

As before, the second-order filters can be selected to produce the required bispectra and cross-bispectra, and the first-order filters are then chosen to yield the correct spectra and cross-spectra. However, it is not clear exactly what spectra and bispectra should be used, except to estimate them from field measurements.

Finally, a perturbation of linear theory through relaxation techniques could be used to produce a « slightly » nonlinear wave theory. A wave field as characterized by the velocity potential $\varphi(x, y, z, t)$ could be produced by linear-statistical-theory simulations, and then $\eta(x, y, t)$ relaxed through several passes over the grid to produce a solution to Laplace's equation which gives a better satisfaction of the Bernoulli equation and the kinematic boundary conditions. This approach is somewhat speculative, but it appears to have some promise of producing meaningful nonlinear wave theory in the hydrodynamic sense.

17. – The above-mean-water-level problem.

A somewhat related difficulty in linear statistical theory and methods which start from linear theory is the « above-mean-water-level » problem. Linear theory assumes the velocity potential extends only up to mean water level. Some researchers have extended the potential field above mean water level by simply substituting the larger Z into the standard linear formulae. Others feel that the linear-theory predictions between sea floor and mean water level should be stretched to the interval between the sea floor and the $\eta(x, y, t)$ sea surface. This, however, leads to velocities which are too large beneath troughs. Perhaps an acceptable modification would be a two-stage procedure. Velocities along a given vertical would first be stretched and then adjusted linearly between the sea floor and the sea surface, so that the surface velocities satisfy Bernoulli's equation. This procedure seems reasonable except it is not clear how to treat the $\partial\varphi/\partial t$ term in Bernoulli's equation.

18. – Conditional simulation of kinematic properties.

Presumably the nonlinear waves are « almost » Gaussian. Hence one simulation scheme would be to introduce the slight nonlinearity through the profile, or sea surface nonlinearities only. Then the kinematics would be simulated conditionally as Gaussian processes, given the nonlinear sea surface previously developed. The theorems for conditional simulations were given previously. The methods are relatively straightforward, but require the manipulation of relatively large covariance matrices.

On the other hand, the time series do not have to extend over more than one or two waves, since the kinematics are keyed to the particular waves desired. One does not have to simulate a long series of waves and then look for larger waves approximating design conditions as is necessary in ordinary unconditional simulation.

Another advantage of the conditional-simulation procedure is that the wave profile at one or several measurement locations from field data can be used as the conditioning information. The kinematics consistent with these

TABLE I. – *Wave profiles from numerical solutions of wave theory* [13]. (H = wave height, η = elevation above mean water level, θ = phase within wave = $2\pi x/L$, where x = distance from

Case	H/H_B	H/L_0	H/d	L/L_0	d/L_0
1-*A*	0.25	0.000390	0.194829	0.119648	0.002000
1-*B*	0.50	0.000779	0.389717	0.128262	0.002000
1-*C*	0.75	0.001169	0.584426	0.137070	0.002000
1-*D*	1.00	0.001564	0.782113	0.146465	0.002000
2-*A*	0.25	0.000974	0.194887	0.186504	0.005000
2-*B*	0.50	0.001946	0.389164	0.199023	0.005000
2-*C*	0.75	0.002925	0.585097	0.210547	0.005000
2-*D*	1.00	0.003884	0.776719	0.222852	0.005000
3-*A*	0.25	0.001948	0.194817	0.259570	0.010000
3-*B*	0.50	0.003886	0.388630	0.276172	0.010000
3-*C*	0.75	0.005821	0.582125	0.291992	0.010000
3-*D*	1.00	0.007753	0.775326	0.308203	0.010000
4-*A*	0.25	0.003902	0.195117	0.358594	0.020000
4-*B*	0.50	0.007772	0.388580	0.379687	0.020000
4-*C*	0.75	0.011678	0.583909	0.401172	0.020000
4-*D*	1.00	0.015553	0.777657	0.422461	0.020000
5-*A*	0.25	0.009752	0.195032	0.541016	0.050000
5-*B*	0.50	0.019505	0.390096	0.566016	0.050000
5-*C*	0.75	0.029163	0.583254	0.597070	0.050000
5-*D*	1.00	0.038997	0.779945	0.627344	0.050000
6-*A*	0.25	0.018312	0.183115	0.718164	0.100002
6-*B*	0.50	0.036631	0.366304	0.743750	0.100002
6-*C*	0.75	0.054927	0.549254	0.783203	0.100002
6-*D*	1.00	0.073041	0.730398	0.824414	0.100002
7-*A*	0.25	0.031267	0.156335	0.899219	0.199999
7-*B*	0.50	0.062490	0.312451	0.931055	0.199999
7-*C*	0.75	0.093785	0.468925	0.981055	0.199999
7-*D*	1.00	0.124492	0.622465	1.035156	0.199999
8-*A*	0.25	0.041995	0.083990	1.013086	0.499998
8-*B*	0.50	0.083974	0.167949	1.059180	0.499998
8-*C*	0.75	0.125988	0.251977	1.125195	0.499998
8-*D*	1.00	0.168087	0.336176	1.193750	0.499998
9-*A*	0.25	0.042615	0.042615	1.017578	0.999996
9-*B*	0.50	0.085197	0.085197	1.065234	0.999996
9-*C*	0.75	0.128025	0.128025	1.132813	0.999996
9-*D*	1.00	0.169650	0.169650	1.210937	0.999996
10-*A*	0.25	0.042602	0.021301	1.017773	1.999993
10-*B*	0.50	0.085218	0.042609	1.065234	1.999993
10-*C*	0.75	0.127534	0.063767	1.134375	1.999993
10-*D*	1.00	0.170401	0.085201	1.222070	1.999993

H_B = breaking wave height, L = wave length, L_0 = deep-water wave length, d = water depth, crest horizontally.)

η/H values								
θ (degrees)								
0	10	20	30	50	75	100	130	180
0.910	0.600	0.199	0.009	−0.080	−0.090	−0.090	−0.090	−0.090
0.938	0.413	0.049	−0.039	−0.062	−0.061	−0.061	−0.062	−0.062
0.951	0.287	0.002	−0.042	−0.050	−0.048	−0.048	−0.049	−0.049
0.959	0.210	−0.014	−0.039	−0.043	−0.040	−0.040	−0.041	−0.041
0.857	0.713	0.424	0.177	−0.060	−0.129	−0.141	−0.143	−0.143
0.904	0.606	0.228	0.031	−0.079	−0.095	−0.096	−0.096	−0.096
0.927	0.470	0.116	−0.014	−0.068	−0.072	−0.072	−0.073	−0.073
0.944	0.341	0.056	−0.027	−0.055	−0.056	−0.056	−0.056	−0.056
0.799	0.723	0.538	0.329	0.025	−0.135	−0.183	−0.199	−0.201
0.865	0.692	0.387	0.153	−0.061	−0.123	−0.133	−0.135	−0.135
0.898	0.596	0.251	0.059	−0.072	−0.099	−0.102	−0.102	−0.102
0.922	0.460	0.154	0.015	−0.064	−0.077	−0.078	−0.078	−0.078
0.722	0.682	0.575	0.431	0.146	−0.089	−0.204	−0.266	−0.278
0.810	0.715	0.506	0.294	0.010	−0.131	−0.174	−0.188	−0.190
0.858	0.667	0.383	0.173	−0.041	−0.119	−0.137	−0.141	−0.142
0.889	0.583	0.284	0.101	−0.055	−0.101	−0.110	−0.112	−0.111
0.623	0.603	0.548	0.465	0.257	0.007	−0.177	−0.334	−0.377
0.716	0.673	0.562	0.420	0.150	−0.077	−0.196	−0.269	−0.284
0.784	0.687	0.498	0.318	0.059	−0.106	−0.175	−0.209	−0.216
0.839	0.582	0.363	0.207	0.012	−0.096	−0.137	−0.156	−0.161
0.571	0.558	0.519	0.458	0.293	0.061	−0.144	−0.360	−0.429
0.642	0.617	0.549	0.452	0.232	−0.009	−0.178	−0.320	−0.358
0.713	0.657	0.530	0.390	0.146	−0.061	−0.180	−0.266	−0.287
0.782	0.594	0.417	0.279	0.079	−0.071	−0.151	−0.205	−0.218
0.544	0.533	0.501	0.450	0.306	0.087	−0.124	−0.370	−0.456
0.593	0.576	0.527	0.453	0.270	0.038	−0.152	−0.345	−0.407
0.653	0.616	0.528	0.420	0.207	−0.010	−0.165	−0.305	−0.347
0.724	0.580	0.443	0.326	0.137	−0.033	−0.147	−0.247	−0.276
0.534	0.524	0.494	0.447	0.310	0.097	−0.116	−0.373	−0.466
0.570	0.555	0.514	0.450	0.285	0.061	−0.138	−0.356	−0.430
0.611	0.586	0.521	0.434	0.243	0.025	−0.150	−0.329	−0.389
0.677	0.572	0.456	0.355	0.177	−0.002	−0.140	−0.278	−0.323
0.534	0.523	0.494	0.446	0.310	0.097	−0.116	−0.373	−0.446
0.569	0.554	0.513	0.450	0.286	0.062	−0.137	−0.356	−0.431
0.609	0.585	0.522	0.436	0.245	0.026	−0.149	−0.330	−0.391
0.661	0.595	0.483	0.375	0.187	−0.002	−0.146	−0.291	−0.339
0.533	0.523	0.494	0.446	0.310	0.097	−0.116	−0.374	−0.467
0.569	0.554	0.513	0.450	0.286	0.062	−0.137	−0.356	−0.413
0.608	0.584	0.521	0.435	0.245	0.026	−0.149	−0.331	−0.392
0.657	0.603	0.496	0.385	0.189	−0.004	−0.148	−0.294	−0.343

profile time series and the imposed directional wave spectrum $S(f, \theta)$ can then be simulated conditionally. Alternatively stream function wave profiles from the work of Dean [13] as summarized in table I provide the nonlinear wave profile.

19. – Summary and conclusions.

1) A variety of procedures for the computer simulation of ocean wave time series have been presented. The methods assume that the wave properties follow a multivariate normal probability law whose parameters are related to the directional wave spectral density in accordance with linear wave theory. Covariance stationarity is also assumed in most of the procedures.

2) Although the frequency domain simulation procedures are more theoretically complicated than the time domain methods, the savings in computer time with the frequency-based procedures justify the extra effort. The frequency domain methods are ten to a hundred times faster than the time domain techniques, primarily because of the speed of the fast-Fourier-transform algorithm. The computer time savings increase with increasing length of the simulation sequences.

3) The time domain procedures are occasionally useful in applications involving short-length simulations or real-time, fluid-structural interactions.

4) A fundamental problem is the simulation of many simultaneous wave properties (*e.g.* all the components of velocity and acceleration at many loading points in a fixed-leg drilling platform). Such problems strain computer capacity in storage and time and require much programmer ingenuity. Simulations of, perhaps up to fifty, simultaneous properties can be performed with frequency domain techniques based on multiplication of normal random numbers by a triangular matrix. Simulations of more properties than that would probably require either the eigenvector matrix method or the random-phase model, both in frequency domain.

5) The methods based on filtered white noise, either in time or in frequency, are very analogous to techniques using the multiplication by a triangular matrix. Computer precision and round-off problems cause severe difficulties if more than fifty simultaneous properties are being simulated. Even for fewer properties, double or triple precision may be necessary to avoid encountering square roots of negative numbers.

6) A major difficulty in the use of simulations in design is that most methods require the simulation of a long wave record, followed by a search for a wave which has the desired design characteristics (height, period, etc.).

New techniques of conditional simulation of short-length records provide a way around this problem. A wave profile from one of the deterministic theories is assumed to be present at a wave staff and the other wave properties are simulated conditionally, given that profile holds and that the specified wave directional spectrum is present. The resulting wave property sequences will illustrate the statistical variability consistent with the presence of that wave profile. Techniques for conditional simulation based on multivariate normal probability theory are outlined. If many wave properties are involved, the eigenvector matrix method in frequency domain appears the most efficient.

7) Nonlinearities of various types may be introduced into the simulations; however, a clear understanding of the types of nonlinearity desired must be developed.

8) All of the statistical procedures outlined for simulating ocean waves are useful within one context or another. There is no single best way. Unavoidably the oceanographer and the engineer will need to study the application and make a choice within that framework.

REFERENCES

[1] V. J. Cardone, W. J. Pierson and E. G. Ward: *J. Petr. Tech.*, **25**, 385 (1976).

[2] L. E. Borgman: *Directional spectra from wave sensors*, in *Ocean Wave Climate*, Vol. **8** (New York, N. Y., 1979), p. 269.

[3] G. Z. Forristall, E. G. Ward, L. E. Borgman and V. J. Cardone: *Storm wave kinematics*, in *Proceedings of the X Technical Conference* (Houston, Tex., 1978).

[4] G. A. Forristall, E. G. Ward, L. E. Borgman and V. J. Cardone: *J. Phys. Oceanography*, **8**, 888 (1978).

[5] G. Birkhoff and S. MacLane: *A Survey of Modern Algebra*, 3rd edition (New York, N. Y., 1965), p. 22.

[6] M. Zelen and N. C. Severo: *Probability functions*, in *Handbook of Mathematical Functions* (New York, N. Y., 1965), p. 927.

[7] E. Scheuer and D. S. Stoller: *Technometrics*, **4**, 278 (1962).

[8] L. E. Borgman: *J. Waterways Harb. Div. Am. Soc. Civ. Eng.*, **95**, 557 (1969).

[9] T. W. Anderson: *An Introduction to Multivariate Statistical Analysis* (New York, N. Y., 1958), p. 19.

[10] L. E. Borgman: *Statistical Properties of Fast Fourier Transform Coefficients Computed from Real-Valued, Covariance-Stationary, Periodic, Random Sequences*, Tech. paper 76-9 (Ft. Belvoir, Va., 1976).

[11] A. G. Journel: *Econ. Geology*, **69**, 673 (1974).

[12] L. J. Tick: *Technometrics*, **3**, 563 (1961).

[13] R. G. Dean: *Evaluation and Development of Water Wave Theories for Engineering Applications*, Spec. Rpt. No. 1 (Fort Belvoir, Va., 1974).

Statistical Precision of Directional Spectrum Estimation with Data from a Tilt-and-Roll Buoy.

L. E. BORGMAN and R. L. HAGAN
University of Wyoming - Laramie, Wy.

T. KUIK
Technisch Physische Dienst - The Netherlands

1. – Introduction.

The tilt-and-roll buoy measures, as a function of time, the elevation of the sea surface above mean water level (η) and the x- and y-components of the slope of the sea surface (η_x and η_y, respectively). In this study, it is supposed that the three time series $\eta(t)$, $\eta_x(t)$ and $\eta_y(t)$ represent the values at a single (x, y) position within an invariant co-ordinate system. That is, orbital motion or twisting of the buoy has been satisfactorily eliminated by instrumentation procedures or by computational corrections. It is also presumed that $\eta(t)$, $\eta_x(t)$, $\eta_y(t)$ behave in accordance with linear statistical wave theory as a stationary trivariate Gaussian set of time series, with each series having mean zero.

Various directional estimates can be computed from the auto- and cross-spectra estimates made from the data for the three time series [1]. With the Gaussian assumption, various mathematical procedures can be used to develop quantitative measures of the bias and error in the directional estimates. The quantitative reliability measures may, thus, be used to investigate bias and error of the fundamental directional statistics under various hypothesized sets of « true » underlying wave conditions.

As an adjunct to these quantitative reliability measures, mathematical procedures are developed for digital simulation of the (η, η_x, η_y) time series for given « population » wave conditions.

2. – Basic data.

The basic data were taken as

$$\eta(n\Delta t) = \text{water level elevation above mean water level,}$$
$$\eta_x(n\Delta t) = x\text{-component of sea surface slope,}$$
$$\eta_y(n\Delta t) = y\text{-component of sea surface slope,}$$

all presumed to have been measured at a fixed (x, y) location for $n = 0, 1, 2, 3, \ldots, N-1$. The quantity Δt is the time increment.

3. – Data Fourier transforms.

The frequency domain versions of η, η_x and η_y are represented by the complex sequences

$$(1)\qquad \begin{cases} U_m - iV_m \;\;= \Delta t \sum\limits_{n=0}^{N-1} \eta(n\Delta t) \exp[-i2\pi mn/N], \\ U_{xm} - iV_{xm} = \Delta t \sum\limits_{n=0}^{N-1} \eta_x(n\Delta t) \exp[-i2\pi mn/N], \\ U_{ym} - iV_{ym} = \Delta t \sum\limits_{n=0}^{N-1} \eta_y(n\Delta t) \exp[-i2\pi mn/N] \end{cases}$$

for $m = 0, 1, 2, 3, \ldots, N-1$. If only one frequency is involved in algebraic manipulations, the subscript m will be deleted.

4. – Population functions.

The time series are assumed to all have mean zero and be stationary. Hence the covariance properties for the three series are

$$(2)\qquad \begin{cases} C_{\eta\eta}(\tau) \;\;= \mathscr{E}[\eta(t)\eta(t+\tau)], \\ C_{\eta_x\eta_x}(\tau) = \mathscr{E}[\eta_x(t)\eta_x(t+\tau)], \\ C_{\eta_y\eta_y}(\tau) = \mathscr{E}[\eta_y(t)\eta_y(t+\tau)], \\ C_{\eta\eta_x}(\tau) \;= \mathscr{E}[\eta(t)\eta_x(t+\tau)], \\ C_{\eta\eta_y}(\tau) \;= \mathscr{E}[\eta(t)\eta_y(t+\tau)], \\ C_{\eta_x\eta_y}(\tau) = \mathscr{E}[\eta_x(t)\eta_y(t+\tau)]. \end{cases}$$

In the above, the symbol $\mathscr{E}[\;]$ denotes the statistical expectation operation.

The population spectral functions are given by

$$(3)\qquad \begin{cases} S(f) \;\;= \int\limits_{-\infty}^{\infty} C_{\eta\eta}(\tau) \exp[-i2\pi f\tau]\, \mathrm{d}\tau, \\ S_x(f) = \int\limits_{-\infty}^{\infty} C_{\eta_x\eta_x}(\tau) \exp[-i2\pi f\tau]\, \mathrm{d}\tau, \\ S_y(f) = \int\limits_{-\infty}^{\infty} C_{\eta_y\eta_y}(\tau) \exp[-i2\pi f\tau]\, \mathrm{d}\tau, \end{cases}$$

$$
\left\{
\begin{aligned}
c_x(f) - iq_x(f) &= \int_{-\infty}^{\infty} C_{\eta\eta_x}(\tau) \exp[-i2\pi f\tau]\,d\tau\,, \\
c_y(f) - iq_y(f) &= \int_{-\infty}^{\infty} C_{\eta\eta_y}(\tau) \exp[-i2\pi f\tau]\,d\tau\,, \\
c_{xy}(f) - iq_{xy}(f) &= \int_{-\infty}^{\infty} C_{\eta_x\eta_y}(\tau) \exp[-i2\pi f\tau]\,d\tau\,.
\end{aligned}
\right. \tag{3}
$$

5. – Spectral estimates.

The estimates for the population spectral functions will be denoted by the same symbols except a ^ will be placed above it. Let

$$\Delta f = (N\,\Delta t)^{-1} \tag{4}$$

be the frequency increment and

$$f_0 = m_0 \Delta f \tag{5}$$

be the frequency associated with the subscript m_0.

Finally let M indicate the set of values of the subscript m, centered at m_0 and containing ν consecutive elements. The spectral estimates may be written

$$
\left\{
\begin{aligned}
\hat{S}(f_0) &= (1/\nu)\sum_M (U_m^2 + V_m^2)/N\Delta t\,, \\
\hat{S}_x(f_0) &= (1/\nu)\sum_M (U_{xm}^2 + V_{xm}^2)/N\Delta t\,, \\
\hat{S}_y(f_0) &= (1/\nu)\sum_M (U_{ym}^2 + V_{ym}^2)/N\Delta t\,, \\
\hat{c}_x(f_0) - i\hat{q}_x(f_0) &= (1/\nu)\sum_M (U_m + iV_m)(U_{xm} - iV_{xm})/N\Delta t\,, \\
\hat{c}_y(f_0) - i\hat{q}_y(f_0) &= (1/\nu)\sum_M (U_m + iV_m)(U_{ym} - iV_{ym})/N\Delta t\,, \\
\hat{c}_{xy}(f_0) - i\hat{q}_{xy}(f_0) &= (1/\nu)\sum_M (U_{xm} + iV_{xm})(U_{ym} - iV_{ym})/N\Delta t\,.
\end{aligned}
\right. \tag{6}
$$

6. – Consequences of statistical linear wave theory.

According to linear statistical wave theory, the three series may be written in pseudointegral form as

$$
\begin{bmatrix} \eta(t) \\ \eta_x(t) \\ \eta_y(t) \end{bmatrix} = 2\int_0^{\infty}\int_0^{2\pi} \sqrt{S(f,\theta)\,d\theta\,df} \begin{bmatrix} \cos\Psi \\ -k\cos\theta\sin\Psi \\ -k\sin\theta\sin\Psi \end{bmatrix}, \tag{7}
$$

where

$$\Psi = kx\cos\theta + ky\sin\theta - 2\pi ft + \Phi(\theta, f)\,, \tag{8}$$

$$(2\pi f)^2 = gk \operatorname{tgh} kd \tag{9}$$

and d is the water depth.

Also Φ is taken as being uniformly distributed on the radian interval $(0, 2\pi)$ and as being independent for each frequency f and direction θ. The function $S(f, \theta)$ in the formula represents the two-sided directional spectral density. That is

$$2\int_0^\infty\int_0^{2\pi} S(f, \theta)\,d\theta\,df = \text{variance of } \eta(t)\,. \tag{10a}$$

The frequency spectral density will be indicated by the same symbol except it will have the one argument f instead of the two arguments (f, θ). Thus

$$S(f) = \int_0^{2\pi} S(f, \theta)\,d\theta\,. \tag{10b}$$

The spreading function, $D(\theta; f)$, is a unit-area function of θ, at each frequency, which gives the distribution of variance over angle for that frequency:

$$D(\theta; f) = S(f, \theta)/S(f)\,. \tag{10c}$$

Usually the argument, f, in $D(\theta; f)$ is dropped to simplify the formulae. It is always recognized as being implicitly present.

It is also convenient to work primarily with the Fourier-series coefficients for $D(\theta)$, since $D(\theta)$ is periodic with period 2π radians. Let a_j and b_j be these coefficients which are also implicitly remembered to be functions of frequency. Hence

$$D(\theta) = \frac{1}{2\pi} + \sum_{j=0}^{\infty} (a_j \cos j\theta + b_j \sin j\theta)\,. \tag{11}$$

As a consequence of (7), it can be shown that theoretically (if $x = y = 0$ is presumed)

$$\begin{bmatrix} C_{\eta\eta}(\tau) \\ C_{\eta_x\eta_x}(\tau) \\ C_{\eta_y\eta_y}(\tau) \\ C_{\eta\eta_x}(\tau) \\ C_{\eta\eta_y}(\tau) \\ C_{\eta_x\eta_y}(\tau) \end{bmatrix} = 2\int_0^\infty\int_0^{2\pi} S(f, \theta) \begin{bmatrix} \cos(2\pi f\tau) \\ k^2\cos^2\theta\cos(2\pi f\tau) \\ k^2\sin^2\theta\cos(2\pi f\tau) \\ k\cos\theta\sin(2\pi f\tau) \\ k\sin\theta\sin(2\pi f\tau) \\ k^2\sin\theta\cos\theta\cos(2\pi f\tau) \end{bmatrix} d\theta\,df\,. \tag{12}$$

The relation between the cross-covariance function between any two random variables (X, Y) and corresponding co- and quad-spectral densities is

$$C_{XY}(\tau) = 2\int_0^\infty c_{XY}(f)\cos(2\pi f\tau)\,\mathrm{d}f + 2\int_0^\infty q_{XY}(f)\sin(2\pi f\tau)\,\mathrm{d}f\,. \tag{13}$$

The uniqueness of Fourier transforms enables (13) to be used with (12) to extract the theoretical cross-spectral relations. They become, after substituting $S(f)D(\theta)$ for $S(f,\theta)$,

$$S_{\eta\eta}(f) = S(f)\,, \tag{14}$$

$$\begin{pmatrix} S_{\eta_x\eta_x}(f) \\ S_{\eta_y\eta_y}(f) \\ c_{\eta\eta_x}(f) - iq_{\eta\eta_x}(f) \\ c_{\eta\eta_y}(f) - iq_{\eta\eta_y}(f) \\ c_{\eta_x\eta_y}(f) - iq_{\eta_x\eta_y}(f) \end{pmatrix} = S(f)\begin{pmatrix} k^2 \\ k^2 \\ -ik \\ -ik \\ k^2 \end{pmatrix}\int_0^{2\pi} D(\theta)\begin{pmatrix} \cos^2\theta \\ \sin^2\theta \\ \cos\theta \\ \sin\theta \\ \sin\theta\cos\theta \end{pmatrix}\mathrm{d}\theta\,. \tag{15}$$

These expressions can be further reduced, after introducing the Fourier coefficients for $D(\theta)$, to

$$S_{\eta_x\eta_x}(f) = S(f)k^2(1+\pi a_2)/2\,, \tag{16}$$

$$S_{\eta_y\eta_y}(f) = S(f)k^2(1-\pi a_2)/2\,, \tag{17}$$

$$c_{\eta\eta_x}(f) = 0\,, \tag{18}$$

$$q_{\eta\eta_x}(f) = S(f)k\pi a_1\,, \tag{19}$$

$$c_{\eta\eta_y}(f) = 0\,, \tag{20}$$

$$q_{\eta\eta_y}(f) = S(f)k\pi b_1\,, \tag{21}$$

$$c_{\eta_x\eta_y}(f) = S(f)k^2\pi b_2/2\,, \tag{22}$$

$$q_{\eta_x\eta_y}(f) = 0\,. \tag{23}$$

For convenience in subsequent algebra, let

$$\begin{cases} A = \pi k a_1\,, \\ B = \pi k b_1\,, \\ C = k^2(1+\pi a_2)/2\,, \\ D = k^2(1-\pi a_2)/2\,, \\ E = \pi k^2 b_2/2\,. \end{cases} \tag{24}$$

Then

$$\frac{1}{S(f)}\begin{pmatrix} S_{\eta_x\eta_x}(f) \\ S_{\eta_y\eta_y}(f) \\ q_{\eta\eta_x}(f) \\ q_{\eta\eta_y}(f) \\ c_{\eta_x\eta_y}(f) \end{pmatrix} = \begin{pmatrix} C \\ D \\ A \\ B \\ E \end{pmatrix}. \tag{25}$$

Another consequence of linear statistical theory is that η, η_x, η_y are multivariate normally distributed. As a result of (1) it can be shown that, for $0 < m < N/2$, U_m, V_m, U_{xm}, U_{ym}, V_{ym} are multivariate normally distributed. Furthermore, these sets of six random quantities are independent for the different m in the interval $0 < m < N/2$, and U_m is independent of V_m, U_{xm} is independent of V_{xm}, and U_{ym} is independent of V_{ym}. In the particular case of the variables derived for the tilt-and-roll buoy, the triple (U_m, V_{xm}, V_{ym}) is also independent of the triple (V_m, U_{xm}, U_{ym}). Both triples are trivariate normal, with mean zero and covariance matrices given by

$$\begin{cases} \operatorname{cov}\begin{pmatrix} U \\ V_x \\ V_y \end{pmatrix} = \dfrac{SN\Delta t}{2}\begin{pmatrix} 1 & A & B \\ A & C & E \\ B & E & D \end{pmatrix}, \\ \operatorname{cov}\begin{pmatrix} V \\ U_x \\ U_y \end{pmatrix} = \dfrac{SN\Delta t}{2}\begin{pmatrix} 1 & -A & -B \\ -A & C & E \\ -B & E & D \end{pmatrix}. \end{cases} \tag{26}$$

The subscript m has been suppressed for notational simplicity, since the above relation holds for any $0 < m < N/2$.

The FFT coefficients $U_{N-m} - iV_{N-m}$, $U_{x,N-m} - iV_{x,N-m}$ and $U_{y,N-m} - iV_{y,N-m}$ are defined as the complex conjugates of the corresponding coefficients $U_m - iV_m$, $U_{xm} - iV_{xm}$ and $U_{ym} - iV_{ym}$. This leaves only the complex FFT coefficients at $m = 0$ and $m = N/2$ unspecified. The imaginary parts at these frequencies are always zero as a consequence of η, η_x, η_y being real-valued:

$$V_0 = V_{x0} = V_{y0} = V_{N/2} = V_{x,N/2} = V_{y,N/2} = 0. \tag{27}$$

The real part at $m = N/2$ is taken as zero, because the analysis is arranged so that no wave energy remains at frequencies as high as the Nyquist frequency. Hence

$$U_{N/2} = U_{x,N/2} = U_{y,N/2} = 0. \tag{28}$$

Finally, it is convenient to arbitrarily use

$$U_0 = U_{x0} = U_{y0} = 0 \tag{29}$$

as a means of forcing the time series (η, η_x, η_y) to have mean zero.

This specifies the random nature of all the FFT coefficients represented by (1) for $0 \leqslant m < N$.

7. – Simulation theory.

There is a variety of ways to simulate Gaussian time series, either in time domain or in frequency domain. The fastest procedures relative to the computer are based on frequency domain operations, followed by a fast Fourier transform to the time domain. The method recommended here produces by matrix multiplication (with a triangular matrix) the two trivariate normals whose characteristics are specified in (26). A mean-zero trivariate normal can be generated from three independent standard normal random numbers (Z_1, Z_2, Z_3) by the operation

$$\begin{pmatrix} X_1 \\ X_2 \\ X_3 \end{pmatrix} = \begin{pmatrix} a_{11} & 0 & 0 \\ a_{21} & a_{22} & 0 \\ a_{31} & a_{32} & a_{33} \end{pmatrix} \begin{pmatrix} Z_1 \\ Z_2 \\ Z_3 \end{pmatrix}. \tag{30}$$

The matrix elements a_{ij} are selected to produce the variances and covariances [2]. Thus

$$\begin{cases} \mathscr{E}(X_1^2) = a_{11}^2 & \text{(solve for } a_{11}), \\ \mathscr{E}(X_1 X_2) = a_{11} a_{21} & \text{(solve for } a_{21}), \\ \mathscr{E}(x_2^2) = a_{21}^2 + a_{22}^2 & \text{(solve for } a_{22}), \\ \mathscr{E}(X_1 X_3) = a_{11} a_{31} & \text{(solve for } a_{31}), \\ \mathscr{E}(X_2 X_3) = a_{21} a_{31} + a_{22} a_{32} & \text{(solve for } a_{32}), \\ \mathscr{E}(X_3^2) = a_{31}^2 + a_{32}^2 + a_{33}^2 & \text{(solve for } a_{33}). \end{cases} \tag{31}$$

For the two independent trivariate normals specified in (26), the procedure gives $(0 < m < N/2)$

$$\begin{pmatrix} U_m \\ V_{xm} \\ V_{ym} \end{pmatrix} = \sqrt{\frac{SN\,\Delta t}{2}} \begin{pmatrix} 1 & 0 & 0 \\ c_1 & c_2 & 0 \\ c_3 & c_4 & c_5 \end{pmatrix} \begin{pmatrix} Z_1 \\ Z_2 \\ Z_3 \end{pmatrix}, \tag{32}$$

$$\begin{pmatrix} V_m \\ U_{xm} \\ U_{ym} \end{pmatrix} = \sqrt{\frac{SN\,\Delta t}{2}} \begin{pmatrix} 1 & 0 & 0 \\ -c_1 & c_2 & 0 \\ -c_3 & c_4 & c_5 \end{pmatrix} \begin{pmatrix} Z_1 \\ Z_2 \\ Z_3 \end{pmatrix}, \tag{33}$$

where

$$\begin{cases} c_1 = A\,, \\ c_2 = \begin{vmatrix} 1 & A \\ A & C \end{vmatrix}^{\frac{1}{2}}, \\ c_3 = B\,, \\ c_4 = \begin{vmatrix} 1 & A \\ B & E \end{vmatrix} \Big/ c_2\,, \\ c_5 = \begin{vmatrix} 1 & A & B \\ A & C & E \\ B & E & D \end{vmatrix}^{\frac{1}{2}} \Big/ c_2\,. \end{cases} \tag{34}$$

The simulation procedure is as follows:

1) Produce six independent mean-zero, variance-one, normal random numbers for each $0 < m < N/2$.

2) Use (32) through (34) to convert these to $(U_m, V_m, U_{xm}, V_{xm}, U_{ym}, V_{ym})$.

3) For $0 < m < N/2$, define

$$(U_{N-m}, V_{N-m}, U_{x,N-m}, V_{x,N-m}, U_{y,N-m}, V_{y,N-m}) = \\ = (U_m, -V_m, U_{xm}, -V_{xm}, U_{ym}, -V_{ym})\,. \tag{35}$$

That is, the first set consists of the complex conjugates of the second set.

4) Set

$$\begin{cases} (U_0, V_0, U_{x0}, V_{x0}, U_{y0}, V_{y0}) = (0, 0, 0, 0, 0, 0)\,, \\ (U_{N/2}, V_{N/2}, U_{x,N/2}, V_{x,N/2}, U_{y,N/2}, V_{y,N/2}) = (0, 0, 0, 0, 0, 0)\,. \end{cases} \tag{36}$$

5) Steps 1) through 4) define the FFT coefficients for $\eta(n\Delta t)$, $\eta_x(n\Delta t)$ and $\eta_y(n\Delta t)$ as defined in eq. (1), for $0 \leq m < N$. The three sequences of complex numbers in frequency domain are processed with the fast Fourier trans-

form to produce the corresponding time series:

$$(37)\qquad \begin{cases} \eta(n\,\Delta t) = \Delta f \sum_{m=0}^{N-1} (U_m - iV_m) \exp[i2\pi mn/N]\,, \\ \eta_x(n\,\Delta t) = \Delta f \sum_{m=0}^{N-1} (U_{xm} - iV_{xm}) \exp[i2\pi mn/N]\,, \\ \eta_y(n\,\Delta t) = \Delta f \sum_{m=0}^{N-1} (U_{ym} - iV_{ym}) \exp[i2\pi mn/N]\,. \end{cases}$$

Alternately, the frequency domain simulations can be substituted in (6) to obtain directly the various spectral estimates which would result from the simulation. The main advantage of reverting completely to the time domain is that this permits any data management procedures (filters, etc.) to be applied to the simulated data just as though it represented measured data. The advantage of staying in frequency domain and going directly to the spectral estimates lies in the greater computational speed associated with this procedure.

8. – Characteristic function for the spectral estimates.

The characteristic function for a random vector $\boldsymbol{X}$ will be specified by $\Phi_X(\boldsymbol{u})$, where $\boldsymbol{u}$ is a parameter vector with the same number of components as $\boldsymbol{X}$. The vector $\boldsymbol{u}$ is the vector of parameters introduced in a type of Fourier transform defined by

$$(38)\qquad \Phi_X(\boldsymbol{u}) = \mathscr{E}[\exp[i\boldsymbol{u}^{\mathrm{T}}\boldsymbol{X}]]\,,$$

where $\boldsymbol{u}^{\mathrm{T}}$ is the transpose of the vector $\boldsymbol{u}$. The joint characteristic functions are developed separately for the two vectors.

From (6), the spectral lines of importance in the present problem are

$$(39)\qquad \begin{cases} \tilde{S} = (U^2 + V^2)/N\,\Delta t\,, \\ \tilde{S}_x = (U_x^2 + V_x^2)/N\,\Delta t\,, \\ \tilde{S}_y = (U_y^2 + V_y^2)/N\,\Delta t\,, \\ \tilde{q}_x = (UV_x - VU_x)/N\,\Delta t\,, \\ \tilde{q}_y = (UV_y - VU_y)/N\,\Delta t\,, \\ \tilde{c}_{xy} = (U_xU_y + V_xV_y)/N\,\Delta t\,. \end{cases}$$

The characteristic function for these six quantities will now be developed [3, 4].

Let

$$\tilde{\boldsymbol{S}} = \begin{pmatrix} \tilde{S} \\ \tilde{S}_x \\ \tilde{S}_y \\ \tilde{q}_x \\ \tilde{q}_y \\ \tilde{c}_{xy} \end{pmatrix}, \qquad \boldsymbol{u} = \begin{pmatrix} u_1 \\ u_2 \\ u_3 \\ u_4 \\ u_5 \\ u_6 \end{pmatrix} \tag{40}$$

be the vector of spectral lines and the parameter vector. Then

$$\begin{aligned} \Phi_{\tilde{\boldsymbol{S}}}(\boldsymbol{u}) &= \mathscr{E}[\exp[i\boldsymbol{u}^{\mathrm{T}}\tilde{\boldsymbol{S}}]] = \\ &= \mathscr{E}[\exp[iu_1(U^2+V^2)/N\Delta t + iu_2(U_x^2+V_x^2)/N\Delta t + \\ &+ iu_3(U_y^2+V_y^2)/N\Delta t + iu_4(UV_x - VU_x)/N\Delta t + \\ &+ iu_5(UV_y - VU_y)/N\Delta t + u_6(U_xU_y + V_xV_y)/N\Delta t]] = \\ &= \mathscr{E}\left[\exp\left[\frac{i(U, V_x, V_y)}{N\Delta t}\begin{pmatrix} u_1 & u_4/2 & u_5/2 \\ u_4/2 & u_2 & u_6/2 \\ u_5/2 & u_6/2 & u_3 \end{pmatrix}\begin{pmatrix} U \\ V_x \\ V_y \end{pmatrix} + \right.\right. \\ &\left.\left. + \frac{i(V, U_x, U_y)}{N\Delta t}\begin{pmatrix} u_1 & -u_4/2 & -u_5/2 \\ -u_4/2 & u_2 & u_6/2 \\ -u_5/2 & u_6/2 & u_3 \end{pmatrix}\begin{pmatrix} V \\ U_x \\ U_y \end{pmatrix}\right]\right]. \end{aligned} \tag{41}$$

For any three-component vector $\boldsymbol{x}$,

$$\int_{-\infty}^{\infty}\int_{-\infty}^{\infty}\int_{-\infty}^{\infty} \exp[-\boldsymbol{x}^{\mathrm{T}} M \boldsymbol{x}/2]\,\mathrm{d}x = \frac{(2\pi)^{\frac{3}{2}}}{\sqrt{|M|}}. \tag{42}$$

The expectation in (41) can be expressed as the product of two such integrals. Let B_1 be the first matrix in (41), while B_2 is the second matrix. Also let A_1 be the covariance matrix for (U, V_x, V_y) and let A_2 be the covariance matrix for (V, U_x, U_y). Then

$$\Phi_{\tilde{\boldsymbol{S}}}(\boldsymbol{u}) = G(\boldsymbol{u})H(\boldsymbol{u}), \tag{43}$$

where

$$\begin{cases} G(\boldsymbol{u}) = \dfrac{\int_{-\infty}^{\infty}\int_{-\infty}^{\infty}\int_{-\infty}^{\infty} \exp[-\boldsymbol{x}^{\mathrm{T}}(A_1^{-1} - (2i/N\Delta t)B_1)\,\boldsymbol{x}/2]\,\mathrm{d}\boldsymbol{x}}{(2\pi)^{\frac{3}{2}}\sqrt{|A_1|}}, \\ H(\boldsymbol{u}) = \displaystyle\int_{-\infty}^{\infty}\int_{-\infty}^{\infty}\int_{-\infty}^{\infty} \exp\left[-\boldsymbol{x}^{\mathrm{T}}\left(A_2^{-1} - \frac{2i}{N\Delta t}B_2\right)\boldsymbol{x}/2\right]\mathrm{d}\boldsymbol{x}. \end{cases} \tag{44}$$

It follows from (42) that

$$\left\{\begin{aligned} G(\boldsymbol{u}) &= \{|A_1|\,|A_1^{-1} - 2iB_1/N\Delta t|\}^{-\frac{1}{2}}\,, \\ H(\boldsymbol{u}) &= \{|A_2|\,|A_2^{-1} - 2iB_2/N\Delta t|\}^{-\frac{1}{2}}\,. \end{aligned}\right. \tag{45}$$

So

$$\Phi_{\tilde{S}}(\boldsymbol{u}) = \{|I - 2iA_1B_1/N\Delta t|\,|I - 2iA_2B_2/N\Delta t|\}^{-\frac{1}{2}}\,, \tag{46}$$

where I denotes the identity matrix.

The spectral estimates were defined in (6) as the average over ν spectral lines of $\tilde{\boldsymbol{S}}$. Let $\hat{\boldsymbol{S}}$ be the sector of such estimates defined by

$$\hat{\boldsymbol{S}} = \frac{1}{\nu}\sum_{M}\tilde{\boldsymbol{S}}_m\,. \tag{47}$$

The usual assumption introduced to obtain approximately the characteristic function for $\hat{\boldsymbol{S}}$ is that the spectral lines are approximately constant over the set $\boldsymbol{M}$. If the spectral lines are linearly varying, the average would be about the same as for an average of statistically stationary lines having the properties associated with the midpoint. As mentioned earlier, the spectral lines, from frequency to frequency, are independent of each other if the original processes are Gaussian. Putting all this together, the characteristic function of $\hat{S}/S$

$$\begin{aligned} \Phi_{\hat{S}/S}(\boldsymbol{u}) &= \left|\exp\left[i\boldsymbol{u}^{\mathrm{T}}\frac{1}{S\nu}\sum_{M}\tilde{\boldsymbol{S}}_m\right]\right| = \prod_{M}|\exp\,[i\boldsymbol{u}^{\mathrm{T}}\boldsymbol{S}_m/\tilde{S}\nu]| = \\ &= \{\Phi_{\tilde{S}}(\boldsymbol{u}/S\nu)\}^{\nu} = \{|I - 2iA_1B_1/\nu SN\Delta t|\,|I - 2iA_2B_2/\nu SN\Delta t|\}^{-\nu/2}\,. \end{aligned} \tag{48}$$

For a 3×3 determinant, it can be shown that

$$\begin{aligned} |I + M| = 1 + m_{11} + m_{22} + m_{33} &+ \begin{vmatrix} m_{22} & m_{23} \\ m_{32} & m_{33} \end{vmatrix} + \\ &+ \begin{vmatrix} m_{11} & m_{13} \\ m_{31} & m_{33} \end{vmatrix} + \begin{vmatrix} m_{11} & m_{12} \\ m_{21} & m_{22} \end{vmatrix} + |M|\,, \end{aligned} \tag{49}$$

where m_{ij} are the elements of $\boldsymbol{M}$. Hence, after some algebra, it is found that

$$\begin{aligned} \Phi_{\hat{S}/S}(\boldsymbol{u}) = \Big\{ & 1 + [-iu_1 - Ciu_2 - Diu_3 - Aiu_4 - Biu_5 - Eiu_6]/\nu - \\ & - \Big[(C - A^2)u_1u_2 + (D - B^2)u_1u_3 + (E - AB)u_1u_6 + \end{aligned} \tag{50}$$

$$+ (CD - E^2) u_2 u_3 + (BC - AE) u_2 u_5 + (AD - BE) u_3 u_4 +$$

$$+ (A^2 - C) \frac{u_4^2}{4} + (AB - E) \frac{u_4 u_5}{2} + (AE - BC) \frac{u_4 u_6}{2} + (B^2 - D) \frac{u_5^2}{4} +$$

$$+ (BE - AD) \frac{u_5 u_6}{2} + (E^2 - CD) \frac{u_6^2}{4} \Bigg] \Big/ \nu^2 + [CD + 2ABE - B^2 C - A^2 D - E^2] \cdot$$

$$\cdot \left[i u_1 u_2 u_3 + i \frac{u_4 u_5 u_6}{4} - i \frac{u_2 u_5^2}{4} - i \frac{u_3 u_4^2}{4} - i \frac{u_1 u_6^2}{4} \right] \Big/ \nu^3 \Bigg\}^{-\nu} .$$

This is the characteristic function for the six spectral estimates after division by the true sea surface spectral density. It forms the basis for expanding the expected values of nonlinear functions in terms of the central moments.

For example, let μ_1 be the expected value of $\hat{q}_x/S$ and μ_2, μ_3 and μ_4 be defined by

$$\mu_k = \mathscr{E}[\{\hat{q}_x/S - \mu_1\}^k] . \tag{51}$$

It can be shown by procedures which will be outlined later that

$$\begin{cases} \mu_1 = A , \\ \mu_2 = (A^2 + C)/2\nu , \\ \mu_3 = A(A^2 + 3C)/2\nu^2 , \\ \mu_4 = [3(A^4 + 6A^2 C + C^2) + 3\nu(A^2 + C)^2]/4\nu^3 . \end{cases} \tag{52}$$

For moderately large ν, the higher-order central moments decrease in importance providing the numerator polynomials are not large. By reference to (24), the numerator terms depend on a_1, b_1, a_2, b_2 and k. The wave number k is quite small for the ocean waves of interest in most engineering considerations. The spreading-function Fourier coefficients are defined by

$$\begin{pmatrix} a_k \\ b_k \end{pmatrix} = \frac{1}{\pi} \int_0^{2\pi} D(\theta) \begin{pmatrix} \cos k\theta \\ \sin k\theta \end{pmatrix} d\theta . \tag{53}$$

Since $|\cos k\theta| \leqslant 1$ and $|\sin k\theta| \leqslant 1$,

$$\begin{cases} |\pi a_k| \leqslant \int_0^{2\pi} D(\theta) \, d\theta = 1 , \\ |\pi b_k| \leqslant \int_0^{2\pi} D(\theta) \, d\theta = 1 . \end{cases} \tag{54}$$

Consequently, the A, B, C, D and E defined in (24) have the bounds

$$\left\{\begin{array}{l} |A|\leqslant k\,, \\ |B|\leqslant k\,, \\ |C|\leqslant k^2\,, \\ |D|\leqslant k^2\,, \\ |E|\leqslant k^2/2\,. \end{array}\right. \tag{55}$$

Thus the expansion of expectations of nonlinear functions in terms of central moments for the spectral estimates appears to be a satisfactory procedure.

9. – The central moments for the spectral estimates.

Let the central differences for the various spectral estimates be designated by

$$\left\{\begin{array}{l} \Sigma \;\;= \{\hat{S} - \mathscr{E}(\hat{S})\}/S\,, \\ \Sigma_x = \{\hat{S}_x - \mathscr{E}(\hat{S}_x)\}/S\,, \\ \Sigma_y = \{\hat{S}_y - \mathscr{E}(\hat{S}_y)\}/S\,, \\ Q_x = \{\hat{q}_x - \mathscr{E}(\hat{q}_x)\}/S\,, \\ Q_y = \{\hat{q}_y - \mathscr{E}(\hat{q}_y)\}/S\,, \\ C_{xy} = \{\hat{c}_{xy} - \mathscr{E}(\hat{c}_{xy})\}/S\,. \end{array}\right. \tag{56}$$

Probably the easiest way to develop the central moments from a characteristic function such as (50) is though the cumulant moments. Let

$$\mu(n_1, n_2, n_3, n_4, n_5, n_6) = \mathscr{E}[\Sigma^{n_1}\Sigma^{n_2}\Sigma^{n_3}Q_x^{n_4}Q_y^{n_5}C_{xy}^{n_6}] \tag{57}$$

denote the central moments. The results developed will be concerned with n_1, n_2, n_3, n_4, n_5 and n_6 all nonnegative and

$$\sum_{j=1}^{6} n_j = n \leqslant 2\,. \tag{58}$$

That is, only moments of second order or less will be examined. The corresponding cumulant moment is defined as

$$c(n_1, n_2, n_3, n_4, n_5, n_6) = \text{coefficient of } \prod_{j=1}^{6}\left[\frac{(iu_j)^{n_j}}{n_j!}\right] \tag{59}$$

in the Maclaurin expansion of $\ln \Phi(\boldsymbol{u})$.

The first-order cumulants are the expectations of each of the random variables. The second- and third-order cumulants exactly equal the corresponding central moments.

10. – Expectations and second-order central moments.

It is useful to list for future reference the expected values and the second-order moments for $\hat{S}/S$ as determined from (50):

$$\mathscr{E}\begin{bmatrix} \hat{S}/S \\ \hat{S}_x/S \\ \hat{S}_y/S \\ \hat{q}_x/S \\ \hat{q}_y/S \\ \hat{c}_{xy}/S \end{bmatrix} = \begin{bmatrix} 1.0 \\ C \\ D \\ A \\ B \\ E \end{bmatrix}, \tag{60}$$

$$\operatorname{cov}\begin{bmatrix} \hat{S}/S \\ \hat{S}_x/S \\ \hat{S}_y/S \\ \hat{q}_x/S \\ \hat{q}_y/S \\ \hat{c}_{xy}/S \end{bmatrix} = (1/\nu)\begin{bmatrix} 1 & A^2 & B^2 & A & B & AB \\ A^2 & C^2 & E^2 & AC & AE & CE \\ B^2 & E^2 & D^2 & BE & BD & DE \\ A & AC & BE & \dfrac{C+A^2}{2} & \dfrac{E+AB}{2} & \dfrac{BC+AE}{2} \\ B & AE & BD & \dfrac{E+AB}{2} & \dfrac{D+B^2}{2} & \dfrac{AD+BE}{2} \\ AB & CE & DE & \dfrac{BC+AE}{2} & \dfrac{AD+BE}{2} & \dfrac{CD+E^2}{2} \end{bmatrix}. \tag{61}$$

The (i, j) element in the covariance matrix gives the expected value of the product of central differences for the i-th variate and the j-th variate. Thus the elements down the main diagonal are the variances, while the off-diagonal elements give the covariances.

11. – Spreading-function characterization.

Estimates for the Fourier coefficients of the spreading function, $D(\theta)$, can be developed from the definitions given in (24) by inserting spectral estimates for the population spectra. Thus (introducing definitions for $A1P$, $B1P$,

$A2P$, $B2P$)

$$(62)\qquad \begin{cases} \pi\hat{a}_1 = \hat{q}_x/k\hat{S} & = A1P\,, \\ \pi\hat{b}_1 = \hat{q}_y/k\hat{S} & = B1P\,, \\ \pi\hat{a}_2 = (\hat{S}_x - \hat{S}_y)/k^2\hat{S} & = A2P\,, \\ \pi\hat{b}_2 = 2\hat{c}_{xy}/k^2\hat{S} & = B2P\,. \end{cases}$$

An estimate for the wave number k can also be determined from

$$(63)\qquad \hat{k} = \{(\hat{S}_x + \hat{S}_y)/\hat{S}\}^{\frac{1}{2}}\,.$$

12. – Centered Fourier coefficients.

It is convenient to use « centered » Fourier coefficients for $D(\theta)$. That is, the co-ordinate system is rotated to make the first harmonic sine coefficient equal to zero. In terms of the new co-ordinate system

$$(64)\qquad D(\theta) = 1/2\pi + (M1P)\cos\theta + (M2P)\cos 2\theta + (N2P)\sin 2\theta + \ldots\,.$$

The definitions in (56) form a natural basis for expansion of estimates. For example consider $\widehat{M1P}$, where the ^ denotes the estimate of the quantity beneath it. From the definition of $\hat{\theta}_0$ as arctg $(\hat{b}_1/\hat{a}_1)$

$$(65)\qquad \cos\hat{\theta}_0 = \widehat{A1P}/(\widehat{A1P} ** 2 + \widehat{B1P} ** 2) ** 0.5\,,$$

$$(66)\qquad \sin\hat{\theta}_0 = \widehat{B1P}/(\widehat{A1P} ** 2 + \widehat{B1P} ** 2) ** 0.5\,.$$

Hence

$$(67)\qquad \widehat{M1P} = \widehat{A1P} * \cos\hat{\theta}_0 + \widehat{B1P} * \sin\hat{\theta}_0 = (\widehat{A1P} ** 2 + \widehat{B1P} ** 2) ** 0.5\,.$$

Thus, with k = wave number,

$$(68)\qquad \widehat{A1P} = \hat{q}_x/k\hat{S} = (Q_x + A)/k(\Sigma + 1)\,,$$

$$(69)\qquad \widehat{B1P} = \hat{q}_y/k\hat{S} = (Q_y + B)/k(\Sigma + 1)\,,$$

$$(70)\qquad \widehat{A2P} = (\hat{S}_x - \hat{S}_y)/k^2\hat{S} = (\Sigma_x + C - \Sigma_y - D)/k^2(\Sigma + 1)\,,$$

$$(71)\qquad \widehat{B2P} = 2\hat{c}_{xy}/k^2\hat{S} = 2(C_{xy} + E)/k^2(\Sigma + 1)\,.$$

If the results are combined, the following formula is obtained:

$$(72)\qquad \widehat{M1P} = \{(Q_x + A)^2 + (Q_y + B)^2\}^{\frac{1}{2}}/k(\Sigma + 1)\,.$$

The corresponding theoretical value for $M1P$ is

$$M1P = (A^2 + B^2)^{\frac{1}{2}}/k\,. \tag{73}$$

Similar equations can be derived for $M2P$ and $N2P$. From trigonometric identities

$$\cos 2\hat{\theta}_0 = \cos^2\hat{\theta}_0 - \sin^2\hat{\theta}_0 = \frac{\widehat{A1P}**2 - \widehat{B1P}**2}{\widehat{A1P}**2 + \widehat{B1P}**2}, \tag{74}$$

$$\sin 2\hat{\theta}_0 = 2\sin\hat{\theta}_0 \cos\hat{\theta}_0 = \frac{2*\widehat{A1P}*\widehat{B1P}}{A1P**2 + B1P**2}. \tag{75}$$

If the various above expressions are substituted into

$$M2P = \widehat{A2P}\cos 2\hat{\theta}_0 + \widehat{B2P}\sin 2\theta_0\,, \tag{76}$$

$$N2P = \widehat{B2P}\cos 2\theta_0 - \widehat{A2P}\sin 2\hat{\theta}_0\,, \tag{77}$$

one gets after some algebra

$$\widehat{M2P} = [\{(\Sigma_x - \Sigma_y) + (C - D)\}\{(Q_x + A)^2 - (Q_y + B)^2\} + \\ + 4(C_{xy} + E)(Q_x + A)(Q_y + B)]/k^2(\Sigma + 1)[(Q_x + A)^2 + (Q_y + B)^2]\,, \tag{78}$$

$$\widehat{N2P} = [2(C_{xy} + E)\{(Q_x + A)^2 - (Q_y + B)^2\} - 2\{(\Sigma_x - \Sigma_y) + \\ + (C - D)\}(Q_x + A)(Q_y + B)]/k^2(\Sigma + 1)\{(Q_x + A)^2 + (Q_y + B)^2\}\,. \tag{79}$$

The « theoretical » values for these parameters are

$$M2P = \frac{(C - D)(A^2 - B^2) + 4ABE}{k^2(A^2 + B^2)}\,, \tag{80}$$

$$N2P = \frac{2E(A^2 - B^2) - 2(C - D)AB}{k^2(A^2 + B^2)}\,. \tag{81}$$

Equations (72), (73), (78)-(81) provide the starting point for a second-order expansion of $M1P$, $M2P$ and $N2P$. The phrase « second-order » as used here refers to an expansion keeping terms of first and second order in Σ, Σ_x, Σ_y, Q_x, Q_y and C_{xy}. The statistical expectation of such expansions involves only means, variances and covariances of the spectral and cross-spectral estimates.

The actual algebra is intricate but straightforward. For $M1P$ in eq. (72), the numerator may be expanded by binomial expansion of

$$\text{numerator} = \sqrt{A^2 + B^2}\left[1 + \frac{2(AQ_x + BQ_y) + Q_x^2 + Q_y^2}{A^2 + B^2}\right]^{\frac{1}{2}}. \tag{82}$$

The denominator is directly

$$(\text{denominator})^{-1} = (1 + \Sigma)^{-1}/k = (1 - \Sigma + \Sigma^2 - \dots)/k\,. \tag{83}$$

The product of these last two equations gives the basic expansion of $\widehat{M1P}$.

The numerators for $\widehat{M2P}$ and $\widehat{N2P}$ in eq. (78) and (79) do not require series expansion. The expressions may be written as first-, second- and third-order products of the centered variables. The denominators for both estimates are identical to each other and may be written

$$(\text{denominator})^{-1} = (A^2 + B^2)(1+\Sigma)^{-1}\left(1 + \frac{2AQ_x + 2BQ_y + Q_x^2 + Q_y^2}{A^2 + B^2}\right)^{-1} \Big/ k^2\,. \tag{84}$$

The second-order expansion for this expression may then be multiplied by the separate numerators to obtain the basic expansions for $\widehat{M2P}$ and $\widehat{N2P}$.

The basic second-order expansions may be algebraically manipulated to obtain second-order expansions of

$$\left\{\begin{array}{l} \widehat{M1P} - M1P\,, \\ (\widehat{M1P} - M1P)^2\,, \\ \widehat{M2P} - M2P\,, \\ (\widehat{M2P} - M2P)^2\,, \\ \widehat{N2P} - N2P\,, \\ (\widehat{N2P} - N2P)^2\,. \end{array}\right. \tag{85}$$

Hence the statistical expectation of the second-order expansion indicated by eq. (85) gives the bias and mean square error for $\widehat{M1P}$, $\widehat{M2P}$ and $\widehat{N2P}$ as expressions of A, B, C, D and E.

The final result of all this algebra may be written as follows for the bias expressions:

$$\mathscr{E}(\widehat{M1P} - M1P) = \frac{B^2C - 2ABE + A^2D}{4k\nu R^3}\,, \tag{86}$$

$$\begin{aligned}\mathscr{E}(\widehat{M2P} - M2P) = \{&(C + D - R^2) - (2E^2 + CD)/R^2 + \\ &+ [A^2D^2 + B^2C^2 + 6ABE(C + D)]/R^4 - \\ &- 4AB[AB(C - D)^2 + 4E(A^2D - ABE + B^2C)]/R^6\}/k^2\nu\,,\end{aligned} \tag{87}$$

$$\begin{aligned}&\mathscr{E}(\widehat{N2P} - N2P) = \\ &= \{-3G(C + D)/2R^4 + 2G(A^2C + B^2D + 2ABE)/R^6\}/k^2\nu\,,\end{aligned} \tag{88}$$

$$R^2 = A^2 + B^2 , \tag{89}$$

$$G = 2E(A^2 - B^2) - 2AB(C - D) . \tag{90}$$

Similarly the root-mean-square errors simplify to the following:

$$\sqrt{\mathscr{E}[(\widehat{M1P} - M1P)^2]} = [\{A^2C + 2ABE + B^2D - R^4\}/2k^2\nu R^2]^{\frac{1}{2}} , \tag{91}$$

$$\begin{aligned}\sqrt{\mathscr{E}[(\widehat{M2P} - M2P)^2]} &= [-2\{(A^2 - B^2)(C - D) + 4ABE\} + \\ &+ \{[(A^2 - B^2)(C - D) + 4ABE]^2 + (A^2 - B^2)^2(C^2 + D^2 - 2E^2) + \\ &+ 8ABE(A^2 - B^2)(C - D) + 8A^2B^2(CD + E^2)\}/R^4 + \\ &+ \{8(A^2 - B^2)[E^2(A^4 - B^4) + A^2B^2(C^2 - D^2) + 2ABE(B^2D - A^2C)] + \\ &+ 16AB[AB(C - D) - E(A^2 - B^2)](B^2C - A^2D)\}/R^6 + \\ &+ 8[AB(C - D) - E(A^2 - B^2)]^2(B^2C - 2ABE + A^2D)/R^8]^{\frac{1}{2}}/\sqrt{\nu}k^2 ,\end{aligned} \tag{92}$$

$$\begin{aligned}\sqrt{\mathscr{E}[(\widehat{N2P} - N2P)^2]} &= \\ &= \langle\{G^2 + 4A^2B^2(D^2 + C^2 - 2E^2) - 8ABE(C - D)(A^2 - B^2) + \\ &+ 2(CD + E^2)(A^2 - B^2)^2\}/R^4 + \{8AB[4ABE + (A^2 - B^2)(C - D)] \cdot \\ &\cdot [-ABC + A^2E + B^2E - ABD] - 4[4ABE + (A^2 - B^2)(C - D)] \cdot \\ &\cdot [A^4D - A^2B^2(C + D) + B^4C]\}/R^6 + \\ &+ 2(A^2D + B^2C - 2ABE)[4ABE + (A^2 - B^2)(C - D)]^2/R^8\rangle^{\frac{1}{2}}/\sqrt{\nu}k^2 .\end{aligned} \tag{93}$$

13. – Bias and error for $\hat{\theta}_0$ and $\hat{S}_0$.

The principal direction for $D(\theta)$ is $\operatorname{arctg}(b_1/a_1)$. From (62)

$$\hat{\theta}_0 = \operatorname{arctg}(\hat{b}_1/\hat{a}_1) = \operatorname{arctg}(\hat{q}_y/\hat{q}_x) . \tag{94}$$

Consequently, the introduction of (56) gives

$$\hat{\theta}_0 = \operatorname{arctg}\{(Q_y + B)/(Q_x + A)\} = \operatorname{arctg}\{(B/A)(1 + Q_y/B)(1 + Q_x/A)^{-1}\} . \tag{95}$$

Since from trigonometric identities

$$\operatorname{arctg} z_1 - \operatorname{arctg} z_2 = \operatorname{arctg}\{(z_1 - z_2)/(1 + z_1 z_2)\} , \tag{96}$$

it follows that the population or « true » value for θ_0 will be arctg (B/A). Hence

$$\hat{\theta}_0 - \theta_0 = \operatorname{arctg}\{(B/A)(1+Q_y/B)(1+Q_x/A)^{-1}\} - \operatorname{arctg}(B/A) = \\ = \operatorname{arctg}\frac{(B/A)(1+Q_y/B)(1+Q_x/A)^{-1} - B/A}{1+(B/A)^2(1+Q_y/B)(1+Q_x/A)^{-1}}. \tag{97}$$

Every function in (97) has a known power series expansion, so the required solution is just a matter of careful hard work.

A measure of the dispersion of the spreading function is given by

$$\hat{S}_0 = [-2\ln(\widehat{M1P})]^{\frac{1}{2}}. \tag{98}$$

The series expansion for $\hat{S}_0$ from (71) is

$$[-2\ln(\widehat{M1P})]^{\frac{1}{2}} = [-\ln(Q_x^2 + Q_y^2 + 2AQ_x + 2BQ_y + A^2 + B^2) - \\ - 2\ln k + 2\ln(1+\Sigma)]^{\frac{1}{2}} = \\ = [S_0^2 - \ln\{1 + (2AQ_x + 2BQ_y + Q_x^2 + Q_y^2)/R^2\} + 2\ln(1+\Sigma)]^{\frac{1}{2}}, \tag{99}$$

where

$$R^2 = A^2 + B^2. \tag{100}$$

Equations (97) and (99) can be expanded to second order in Σ, Q_x and Q_y and the expectation operator applied to give second-order approximations to $\mathscr{E}[\hat{\theta}_0 - \theta_0]$, $\mathscr{E}[(\hat{\theta}_0 - \theta_0)^2]$, $\mathscr{E}[\hat{S}_0 - S_0]$ and $\mathscr{E}[(\hat{S}_0 - S_0)^2]$. These are measures of the bias and root-mean-square error for the estimates $\hat{\theta}_0$ and $\hat{S}_0$.

Strictly speaking, both $\hat{\theta}_0$ and $\hat{S}_0$ are angular quantities and should, in general, be treated as such. However, if the errors $\hat{\theta}_0 - \theta_0$ and $\hat{S}_0 - S_0$ are small, the expectations of the differences and squared differences are quite good measures. Values from computer simulations will be given later which will support this conclusion.

The results of the second-order expansions for $\hat{\theta}_0 - \theta_0$ and $\hat{S}_0 - S_0$ are

$$\mathscr{E}(\hat{\theta}_0 - \theta_0) = [AB(C-D) + E(B^2 - A^2)]/2\nu R^4, \tag{101}$$

$$\sqrt{\mathscr{E}[(\hat{\theta}_0 - \theta_0)^2]} = \sqrt{(A^2D - 2ABE + B^2C)/2\nu R^4}, \tag{102}$$

$$\mathscr{E}(\hat{S}_0 - S_0) = \{S_0^2[(A^2 - B^2)(C-D) - R^4 + 4ABE] + \\ + [R^4 - A^2C - B^2D - 2ABE]\}/4S_0^2R^4\nu, \tag{103}$$

$$\sqrt{\mathscr{E}[(\hat{S}_0 - S_0)^2]} = \sqrt{(A^2C + B^2D + 2ABE - R^4)/2S_0^2R^4\nu}. \tag{104}$$

TABLE I. – *Bias and r.m.s. error for second-order formula for averages of* 100 *simulations with* 40 *degrees of freedom and a theoretical wave number of* 0.012431. (Angles in radians.)

	$\hat{\theta}_0$	$\hat{S}_0$	$\widehat{M1P}$	$\widehat{M2P}$	$\widehat{N2P}$
Case 1. $\pi a_1 = -0.923$, $\pi b_1 = 0$, $\pi a_2 = 0.725$, $\pi b_2 = 0$ (symmetric unimodal)					
bias:					
second-order theory	0	-0.0071	0.0019	0.0016	0
simulations with	0.0063	-0.0107	0.0031	0.0018	0.0008
theory wave number	-0.0034	-0.0047	0.0009	0.0032	-0.0017
simulations with	-0.0075	-0.0049	0.0005	0.0043	-0.0026
estimated wave number	-0.0057	-0.0043	-0.0033	-0.0092	-0.0004
r.m.s.:					
second-order theory	0.0635	0.0440	0.0163	0.0516	0.0201
simulations with	0.0650	0.0475	0.0172	0.0553	0.0204
theory wave number	0.0626	0.0471	0.0167	0.0485	0.0226
simulations with	0.0615	0.0582	0.0222	0.0779	0.0197
estimated wave number	0.0635	0.0674	0.0262	0.0917	0.0243
Case 2. $\pi a_1 = -0.462$, $\pi b_1 = -0.462$, $a'_2 = 0$, $b'_2 = 0$ (symmetric bimodal)					
bias:					
second-order theory	0	-0.0163	0.0096	0.0287	0
simulations with	-0.0112	-0.0250	0.0145	0.0356	-0.0126
theory wave number	-0.0073	-0.0052	0.0028	0.0365	0.0034
simulations with	-0.0125	0.0062	-0.0042	0.0068	-0.0089
estimated wave number	0.0541	-0.0058	0.0029	0.0327	0.0441
r.m.s.:					
second-order theory	0.1711	0.0709	0.0428	0.1581	0.1581
simulations with	0.1850	0.0935	0.0556	0.1766	0.1936
theory wave number	0.1776	0.0780	0.0466	0.1683	0.1719
simulations with	0.1532	0.1031	0.0616	0.1568	0.1361
estimated wave number	0.1785	0.1141	0.0679	0.1737	0.1523
Case 3. $\pi a_1 = -0.839$, $\pi b_1 = -0.084$, $\pi a_2 = 0.593$, $\pi b_2 = 0$ (nonsymmetric bimodal)					
bias:					
second-order theory	0.0021	-0.0109	0.0031	0.0054	-0.0007
simulations with	0.0049	-0.0030	-0.0010	-0.0041	0.0001
theory wave number	-0.0039	-0.0050	0.0011	0.0048	0.0029
simulations with	0 0121	-0.0059	0.0016	0.0067	0.0086
estimated wave number	-0.0017	-0.0116	0.0030	0.0032	0.0047
r.m.s.:					
second-order theory	0.0858	0.0906	0.0446	0.1023	0.0566
simulations with	0.0815	0.0938	0.0451	0.0968	0.0563
theory wave number	0.0828	0.0707	0.0351	0.1060	0.0501
simulations with	0.0778	0.0675	0.0330	0.0978	0.0474
estimated wave number	0.0799	0.0959	0.0451	0.1086	0.0484

14. – Accuracy of the second-order approximations.

As a means of testing the accuracy of the second-order approximations, a number of computer simulations for various spreading functions were run. The bias and r.m.s. error as computed from a sequence of 100 estimates of θ_0 and S_0 were compared with the second-order predictions of the bias and r.m.s. error. Each simulation produced spectra and cross-spectra for η, η_x and η_y with 40 degrees of freedom. Theoretical conditions corresponding to a wave number of 0.012 431 (or frequency of 0.1 Hz for waves in 200-foot water depth).

Three spreading functions were used. Case I was a unimodal, symmetric spreading function. Both case II and case III were bimodal. Case II was symmetric, while case III was asymmetric. In addition, estimates for some runs were made with the theoretical wave number for a 0.1 Hz frequency and, for others, the wave number was computed from the simulated data using eq. (63). The results of these runs are given in table I. The agreement between the simulations and the second-order approximations for bias and r.m.s. appear reasonable and support the adequacy of the formulae as measures of accuracy for data management.

15. – Conclusions.

This study is part of an ongoing investigation of procedures for analyzing both unimodal and bimodal spectral spreading functions. The second-order estimates of bias and r.m.s. error are useful in themselves. They also provide a basis for studying the more complicated problem of the discrimination between bimodal and unimodal conditions from tilt-and-roll wave buoys.

* * *

The research reported here was funded by the Technical Physische Dienst (Institute of Applied Physics) of the Netherlands. Their support is sincerely appreciated.

REFERENCES

[1] M. S. LONGUET-HIGGINS, D. E. CARTWRIGHT and N. D. SMITH: *Observations of the directional spectrum of sea waves using the motions of a floating buoy*, in *Ocean Wave Spectra* (Englewood Cliffs, N. J., 1963), p. 111.

[2] E. SCHEUER and D. S. STOLLER: *Technometrics*, **4**, 278 (1962).

[3] L. E. BORGMAN: *Statistical properties of fast Fourier transform coefficients*, Tech. Paper No. 76-9 (Fort Belvoir, Va., 1976).

[4] N. R. GOODMAN: *On the joint estimation of the spectra, cospectrum, and quadrature spectrum of a two-dimensional stationary Gaussian process*, Scientific Paper No. 10, New York University (1957).

Extremal Statistics in Wave Climatology.

L. E. BORGMAN

University of Wyoming - Laramie, Wy.

D. T. RESIO

Wave Dynamics Division, Hydraulics Laboratory, U.S.A.E.W.E.S.
Vicksburg, Miss.

1. – Introduction.

Two basic problems in extremal theory will be presented together with the relevant solutions and formula. These, in turn, will be used to motivate a general outline of the techniques used in the application of the statistics of extremes to oceanographic engineering. A third problem related to the use of second-largest, third-largest and other near-extremes to strengthen predictions based on the observed largest values will be introduced later.

Some basic problems related to the largest wave height will be discussed. These will include the joint distribution of the maximum wave height and associated wave period, and probability laws for the maximum wave height in time-varying storms.

There is considerable controversy concerning the best procedures for estimating the largest wave height in multiyear intervals. Engineering literature contains many conflicting procedures and often arbitrary justification for the procedures. It seems appropriate at this time to take a close look at the various estimation methods and to try to provide guidelines for their use. Most of the methods are reasonably valid within certain data situations, but may be incorrect elsewhere. A careful catalog of suggestions and cautions would certainly help the working engineer make judgements concerning which procedures are appropriate for a given situation.

Extremal prediction procedures can be divided into two basic classes. In the first, measured data are graphed, after choosing some initial fundamental probability law. Typically, the graph paper is chosen so that the data lie along a straight line if the initial probability law is valid. A curve-fitted line may then be extended to provide predictions of the extreme «expected» in substantially longer time intervals. In the second basic class, the measured data are used to calibrate a statistical or physical model of the phenomena producing the waves. The probabilities of joint occurrence of various statistical or meteorological combinations are developed from historical records and used to extend the model to produce extremal estimates.

Both basic classes can be approached in a variety of ways. In most cases, different approaches lead ultimately to results which are not significantly different from one another. However, some of the procedures are critically dependent and sensitive to initial assumptions. For this reason, it is desirable to examine in detail the variety of possible approaches.

In the first class, the extrapolation is most commonly based on order statistics combined with an initial distribution such as the Rayleigh, the Weibull, the extremal types, the Pearson curve family, the logarithmic normal, or other distributions. It is important that some reliability measure be estimated for the final extrapolation. This may be based on simulations or on statistical formulae. The method of curve fitting may also affect the actual extrapolated value produced. However, the time extent of the data base is probably the most critical factor.

In the second basic class, that of model building, a wide variety of methods of attack are possible. The classic approach is based on hindcasts of significant wave heights for historical storms, followed by some type of joint-probability weighting of storm parameters to extrapolate the hindcast results to give long-term extremes at a given site. The site-dependent results may subsequently be « smoothed » in some probabilistic way to give site-independent extremal estimates. Subjective judgements are incorporated in these procedures in a variety of ways.

Besides the hindcast approach, model-building methods also include techniques based on the estimation from data of the parameters for the initial population probability law. The fitted initial population distribution is then statistically manipulated in predicting the distribution for long-term extremes.

Reliability of extremal statistics in the model-building context are difficult to determine because of the very subjective nature of the choices in the model construction. Sometimes simulation can be used; sometimes parameter estimate variability can be extended on through the formulae to give the variability of the extremal estimates.

The model method is simultaneously the most promising procedure and the one most likely to lead to gross errors. The extrapolation method is quite reliable for short extensions beyond the data base, but becomes increasingly unreliable for extrapolation longer than twice the data extent. Both basic classes fail miserably if the extremes arise from a different basic population than the measured data (*i.e.* data from winter storms and extremes from hurricanes). Data arising from a population of mixtures are particularly difficult to use in the extrapolation methods, but may be usable in model building if the mixture can be characterized and parameterized.

Ultimately, there is no really satisfying answer for estimating long-term extremes from short-term data bases. Every approach has its faults and a variety of answers is possible. This does not mean that long-term extremal estimates should not be made, but only that careful, well-considered engineering judgement is a necessary component in the preparation of extremal estimates.

Much of the material in this summary of extremal procedures in ocean wave studies is combined from two earlier papers by the authors in various conference proceedings [1, 2].

2. – Problem No. 1.

In its simplest form, the extreme-value problem can be stated as follows. Let $X_1, X_2, \ldots, X_n$ be a set of observations, with the i-th observation being taken from a population with a probability law

$$F_{X_i}(x) = P[X_i \leqslant x]\,. \tag{1}$$

Suppose that $X_{(n)}$ is defined as max $(X_1, X_2, \ldots, X_n)$. What is the probability $P[X_n \leqslant x] = F_{X_{(n)}}(x)$?

The answer may be stated in increasingly simpler form if additional restrictions are applied:

$$F_{X_{(n)}}(x) = \begin{cases} P[X_1 \leqslant x, X_2 \leqslant x, \ldots, X_n \leqslant x]\,, & \text{in general}\,, \quad (2A) \\ \prod_{i=1}^{n} F_{X_i}(x)\,, & \text{if observations are independent}\,, \quad (2B) \\ F_X^n(x)\,, & \text{if } F_{X_i}(x) = F_X(x) \text{ for } i = 1, 2, \ldots, n\,. \quad (2C) \end{cases}$$

The proof of this relation is straightforward. The largest of a set of random variables will be less than or equal to a number x if and only if every one of the random variables is less than or equal to x. That is, the two sets are equivalent

$$[X_{(n)} \leqslant x] \rightleftarrows [X_1 \leqslant x, X_2 \leqslant x, \ldots, X_n \leqslant x]\,. \tag{3}$$

Therefore, the probability of one equals the probability of the other. This is the first line of eq. (2).

For nonstationary, intercorrelated sequences of random variables, the general relationship is the natural starting point. The joint distribution for the total sample, with all arguments set equal to the common value x, is the distribution function for the extreme.

If the observations are nonstationary but independent of each other, the basic relationship that the joint probability for independent events is equal to the product of the probabilities for the individual events can be introduced. This gives the second line of eq. (2). Finally, if the sequences of observations are stationary, *i.e.* all have the same probability law, the third line of eq. (2) results.

In a given application, the fundamental relation to be used as the starting point for further derivation depends on which assumptions are justifiable.

3. – Problem No. 2.

A slightly more involved extremal problem is the following. Suppose that a random number N of independent events occur during the time interval of interest and that each event has associated with it a random intensity X_i. What is the probability law for the largest intensity occurring during the time interval? The solution to this problem may be stated most easily in terms of the « so-called » probability-generating function, $G(s)$, for the random variable N. Let

$$P_N(n) = P[N = n] \tag{4}$$

specify the probability law for N. The probability-generating function for N is defined as

$$G(s) = \sum_{n=0}^{\infty} P_N(n)\, s^n \,. \tag{5}$$

The function $G(s)$ has a simple formula for many of the discrete-probability laws commonly used. For example, the Poisson and binomial probability laws yield

Poisson:

$$\begin{cases} P_N(n) = \exp[-\lambda L]\, \lambda^n / n!\,, \\ G(s) \;\;\; = \exp[-\lambda L(1-s)]\,; \end{cases} \tag{6}$$

binomial:

$$\begin{cases} P_N(n) = \begin{pmatrix} m \\ n \end{pmatrix} \theta^n (1-\theta)^{m-n}\,, \\ G(s) \;\;\; = (1-\theta+\theta s)^m\,. \end{cases} \tag{7}$$

The solution to problem No. 2 may be stated as

$$F_{X_{(N)}} = G(F_X(x))\,. \tag{8}$$

The right-hand side of this equation is the probability-generating function with its argument, s, replaced by $F_X(x)$.

The proof for eq. (8) is relatively simple and is based on the two fundamental probability relations

$$P(A \text{ and } B) = P(A, \text{ given } B)P(B) \tag{9}$$

and, if $A_1, A_2, \ldots, A_n$ are mutually exclusive,

$$P(A_1 \text{ or } A_2 \text{ or } \ldots \text{ or } A_n) = \sum_{j=1}^{n} P(A_j)\,. \tag{10}$$

In problem No. 2, A_j is the event that j observations occur and the largest observation is less than or equal to x. Thus

$$\begin{aligned} P(\text{largest } X \leqslant x) &= \sum_{j=0}^{\infty} P[N = j \text{ and } X_{(N)} \leqslant x] = \\ &= \sum_{j=0}^{\infty} P[X_{(N)} \leqslant x, \text{ given } N = j]P[N = j] = \sum_{j=0}^{\infty} (F_X(x))^j P_N(j) = G(F_X(x))\,. \end{aligned} \tag{11}$$

4. – Return period and encounter probability.

Two commonly used measures of risk are the nonencounter probability and the return period for a given intensity of event. The nonencounter probability, $\mathrm{NE}(x)$, for a time interval or life of operation, L, is the probability that the largest intensity occurring during that time interval is less than or equal to x. A careful reading of this definition will show that

$$\mathrm{NE}(x) = \begin{cases} F_{X_{(n)}}(x)\,, & \text{problem No. 1,} \\ G(F_X(x))\,, & \text{problem No. 2,} \end{cases} \tag{12}$$

where the data set $X_1, X_2, \ldots, X_n$ covers the life L. Thus the nonencounter terminology just draws attention to one particular interpretation of the extremal distribution function.

A useful formulation of this problem, often used in ocean engineering, arises from eq. (11) combined with eq. (6):

$$\mathrm{NE}(x) = \exp\left[-\lambda L(1 - F_X(x))\right], \tag{13}$$

where λ = average number of events occurring per unit time.

The return period is defined as the average waiting time between exceedences of x. In problem No. 1, this would be the average number of observations between exceedences. This will be derived for the independent, identically distributed situation (3rd line, eq. (2)). Let W be the random number of observations preceding and including the first exceedence of x. Then, if $p = F_X(x)$,

$$P(W = w) = p^{w-1}(1 - p) . \tag{14}$$

The average, or expected, value of W would be the return period R, or

$$\begin{aligned} R &= \sum_{w=1}^{\infty} w p^{w-1}(1 - p) = (1 - p)(\mathrm{d}/\mathrm{d}w)\Big(\sum_{w=0}^{\infty} P^w\Big) = \\ &= (1 - p)(\mathrm{d}/\mathrm{d}w)\big(1/(1 - p)\big) = 1/(1 - p) = 1/\big(1 - F_X(x)\big) . \end{aligned} \tag{15}$$

A similar derivation for the Poisson case in problem No. 2 is as follows

$$P(W > w) = P\big(\text{no exceedences of } x \text{ occur during } (0, w)\big) . \tag{16}$$

But the right-hand side of the above equation is just the nonencounter probability for the life w. Hence from eq. (13)

$$P(W > w) = \exp\big[-\lambda w\big(1 - F_X(x)\big)\big] , \tag{17}$$

$$P(W \leqslant w) = 1 - \exp\big[-\lambda w\big(1 - F_X(x)\big)\big] . \tag{18}$$

In the above formulae, w is the variable, and x is held fixed. The probability density for W would be

$$f_W(w) = (\mathrm{d}/\mathrm{d}w)\,P(W \leqslant w) = \theta \exp[-\theta w] , \tag{19}$$

where

$$\theta = \lambda\big(1 - F_X(x)\big) . \tag{20}$$

The mean value of W is

$$R = \int_0^{\infty} \theta w \exp[-\theta w]\,\mathrm{d}w = 1/\theta = 1/\lambda\big(1 - F_X(x)\big) . \tag{21}$$

The relation between R and NE follows from eqs. (12), (13), (15) and (21) as

$$\mathrm{NE}(x) = \begin{cases} (1 - 1/R)^n , & \text{problem No. 1,} \\ \exp[-L/R] , & \text{problem No. 2, Poisson case,} \end{cases} \tag{22}$$

where n is the life relative to problem No. 1 and L is the life for problem No. 2

5. – Asymptotic extremal probability laws.

Two distribution functions $F_1(x)$ and $F_2(x)$ are said to be of the same type if there exist constants $a > 0$ and b such that

$$F_1(ax + b) = F_2(x) . \tag{23}$$

In terms of this concept, it is natural to seek a function $\Phi(x)$ such that for some sequences $a_1, a_2, a_3, \ldots$ and $b_1, b_2, b_3, \ldots$ (with $a_j > 0$)

$$\lim_{n\to\infty} F^n(a_n x + b_n) = \Phi(x) \tag{24}$$

for every continuity point of $\Phi(x)$. If such a $\Phi(x)$ can be found, then, if n is quite large,

$$F^n(a_n x + b_n) \approx \Phi(x) . \tag{25}$$

Hence for large, but fixed, n

$$F^n(x) \approx \Phi[(x - b_n)/a_n] \tag{26}$$

and $\Phi(x)$ characterizes the distribution of the largest, except for a linear transformation on the argument.

It has been shown that, if one excludes improper distribution functions of the form

$$\Phi(x) = \begin{cases} 1.0 & \text{for } x > c, \\ 0 & \text{for } x < c, \end{cases} \tag{27}$$

there are only three possible limiting types [3-5]

$$\Phi(x) = \begin{cases} (A) & \exp[-\exp[-x]] & \text{for } -\infty < x < \infty, \\ (B) & \begin{cases} \exp[-(c/x)^k] & \text{for } x \geqslant 0, \\ 0 & \text{for } x < 0, \end{cases} \\ (C) & \begin{cases} 1 & \text{for } x \geqslant w, \\ \exp[-c(w-x)^k] & \text{for } x < w. \end{cases} \end{cases} \tag{28}$$

Asymptote (A) is quite commonly used in extremal applications. However, asymptote (C), which fits situations for variables limited by some upper bound, deserves more attention than it usually gets. This would be a natural possibility for wave heights limited by breaking, for example.

6. – Plotting position formula.

Utilization of any one of these asymptotes is usually by plotting on graph paper. Let $X_1, X_2, \ldots, X_n$ be a set of independent, identically distributed random variables and suppose these are re-ordered into the order statistics

$$X_{(1)} < X_{(2)} < \ldots < X_{(n)} .$$

The probability law for $X_{(k)}$ can be derived from the binomial probability law:

$$P(X_{(k)} \leqslant x) = P[k \text{ or more of the } X_i \text{ are less than } x] . \tag{29}$$

Let

$$r = P[X_i \leqslant x] = F_X(x) . \tag{30}$$

Then, in binomial terminology,

$$P(X_{(k)} \leqslant x) = P(k \text{ or more successes in } n \text{ trials}) , \tag{31}$$

where an observation being less than or equal to x is called a success. Hence

$$P(X_{(k)} \leqslant x) = \sum_{j=k}^{n} \binom{n}{j} r^j (1-r)^{n-j} . \tag{32}$$

The summation may be expressed in terms of the incomplete beta-function (ref. [6], p. 944-945, eqs. (26.5.1), (26.5.24)):

$$P(X_{(k)} \leqslant x) = [n\,!/(k-1)\,!\,(n-k)\,!] \int_0^r t^{k-1}(1-t)^{n-k}\,\mathrm{d}t . \tag{33}$$

Therefore, the probability density function for $X_{(k)}$ is

$$f_{X_{(k)}}(x) = [\mathrm{d}/\mathrm{d}x] P(X_{(k)} \leqslant x) = k \binom{n}{k} [F_X(x)]^{k-1} [1 - F_X(x)]^{n-k} f_X(x) . \tag{34}$$

The theoretical average value for $r = F_X(x)$ is

$$E[F_X(x)] = k \binom{n}{r} \int_0^1 r r^{k-1} (1-r)^{n-k}\,\mathrm{d}r = k/(n+1) . \tag{35}$$

Reference [5], p. 15, eq. (17) argues persuasively for the use of $\hat{F}_k = k/(n-1)$ as an unbiased estimate of $F_X(X_{(k)})$. Thus a plot of $\hat{F}_k$ *vs.* the observed $X_{(k)}$ gives a prediction of the distribution function from which the observations $X_1, X_2, \ldots, X_n$ came.

Other plotting formulae have been proposed [7-10], but the $k/(n+1)$ formula is simple and adequate for most purposes.

7. – The use of plotting paper.

Let $\Phi(x)$ represent the family of distribution functions presumed to hold for given observed data. This means that, for some $a > 0$ and b, the distribution function for the data is $\Phi(ax + b)$. By the previous section

$$\hat{F}_k = k/(n+1) \approx \Phi(aX_{(k)} + b)\,. \tag{36}$$

Hence, if $\Phi^{-1}(p)$ is the inverse function to $\Phi(x)$,

$$\Phi^{-1}(\hat{F}_k) = aX_{(k)} + b \tag{37}$$

and a plot of $\Phi^{-1}(k/(n+1))$ *vs.* $X_{(k)}$ should give a straight line with intercept b and slope a. For asymptote (A) in eq. (28), for example, a plot of $-\ln\left[-\ln\left(k/(n+1)\right)\right]$ *vs.* $X_{(k)}$ should lead to a straight line from which a and b can be estimated if that asymptote holds for a data set each member of which is a largest value from n observations.

The return period for the number of observations between exceedences of the observed $X_{(k)}$ can be estimated from $\hat{F}_k$ by analogy to eq. (15) as

$$\hat{R}_k = 1/(1 - \hat{F}_k)\,. \tag{38}$$

The straight-line formula given by eq. (37) can be restated in terms of R as

$$\Phi^{-1}(1 - 1/\hat{R}) = aX_{(k)} + b\,. \tag{39}$$

Nonlinear scales with division marks at equal increments of either $\hat{F}_k$ or $\hat{R}$ are often drawn on the margins of the graph paper to assist in the plotting routine. The linear distance between increment marks are determined from

$$y = \Phi^{-1}(\hat{F}_k) \tag{40}$$

or

$$y = \Phi^{-1}(1 - 1/\hat{R}) \tag{41}$$

for the two cases.

In addition to the extremal asymptotes, other probability distribution functions are often fitted by straight-line plots. Normal probability paper is, of course, a common example. Another is the Weibull distribution, for which

$$p = \Phi(x) = \begin{cases} 1 - \exp[-(ax+b)^c] & \text{for } ax+b>0\,, \\ 0 & \text{for } ax+b<0\,, \end{cases} \tag{42}$$

where $a>0$ and $c>0$. The inverse function, for this case, is

$$x = \Phi^{-1}(p) = \{[-\ln(1-p)]^{1/c} - b\}/a\,. \tag{43}$$

If c is known, a plot of $[-\ln(1-F_k)]^{1/c}$ or of $[\ln \hat{R}_k]^{1/c}$ *vs.* $X_{(k)}$ should give straight-line estimates of a and b.

If c is incorrectly chosen, the graphed points will show curvature. This suggests a method for estimating c. Suppose that u_k is defined by

$$u_k = -\ln(1-\hat{F}_k) = \ln \hat{R}_k\,. \tag{44}$$

If c is correct, the graph of $u_k^{1/c}$ *vs.* $X_{(k)}$ will be a straight line. The correlation coefficient ϱ, defined as

$$\varrho = \sum_{k=1}^{n} (X_{(k)} - \bar{X})(u_k^{1/c} - \bar{u}^{1/c}) \bigg/ \left[\sum_{k=1}^{n} (X_{(k)} - \bar{X})^2 \sum_{k=1}^{n} (u_k^{1/c} - \bar{u}^{1/c})^2\right]^{\frac{1}{2}}, \tag{45}$$

where

$$\bar{X} = (1/n)\sum_{j=1}^{n} X_j\,, \qquad \bar{u}^{1/c} = (1/n)\sum_{k=1}^{n} u_k^{1/c}\,, \tag{46}$$

measures the degree of straightness or linearity between $u_k^{1/c}$ and $X_{(k)}$. Hence a good estimate of c would be the value which maximizes ϱ. A variety of routines are available by which a digital computer may be used to search for the c value which maximizes ϱ relative to the data available.

8. – Extrapolation to longer return periods.

Suppose that data are available for the largest wave height occurring in each month of a four-month « stormy » season for a ten-year interval. Thus $n = 40$ observed monthly extremes are available. Suppose one convinces oneself that the monthly extremes are approximately independent and identically distributed. It would then seem reasonable to try straight-line plots for the asymptotic extremal distributions or attempt experimental plots with other distributions like the Weibull or the logarithmic normal. If one of these yields an acceptable straight line, the line could be extended on past the data. The extrapolation would provide an estimate of the right-hand « tail » of the distribution function for monthly maximum wave heights during the « stormy » season.

A 20-year (or 80-stormy month) return period would be associated with monthly maximum values by the formula

$$F_{\text{mo max}}(x) = 1 - 1/R = 1 - 1/80 = 0.9875\,. \tag{47}$$

Hence the value x associated with $F_{\text{mo max}}(x) = 0.9875$ is the 20-year wave. The extrapolation becomes increasingly uncertain the further it is carried to the right beyond the points. As a rule of thumb, it is dangerous to extend it more twice the data base beyond the largest observation. An application of this general method to waves in the North Sea is given by DRAPER [11]. A study of extrapolation uncertainty for this problem was made by PETRAUSKAS and AAGAARD [12].

9. – The right-tail function.

As before, let $F_X(x)$ be the distribution function for the basic variable of interest. Define

$$W_n(x) = \begin{cases} n(1 - F_X(x)) & \text{for problem No. 1,} \quad (48) \\ \lambda L(1 - F_X(x)) & \text{for problem No. 2.} \quad (49) \end{cases}$$

A reasonable approximation to the distribution function for the largest value in n observations (problem No. 1) or the largest value during the operation life L (problem No. 2) where λ is the average number of events per unit time is (ref. [13], p. 370; ref. [14])

$$F_{\max X}(x) = \exp[-W_n(x)]\,. \tag{50}$$

This equation is exactly true for problem No. 2 in the Poisson case (see eqs. (6), (8)). It is a good approximation for problem No. 1, because

$$F_{X_{(n)}}(x) = [F_X(x)]^n = [1 - W_n(x)/n]^n \to \exp[-W_n(x)] \qquad \text{as } n \to \infty. \tag{51}$$

10. – Problem No. 3.

Consider the following problem. The largest, second-largest and third-largest daily maximum wave height for each of the four « stormy » months of the year have been recorded for ten years. Thus there are 40 monthly maximum waves which are the second-largest daily maximum for the month in which they occurred, and 40 waves which represent the third-largest daily maximum for each month. There would be no problem in using the 40 largest values in straight-line plotting by the methods previously discussed. However, how would the second-largest and third-largest be used to strengthen the determination?

By eq. (33), the distribution functions for the largest, second-largest and third-largest of n values are, respectively,

$$F_1(x) = P(X_{(n)} \leqslant x) = r^n , \tag{52}$$

$$F_2(x) = P(X_{(n-1)} \leqslant x) = n(n-1) \int_0^r t^{n-2}(1-t)\, dt , \tag{53}$$

$$F_3(x) = P(X_{(n-2)} \leqslant x) = [n(n-1)(n-2)/2] \int_0^r t^{n-3}(1-t)^2\, dt . \tag{54}$$

Let $X_{(1,k)} < X_{(2,k)} < X_{(3,k)} < \ldots < X_{(n,k)}$ be the ranked set of values for the k-th largest. In the above example, $n = 40$ and $k = 1, 2, 3$.

As before, $j/(n+1) = \hat{F}_{jk}$ is an unbiased estimate of $F_k(X_{(j,k)})$. Rather than plotting $\hat{F}_{jk}$ *vs.* $X_{(j,k)}$, it is more useful to set $\hat{F}_{jk}$ equal to $F_k(x)$ in eqs. (52) through (54) and solve for r (called $\hat{r}$, in this context). This is easy to do for eq. (52), where $r = [F_1(x)]^{1/n}$, but it is more difficult for the other equations. The process involved is the inversion of the incomplete beta-function and computer routines are available to solve for r, given $F_k(x)$ [15]. Alternately, tables for the upper five extremes may be used [14, 15].

$$\hat{r} = \hat{F}_X(X_{(j,k)}) \tag{55}$$

can be plotted *vs.* $X_{(j,k)}$ for each of the 120 values in the example to delineate the right-hand tail of $F_X(x)$. Alternately, the right-hand tail function $W_{40}(x)$ given by $40(1-r)$ could be plotted. In either case, the resulting graph characterizes the right-hand tail of the basic population distribution function. This function could then be used for extremal predictions either by incorporating it into a model or by using formulae for the type given in eq. (2) or eq. (8).

If the larger waves were breaking in the statement of problem No. 3, the graphs of $\hat{F}_X$ would specify the right-hand tail of the wave height distribution and perhaps indicate the deviations from the Rayleigh distribution for breaking waves.

11. – Probabilities for the highest wave and associated variables.

Distribution theory is intimately involved in the model-building approach to extremal statistics. The k-th largest of a set of n random variables is called an order statistic. The distributional properties of such quantities under various chosen assumptions are studied as a part of mathematical probability generally referred to as « Order Statistics ».

As an illustration of the use of the distribution theory for order statistics in ocean engineering, the following problem will be considered. Suppose that n random wave heights, H_i, and their associated periods, T_i, occur according to the probability density $f_{H,T}(h_i, t_i)$ and distribution function $F_{H,T}(h_i, t_i)$. Suppose that each height-period pair is independent of the other height-period pairs and all follow the common probability law. What is the joint-probability law for the maximum H_i value and the T_i associated with it? If T_A is defined as the period associated with whatever wave happens to have the largest height in the n waves, the problem may be expressed as « What is $P(H_{\max} \leqslant h$ and $T_A \leqslant t)$? ».

The answer to the above problem is

$$P[H_{\max} \leqslant h,\ T_A \leqslant t] = \int_0^t \int_0^h f_{H,T}(h_1, t_1)\, n F_H^{n-1}(h_1)\, \mathrm{d}h_1\, \mathrm{d}t_1\,. \tag{56}$$

The proof is as follows. Let Q denote the left-hand side of eq. (56):

$$Q = P\left[\bigcup_{i=1}^{n} (H_i \leqslant h,\ T_i \leqslant t,\ H_i \text{ is the largest wave})\right]. \tag{57}$$

The events in the union on the right-hand side of eq. (57) are mutually exclusive and exhaustive. Furthermore, each of them has the same probability by symmetry. Hence

$$Q = \sum_{i=1}^{n} P(H_i \leqslant h,\ T_i \leqslant t,\ H_i \text{ is the largest wave}) = nP(H_1 \leqslant h,\ T_1 \leqslant t,\ H_1 \text{ is the largest wave})\,. \tag{58}$$

Continuing the derivation,

$$Q = nP(H_1 > H_2,\ H_1 > H_3, \ldots,\ H_1 > H_n,\ H_1 \leqslant h,\ T_1 \leqslant t)\,, \tag{59}$$

$$Q = n \int_0^t \int_0^h \int_0^{h_1} \ldots \int_0^{h_n} f_{H,T}(h_1, t_1)\, f_{H_2}(h_2) \ldots f_{H_n}(h_n)\, \mathrm{d}h_n \ldots \mathrm{d}h_2\, \mathrm{d}h_1\, \mathrm{d}t_1\,. \tag{60}$$

The integration is over the part of the $H_1, H_2, \ldots, H_n, T_1$ space where the event on the right-hand side of eq. (59) holds. Also the pairwise independence is introduced here. Since

$$F_H(h) = \int_0^h f_H(h)\,\mathrm{d}h\,, \tag{61}$$

it follows that

$$Q = \int_0^t \int_0^h f_{H,T}(h_1, t_1)\, n F_H^{n-1}(h_1)\,\mathrm{d}h_1\,\mathrm{d}t_1\,, \tag{62}$$

which is the final result.

A similar probability distribution function for maximum wave height, associated period and associated crest elevation η_c may be stated by analogy

$$P[H_{\max} < h,\ T_A < t,\ \eta_c < y] = \int_0^y \int_0^t \int_0^h f_{H,T,\eta_c}(h_1, t_1, y_1) n F_H^{n-1}(h_1)\,\mathrm{d}h_1\,\mathrm{d}t_1\,\mathrm{d}y_1\,. \tag{63}$$

Of course, in order to use these formulae, the joint densities involved are needed. Some work has been done [16-20] and a number of studies of height-period probability laws have been made [21-26].

The equations for extreme wave height and associated period may be extended to waves in storms with time-varying storm intensities. Four basic theories related to this situation may be stated.

Theorem 1. Let (H_i, T_i), $i = 1, 2, \ldots, n$, be independent pairs of random variables with a continuous and differentiable joint-distribution function. Let $f_{H_i,T_i}(h, \tau)$ be the probability density function for (H_i, T_i), $f_{H_i}(h)$ be the probability density function for H_i and $F_{H_i}(h)$ be the distribution function for H_i. Then the probability density for the maximum value of H_i and the value of T associated with the maximum is

$$f_{H\max,T\mathrm{assoc}}(h, \tau) = \left(\prod_{j=1}^n F_{H_j}(h)\right)\left(\sum_{i=1}^n [f_{H_i,T_i}(h, \tau)/F_{H_i}(h)]\right). \tag{64}$$

Proof.

$P[\max H \leqslant h$ and $T\,\mathrm{assoc} \leqslant \tau] =$

$$= P\left[\bigcup_{i=1}^n H_i \leqslant h,\ T_i \leqslant \tau \text{ and } \max H \text{ occurs on the } i\text{-th pair}\right] =$$

$$= \sum_{i=1}^n P\{H_i \leqslant h,\ T_i \leqslant \tau \text{ and } \max H \text{ occurs on the } i\text{-th pair}\} =$$

$$= \sum_{i=1}^n P\{H_i \leqslant h,\ T_i \leqslant \tau \text{ and } H_i > H_j \text{ for } 1 \leqslant j \leqslant n,\ j \neq i\} =$$

$$= \sum_{i=1}^{n} \int_0^h P\left(\bigcap_{j=1, j \neq i}^{n} [H_j < x | H_i = x] \cap \{P[T_i \leqslant \tau | H_i = x]\} f_{H_i}(x) \right) dx =$$

$$= \sum_{i=1}^{n} \int_0^h \left[\prod_{j=1}^{n} F_{H_j}(x) \right] F_{T_i | H_i = x}(\tau) f_{H_i}(x) \, dx =$$

$$= \sum_{i=1}^{n} \int_0^h \left[\sum_{j=1}^{n} F_{H_j}(x) \right] [F_{T_i | H_i = x}(\tau) f_{H_i}(x) / F_{H_i}(x)] \, dx .$$

Now

$$f_{H\max, T\mathrm{assoc}}(h, \tau) = [\partial^2 / \partial h \, \partial \tau] P\{\max H \leqslant h \text{ and } T \,\mathrm{assoc} \leqslant \tau\} =$$

$$= [\partial^2 / \partial h \, \partial \tau] \left[\sum_{i=1}^{n} \int_0^h \left(\prod_{j=1}^{n} F_{H_j}(x) \right) \{F_{T_i | H_i = x}(\tau) f_{H_i}(x) / F_{H_i}(x)\} \, dx \right] =$$

$$= \sum_{i=1}^{n} \left(\prod_{j=1}^{n} F_{H_j}(h) \right) (f_{H_i, T_i}(h, \tau) / F_{H_i}(h)) = \left(\prod_{j=1}^{n} F_{H_j}(h) \right) \left(\sum_{i=1}^{n} [f_{H_i, T_i}(h, \tau) / F_{H_i}(h)] \right) =$$

$$= \left(\prod_{j=1}^{n} F_{H_j}(h) \right) \left(\sum_{i=1}^{n} [f_{H_i, T_i}(h, \tau) / F_{H_i}(h)] \right) .$$

Theorem 2. Under the same assumptions as theorem 1, the probability density for H max is

$$f_{H\max}(h) = \left(\prod_{j=1}^{n} F_{H_j}(h) \right) \sum_{i=1}^{n} [f_{H_i}(h) / F_{H_i}(h)] . \tag{65}$$

Proof.

$$f_{H\max}(h) = \int f_{H\max, T\mathrm{assoc}}(h, \tau) \, d\tau = \left(\sum_{j=1}^{n} F_{H_j}(h) \right) \left(\sum_{i=1}^{n} [f_{H_i}(h) / F_{H_i}(h)] \right) .$$

Theorem 3. Under the same assumptions as theorem 1, the conditional-probability density for the associated period, given $H \max = h$, is

$$f_{T\mathrm{assoc} | H\max = h}(\tau) = \left\{ \sum_{i=1}^{n} [f_{H_i, T_i}(h, \tau) / F_{H_i}(h)] \right\} \Big/ \left\{ \sum_{i=1}^{n} [f_{H_i}(h) / F_{H_i}(h)] \right\} . \tag{66}$$

Proof.

$$f_{T\mathrm{assoc} | H\max = h}(\tau) = f_{H\max, T\mathrm{assoc}}(h, \tau) / f_{H\max}(h) .$$

After substitution from theorems 1 and 2, the result follows.

Corollary to theorem 3. Under the same assumptions

$$f_{T\mathrm{assoc} | H\max = h} = \left(\sum_{i=1}^{n} \{f_{H_i}(h) / F_{H_i}(h)\} \{f_{T_i | H_i = h}(\tau)\} \right) \Big/ \left(\sum_{i=1}^{n} \{f_{H_i}(h) / F_{H_i}(h)\} \right) = \tag{67}$$

$$= \sum_{i=1}^{n} w_i f_{T_i | H_i = h}(\tau) \Big/ \sum_{i=1}^{n} w_i ,$$

where

$$w_i = f_{H_i}(h)/F_{H_i}(h) . \tag{68}$$

Theorem 4. Suppose $\{H_i, i = 1, 2, ..., n\}$ are independent, identically distributed continuous random variables with probability density $f_{H_i}(h) = f_H(h)$; $\{T_i, i = 1, 2, ..., n\}$ are also independent, identically distributed continuous random variables with probability density $f_{T_i}(\tau) = f_T(\tau)$. Let $f_{H,T}(h, \tau)$ be the joint-probability density function for (H, T). Then:

1) The probability density for the maximum value of H_i and the value of T associated with the maximum is

$$f_{H\max, T\text{assoc}}(h, \tau) = hF_H^{n-1}(h) f_{H,T}(h, \tau) . \tag{69}$$

2) The probability density for H max is

$$f_{H\max}(h) = nF_H^{n-1}(h) f_H(h) . \tag{70}$$

3) The conditional-probability density for the associated period, given H max $= h$, is

$$f_{T\text{assoc}|H\max=h}(\tau) = f_{H,T}(h, \tau)/f_H(h) . \tag{71}$$

12. – An example of extremal estimates by extrapolation.

PETRAUSKAS and AAGAARD [12] give characteristic « largest » wave heights as hindcast for 22 frontal-system storms relative to a specific West Coast location. The values represent all frontal-system storms generating a « largest » wave height that would be obtained in a 60-year interval if each frontal-system storm (with exceedence of 12.0 feet largest wave height) were hindcast by the same basic procedure. If we assume that the storms continue at the same rate, there would be 44 such storms in 120 years. Thus the estimation process can be divided into two parts. First the basic distribution function for the « largest » hindcast wave height in a single storm with exceedence of the 12.0 foot lower limit must be estimated. Then the basic distribution is manipulated to obtain the distribution function for the 120-year maximum hindcast « largest » height.

For the first part, the 22 hindcast values may be ranked in order of increasing size. Let $x_{(k)}$ denote the value with rank k. That is, $x_{(1)} \leqslant x_{(2)} \leqslant x_{(3)} \leqslant ... \leqslant x_{(22)}$. The basic distribution function may be estimated [5] by

$$\hat{F}_X(x_{(k)}) = k/(n+1) , \tag{72}$$

where $n = 22$ in the present case. (This is not the only possible way to make this estimate, but it is a fairly good way and will be used here.)

For the second part, if we assume that all the hindcast values are independent of each other and identically distributed, it follows from eq. (2) that

$$F_{X_{\max(120\,\mathrm{y})}}(x) = [\hat{F}_X(x)]^{44}\,. \tag{73}$$

Table I summarizes these computations.

A difficulty arises immediately. Even if one only wanted the median 120-year maximum height, it is substantially larger than the largest hindcast height during the 60 years. One has to extrapolate the fourth column in table I in order to estimate the median 120-year maximum which is the 0.984 fractile for the basic distribution.

TABLE I. – *Distribution function for the* 120-*year maximum hindcast of largest wave height in frontal-system-type storms.*

Rank k	Hindcast « largest » wave height	$\hat{F}_X(x_{(k)})$	$F_{X_{\max(120\,\mathrm{y})}}(x_{(k)})$	$-\ln[-\ln \hat{F}]$
1	12.00	0.043	0	−1.143
2	12.00	0.087	0	−0.893
3	12.30	0.130	0	−0.711
4	13.30	0.174	0	−0.559
5	15.90	0.217	0	−0.423
6	16.50	0.261	0	−0.295
7	17.60	0.304	0	−0.174
8	19.00	0.348	0	−0.055
9	19.40	0.391	0	0.064
10	19.80	0.435	0	0.183
11	19.80	0.478	0	0.304
12	22.40	0.522	0	0.430
13	22.60	0.565	0	0.561
14	23.30	0.609	0	0.700
15	23.90	0.652	0	0.850
16	24.00	0.696	0	1.014
17	24.90	0.739	0	1.196
18	28.10	0.783	0	1.406
19	29.00	0.826	0	1.655
20	30.90	0.870	0.002	1.968
21	34.70	0.913	0.018	2.397
22	35.00	0.957	0.141	3.113

Curvilinear extrapolation for any distance is difficult. Consequently most investigators try to surmise the functional form for $F_X(x)$ and make a transformation of variables which produces a straight-line plot. For example, suppose one assumes that the hindcast « largest » wave heights have the extremal form (eq. (28A))

$$F_X(x) = \exp\left[-\exp\left[-(ax+b)\right]\right]. \tag{74}$$

Then

$$-\ln\left[-\ln F_X(x)\right] = ax + b \tag{75}$$

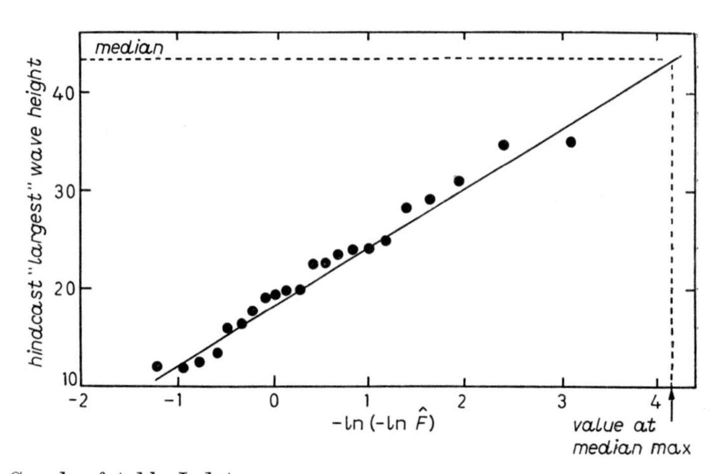

Fig. 1. – Graph of table I data.

and a graph of $-\ln[-\ln F_X(x)]$ *vs.* x would be a straight line. It would be expected that, by analogy, a plot of $-\ln[-\ln \hat{F}_X(x_{(k)})]$ *vs.* $x_{(k)}$ would also be a straight line. The values of $-\ln[-\ln \hat{F}]$ are listed in the fifth column of table I and are graphed as fig. 1.

Since the median 120-year maximum is the 0.984 fractile as referenced to the basic distribution function, the median 120-year maximum will correspond to a $-\ln[-\ln F]$ value of 4.151. From fig. 1, then, the median 120-year maximum would be 43.2 feet.

The interquartile range for the 120-year maximum would be obtained as follows. The 0.25 fractile and the 0.75 fractile for the 120-year maximum would be the $0.25^{1/44} = 0.969$ and the $0.75^{1/44} = 0.993$ fractile for the basic distribution function given in fig. 1. Consequently, the corresponding $-\ln(-\ln \hat{F})$ values would be 3.458 and 5.030. Thus the 0.25 and 0.75 fractiles for the 120-year maximum would be 39.2 and 48.7 feet, respectively. The interquartile range for the 120-year maximum is thus $48.7 - 39.2 = 9.5$ feet.

The example was selected to bring into sharp relief some of the assumptions and difficulties in the extrapolation method. It is primarily an illustration and the median 120-year maximum value of 43.2 feet produced should not be accepted uncritically.

From a purely logical viewpoint, what are some of the critical aspects of the method of extremal estimation by extrapolation? Various of these aspects will be discussed in the following.

13. – Plotting and distribution formulae.

Clearly the extrapolation procedure is significantly affected by the choice of the plotting position formula. The $k/(n+1)$ computation is just one possible selection. Indeed, PETRAUSKAS and AAGAARD used another, somewhat more complicated and a little more statistically justifiable, choice due to GRINGORTEN [10] and obtained a somewhat different graph. It is really not a question of which is right or which is wrong. Each plotting procedure has some statistical or logical justifications.

The selection of a formula for the basic distribution function is similarly arbitrary. Sometimes physical reasoning may be used to support a particular choice. Unfortunately, several formulae usually fit the data equally well, so that the data themselves cannot be used to specify which formula is « best ». PETRAUSKAS and AAGAARD study a variety of different choices for fitting the table II data and demonstrate the complexity of the decision procedure in this situation.

TABLE II. – *Comparison of hurricane Carla fitted $H_{\text{r.m.s.}}$ and that predicted by sea surface variance.*

Date	Hour	σ^2 (a)	$\sqrt{8\sigma^2}$	$H_{\text{r.m.s.}}$ (b)	Departure
September 8, 1961	0600	9.58	8.7	8.0	—
	1200	11.48	9.6	7.2	—
	1800	13.77	10.5	17.2	+
September 9, 1961	0	13.16	10.3	14.0	+
	0600	18.13	12.0	12.8	+
	0620	15.60	11.2	10.8	—
	1200	25.38	14.3	14.1	—
	1500	26.75	14.6	15.6	+
	1800	17.74	11.9	13.3	+
	2100	20.63	12.8	17.7	+
September 10, 1961	0	22.64	13.5	12.1	—
	0100	22.04	13.3	16.4	+
	0300	23.57	13.7	18.3	+

(a) From wave recorder.
(b) From 19 largest waves.

The critical property of the basic distribution function relative to its effect on the extremal statistics is its behavior in the right-hand tail of $F_X(x)$. One should be very careful to ensure that the basic distribution function adequately reflects any physical limiting processes which may impose restrictions on extremal events. Wave breaking is a case in point. Evidence is accumulating [27-30] that the Rayleigh distribution must be modified in the far right tail to decrease the higher fractiles of wave height. Figure 2 shows this behavior. Presumably this departure from the Rayleigh distribution is caused by waves breaking and reforming when the height exceeds certain critical levels.

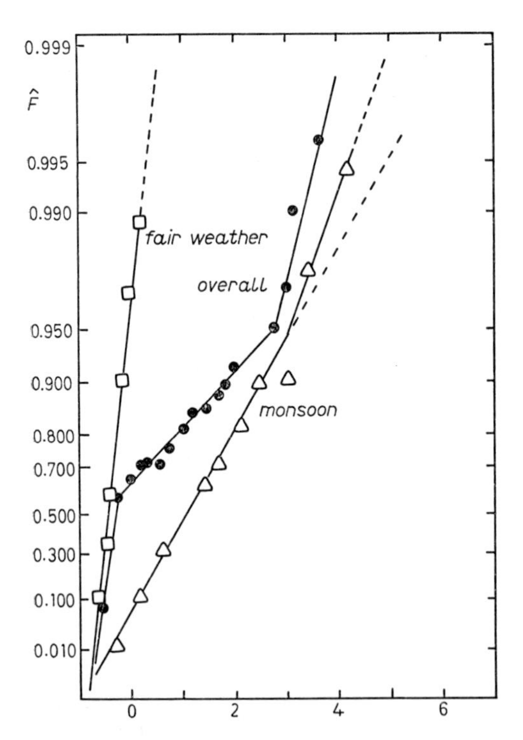

Fig. 2. – Theoretical and observed wave height distributions. Observed waves from 72 individual 15-minute observations, containing a total of 11 678 waves (15 364 calculated waves), from several Atlantic coast wave gages are superimposed on the Rayleigh distribution curve (fig. 3-4, [29]).

14. – Sample variability.

Even if one has a perfect plotting position formula and « knows » precisely what the proper distribution function is, estimate inaccuracies arise due to the random accidents of the particular data measured. One data set may possess lower values overall than another data set drawn from the identical basis population. In the extrapolation method for extremal estimation, however, the slope of the line produced by the data set with the accidentally lower values will be less than the slope for the data set with the slightly larger values. This difference leads to substantial discrepancies if the lines are extrapolated a considerable distance.

As an example, 10 sets of 50 random numbers each were taken from a normal random-number table where the numbers have zero mean and unit standard deviations. The largest number for each set of 50 was recorded and the 10 values, each the maximum of 50 normals, were used to predict the median of the maximum of 500 such normal random numbers. The plotting position formula due to GRINGORTEN [10]

$$\hat{F}_X(x_{(k)}) = (k - 0.44)/(n + 0.12) \tag{76}$$

was used. This procedure was repeated three times with a different collection of random numbers. The true median for the maximum of 500 numbers drawn from the normal population would be the 0.998 61 fractile or 2.99.

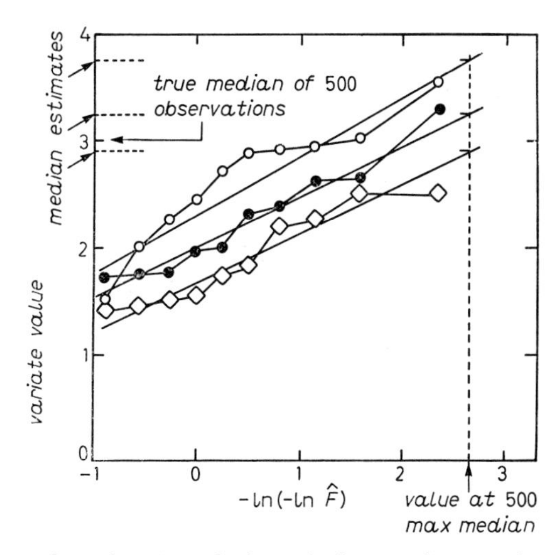

Fig. 3. – Three examples of extrapolation of the maximum of 50 observations from a $N(0, 1)$ population.

Figure 3 shows the three extrapolations. As can be seen, one comes out fairly near the true value, while the other two are substantially different. There is really nothing one can do about this intrinsic variability from data set to data set except to be aware of it in the decision-making process. A fairly good rule of thumb is to not extrapolate more than twice the extent of the data base. That is, 50 years might be pushed to make predictions relative to 100-year periods and so forth. Unfortunately, engineering deadlines and nonpostponable decisions may force one to make extrapolations on inadequate data, just because guidance from these figures is better than nothing. A bit of skepticism and hedging together with an attempt to avoid truly disasterous consequences is about your only option in such circumstances.

The paper by PETRAUSKAS and AAGAARD [12] presents the results from simulation studies which show reasonable ranges for the variation of slope in the extrapolation process due to the intrinsic randomness of the data.

15. – Population assumptions.

In deriving eq. ($2B$), it was casually assumed that the data were independent and identically distributed. These two assumptions make the analysis much simpler. Unfortunately, they are often not appropriate.

If the data are not independent, it is necessary to retreat to eq. ($2A$). Statistically, it is very difficult to work with multivariate probability laws. There are many unresolved research questions.

Several studies have shown [31, 32] that limited interdependence does not seriously change the basic extremal formulae. There may, of course, be some changes in parameter values. Most developments of extremal methods, therefore, proceed on the assumption that independence holds and introduce calibration procedures at some point in the calculations.

If independence is accepted but the data are not identically distributed, eq. ($2B$) must be the basis for analysis. A good example of this is the sequence of wave heights during the onset, peak and tail end of a hurricane. One model for this situation was developed from manipulations of eq. ($2B$) as

$$F_{\max H}(h) = \exp\left[\int_a^b \ln\left[1-\exp\left[-2h/H_s(t)\right]\right] \mathrm{d}t/T(t)\right], \tag{77}$$

where $H_s(t)$ and $T(t)$ are the significant wave height and wave period, respectively, as functions of time [33]. Other similar formulae have been proposed [34-37].

16. – The greatest danger.

During September, 1976, hurricane Kathleen crossed Mexico from the Gulf of Mexico into the Pacific and then turned north, crossing the Southern California desert. Some desert areas of San Diego County received more than 10 inches of rain in slightly less than 24 hours. The mountains west of Palm Desert received eight inches, while the floor of Coachella valley received better than four inches of precipitation in this same period [38].

Any extremal predictions based on routine yearly precipitations would not encompass the hurricane precipitation, since it is an entirely different phenomenon than the usual thunderstorm or cold-front-type storms. Data for one type of phenomena cannot be used to estimate extremes for a different type of phenomena. By similar argument, East Coast wave data which by accident do not contain hurricane-induced waves could not logically be used to predict the extreme wave heights which would result when a hurricane does reach that site.

Another more insidious example of such confusion of phenomena occurs when the data arise from a mixture of several populations. This can happen unknowingly.

An excellent example can be found in the analysis of wave records off Mangalore Harbor in India [39, 40]. In this area, the year can be subdivided into two distinct seasons: fair weather and monsoon. Figure 4 shows a plot of $F_{(x)}$ against H_{max} for each season as well as for the entire year. Clearly the stratification of wave heights into seasons leads to significantly different results than the treatment of the year as a whole. Since there is no *a priori* reason to expect the distribution of wave heights in the fair-weather portion of the year to markedly affect the distribution of wave heights during the monsoon

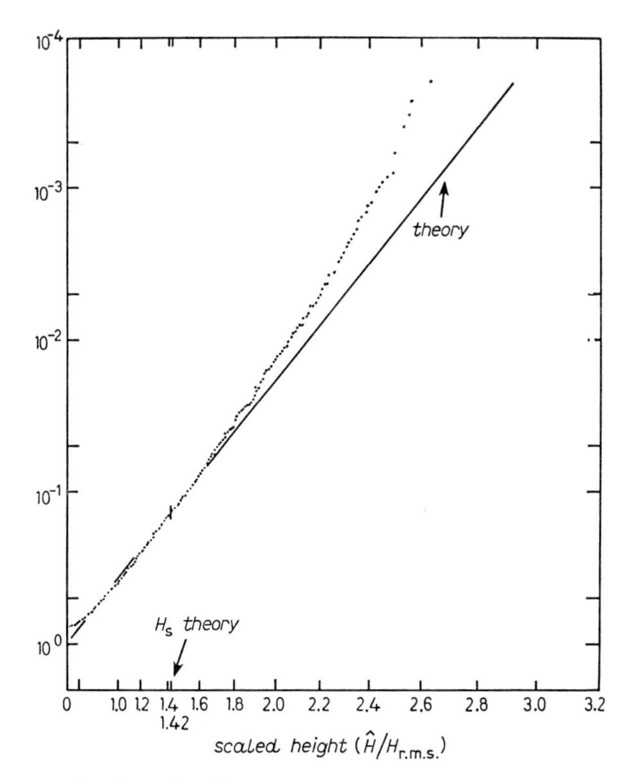

Fig. 4. – An example given by GOURLAY [40].

season, extrapolations from the stratified monsoon sample probably give a better estimate of expected extremes. It is also interesting to note that, even in the data for only the monsoon season, there is a sharp discontinuity in the straight-line plot. Such a discontinuity can be indicative of the presence of still another population within the monsoon season. Possibly a limiting factor such as energy loss caused by increased wave breaking due to depth becomes important for wave heights above four meters, or possibly this region represents individual wave heights which are affected by the deviation from the Rayleigh distribution, as shown in fig. 2. At any rate this example well demonstrates some difficulties to be expected in using extrapolations from mixed populations. In the United States, such differences might occur if particularly stormy seasons (typically winter) are not treated separately from the rest of the year.

If only one observation is from the more intense phenomenon, the one value will plot far off the straight line from the rest of the data. Unfortunately, anomalous behavior by one point can also be due to accidental gross errors in recording or measuring data. One is always concerned whether the outlier is an observer error or is physically real. In the first case, the point should be deleted. In the second, the unusually large value is perhaps the only datum point present arising from the phenomena which will produce the long-term extreme. The inclusion of this other into a sample can severely distort the parameter estimates as well as influence the type of parent curve presumed to represent the distribution of extremes.

17. – Extremal prediction by model building.

If the data in column 2 of table II were actual measurements of the largest wave height occurring at a particular site during the frontal-type storm, the extrapolation method would produce a pure data-based extremal estimate. The use of hindcast data, instead of measurements, gives a hybrid model and extrapolation procedure. The hindcast methods represent a mathematical modeling of the physical processes involved [41, 42].

In many cases, the actual data set available is insufficient for the extrapolation method. For example, the number of hurricanes passing near a given site may be quite limited. It would be hazardous to extrapolate to long-time intervals by the hindcast-and-line-fit method if only the properties of the local hurricanes were used. Therefore, the hurricanes passing near the site are assumed to be a random sample drawn from the hurricanes affecting a larger region around the site. The number of hurricanes expected to pass near the site per year may also be taken as random, perhaps behaving as a Poisson random variable

$$P[N = n] = \exp[-\lambda Lt]\,(\lambda Lt)^n/n\,! \tag{78}$$

Here λ denotes the number of storms per mile of coastline per year, t is the prediction time interval and L is length interval about the site.

The larger set of hurricanes occurring in the region and the number of storm coastal crossings up and down the coast may be used to strengthen the extremal prediction for the site. The probabilistic assumptions introduced and the resulting formulae constitute a process called statistical modeling. The construction of statistical models may require substantial statistical and mathematical sophistication and may involve many subjective judgements on the part of the designer.

The combination of hindcasts and a probabilistic framework join together both statistical and mathematical aspects of modeling to produce a probabilistic-physical model. Statistical formulae can then be used to predict the extremal behavior of various aspects of the model.

Returning to the hurricane example, one can proceed to make an extremal estimate by hindcasting the wave property of interest for each hurricane and then ranking the estimates. These are graphed with some subjective choice of a plotting position formula to produce an estimate of the distribution function $\hat{F}_X(x)$ for the property relative to its regional occurrence. The maximum occurrence at the site in t years can then be shown to have the distribution function

$$F_{\max X}(x) = \exp\left[-\lambda L t[1 - \hat{F}_X(x)]\right] \tag{79}$$

subject to the error of estimation in $\hat{F}_X(x)$.

Various sophistications can be introduced into this simple structure. The path and direction of travel of the hurricane can be made probabilistic. The distribution function for the wave property in each storm can be hindcast instead of a single value. Probability laws can be developed for the regional storm parameters as an alternative to the utilization of the actual occurring storms.

Each sophistication adds more generality to the model, but also increases the computation complexity and enlarges the data base needed to adequately calibrate the model. Obviously a balance needs to be maintained which successfully represents the real world, is computationally feasible, and does not require unavailable data.

18. – The joint-probability method.

The hurricane example just discussed is one illustration of a general model structure which has come to be labeled « the joint-probability method ». It is worthwhile to define this basic structure more formally. The joint-probability method has the following aspects.

1) The damaging phenomenon is produced by well-defined entities or events (storms, earthquakes, etc.) which can be parameterized (storm radius, maximum wind, etc.).

2) A physical model can be developed which predicts the magnitude (and/or probability law) of the damaging phenomenon from the parameter values for the event.

3) Historical data are used to predict the probabilities associated with the occurrence of various event intensities. This is usually expressed as the multivariate probability law for the parameters characterizing the event.

4) The predictions from item 2) are combined with the probabilities from item 3) to obtain extremal estimates for the site.

The joint-probability method has been used extensively by NOAA [43] in recent years to predict coastal flooding. A computerized surge model is used to predict the surge height for a range of different storm parameters, with long-term climatology providing the probabilities for each combination of storm parameters. Statistical formulae are then used to obtain risk criteria associated with various intensities of flooding.

19. – State of art in hindcast procedures.

The accuracy associated with various hindcast procedures is obviously very important in many methods for preparing extremal estimates of coastal phenomena. If the hindcast procedure has a tendency to overestimate or underestimate the quantity being predicted, these deviations will be exaggerated in the computation of extremes.

It is worthwhile then to briefly examine the current state of the art in making hindcasts for waves or flooding. A technique for obtaining unbiased wind estimates over the region of wave generation is essential to reliable hindcasts. Extensive studies [44, 45] have indicated that the surface wind speed is not a simple constant times the geostrophic wind speed. The use of winds estimated in this way can result in serious overestimation of extreme wave heights.

Once the wind speeds are calibrated over the water or a previously calibrated method is chosen, several wave prediction methods are available. These range from the simplicity and directness of empirical significant-wave techniques to the complexity of a numerical model for wave spectra. Recent efforts would seem to suggest that most of the methods give fairly comparable results in simple wind fields over deep water and away from any coasts. Near the coast, however, the significant-wave methods can produce overestimates of wave heights due to failure to consider the directional spreading of the waves. Of these significant-wave methods, that due to WILSON [46] has the form for growth of wave height with fetch which agrees with recent field efforts [47, 48]

$$H_{\frac{1}{3}} \sim c\sqrt{F}\,, \tag{80}$$

where F is the fetch for wave generation and c is an empirical constant.

Extensive numerical models based on the solution to the radiative-transfer equation for two-dimensional wave spectra [49-51] require considerable computer time, but appear to give results in comparison with observed data. A multiparametric model for the radiative-transfer equation has been advanced by HASSELMANN [47] and a simplified one-parameter model for this has been proposed by HASSELMANN [52]. There is little current agreement regarding the accuracy of the results from these parametric models other than in the case of fetch-limited wave spectra.

A limited comparison of the relative performances of several numerical models can be found in ref. [53]. Major factors to consider in choosing which hindcast method to use in hindcasting storm waves are the requisite accuracy and the relative economies of different methods. When combined with reasonable probability models for storms or long historical records of storm wind fields, these hindcast models all give estimates of extremes that appear adequate for most design purposes.

Maximum surge levels from hurricanes are estimated by some form of a solution to the equations of motion for an incompressible fluid with prescribed boundary conditions. These equations can be solved by either finite-difference techniques or finite-element techniques. The finite-difference method has been applied to rectilinear grid systems and more recently has been extended to curvilinear grid systems. The ability of the curvilinear system to represent the land-sea interface and to produce a reduction in program logic makes these models presently the best available. JELESNIANSKI [54] developed a model in nonorthogonal curvilinear co-ordinates, but encountered considerable problems related to cross-terms in the co-ordinates. More recently REID and WANSTRATH [55] developed a model in orthogonal, curvilinear co-ordinates which appears to eliminate many of these problems.

The finite-element technique has the advantage of fitting coastlines very precisely and accommodating highly variable element sizes. However, some problems in integrating the source terms over a time step make these models somewhat more costly at present than the Reid-Wanstrath model. PAGENKNOPF and PEARCE [56], in a comparison of finite-difference and finite-element methods, conclude that there does not appear to be any significant advantage to the finite-element method.

The largest source of errors in the surge models is most likely to come from the lack of precise parameterizations for the coefficient of drag and the indeterminate influence of land on the hurricane parameters at landfall. However, as with hindcast models, the surge models, when combined with reasonable probabilities of storm parameters, give estimates of extremes that appear adequate for most design purposes.

20. – Sources of error in modeling.

The method of extremal prediction with statistical-physical models clearly possesses a great degree of flexibility and allows the the introduction of numerous special features in any particular problem. This is not an unmixed blessing, however, and mathematical modeling is not a panacea. The great flexibility has associated with it a major dependence on the judgement and mathematical maturity of the investigator. Many subjective decisions are involved. Paradoxically, extremal estimates with statistical-physical models may provide the best possible estimate or an abysmal failure.

Cross-checking of the model prediction against data is extremely important. Often the data may be artificially divided into two portions at random. The first part is used to calibrate the model through parameter estimation. The second data set is then used to test the prediction accuracy.

The previous section on hindcast methods and their reliability is, of course, relevant to the prediction accuracy of models. Hindcasting is an important component of most models. Questions of interdependence and true probability behavior also may seriously affect prediction accuracy.

The joint probabilities for various combinations of storm parameters are usually estimated by a multivariate frequency histogram. This has some disadvantages as compared to the method of fitting a parameterized family of surfaces. The histogram is arbitrarily truncated about the values which historically occurred. Thus the possibility of future storms with decidedly different parameter values is ignored. The family of surfaces, on the other hand, tails out around the histogram edges introducing small but nonzero probabilities of events outside the historical data set. The joint-probability density can also be estimated by smoothed-histogram methods [57] and the same advantages obtained without assuming particular density formulae. If the method of parameterized density families is used, the usual questions concerning parameter estimate reliability must be studied.

It is difficult to be very specific in the enumeration of error sources in modeling, since a point-by-point examination of each particular model is necessary to prepare such a list. However, it is important to recognize that major sources for error do exist, so that the decision-maker is alerted to the necessity for appropriate cross-checking before the predictions are used in engineering planning.

21. – Possible future improvements.

All of the previous methods reviewed in this paper have treated a single source of information, model outputs or measured data, and have brought the engineer to the point where he still must select some subjective risk level. However, methods using Bayesian analyses and the framework of decision theory are capable of overcoming these shortcomings. Given a set of initial information, I, say from model outputs or the joint-probability method, a set of historical data, Q, can be used to update this information by the application of the Bayes theorem. If the prior probability density function of the extremal parameter set, θ, is expressed as $f'(\theta|I_0)$, then the updated distribution of θ is given by

$$f''(\theta|Q, I_0) = f(Q|\theta)\cdot f'(\theta|I_0)/k\,, \tag{81}$$

where $f''(\theta|Q, I)$ is the posterior density function for θ based on both the initial information and the historical data, $f(Q|\theta)$ is the probability of the observations given the parameter set (or the sample likelihood function of the parameter set, conditional upon the observation), and k is a normalizing constant.

This parameter uncertainty can be included into the estimation of the density function of wave heights (or surge heights) by integrating the conditional probability of wave heights given a parameter set over all parameter values

$$f(H) = \int_\theta f(H|\theta)\cdot f''(\theta)\,\mathrm{d}\theta\,. \tag{82}$$

If one can determine a loss function, l, for different wave heights, then the expected loss for a particular design height can be computed from

$$E(l|H) = \int_H l(H)\,f(H)\,\mathrm{d}H\,. \tag{83}$$

WOOD and RODRIGUEZ-ITURBE [58] show that application of such Bayesian methods can lead to significantly different choices of design for hydrologic structures than those based on point estimators and they provide details permitting the quantification of factors such as preferences and their introduction into the decision process.

22. – Summary.

The extensive discussion of problems and difficulties in making predictions of extreme wave heights is not intended to scare the design engineer away from making such estimates. Rather it is offered as a check list of things to examine before seriously using an extremal estimate. Despite their inexact nature, numerical values for wave extremes are necessary in most coastal design procedures. They cannot really be avoided. One can only strive to use them intelligently and with a full understanding of their inexact nature.

In the extrapolation method, one should consider

1) plotting position formula effects,

2) parent population choices and the right-tail behavior of the distribution function for the population,

3) data interdependence,

4) lack of identically distributed data,

5) effects of sample randomness,

6) the possibility that extremes arise from different phenomena than the data,

7) mixtures in the data.

In model building, the list of precautions is more vague. However, one should carefully

1) examine the appropriateness of the interdependence assumptions and probability laws chosen for the statistical structure,

2) review the wave or surge prediction methods to make sure the physical assumptions are satisfied in the application,

3) cross-check the predictions against data if at all possible.

In both general procedures, it is important to make an effort to estimate the reliability or range of possible variation of the extremal events. The statement of a single value, say the median, for the estimated extreme can be very misleading. One needs to also state some measure of spread, say the interquartile range, which alerts those who may later use the figures to the random nature of the future extreme event and the range of values over which it may actually occur.

This has been an expository paper reviewing the field of extremal statistics and its use in ocean engineering from a particular personal viewpoint. The field is large and the length of the paper was necessarily limited. Many significant topics and investigations were unavoidably not included.

However, it is hoped that the general presentation given will permit engineers to get a better grasp of the statistical methods used in treating extremal problems. Certainly there are many questions concerning reliability and risk analyses, for engineering design in the ocean environment, which require a working knowledge of this field.

* * *

The authors wish to gratefully acknowledge the assistance provided by Mr. A. YFANTIS, Mrs. F. PRYOR and Miss L. BOSWELL, in the preparation of this paper.

REFERENCES

[1] L. E. BORGMAN: *Extremal statistics in ocean engineering*, in *Proc. Cicil Eng. Oceans*, Vol. **3** (New York, N. Y., 1975), p. 117.
[2] L. E. BORGMAN and D. T. RESIO: *Extremal prediction in wave climatology*, in *Proceedings Ports 1977*, Vol. **1** (New York, N. Y., 1977), p. 394.
[3] R. A. FISHER and L. H. C. TIPPETT: *Proc. Cambridge Philos. Soc.*, **24**, part 2, 180 (1928).
[4] B. GNEDENKO: *Ann. Math.*, **44**, 423 (1943).
[5] E. J. GUMBEL: *Statistical Theory of Extreme Values and Some Practical Applications*, NBS Appl. Math. Ser. 33 (Washington, D. C., 1954).
[6] M. ZELEN and N. C. SEVERO: *Probability functions*, in *Handbook of Mathematical Functions*, NBS Appl. Math. Ser. 55 (Washington, D. C., 1964).
[7] L. BEARD: *Trans. Am. Soc. Civ. Eng.*, **108**, 1110 (1943).
[8] G. BLOM: *Statistical Estimates and Transformed Beta-Variables* (New York, N. Y., 1958).
[9] M. A. BENSON: *J. Hydraulics Div. Am. Soc. Civ. Eng.*, **88**, HY6, part 1 (1962).
[10] I. I. GRINGORTEN: *J. Geophys. Res.*, **68**, 813 (1963).
[11] L. DRAPER: *Proc. Inst. Civ. Eng.*, **26**, 271 (1963).
[12] C. PETRAUSKAS and P. M. AAGAARD: *Soc. Petr. Eng. J.*, **11**, 23 (1971).
[13] H. CRAMER: *Mathematical Methods of Statistics* (Princeton, N. J., 1946), p. 370.
[14] L. E. BORGMAN: *J. Geophys. Res.*, **66**, 3295 (1961).
[15] L. E. BORGMAN: *The frequency distribution for the m-th largest of n observations*, M. S. Thesis, University of Houston (1959).
[16] C. L. BRETSCHNEIDER: *Estuary and Coastline Hydrodynamics*, edited by A. T. IPPIN, Chapt. 3 (New York, N. Y., 1966).
[17] M. D. EARLE, C. C. EBBESMEYER and D. J. EVANS: *J. Waterways Harb. Coast. Eng. Am. Soc. Civ. Eng.*, **100**, 257 (1974).
[18] R. C. GOODKNIGHT and T. L. RUSSEL: *J. Waterways Harb. Div. Am. Soc. Civ. Eng.*, **89**, 29 (1963).
[19] M. S. LONGUET-HIGGINS: *J. Marine Res.*, **11**, 245 (1952).
[20] J. H. NATH and F. L. RAMSEY: *Probability distributions of breaking wave heights*, in *Proceedings of the International Symposium on Ocean Wave Measurement and Analysis* (1974), p. 379.
[21] E. CHEN: *Statistical distributions of wave heights and periods for hurricane waves*, Ph. D. Thesis, University of Wyoming (1979).
[22] E. CHEN, L. E. BORGMAN and E. YFANTIS: *Height and period distribution for hurricane waves*, in *Proc. Civil Eng. Oceans*, Vol. **4** (San Francisco, Cal., 1979), p. 321.

[23] A. CAVANIE, M. ARHAN and R. EZRATY: *A statistical relationship between individual height and periods of storm waves*, BOSS 76 (Trondheim, 1976).
[24] R. EZRATY, M. LAURENT and M. ARHAN: *Comparison with observations at sea of period and height dependent sea state parameters from a theoretical model*, OTC 2744, 149 (1977).
[25] Y. GODA: *The observed joint distribution of periods and heights of sea waves*, in *Proceedings of the XVI Conference on Coastal Engineering* (Hamburg, 1978).
[26] M. S. LONGUET-HIGGINS: *J. Geophys. Res.*, **80**, 2688 (1975).
[27] R. E. HARING and J. C. HEIDEMAN: *Gulf of Mexico rare wave return periods*, in *Proceedings of the Offshore Technical Conference* (1978), p. 1537.
[28] R. A. HARING, A. R. OSBORNE and L. P. SPENCER: *Extreme wave parameters based on continental shelf storm wave records*, in *Proceedings of the Conference on Coastal Engineering* (Hawaii, 1976).
[29] U. S. ARMY CORPS ENGINEERS: *Shore Protection Manual* (Fort Belvoir, Va., 1973).
[30] G. A. FORRISTALL: *J. Geophys. Res.*, **83**, 2353 (1978).
[31] S. M. BERMAN: *Ann. Math. Stat.*, **33**, 894 (1962).
[32] G. S. WATSON: *Ann. Math. Stat.*, **25**, 789 (1954).
[33] L. E. BORGMAN: *J. Waterways Harb. Coast. Eng. Am. Soc. Civ. Eng.*, **99**, 187 (1973).
[34] J. A. BATTJES: *Long-Term Wave Height Distribution at Seven Stations around the British Isles*, No. A.44 (England, 1970).
[35] P. J. DURNING: *Prediction of maximum wave height from historical data*, in *OTC Conference* (Houston, Tex., 1971).
[36] K. G. NOLTE: *Statistical methods for determining extreme sea states*, in *Proceedings of the II International Conference on Port and Ocean Engineering and Artic Conditions* (Iceland, 1973), p. 705.
[37] C. Y. YANG, A. M. TAYFUN and G. C. HSIAO: *Extreme wave statistics for Delaware coastal waters*, in *Proceedings of the International Symposium on Ocean Wave Measurement and Analysis* (1974), p. 352.
[38] G. VARGAS and M. VARGAS: *Desert Mag.*, **40**, 40 (1977).
[39] J. DATTABRI: *J. Waterways Harb. Coast. Eng. Am. Soc. Civ. Eng.*, **99**, 39 (1973).
[40] M. R. GOURLAY: *J. Waterways Harb. Coast. Eng. Am. Soc. Civ. Eng.*, **100**, 54 (1974).
[41] V. J. CARDONE, W. J. PIERSON and E. G. WARD: *Hindcasting the directional spectra of hurricane generated waves*, in *OTC Conference*, paper 2332 (Houston, Tex., 1975).
[42] E. G. WARD, L. E. BORGMAN and V. J. CARDONE: *J. Petr. Tech.*, **31**, 632 (1979).
[43] V. A. MYERS: *Joint Probability Method of Tide Frequency Analysis Applied to Atlantic City and Long Beach, New Jersey*, WBIM Hydro II (1970).
[44] V. J. CARDONE: *Specification of the Wind Distribution in the Marine Boundary Layer for Wave Forecasting*, Rpt. 69-1 (1969).
[45] D. T. RESIO and C. L. VINCENT: *J. Waterways Harb. Coast. Eng. Am. Soc. Civ. Eng.*, **103**, 265 (1977).
[46] B. W. WILSON: *Trans. Am. Soc. Civ. Eng.*, **128**, 104 (1963).
[47] K. HASSELMANN *et al.*: *Measurements of wind-wave growth and swell decay during the Joint North Sea Wave Project (JONSWAP)*, *Dtsch. Hydrogr. Z., Suppl. A*, Vol. **8** (1973).
[48] H. MITSUYASU: *On the growth of the spectrum of wind-generated waves*, No. 16, 67, Rep. Res. Inst. Appl. Mech., Kyushu University (1968).
[49] T. INONE: *On the Growth of the Spectrum of a Wind-Generated Sea According to a Modified Miles-Phillips Mechanism and its Application to Wave Forecasting*, Rpt. 67-5 (New York, N. Y., 1967).

[50] T. P. Barnett: *J. Geophys. Res.*, **73**, 513 (1968).
[51] J. A. Ewing: *Dtsch. Hydrogr. Z.*, **24**, 241 (1971).
[52] K. Hasselmann, D. B. Ross, P. Muller and W. Sell: *J. Phys. Oceanogr.*, **6**, 200 (1976).
[53] P. Dexter: *J. Phys. Oceanogr.*, **4**, 635 (1974).
[54] C. Jelesnianski: *Co-ordinate system for storm surge equations of motion with a mildly curved coast*, in *Proces-Verbauz* 14 (Grenoble, 1975).
[55] J. J. Wanstrath: *Storm surge simulation in transformed co-ordinates*, Ph. D. Thesis, Texas A&M University (1975).
[56] J. R. Pagenknopf and B. R. Pearce: *Evaluation of Techniques for Numerical Calculation of Storm Surges*, Rpt. 199, M.I.T. (1975).
[57] E. Parzen: *Ann. Math. Stat.*, **33**, 520 (1962).
[58] E. F. Wood and I. Rodriguez-Iturbe: *Water Resour. Res.*, **11**, 533 (1975).

Experimental Characteristics of Wind Waves.

L. CAVALERI

Istituto per lo Studio della Dinamica delle Grandi Masse, C.N.R.
S. Polo 1364, 30125 *Venezia, Italia*

Wind waves are a natural phenomenon that most people have experienced in their lives, in a more or less happy situation. Some characteristics like wave height, length, water velocity have been known to coastal populations since long ago, but the further step of trying to quantify these variables in a precise way was not done till a quite recent time. As usual, the push arrived when the waves became a factor strongly influencing some human activity, and at the same time the theory flourished. Wind waves have the characteristics of being possibly described both in some very simple way, a sort of thumb rule of the sailor, as by means of highly sophisticated mathematics, when one tries to describe the phenomenon in detail. Mainly in an unexplored field, like that of the sea till the last war, theory and experiment must go side by side, correcting and adapting each other, with continuous reciprocal comparison. It is my intention to devote this lecture to a survey of the characteristics of wind waves as they are actually measured in nature, comparing them with the corresponding values foreseen by the theory. It is worth-while to say that to measure waves is not an easy job, and, before being accepted, each output number should be carefully weighted. Notwithstanding this fact, the boomed economic interest in the sea has led to a very large amount of experimental data, and I will give only a few samples of them, according to the different aspects of the problem.

It is obvious that the first natural description of a sea concerns its visible part, the surface, and this is going to be the first subject we will discuss. Later we will shift below the surface, where the intuition, less supported by the direct experience, is of less help.

The first idea of a wave is quite close to that of a sinusoidal profile

$$y = a \cos (kx - \sigma t) \tag{1}$$

characterized by amplitude a, wave number $k = 2\pi/L$ with L the wave length, $\sigma = 2\pi f$ with f the frequency, x and t spatial and temporal co-ordinates, respectively. This simple representation is perhaps close to the idea of the sea in the mind of a sailor, but it has little resemblance with what we actually see on the sea surface. A more realistic representation is obtained by superposing a certain number of waves like (1), with different amplitude and period, so having

$$y = \sum_{n=1}^{N} a_n \cos (k_n x - \sigma_n t + \varepsilon_n) . \tag{2}$$

The random phase ε_n makes the components uncorrelated among themselves. This representation turns out to be much more satisfactory as, with a sufficient number of components, say 10 or more, we can build a diagram well resembling the actual shape of the sea surface (fig. 1). Besides we know from

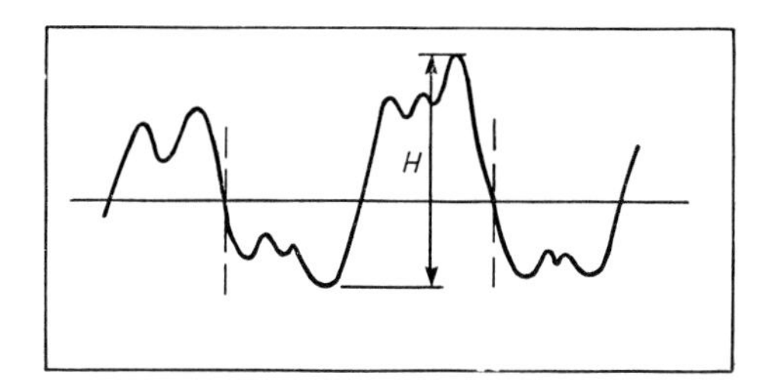

Fig. 1. – Profile of a sea surface. Wave height is defined as the difference between maximum and minimum between two successive down-crossings of the zero level.

Fourier's theorem that, provided N is large enough, any such signal can be decomposed in a series like (2), which makes us confident about its practical use. This idea was formalized in the 50's by LONGUET-HIGGINS [1] and PIERSON [2] introducing the concept of a wave spectrum $E(f)$ (fig. 2), describing

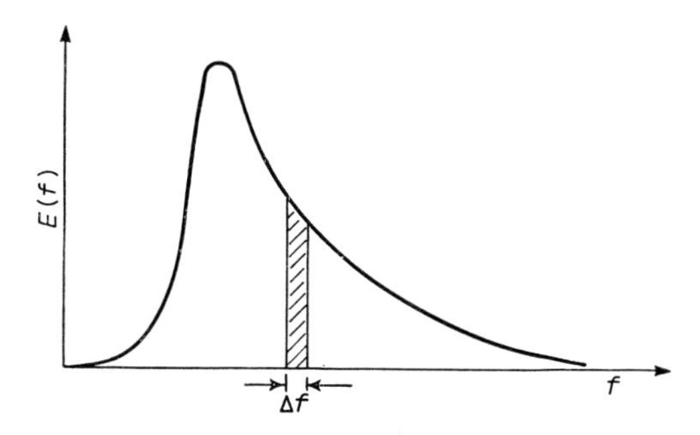

Fig. 2. – Spectrum of the sea surface. The shadowed area is proportional to the square of the amplitude of the related discrete component.

the continuous distribution of energy on frequency. If one resolves the continuous spectrum into discrete components of width Δf, $E(f)$ is related to the amplitude of the single component by

$$E(f)\Delta f = \tfrac{1}{2}a^2 . \tag{3}$$

There are some basic assumptions under (2). The first one is linearity, that allows us to superimpose the different wave components. If you further assume that the sea surface irregularities are locally homogeneous, stationary and Gaussian, you can show [1, 2] that the average quantities are invariant under translation in space and time, provided, as it is usually the case, that the processes which are changing the sea state have scales which are large compared with the wave scales. This is an extremely useful point as it allows us to stick at a fixed point and to measure waves there, later interpreting the results as characterizing the sea surface profile at a given instant.

A number of consequences can be derived from (2). It can easily be seen that (2) provides an evenly distributed profile with respect to the mean sea level, and, under the hypothesis of a sufficiently narrow spectrum, LONGUET-HIGGINS [1] proved that the wave heights should be described by a Rayleigh distribution. Here the wave height is defined (fig. 1) as the difference between maximum and minimum values between two successive down-crossings of the zero level. A mathematical expression for the Rayleigh distribution is

$$P(H > H_0) = \exp[-H_0^2/8m_0] , \tag{4}$$

where the left-hand side denotes the probability that a wave height H is greater than H_0, and m_0 is the total variance of the spectrum. The distribution (4) is of strong practical importance, mainly in its far right tail, as it gives the probability of appearance of the very large waves, so crucial for the design of the offshore structures. Extensive studies of experimental data indicate that (4) overpredicts the actual probability of appearance of the largest waves.

Figure 3 reports a recent result by FORRISTALL [3] obtained from the analysis of several storms, including some very high hurricane waves in the Gulf of Mexico. It is clear that the Rayleigh distribution (continuous line) overpredicts the experimental data, fitted by an empirical distribution (broken line), and that the error increases with the wave height. The figures reported by FORRISTALL are that the significant wave height is 0.942 times that calculated from the Rayleigh distribution and the expected value of the maximum wave in 1000 is 0.907 the height calculated from the Rayleigh distribution. This concerns the wave height. A similar check can be done on the level distributions, that we have previously assumed to be Gaussian. This check does not turn out to be so negative. Figure 4 shows the actual distribution of the surface, normalized with respect to m_0, and fitted by a Gram-Charlier distribution that is fairly close to the originally assumed Gaussian distribution. Actually this fitness is perhaps more than we expected after the straightforward assumption of linearity. There is a slight excess of negative levels, deriving from some degree of nonlinearity in the actual sea. This is connected to the known fact that the crests are higher and shorter with respect

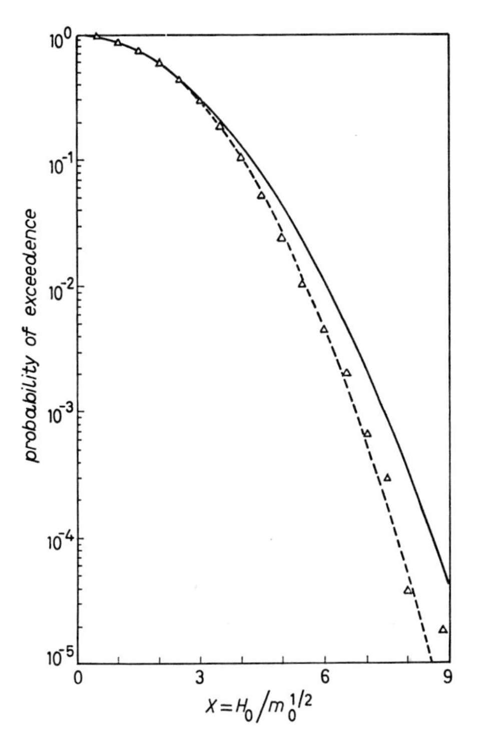

Fig. 3 – Probability of exceeding a given nomalized wave height. The triangles are data, the solid line is the Rayleigh distribution, and the dashed line is the empirical Weibull distribution (after [3]).

to the sinusoidal profile given by (1), while, on the contrary, the troughs are rounder and longer. Nevertheless, we see that the linear theory, even if not fully correct, can certainly give some satisfactory result.

I have so far spoken about wave height and level distributions. These are sophisticated means, and it is obvious that a more compact way to describe the sea conditions at a given instant is required, both for practical communication purposes as for use by unexperienced people. This has been identified in the significant wave height H_s, defined as the mean of the highest one third of the actual wave height distribution, and indicated also with $H_{\frac{1}{3}}$. Similar intuitive definitions have been introduced for $H_{\frac{1}{2}}$ or $H_{\frac{1}{10}}$, but it has been found (for instance by GODA [4]) that $H_{\frac{1}{3}}$ is much more stable than the other parameters in the statistical analysis of a wave record. Besides the distinctive advantage of $H_{\frac{1}{3}}$ (which was also a reason for choosing it) is that it is close to the actual wave height estimated by a sailor's eye.

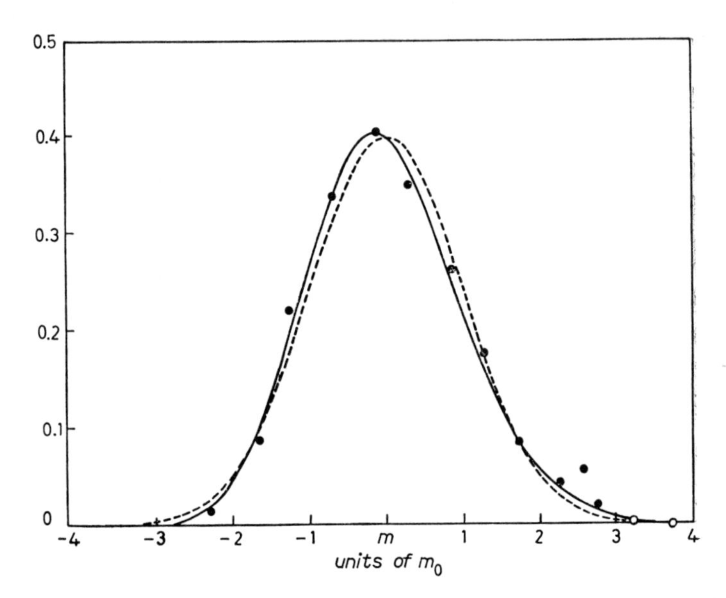

Fig. 4. – Actual distribution of sea level fitted by a Gram-Charlier distribution (———), compared to that due to the Gaussian assumption (— — —).

Starting from (4) it is possible to show that the significant wave height is related to the spectrum by

$$H_{\frac{1}{3}} = 4.005\, m_0^{\frac{1}{2}}. \tag{5}$$

However, similarly to the Rayleigh distribution, this expression has been found to overpredict the experimental values. FORRISTALL [3] finds a better agreement for

$$H_{\frac{1}{3}} = 3.77\, m_0^{\frac{1}{2}}. \tag{6}$$

The coefficient is quite close to that previously found by GODA [5], equal to 3.79, for deep-water conditions. The wave height distribution changes anyhow when we approach the shore and the waves are deformed by the presence of the bottom. In these conditions GODA [5] found good agreement between (5) and his experimental results.

When we look at the sea, our attention is usually focused on the largest waves. But if we analyse a wave in detail or wonder about its generation by wind, we sooner or later realize the importance of its secondary aspects. The right tail of a wave spectrum (fig. 2) is associated to the high-frequency waves, short hence not very high, and it is intuitive that their contribution to $H_{\frac{1}{3}}$ is not substantial at all. Anyhow there is a number of phenomena that are strictly related to the shape of this part of the spectrum. It is a remarkable point that PHILLIPS [6] succeeded in giving its mathematical expression basing himself only on dimensional arguments. Phillips' expression is

$$E(f) = \alpha g^2 \sigma^{-5} \tag{7}$$

and it has later been confirmed by a large number of experiments. Uncertainty exists on the actual value of α. Using the few data existing at the time, PHILLIPS had suggested the value 0.073, but, mainly during the very extensive and accurate JONSWAP experiment in the North Sea [7], it was found that α is actually a function of the adimensional fetch $\tilde{x}$ defined as

$$\tilde{x} = gx/U_{10}^2 , \tag{8}$$

where g is gravity acceleration, x is actual fetch, U_{10} is wind speed measured at a 10 m height above sea level. The actual expression for α is

$$\alpha = 0.076\, \tilde{x}^{-0.22} . \tag{9}$$

More precisely (7) gives a curve that cannot be overpassed by the spectrum, and it is usually reached on the right side of the spectrum when wind is actively blowing on the sea surface.

It is now time we breath deeply and we shift below the surface. It is a surprising but common belief that, below a narrow layer, water particles are not affected by the eventual waves present at the surface. It is anyhow enough to try to dive once in stormy conditions to change the opinion about this point. There are two variables on which we can focus our attention to observe the underwater implications of a wave: pressure and particle velocity. Once a surface wave (1), running over a bottom depth h, is given, we can consider an intermediate point at depth z, supposed negative downwards from the surface. The linear theory then supplies the expressions for pressure p and horizontal and vertical velocity u, w, respectively,

$$p = \varrho g a \frac{\cosh k(z+h)}{\cosh kh} \cos(kx - \sigma t) , \tag{10}$$

$$u = a\sigma \frac{\cosh k(z+h)}{\sinh kh} \cos(kx - \sigma t) , \tag{11}$$

$$w = a\sigma \frac{\sinh k(z+h)}{\sinh kh} \sin(kx - \sigma t) . \tag{12}$$

Here ϱ is water density. w, like z, is positive upwards. Note that in a given depth a wave with a certain wavelength has a well-defined phase speed, hence T and L, *i.e.* σ and k, are strictly related. This is expressed by the dispersion relationship

$$\sigma^2 = gh \operatorname{tgh} kh \,, \tag{13}$$

so called because waves with different period proceed with different phase speed and consequently disperse. Several things should be noticed from (10)-(12). p and u are clearly in phase with the surface wave (1), in the sense that their maximum values are reached simultaneously at the maximum elevation of y. On the contrary, w is out of phase of 90°, and it is, therefore, an orthogonal function with respect to u. This tells us something about the actual path of a water particle. Two harmonic motions like (11) and (12), orthogonal to each other, produce an ellipse, and, because $\cosh x > \sinh x$, this ellipse has its major axis on a horizontal plane. Expressions (10)-(12) are simplified in deep water, *i.e.* when $h \to \infty$, producing

$$p = \varrho g a \exp[kz] \cos(kx - \sigma t) \,, \tag{14}$$

$$u = a\sigma \exp[kz] \cos(kx - \sigma t) \,, \tag{15}$$

$$w = a\sigma \exp[kz] \sin(kx - \sigma t) \,. \tag{16}$$

Not squeezed by the bottom, the ellipse has now turned into a circle, which is the actual path a particle follows in deep water. There is another interesting point about (14)-(16). If we forget for a moment about the terms ϱg and $\exp[kz]$, the latter present in all three, we end up with three expressions that represent the vertical profile of a wave and the velocity component of a particle at the surface. This just tells us that, if we consider a horizontal layer at a given depth, under wave action this behaves like a surface wave. In other words, the kinematics of a wave is such that, at least in deep-water conditions, if we could take off a surface layer of water, we would be left with a perfect wave as the original one, with only reduced amplitude, the reduction ratio being given from (14) as $\exp[kz]$.

What shall we do with a real sea, whose profile is given by (2)? Well, the big advantage of the linear hypothesis we have originally assumed is that the sea can be split into a series of sinusoids, and each of them is acting independently on the others. This implies that we can apply (10)-(12) to each single component, and that the actual pressure and velocity at a given point will be just the sum of those due to each single component. Combining (2) and (10)-(12), we have therefore the expression for the pressure and velocity at a given depth

z under a complicated sea:

$$p = \varrho g \sum_{n=1}^{N} a_n \frac{\cosh k_n(z+h)}{\cosh k_n h} \cos(k_n x - \sigma_n t)\,, \tag{17}$$

$$u = \sum_{n=1}^{N} a_n \sigma_n \frac{\cosh k_n(z+h)}{\sinh k_n h} \cos(k_n x - \sigma_n t)\,, \tag{18}$$

$$w = \sum_{n=1}^{N} a_n \sigma_n \frac{\sinh k_n(z+h)}{\sinh kh} \sin(k_n x - \sigma_n t)\,. \tag{19}$$

Surface level, pressure and particle velocity can be measured, and we can, therefore, try to verify if (17)-(19) are correct. In particular, we will focus our attention on the so-called attenuation coefficients k_p, k_u, k_w, obtained from (1) and (10)-(12), and expressing the transfer functions between signals at depth z and the surface profile [2]

$$k_p = \frac{\varrho g \cosh k(z+h)}{\cosh kh}\,, \tag{20}$$

$$k_u = \frac{g \cosh k(z+h)}{c \sinh kh}\,, \tag{21}$$

$$k_w = \frac{g \sinh k(z+h)}{c \sinh kh}\,. \tag{22}$$

c is phase speed $= L/T = \sigma/k$.

Standard analysis proceeds via the power spectrum, and, when we analyse a record of the four signals y, p, u, w, we end up with four spectra E_y, E_p, E_u, E_w. From (3), for each frequency component E is proportional to a^2, and, therefore, once a proper calibration of the signals has been done, the experimental attenuation coefficients are given by

$$k_p'^2 = \frac{E_p}{E_y}\,, \tag{23}$$

$$k_u'^2 = \frac{E_u}{E_y}\,, \tag{24}$$

$$k_w'^2 = \frac{E_w}{E_y}\,. \tag{25}$$

Note that, even if written in a compact way for clarity, (23)-(25) refer to each single component. We have now theoretical (20)-(22) and experimental (23)-(25) results, and the obvious thing is to compare them. I will first speak about pressure.

Pressure has never received much attention, probably because the pressure under a running wave does not have big consequences on an engineering structure. Anyhow the measurement of waves by a submerged pressure transducer is a common and useful technique, and it is, therefore, worth-while to check the correctness of (20). The first who did this was DRAPER [8], who used a transducer fixed on a pole relying on the tide for variations in depth. He found an attenuation 16 % greater than predicted by theory. Now this turns out to be surprising, as, if some nonlinearity was present, I would expect underwater pressure to be larger than foreseen by linear theory, as nonlinear processes usually work in the sense of increasing the effect. KAWANABE and TAGUCHI [9], using three pressure transducers at different levels, found that, once corrected to the surface, the three results differed among themselves as much as 10 %, but, contrarily to DRAPER, they found this percentage to vary with frequency. Now this looks quite sensible. If a difference from the theory exists, I expect it to vary with depth, more exactly with the relative depth evaluated with respect to the actual wavelength. It is then evident, as wavelength depends on frequency, that at a given point the relative depth, hence the difference from theory, must depend on the frequency we are considering. The point is that the discrepancy between linear theory and experiments now reported is fairly close to the accuracy of the data, and, therefore, no definitive conclusion can be drawn. What I believe to be a more accurate measurement has been done by two other researchers and myself [10], taking advantage of the oceanographic tower of C.N.R. I will use quite a bit of results obtained from this place, and it is, therefore, worth-while to devote a few words to describe it. The tower is shown in fig. 5 and a sketch of the main instrumental arrangement is shown in fig. 6. Basically we measure surface level, pressure at two different depths and the three-dimensional velocity field. The underwater set can be fixed at any desired depth. The standard technique is to carry out several records at different depth during the same storm, so as to explore the whole water column. Local depth is 16 m. There is a large number of records out of the tower, but, as far as pressure is concerned, all of them show the same behaviour. The way we have analysed the data is to shift down the surface signal by means of (20), the theoretical attenuation factor, and to compare it with the actual output of the transducer. In other words, we compare (23) with (20). The ratio k_p'/k_p is shown in fig. 7 for 8 different records. These have been grouped together two by two, according to the depth of the cart. Peak frequency during these records varied between 0.17 and 0.20 Hz. It is soon evident that, at least in the energetic part of the spectrum, the waves are more attenuated than foreseen by linear theory ($R < 1$). We should anyhow be careful about the possibility of contamination of data by noise. A good method to check this is to evaluate the coherence between two signals. It is obvious that, if two signals show a high coherence, any random noise superimposed on them must have a very

Fig. 5. – The oceanographic tower of C.N.R. (see location on map, upper left-hand corner). The cart holding the instruments (sketched in fig. 6) slides on two vertical wires vertically tensed between the horizontal platform and a heavy flat ballast on the bottom. The light frame with a man above allows easy handling of the instruments (after [10]).

limited value. The range of frequency with squared coherence greater than 0.98 is shown for each record in the figure. Also shown are the 95 % confidence limits. R seems to vary linearly with frequency, and the slope of the line becomes smaller the closer we go to the surface. I must say that no definite ex-

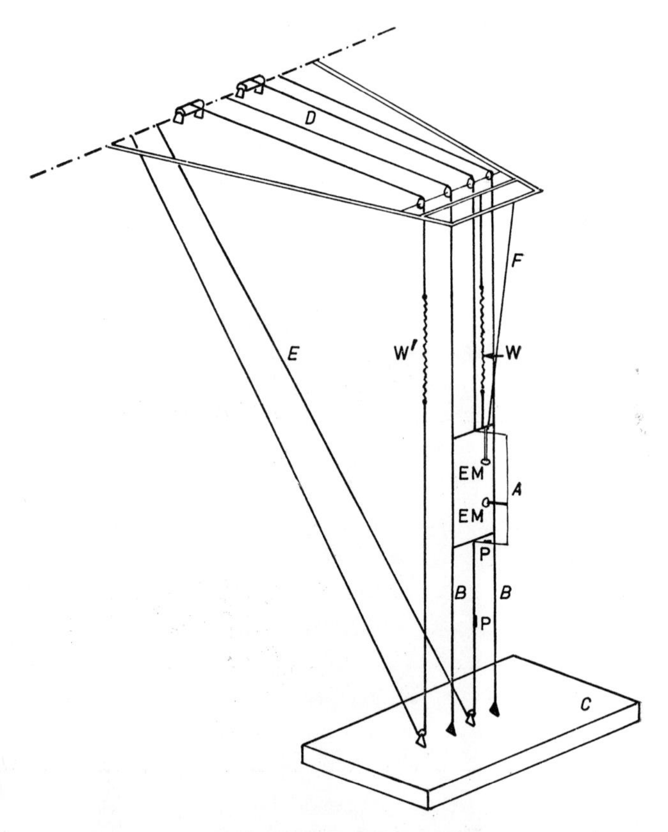

Fig. 6. – Schematic diagram of the instrumental arrangement on the tower (see fig. 5). Instruments are indicated by W (wave staff), EM (electromagnetic currentmeters) and P (pressure transducer). W′ is the reserve position of the wave staff when the cart is operated very close to the surface (after [10]).

planation of this behaviour has yet been proposed. Clearly some sort of non-linearity is working here. We have considered the possible effect of the bottom, but we had peak frequency till 0.20 Hz, and a 5 s wave has a wavelength of 40 m in deep water, with respect to which 16 m are not really shallow conditions. One would expect some consequence from the finite amplitude of waves (linear theory assumes infinitesimal amplitude). To check this we put together all the records taken more or less at the same depth, independently of the value of H_s. During a storm this could vary from 0.50 to $(2 \div 3)$ m, but, notwithstanding this fact, no difference was evident in the values of R

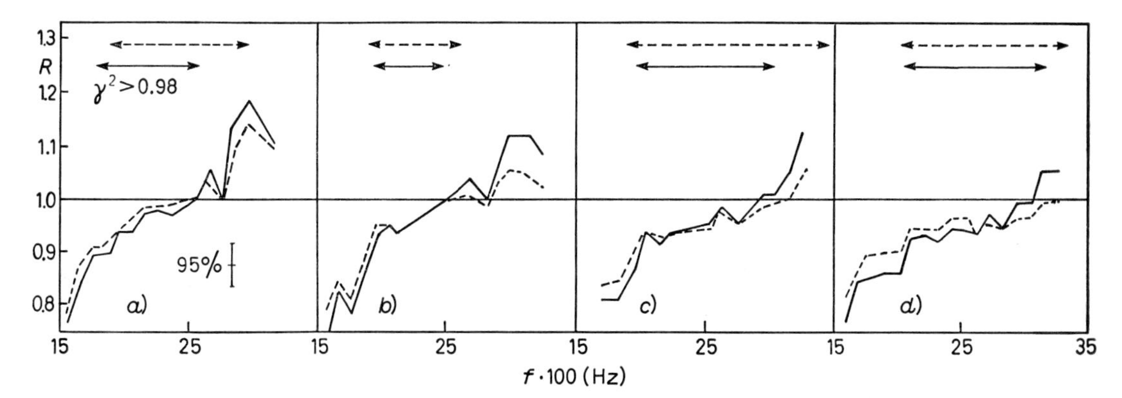

Fig. 7. – Ratio R between experimental and theoretical pressure amplitude response functions at various depths: *a*) ——— 11.3, ——— 14.5; *b*) ——— 8.5, ——— 11.7; *c*) ——— 6.2, ——— 9.4; *d*) ——— 4.5, ——— 7.7. Two records are shown in each diagram. $R < 1$ means an attenuation greater with respect to linear theory. Coherence (> 0.98) with surface amplitude and 95% confidence limits are shown. Note that the average slope of the lines decreases while measurements approach the surface (after [10]).

out of the different records. Therefore, we concluded the trend of R not to be a finite-amplitude effect.

I should point out at this stage that the differences between linear theory and experiments are not disastrous. A 10% error, probably less in the estimation of H_s, is something that can be accepted for many engineering purposes. If we think of its original assumptions, it is really surprising that the linear theory works so well. About this it is worth-while to quote some recent results by GUZA and THORNTON [11], who made a comparison between bottom pressure and surface wave staff measurements in very shallow water, close to the shore. They found that, working in 5.5 m of depth and with a 14 s wave (deep-water wavelength $\simeq$ 300 m), hence in very shallow conditions, the error due to linear theory was less then 20% for each single frequency.

Up to now we have not mentioned the direction the waves are moving. This was unessential as far as pressure is concerned, as pressure is a scalar. Conditions are different when we start considering particle velocity. In a wave the water particle moves in a plane normal to the wave crest, and, therefore, at least for the horizontal component, the direction is a basic information. If we think a bit, we see that the description given by (2) is not fully satisfactory. Direction is not taken into account. Now first of all we can have waves coming from different directions, but besides there is no reason to exclude that two wave systems with the same period are travelling in different directions. For instance, the consequences of two different storms can reach a third position at the same time. More generally, we must expect the energy present at a certain frequency to have a continuous distribution over all the

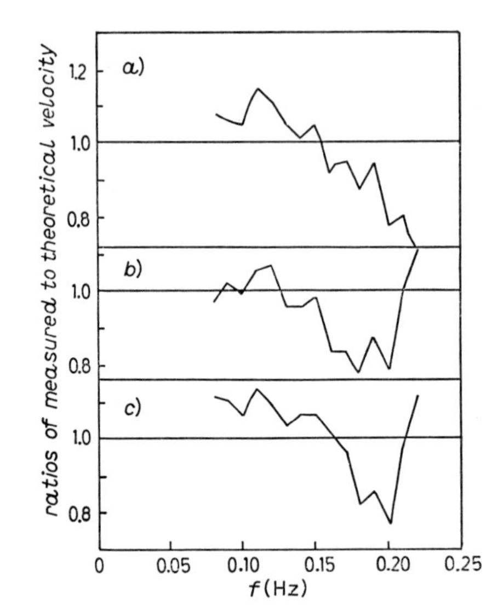

Fig. 8. – Ratio of measured to theoretical velocity at 4, 10, 17 m, in 20 m of depth (after [12]): *a*) currentmeter 1, *b*) currentmeter 2, *c*) currentmeter 3.

horizontal directions. Choosing discrete values of the direction similarly to what was done for frequency, we end up with the concept of the bidimensional spectrum $F(f, \theta)$, where energy is specified for frequency and direction. For each frequency, circular integration over direction supplies the monodimensional spectrum $E(f)$ we had previously defined. This means that we have to think of the sea surface as the ensemble of an infinite (very large in practice) number of component waves, with different frequency, amplitude and direction. (2) is, therefore, generalized to

$$y = \sum_{n=1}^{N} \sum_{m=1}^{M} a_{n,m} \cos (\boldsymbol{k}_{n,m} \cdot \boldsymbol{x} - \sigma_n t + \varepsilon_{n,m}) , \tag{26}$$

where $\boldsymbol{k}_{n,m}$ and $\boldsymbol{x}$ are horizontal vectors. There are several means to determine experimentally a bidimensional spectrum, which follow basically two main lines. The first group obtains the information by measuring the sea surface elevation or slope at the same time at different close locations. The second is concerned with the horizontal components of water particles. I will devote the last part of the lecture to the latter point.

Starting from a two-dimensional spectrum we have to deal with two horizontal velocity components u, v and with their spectra E_u, E_v. The overall energy associated to a horizontal velocity spectrum at a given frequency is $E_U = E_u + E_v$, and (24) must now be corrected to

$$k_u'^2 = \frac{E_U}{E_y} . \tag{27}$$

It is natural that the first check we do is on the validity of (27) and (25), quite similarly to what we have previously done for pressure. Figure 8 reports the results obtained by FORRISTALL *et al.* [12] from a tower in the Gulf of Mexico during hurricane Delia. Besides surface level, horizontal velocity was measured at three different depths, 4, 10, 17 m, respectively, in a 20 m depth. Significant wave height was 4.5 m, peak period 8 s. We see a tendency towards underpredicting at low frequencies. However, the agreement between experiment and prediction of linear theory was better than 10 % in the frequency range in which energy was concentrated. This discrepancy was of the same order of magnitude of that found from the C.N.R. tower [10], and whose results for a single record are reported in fig. 9. Coherence and confidence limits are reported in the figure. Here also the vertical attenuation factors k_w and k'_w are shown, and their discrepancy is similar to that found for the horizontal velocity.

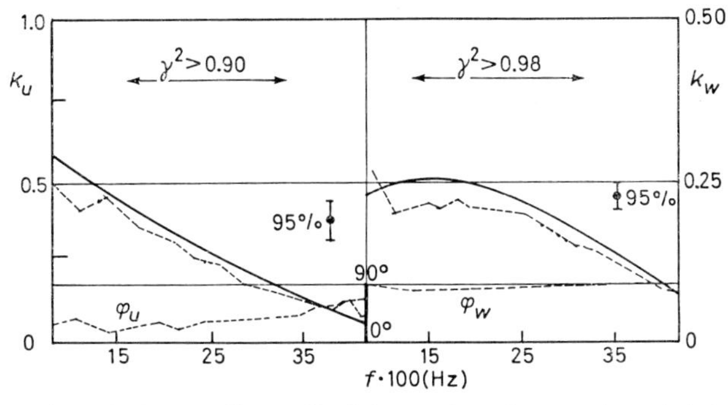

Fig. 9. – Comparison between theoretical (———) and experimental (— — —) transfer functions for combined horizontal (u) and vertical (w) velocity components. Coherence and confidence limits are shown. Note the phase angles φ_u and φ_w with respect to the surface elevation.

At this stage we might wonder if higher-order theories could provide a better agreement between theory and experiment. FORRISTALL *et al.* [12] tried this way, but the results turned out to be much worse than those out of linear theory. The explanation of this lies in the unidirectionality of waves assumed by high-order theories, whose consequences are illustrated in fig. 10.

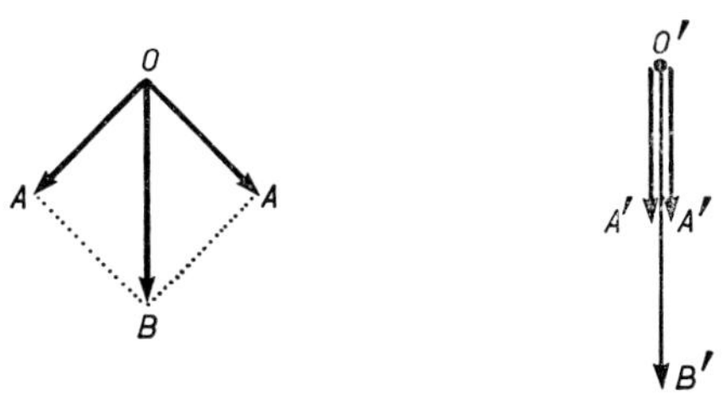

Fig. 10. – The sum of two vectors (horizontal velocity due to two wavelets) decreases the more spread the two vectors are in direction.

Consider two vectors, representing the horizontal components due to two wavelets. For the same amplitude, the more spread they are in direction, the smaller is the modulus of the resulting vector. If symmetrical, the transversal components cancel each other. This is, of course, extensible to a continuous directional distribution, and it is, therefore, evident that, if we do not take directional spreading into account, we will be led to overpredict the actual horizontal velocity. FORRISTALL *et al.* concluded that to take directionality into account was much more important than to consider nonlinear effects.

The bidimensional horizontal motion is well shown by a sample of the time history of the horizontal velocity measured at the upper currentmeter, still from [12], and shown in fig. 11.

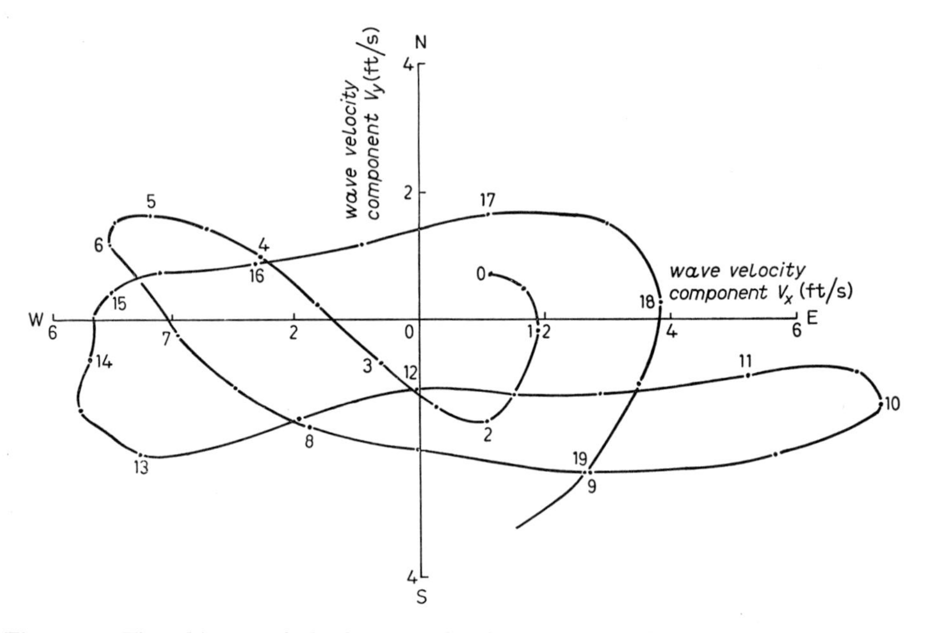

Fig. 11. – Time history of the horizontal velocity measured at the top currentmeter during the largest wave. Numbers on the curve give the time in seconds (after [12]).

Just because we have seen that the horizontal velocity depends on the spreading of the spectrum, it is of interest to see how spread this actually is. This is not a simple job, because the information we can extract from a standard set of instruments, say surface elevation plus two horizontal velocity components, like those used in [10, 12], is not sufficient for a full description of the distribution. What we are left with is the actual distribution filtered by some smoothing function depending on the instrumental set that has been used. Once a certain kind of distribution has been accepted, say for instance $\cos^n(\Phi - \Phi_0)$ (setting to zero the values for $|\Phi - \Phi_0| > \pi/2$), the data analysis will provide the value of Φ_0, mean wave direction, and the exponent n, that is simply a measurement of the peakedness of the distribution. Practically all the measurements available in the literature coincide in saying the distribution is much narrower close to the spectral peak and more spread laterally on both sides of the spectrum, especially towards the high frequencies. There is also a suggestion, first by MITSUYASU *et al.* [13], that the distribution should depend on fetch, but no definitive conclusion has been reached on this point.

I want now to go back to fig. 9 that shows the results of the analysis of the velocity data out of the C.N.R. tower. There is a striking feature in these results. The lowest part shows the phase difference between the surface elevation and u and w, respectively. Now φ_w is close to the theoretical value of 90°, but φ_u shows a well-definite departure from the expected value of 0°. This simply means that the maximum horizontal velocity is reached well after the passage of the crest. This also means that u and w are no longer orthogonal functions, and that the ellipse described by a particle has now its main axis obliquely oriented downwards. Now the expression

$$\frac{1}{T}\,\varrho\int_0^T uw\,\mathrm{d}t \tag{28}$$

is a measure of the downward flux of horizontal momentum, connected to the wave development under the wind action. Table I reports from [10] various characteristics concerning a sample of records. The vertical flux of horizontal momentum is given by $\overline{uw}$, expressed in dyn/cm². From the wind speed we can estimate the atmospheric stress at the surface to range from about 0.5 to 2 dyn/cm², orders of magnitude smaller than the value actually found.

TABLE I.

Record	Time (GMT) 10, 11 March 1976	Depth of current meter C_2 (m)	Mean wind speed (m/s) and direction (degrees)	Significant wave height (m)	$\overline{u^2}$ (cm/s)²	$\overline{v^2}$ (cm/s)²	$\overline{w^2}$ (cm/s)²	$\overline{uw}$ (cm/s)²	$\overline{vw}$ (cm/s)²
5	2238	3.7	15; 080	1.74	559	90	662	172	18
6	2330	10.4	14; 080	1.63	101	25	52	22	6
7	0031	7.9	14; 080	1.84	204	36	188	71	22
8	0122	5.6	13; 080	1.56	260	42	277	85	11
9	0620	10.6	11; 070	1.28	40	6	23	9	1
10	0848	3.8	8; 070	1.07	210	33	247	55	14
11	0949	8.0	8; 070	0.90	33	5	31	10	3
12	1032	5.6	7; 070	0.79	51	12	62	15	2

$\overline{uw}$ seems to decrease while increasing the depth of the point of measurement, but it always keeps much larger than the surface wind stress. Also on this point there has been no satisfactory explanation. The only hint is that all these measurements were made under active wind conditions, with frequent breaking waves. Breaking waves mean turbulence injected into the surface layers and superimposed to the velocity field due to waves. We think, therefore, that a possible explanation for the above discrepancy lies in the presence of turbulence influencing the results, but no conclusion about this point has been reached up to now.

I believe it worth-while to conclude pointing out the possible consequences of the instruments themselves on the actual final results. When an instrument is put at a point to measure something there, its presence itself will influence the local characteristics. We can only hope to decrease this influence to a minimum. A consequence of this can be recognized in fig. 9. I mentioned before, and we could see in fig. 6, that we measure the three-dimensional velocity field by means of two electromagnetic currentmeters at cross angles. Each currentmeter consists basically of a disk measuring velocity on its plane. The holding arm can be oriented so as to align the u-w measuring meter in the main wave direction. The u-v measuring unit is on the contrary permanently connected to the horizontal plane, and, when strong vertical velocities are present, we must expect vortex shedding to be present around it. In effect, we found a much higher noise level for this currentmeter, and we were forced to accept a lower coherence between $\boldsymbol{U} = \boldsymbol{u} + \boldsymbol{v}$ and the surface level, 0.90, with respect to the value 0.98 accepted for all the other couples of variables.

The main conclusion that can be drawn from this lecture is the validity of the linear theory. But be careful, this strictly depends on the accuracy you want from your measurements. For many engineering purposes a 10% accuracy is enough, but there are cases in which a smaller error is highly desirable. In those cases the linear theory will not suffice any more, nevertheless do not forget that often, to open one way, you will close another one worsening the result. Also the accuracy of the instruments must be taken into account, and at this level we are often playing at the limit of reliability of most of the commercial instruments. Sometimes the best thing to do is to recognize our actual limits and those of the results we are providing.

REFERENCES

[1] M. S. Longuet-Higgins: *J. Mar. Res.*, **11**, 245 (1952).

[2] W. J. Pierson jr.: *A Unified Mathematical Theory for the Analysis, Propagation and Refraction of Storm Generated Ocean Surface Waves*, Part I and II, N.Y.U. Coll. of Eng. (New York, N. Y., 1952).

[3] G. Z. Forristall: *J. Geophys. Res. C*, **83**, 2353 (1978).

[4] Y. GODA: *Report of the Port and Harbour Research Institute*, Vol. **18**, No. 1 (1979).
[5] Y. GODA: *Proceedings of the International Symposium on « Ocean Wave Measurements and Analysis »*, Vol. **1** (New Orleans, La., 1974), p. 320.
[6] O. M. PHILLIPS: *J. Fluid Mech.*, **4**, 426 (1958).
[7] K. HASSELMANN, T. P. BARNETT, E. BOUWS, H. CARLSON, D. E. CARTWRIGHT, K. ENKE, J. A. EWING, H. GIENAPP, D. E. HASSELMANN, P. KRUSEMAN, A. MEERBURG, P. MÜLLER, D. J. OLBERS, K. RICHTER, W. SELL and H. WALDEN: *Ergeb. Dtsch. Hydr. Z., Reihe A* (8°), No. 12 (1973).
[8] L. DRAPER: *Houille Blanche*, **12**, 926 (1957).
[9] Y. KAWANABE and K. TAGUCHI: *On simultaneous observations of waves with three pressure type wave recorders suspended from a single buoy, Kobe Marine Observatory* (1968), p. 113.
[10] L. CAVALERI, J. A. EWING and N. D. SMITH: NATO *Symposium on Turbulent Fluxes through the Sea Surface, Wave Dynamics and Prediction*, edited by A. FAVRE and K. HASSELMANN (New York, N. Y., 1978), p. 257.
[11] R. T. GUZA and E. B. THORNTON: *J. Geophys. Res. C*, **85**, 1524 (1980).
[12] G. Z. FORRISTALL, E. G. WARD, V. J. CARDONE and L. E. BORGMANN: *J. Phys. Ocean.*, **8**, 888 (1978).
[13] H. MITSUYASU, F. TASAI, T. SUHARA, S. MIZUNO, M. OHKUSU, T. HONDA and K. RIKIISHI: *J. Phys. Ocean.*, **5**, 750 (1975).

Generation and Dissipation of Wind Waves.

L. CAVALERI

Istituto per lo Studio della Dinamica delle Grandi Masse, C.N.R.
S. Polo 1364, 30125 *Venezia, Italia*

Wind waves are probably, with the tide, the most spectacular effect of the ocean. They grow, evolve, break, mix, never the same, exciting the fantasy and the imagination of the observer. Nevertheless, till a few decades ago the only persons involved with waves were the sailors and the fishermen. As we said in the previous lecture, this left the field practically unstudied till a quite recent time. To understand the stage of the wave study in the nineteenth century it is enough to say that I have personally come across an Italian book (in an English library) where the author, a supposed expert, claimed it had been mathematically proved no wave could physically exist higher than 3.72 m! As a matter of fact some theoretical study did exist. After all the simple sinusoidal wave can be applied to quite a number of arguments. But little connection with the real sea was available, and seafarers were carrying on with some « thumb rule » dictated by previous experience. I have some personal experience with these people and I can tell you they can do some extremely good guesswork. This method has anyhow two disadvantages: first it works in what we can call an average or standard situation, second it does not tell anything about the physics of what is going on. From a certain point of view we can call it a statistical method. Well, if we now want to study generation and dissipation of wind waves, we must accept to start again from the beginning, knowing we are going to begin with probably worse results, hopefully later improving them beyond the actual point. Again, like for the kinematics of a wave, we have to proceed side by side with theory and experiment, observing the sea, imagining a theory and devising an experiment to verify and correct it.

It is an immediate experience that wind, blowing on a flat, calm sea, produces waves. It is much less intuitive how this happens, and I must soon say that the problem is far from being satisfactorily understood. The first one who speculated and wrote something about the generation of waves by wind was JEFFREYS [1], who proposed the « sheltering hypothesis ». He made the plausible suggestion that wind blowing over a pre-existing wave crest creates an asymmetrical distribution of pressure along the wave surface, mainly due to the supposed boundary layer separation that should occur on the downwind side of the wave crest. The component of air pressure in phase with the wave slope (the vertical velocity of water surface) does work on the waves and makes them grow. This theory lost support when laboratory experiments showed that the resulting pressures were too small to justify the actual wave growth.

The next two important steps were done simultaneously by PHILLIPS [2] and MILES [3]. Phillips' argument was based on the pressure distribution on the sea surface due to wind. Pressure is not uniform and its longitudinal distribution can be decomposed in a Fourier series with components of different wavelength, transported along with the average wind speed U. Each pressure wave, acting on the surface, tends to deform the sea surface creating a surface wave with the same wavelength L. Given its length L, the phase speed c of a wave is well established, and in general we will find $c \neq U$. The pressure wave and the surface wave will soon run out of phase cancelling the previous effect and, averaging in time, we find a null effect. For the special case $c = U$ the pressure wave will continue to feed energy into the surface wave, making, therefore, its energy grow at a constant rate. Phillips' mechanism is not believed anyhow to be the major cause of wave growth, as experiments indicate a much greater rate than foreseen through pressure fluctuations, and this rate turns out to be exponential in time. An explanation for this behaviour was proposed by MILES [3]. He argued that any surface wave must introduce a disturbance on the air flow, and he was able to show that the part of the induced pressure disturbance which is in phase with the wave slope does work on the wave and causes it to grow. This feedback mechanism, with an effect depending on the already existing value, results in an exponential growth rate for the wave energy.

It is remarkable that the two Phillips' and Miles' mechanisms, proposed more or less simultaneously, complete each other. As mentioned above, PHILLIPS is unable to explain the actual growth rate, but, on the other hand, to operate, MILES requires the presence of an already existing disturbance on the surface. On the contrary, with a growth rate proportional to the actual energy, MILES would indicate a null input of energy. So Phillips' is actually the trigger mechanism that allows the Miles' wind-wave interaction to operate. If we indicate with $E = E(f)$ the energy at a given frequency, the differential equation for the growth of wave energy can now be written

$$\frac{\partial E}{\partial t} = \alpha + \beta E\,, \tag{1}$$

where the coefficients α and β refer to Phillips' and Miles' mechanisms, respectively. MILES evaluated β on the basis of his theory, but the experimental results suggested that he had underestimated the true value by an order of magnitude. A remarkable experiment was conducted by BARNETT and WILKERSON [4], who measured from an airplane how waves due to an offshore wind grew with the distance from the coast. On the basis of these results they presented a modified expression for β,

$$\beta = 5sf\left(u\,\frac{\cos\delta}{c} - 0.90\right). \tag{2}$$

Here s is ratio of density of air to that of water, f is frequency, δ is angle between wind and wave direction, c is phase speed. The term 0.90 allows for the experimental evidence of waves running faster than wind. Whatever the expression of β, it is anyhow evident that the exponential growth out of (1) cannot last indefinitely. Waves must sooner or later reach some limit height, after which the extra energy present at each frequency must be lost by other means, conceivably by breaking. It is obvious that the maximum height depends on wavelength, hence on frequency. The actual limit value of energy at each frequency, *i.e.* the limit shape of the spectrum, was given by PHILLIPS [5] as

$$E(f) = \alpha g^2 \sigma^{-5} \tag{3}$$

on purely theoretical arguments. (3) has been confirmed by innumerable experiments, in particular also by BARNETT and WILKERSON [4]. They found that, after a first linear and then exponential growth, once reached the corresponding limiting value given by (3), the energy effectively adjusts itself to this value without any further growth.

Expression (2) has been widely used for the practical estimate of wave energy, but there is clearly a good amount of empiricism in it. The feeling we have at this point is that the Phillips' and Miles' mechanisms, mainly the latter, do certainly exist, but there must be something else working on waves. We are stretching the β coefficient to force in it something due to some other phenomenon. A bright explanation, from a completely unexpected direction, came from PHILLIPS [6] and HASSELMANN [7]. PHILLIPS discovered that under certain conditions there was the possibility of a continuous exchange of energy among the different frequencies of a spectrum, and HASSELMANN readily applied his theory to the wave spectrum. The « resonant conditions » under which such an exchange is possible were found to be

$$\sigma_1 \pm \sigma_2 \pm \sigma_3 \pm \sigma_4 = 0\,, \tag{4}$$

$$\boldsymbol{k}_1 \pm \boldsymbol{k}_2 \pm \boldsymbol{k}_3 \pm \boldsymbol{k}_4 = 0\,, \tag{5}$$

where σ_i and $\boldsymbol{k}_i$ are the circular frequencies and the wave number vectors, respectively. (4) and (5) tell us that the energy exchange takes place among quadruplets of frequencies. To evaluate the energy exchange in a spectrum due to these nonlinear wave-wave interactions, we have to check all the possible combined quadruplets of frequencies to see if (4) and (5) are verified. Then the actual, positive or negative, energy fed into a certain component is given by the Boltzmann-integral expression [8]

$$\frac{\partial N_4}{\partial t} = \int T(N_1 N_2 N_3 + N_1 N_2 N_4 - N_1 N_3 N_4 - N_2 N_3 N_4) \cdot \delta(\sigma_1 + \sigma_2 - \sigma_3 - \sigma_4)\, \mathrm{d}\boldsymbol{k}_1\, \mathrm{d}\boldsymbol{k}_2 , \tag{6}$$

where $N_j = N(\boldsymbol{k}_j) = F(\boldsymbol{k}_j)/\sigma_j$, $\sigma_j = \sqrt{gk_j}$, $F(\boldsymbol{k})$ is the two-dimensional wave spectrum with respect to wave number, and the transfer coefficient T is a quadratic homogeneous function of the four wave numbers $\boldsymbol{k}_1, \ldots, \boldsymbol{k}_4$. The calculation necessary for the actual evaluation of (6) for all the components is very heavy. Together with the lack of complex measurements required to experimentally verify it, this delayed the general acceptance and application of nonlinear interactions to wave evaluations. An end to the discussions was made by the JONSWAP experiment [8], in which several different institutes collaborated together for an extensive series of wave measurements under ideal field conditions. The comparison of spatially displaced spectra taken under a constant offshore blowing wind proved the active role played by the nonlinear interactions in the evolution of the wave spectrum. Moreover, they proved to be the dominant factor, forcing the spectrum of waves growing under an active generating wind to assume permanently a well-defined shape. For obvious reasons this was called JONSWAP (J) spectrum (fig. 1), and it is given [8] by

$$E(f) = \alpha g^2 (2\pi)^{-4} f^{-5} \exp\left[-\frac{5}{4}\left(\frac{f}{f_m}\right)^{-4}\right] \gamma \exp\left[\frac{-(f-f_m)^2}{2\sigma^2 f_m^2}\right], \tag{7}$$

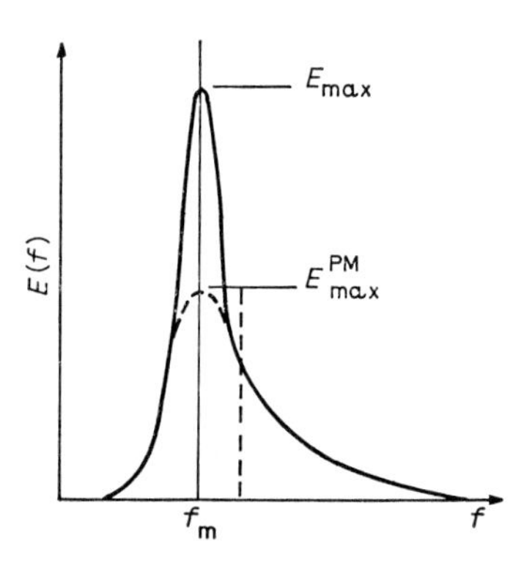

Fig. 1. – JONSWAP spectrum (continuous line) as compared to its limit from Pierson-Moskowitz (broken line); $\gamma = E_{\max}/E^{\mathrm{PM}}_{\max}$.

where $\sigma = \sigma_a$ for $f \leqslant f_m$ and $\sigma = \sigma_b$ for $f > f_m$. This expression contains 5 free parameters f_m, α, γ, σ_a and σ_b. f_m is the frequency at the peak of the spectrum, α is the usual Phillips constant (the spectrum approaches the usual f^{-5} power law for large f/f_m). The remaining three parameters define the shape of the spectrum: γ is the ratio of the maximal spectral energy to the maximum of the corresponding Pierson-Moskowitz (PM) spectrum [9] that, as modified in [8], is given by

$$E_{\text{PM}}(f) = \alpha g^2 (2\pi)^{-4} f^{-5} \exp\left[-\frac{5}{4}\left(\frac{f}{f_m}\right)^{-4}\right]. \tag{8}$$

We will soon come back to the physical meaning of this spectrum. σ_a and σ_b give the width of the left and right side of the spectrum, respectively.

Let us now forget for a moment about expressions (7) and (8). Consider a constant wind blowing for a very long time on an unlimited sea. It is reasonable to expect that after a certain period of time, depending on the actual wind speed, waves will achieve some equilibrium state, a sort of equilibrium among the different acting forces, when they do not grow any more, or at least the evolution is so slow that we can neglect it for all practical purposes. These are called « fully developed conditions », and PIERSON and MOSKOWITZ [9] found experimentally the corresponding spectrum to be well fitted by (8). Its shape is compared to the J one in fig. 1. The physics of wave growth that came out of J was that the main energy input by wind takes place in the frequency range just to the right of the spectral peak, and most of this energy is then brought to lower frequencies by the nonlinear wave-wave interactions. This allows a shift of energy, hence of spectral peak, towards lower frequencies with increasing time and fetch. This is coincident with what we find in the sea, that the wave period grows with the distance from which the waves are coming. The shape of the J spectrum was chosen in such a way to fit the experimental results, but to relax into the PM one once the fully developed conditions were reached. The key parameter is the γ peak enhancement factor (fig. 1), found to have an average value 3.3 in the generating area, and to lower to 1 when generation is over, so making (7) turn into (8).

The Hasselmann theory had a number of consequences. For instance, it is now suggested that the original Miles' estimate of the coefficient β is much closer to the truth than previously supposed. We do not need any longer very large values of β to justify the experimental wave growth, as the continuous energy input on the high-frequency side of the spectrum, coupled with the nonlinear wave-wave interactions, can fully explain the low-frequency energy growth. This is now a very active field of research, and a number of experiments have been carried out on it. One of the most interesting hypotheses is that the energy transfer from wind to waves does not take place wave by wave, but it is mainly concentrated on single bursts. BANNER and MELVILLE [10] have experimentally found that the wind drag on the wave, on which the energy transfer depends, is increased by more than an order of magnitude over breaking waves. This suggests again the sheltering hypothesis of Jeffreys [1]. So we see that the process of wave generation, even if miles ahead of where it was not many years ago, is far from being completely understood. The difficulty lies in that there are many different phenomena acting together and concurring with wave growth. These phenomena interact and influence each other, in ways difficult to envisage, and lead to the strong nonlinearity of the problem. In my opinion it will take still some time to reach what could be called a satisfactory stage of the problem.

So far I have not spoken about water depth. In effect all the results reported above implicitly assumed deep-water conditions, and the experiments on wave growth are usually conducted in deep water to avoid the further factor of the water depth. Anyhow the eventual limited depth of the bottom does not modify the basic physics of generation. Rather it introduces a number of phenomena of conservative and dissipative type. We will be now concerned with them.

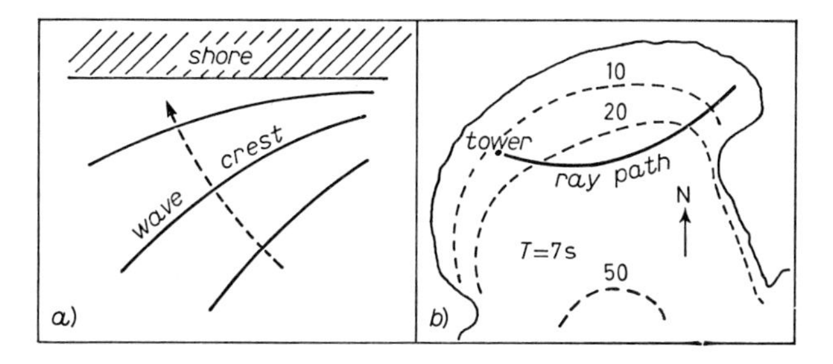

Fig. 2. – *a*) Refraction of a wave approaching obliquely a shore. *b*) Path of a wave ray in the shallow water of the Northern Adriatic Sea.

There are two conservative processes acting when a wave approaches the shore. The first one is refraction. The phase speed c of a wave depends on period and depth. If a long crested wave (fig. 2*a*)) approaches obliquely a beach, different points of the crest will experience different depth, hence different c. The wave is, therefore, forced to turn from its original direction, moving usually almost perpendicular to the shore. More generally a wave packet, that would proceed in a straight direction in deep-water conditions, in shallow water will experience a path depending on the actual bottom topography (fig. 2*b*)). The process is fairly similar to the optical refraction, and the theory is well established.

The second conservative process is shoaling. Physically it depends on the conservation of energy of a single wave while this approaches the shore. During the approach the kinematics of a wave varies, but the overall energy must remain the same. The result is first a decrease, then an increase of its height (fig. 3), that ends up in what we know as surf on the beach, appearing when the shoaling increased wave height reaches a limit value with respect to its wavelength and depth.

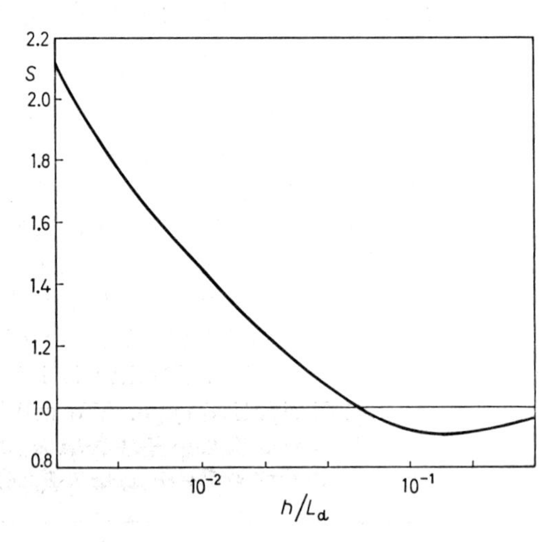

Fig. 3. – Shoaling coefficient S, as ratio of wave heights, with respect to deep-water conditions. h is depth, L_d is deep-water wavelength.

Apart from the very last moment these two processes are conservative, in the sense that the overall energy does not vary. Parallel to them we find another series of processes that act to extract energy from the wave field. Those envisaged up to now are percolation, wave-bottom interaction, bottom friction and bottom scattering. They will be now singularly analysed.

Percolation. We know from the previous lecture that a wave travelling on the sea surface has an underwater pressure field associated with it. In particular on a limited depth the pressure wave on the bottom is given by

$$p = \varrho g a \frac{1}{\cosh kh} \cos (kx - \sigma t) \tag{9}$$

with the usual meaning of symbols. We find, therefore, a different pressure from place to place, with maximum difference $\Delta p = 2\varrho g a/\cosh kh$. If the bottom is permeable, the pressure differences will force water through the bottom material from zones of high to those of low pressure. The water motion through the bottom is not certainly without loss of energy, and this energy can be extracted only from waves. For an isotropic sand and a deep sand layer (greater than 0.3 wavelength) SHEMDIN *et al.* [11] found the loss of energy to be expressible as

$$\frac{\partial E}{\partial t} = -\frac{k\alpha}{\cosh^2 kh} E, \tag{10}$$

where α is the percolation coefficient. Values of α for the different sand size have been provided by laboratory experiments [12] by SLEATH. For a sand diameter 1.3 mm and 0.38 mm he found α to be 1.15 and 0.124. Correspondingly SHEMDIN *et al.* [11], defining from (10) a damping coefficient

$$\gamma_{p} = -\frac{1}{E}\frac{\partial E}{\partial t}, \tag{11}$$

have found, respectively, $\gamma_p = 2.55 \cdot 10^{-4}$ and $\gamma = 0.69 \cdot 10^{-4}$. From the practical point of view percolation has an appreciable influence only for coarse sand. When grains are small, say 0.2 mm or less, it can be neglected without any practical consequence.

Wave-bottom interaction. This process depends again on the wave pressure on the bottom, and it is connected to its possible elasticity. If the bottom material is deformable under the action of pressure, it will respond to wave action with a vertical oscillation opposite in phase to the surface wave. If the material is fully elastic, it would act like a spring absorbing and returning energy without any loss. If, as is usually the case, a percentage of nonelasticity exists, part of the energy absorbed during each wave cycle will be absorbed and lost by hysteresis. This will, of course, cause an attenuation of wave energy. For a sandy bottom this effect is usually quite limited and ROSENTHAL [13] has estimated a damping coefficient $\gamma_{bi} = 1.47 \cdot 10^{-5}$ (the definition is similar to that of γ_p). γ_{bi} can anyhow rise to very large values when the bottom sediment is composed of soft material such as mud or decomposed organic matter. These conditions prevail for instance in the Gulf of Mexico and TUBMAN and SUHAYDA [14] have reported extremely high spatial attenuation rates in this area. It is believed that a real mud wave, whose height can be of the order of metres, is following on the bottom the surface wave, consuming energy at a high rate because of the viscous characteristics of the sediments. HSIAO and SHEMDIN [15] have developed a detailed formulation and solution for the coupled two-layer problem, and found their theory consistent with the experimental results. The decay rate due to bottom motion is at least two orders of magnitude greater than estimated rates based on either friction or percolation mechanisms.

Bottom friction. This mechanism was the first to be recognized and it is probably the most studied one. Its action is quite intuitive. A surface wave implies a motion of the water particles below it. This motion reduces to a longitudinal oscillation close to the bottom, and the water particle is forced to move against the drag exerted by the bottom. Similarly to the drag for a current, we can assume a quadratic friction law

$$\boldsymbol{\tau} = -\varrho c_f |u| \boldsymbol{u}, \tag{12}$$

where ϱ is water density, c_f is the drag coefficient, $\boldsymbol{u}$ is the instantaneous velocity vector. Of course, the friction $\boldsymbol{\tau}$ on the water particle acts in the direction opposite to $\boldsymbol{u}$. The product $\boldsymbol{\tau} \cdot \boldsymbol{u}$ gives the power absorbed by friction, and, for a single sinuoidal wave, averaging over a wave cycle, we find [16]

$$\gamma_{bf} = -\frac{4}{3} \frac{c_f \sigma^3 H}{g \sinh^3 kh} . \tag{13}$$

H is wave height. A complication arises because (12) is a quadratic law, hence nonlinear, and waves of different frequency interact with each other. This aspect has been attacked by HASSELMANN and COLLINS [17]. Under the hypothesis of a Gaussian wave field, they derived an equation for the rate of energy dissipation due to bottom friction in a random sea, and the damping rate was estimated to be

$$\gamma_{bf} = -\frac{c_f g k v}{2\pi\sigma^2 \cosh^2 kh} \langle u \rangle , \tag{14}$$

where v is the group velocity and $\langle u \rangle$ denotes an ensemble average for the wave orbital velocity at the bottom. $\langle u \rangle$ is related to the frequency energy spectrum $E(f)$ [18] by

$$\langle u \rangle = \left(\sum_f E(f) \frac{g^2 k^2}{\sigma^2 \cosh^2 kh} \Delta f \right)^{\frac{1}{2}} . \tag{15}$$

We realize now that we need the knowledge of the whole spectrum to evaluate the damping rate at each single frequency.

The problem is now left to the estimate of the friction coefficient c_f. This depends on the characteristics of the boundary layer on the bottom and on the bottom roughness height. It is intuitive that bottoms made of thin smooth sand, large grains, or with sand ripples must offer highly different resistances to water motion. Theoretical and experimental studies have shown that c_f can easily vary of two orders of magnitude according to the bottom influence and to the flow conditions. The latter have a direct influence, because the characteristics of a sandy bottom (smooth, ripples, or incoherent sand) depend in turn on the intensity of motion of the water particles. The problem is, therefore, closed on itself. To know the damping rate we need the knowledge of the friction coefficient, but this is not known until when the wave conditions are given. Of course, the problem is solved step by step, and a general procedure for the evaluation of c_f has been proposed in [15].

From the quantitative point of view bottom friction is generally the dominant factor when the sediment is composed of fine sand, with diameter less than 0.4 mm. In comparison with the above-reported values for percolation and bottom interaction SHEMDIN *et al.* [11] have estimated the damping rate due to bottom friction up to values of $10^{-2} \div 10^{-1}$. Figure 4, taken from [11], shows the different damping coefficients associated with percolation, bottom motion and friction mechanisms for different value of the product kh.

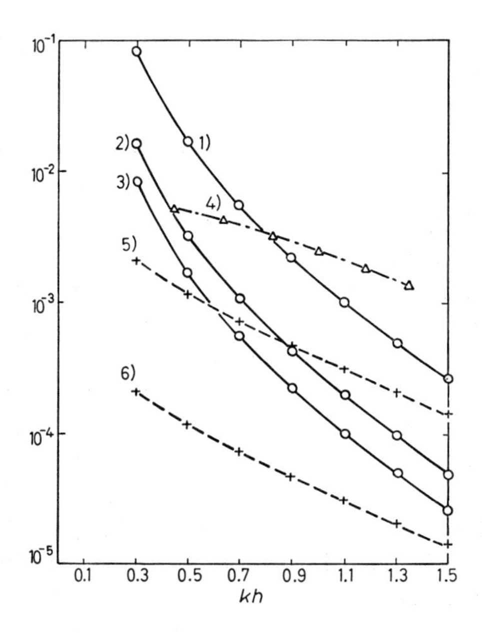

Fig. 4. – Damping coefficients associated with percolation, bottom motion and friction mechanisms for different kh (after [11]): curve 1) friction, $c_f = 0.05$, $a = 1$ m; curve 2) friction, $c_f = 0.01$, $a = 1$ m; curve 3) friction, $c_f = 0.005$, $a = 1$ m; curve 4) bottom motion, $\nu = 0.32$ m²/s, $G/\varrho = 10$ m²/s²; curve 5) percolation, coarse sand, $\alpha = 1$ cm/s; curve 6) percolation, fine sand, $\alpha = 0.1$ cm/s; for all curves $T = 7.75$ s.

Bottom scattering. This process arises from the following phenomenon. Consider a monochromatic wave field propagating in a given direction. Suppose besides that the bottom is not flat but that it presents sinusoidal oscillations, with the crests parallel to the surface ones. It turns out that, if the bottom wavelength is half the surface one, there is a gradual reflection of the surface energy in the backward direction. Different patterns arise when the bottom crests have a longitudinal component. As for the surface, we can define a bidimensional spectrum for the bottom, and there is, therefore, a continuous scattering of surface energy by the oscillations of the bottom. This mechanism has been explored by LONG [19], who has found it to be potentially a dominant process, but an experimental verification of it has not yet been possible. For an inshore coming wave field, the bottom scattering mechanism predicts an offshore going energy increasing in the offshore direction. To verify this we should succeed in measuring at at least two different positions the directional energy distribution for each frequency and to be able to resolve in it two peaks going in opposite directions. With a complicated array system this is in principle possible, but it has not yet been achieved. Hence we are still left with the question if the bottom scattering does really work or not. Would the reply be positive, the overall argument of bottom dissipation would become much more complicated, for the requirement of knowing the bottom spectrum and the heavy computations involved.

The ones listed above are those that, at the actual stage of the theory, are thought to be the main mechanisms concurring with the generation and dissipation of wind waves. I believe it to be useful to make a distinction among them. Phillips', Miles', percolation and bottom interaction are linear mechanisms, in the sense that the input or output at each frequency does not depend on the remaining values of the spectrum. Nonlinear wave-wave interactions, breaking, bottom friction and bottom scattering are highly nonlinear processes, involving the whole spectrum in their evaluation.

I mentioned above that a satisfactory stage for the theory of wave generation has not yet been reached. Much the same can be said about dissipation. We cannot certainly be happy about a situation in which we do not know if what could be the potentially basic mechanism is actually effective or not! Probably most of these doubts will be clarified in the not too distant future through some large-scale extensive and sophisticated experiments in the field. My personal feeling anyhow is that, the farther ahead we go, the more complications we will find in the overall phenomenon. We will have to decide at which stage to stop to be able to provide some compact laws useful for practical application.

REFERENCES

[1] H. Jeffreys: *Proc. R. Soc. London Ser. A*, **110**, 341 (1952).
[2] O. M. Phillips: *J. Fluid Mech.*, **2**, 417 (1957).
[3] J. W. Miles: *J. Fluid Mech.*, **3**, 185 (1957).
[4] T. P. Barnett and J. C. Wilkerson: *J. Marine Res.*, **25**, 3 (1967).
[5] O. M. Phillips: *J. Fluid Mech.*, **4**, 426 (1958).
[6] O. M. Phillips: *J. Fluid Mech.*, **9**, 193 (1960).
[7] K. Hasselmann: *J. Fluid Mech.*, **12**, 481 (1962).
[8] K. Hasselmann, T. P. Barnett, E. Bouws, H. Carlson, D. E. Cartwright, K. Enke, J. A. Ewing, H. Gienapp, D. E. Hasselmann, P. Kruseman, A. Meerburg, P. Müller, D. J. Olbers, K. Ritcher, W. Sell and H. Walden: *Ergeb. Dtsch. Hydr. Z., Reihe A* (8°), No. 12 (1973).
[9] W. J. Pierson jr. and L. Moskowitz: *J. Geophys. Res.*, **69**, 5181 (1964).
[10] M. L. Banner and W. K. Melville: *J. Fluid Mech.*, **77**, 825 (1976).
[11] O. Shemdin, K. Hasselmann, S. V. Hsiao and K. Herterich: NATO *Symposium*

on Turbulent Fluxes through the Sea Surface, Wave Dynamics and Prediction, edited by A. FAVRE and K. HASSELMANN (New York, N. Y., 1978), p. 347.
[12] J. F. H. SLEATH: *J. Hydr. Div., HY2, Am. Soc. Civ. Eng.*, 367 (1970).
[13] W. ROSENTHAL: *J. Geophys. Res.*, **83**, 1980 (1978).
[14] M. W. TUBMAN and J. N. SUHAYDA: *Proceedings of the XV International Conference on Coastal Engineering* (Honolulu, Hawaii, 1976).
[15] S. V. HSIAO and O. SHEMDIN: *Coastal Engineering Conference, Am. Soc. Civ. Eng.* (Hamburg, 1978), p. 434.
[16] J. A. PUTMAN and W. JOHNSON: *Trans. Am. Geophys. Un.*, **30**, 67 (1949).
[17] K. HASSELMANN and J. I. COLLINS: *J. Marine Res.*, **26**, 1 (1968).
[18] J. I. COLLINS: *J. Geophys. Res.*, **77**, 2693 (1972).
[19] R. B. LONG: *J. Geophys. Res.*, **78**, 7861 (1973).

Mathematical Models for Wave Forecasting.

L. Cavaleri

Istituto per lo Studio della Dinamica delle Grandi Masse, C.N.R.
S. Polo 1364, 30125 *Venezia, Italia*

In the two previous lectures I have first spoken about the characteristics of wind waves, as they are found in nature. Then I have shifted to the dynamics through which waves are created and destroyed. To do this we have gone into details, analysing the various known phenomena that concur with the wave energy budget. There are usually two different ways to look at a physical process. The first one is that of the physicist, who studies the process in itself, without any necessary interest in the further application of his findings. The second one is characteristic of the engineer, who faces a practical problem, and devises some way to apply the theoretical results to his practical purposes. The natural development of the second lecture is to see how the theories concerned with wave generation and dissipation are applied to the practical evaluation of the characteristics of a wave field. This will be the subject of this lecture.

Until 1942, the study of a mathematical wave and that of an ocean wave had no practical point of contact. The sea looked too rough and disorganized to be attractive for the neat constructions of the mathematicians. Nevertheless, the war was in progress and the Allies were planning a return to Europe through the sea. This was the starting push that triggered the mechanism of the models for wave forecasts. To build a model anyhow requires a theory, experimental data and time. And do not forget that the fast digital computers were still a dream in the mind of a few people. Nevertheless, the problem was there, and it required to be solved fast. Sverdrup and Munk (SM) [1] submitted a solution in September of 1943, and this was later published in 1947, after the end of the war. Their method of attack was based simply on the few equations describing the behaviour of a wave on the ocean surface, concerning its overall energy, the phase speed, the speed of propagation of energy, on a few sparse experimental data and on estimated rate of energy dissipation. With this material they were capable of supplying a few relationships between wind speed, fetch, time duration, wave height and period. These relationships were transformed in an abacus of immediate and practical use. The practical results were quite appreciable and the method remained in use till very recent times. As a matter of fact, it reveals itself useful even today if you need a fast guess for practical purposes.

The SM method is clearly parametric. There is no connection with the physics of air-sea interaction. The only things we need to know are the wind field and the geometry of the basin, the minimum information required for any application.

It took 10 years before the physical models came into the game. This happened when GELCI proposed his balance energy equation

$$\frac{\partial E}{\partial t} = c_g \cdot \nabla E + S . \tag{1}$$

Here E is the energy at a certain frequency, c_g is group velocity, coincident with the energy propagation speed, ∇ is divergence, and S is the energy input from external sources. This equation simply expresses the physical fact that the overall energy budget, for a given frequency and on a given area, is established by the input and output from the borders plus the, positive or negative, contribution from external means. It is clear that, if, from our previous study of the physical processes, we are capable of expressing S in mathematical form, we can then integrate numerically (1) and end up with an estimate of E. This was done by GELCI and, together with CAZALE and VASSAL [2], he proposed the first numerical method for wave forecasts. It is interesting to know that, with many revisions, the method is still in use by the French Meteorological Service.

The source function in [2] was rather empiric. After all Phillips' [3] and Miles' [4] mechanisms were both proposed in the same year. After some intermediate attempts, a definitive progress was made by BARNETT [5]. By 1968 HASSELMANN [6] had already revealed the role of the nonlinear wave-wave interactions, and the Phillips' and Miles' mechanisms had received their experimental verifications. Taking advantage of his experimental results [7], BARNETT was able to tune his model much better than anyone else before. His source function included both the Phillips' and Miles' processes with the standard form $\alpha + \beta E$. The big step ahead of BARNETT was done through the introduction of the nonlinear wave-wave interactions. These are too complicated and time consuming to be correctly evaluated, and BARNETT resorted to a parametrized form for them.

The results were quite good, and a few years later EWING [8] proposed his model for the North Atlantic. I will spend a bit of time on this, as it represents the temporary conclusion of the physical approach to the problem of forecast in deep water. The grid and the area covered by it are shown in fig. 1. The mesh size was 120 nautical miles. The source function in (1) was written as

$$S = (\alpha + \beta E)[1 - (E/E_s)^2] + A + BE . \tag{2}$$

The model is fairly similar to that of Barnett [5], but it has a better definition of the generative terms α and β. The nonlinear interaction terms, *i.e.* $A + BE$, are present in a different way. At each integration step the actual spectrum at each grid point is compared and best fitted to a family of spectra representing a generalization of that of Pierson-Moskowitz. The integration of these spectra for the evaluation of the nonlinear terms had been previously carried out and stored in memory in the form of the coefficients A and B, so

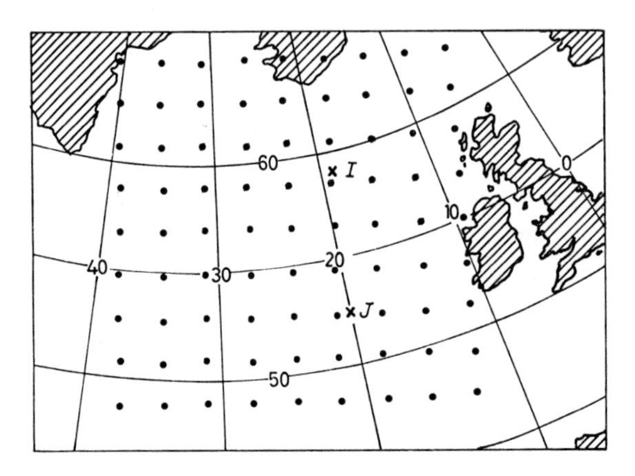

Fig. 1. – Spatial grid used to define wave and wind fields shown on a stereographic projection (after [8]).

to be readily accessible during the numerical integration. Breaking is taken into account through the term $c = 1 - (E/E_s)^2$. E_s represents the saturation energy at each frequency corresponding to the Phillips' limit spectrum [9]. The term c, multiplied by the input energy $\alpha + \beta E$ in (2), makes the energy approach asymptotically its limit value.

Ewing considered a directional spectrum with 8 frequencies and 12 directions, *i.e.* 96 components. Angular resolution was 30 degrees, frequency was resolved every 0.02 Hz. At each time step t, all the 96 components, *i.e.* the directional spectrum, are defined at each grid point. Integration of (1) and (2) over the time interval Δt provides the new values of the spectrum corresponding to time $t' = t + \Delta t$. The procedure is repeated along the whole period of interest.

Figure 2 shows a comparison between the measured and computed significant wave heights at two stations in the North Atlantic. Figure 3 shows an analogous comparison for the spectra.

We mentioned above that Ewing's model was the temporary conclusion of the physical approach to the problem. The reason for this was that there were two basic uncertainties affecting the final results. A basic one was the wind input. Clearly this is a key information for every model. The wind field is usually evaluated starting from the pressure field, but large uncertainties did exist (and still exist!) on its dependence on other factors, the main one being the air stability conditions. An error of 20 % on the estimated wind field does not seem too bad, but think that the energy of a fully developed sea depends on the 4th power of it. This takes the corresponding error up to 100 %! Besides, by 1973, the year of JONSWAP [10], people had fully realized the key role played by the nonlinear wave-wave interactions, but also the prac-

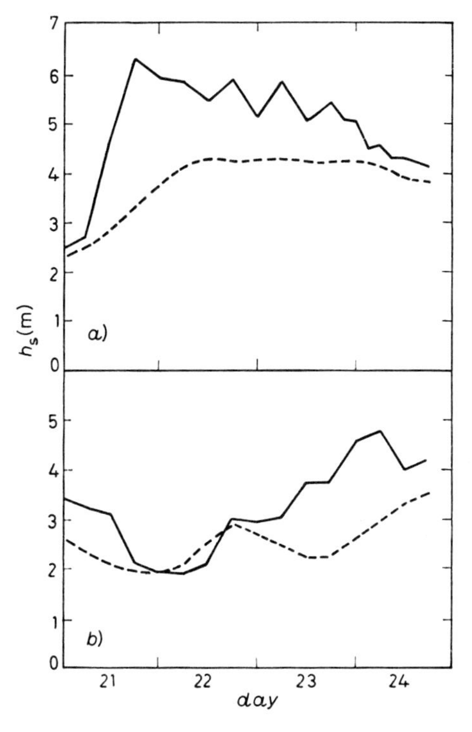

Fig. 2. – Comparison of measured (continuous) and computed (dashed line) significant wave heights during November 1966 (after [8]): *a*) station India, *b*) station Juliett.

tical impossibility of taking them correctly into account during the numerical integration of (1), this for the extremely heavy computations associated with them. So there was clearly no point in pushing further the physical models, as one was in any case left with these large uncertainties.

A completely different approach was, therefore, proposed by HASSELMANN *et al.* [11]. The JONSWAP (J) experiment had shown that, under active wind conditions, the wave spectrum was permanently linked to a fixed shape, namely the J spectrum we have mentioned in the previous lecture. If we assume this as a starting point, to define completely a spectrum it is not necessary any longer to specify the value of energy frequency by frequency. It is now enough to specify the value of the 5 parameters defining a J spectrum, namely f_m,

α, γ, σ_a, σ_b. For what concerns the directional distribution of energy, we can assume this to be centred in the wind direction Φ_0, and some law, like, *e.g.*, $\cos^2(\Phi - \Phi_0)$, for the directional spreading around it. Of the 96 variables we had at each point, we are now left with 5, which implies a high gain in com-

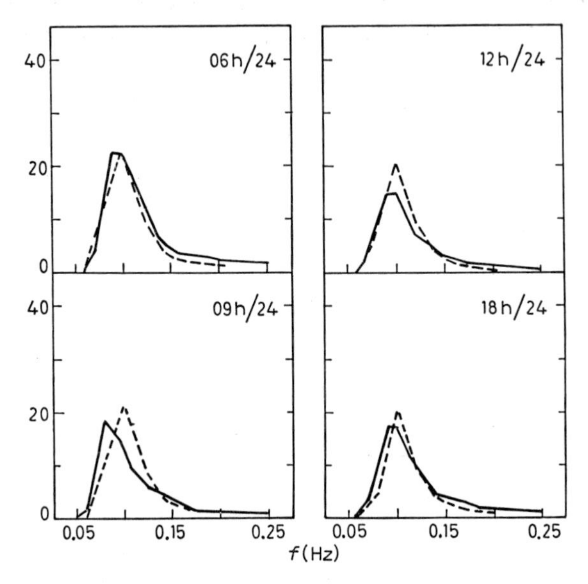

Fig. 3. – Comparison of measured (continuous) and computed (dashed line) one-dimensional spectra during November 1966 (after [8]), station India. The vertical scale units in the graphs are m^2 s (after [8]).

puter time. Moreover, the values of σ_a and σ_b being highly scattered, it was decided to fix them at their mean values, so reducing the number of variables to 3. This seems a quite crude approach, but it is very effective. The point is that the dominant role of the nonlinear wave-wave interactions is implicitly taken into account by the shape of the spectrum. The mathematics is not obvious. It implies a projection of the energy balance equation (1) into the parameter space and the assumption of some functional form for the energy input by wind. The results are 3 differential equations, one for each parameter, to integrate in time. The model needs a calibration, and this has been obtained by using the recorded data out of a number of storms in the North Sea.

Now it is worth-while to stop for a moment and think. We had started 30 years before with an empirical model (SM). Then the physical models came in, till when it was realized that the physics was too complicated to be handled in the model, and we resorted again to parametric models. Clearly there is a good deal of difference between the actual parametric approach and the original one by SM. HASSELMANN *et al.* [11] have embedded a fairly good amount of physics into their few and simple equations. But there are a couple of points that must be clarified. The SM method was empirical, but fitted to experimental evidence, without restriction. The physical models, with their now clear limitations, to be operated simply required the specification of the wind field and the coastal topography. Now the recent parametric model, more physical than the SM model, simpler than the physical one, has a strong limitation: it is valid only in a generating area, as it is only in these conditions that the wave spectrum assumes the J shape. Besides swell is not considered, nor is any energy that is not going in the wind direction. This has resorted GÜNTHER *et al.* [12] to develop a hybrid model where the spectrum is divided into two sections. A high-frequency section, variable according to the wind speed, is dealt with with the pure parametrical approach of Hasselmann *et al.* [11]. The remaining low-frequency part does not receive any external input, and it is simply carried along as an energy packet. Figures 4 and 5 show the measured and computed significant wave heights and zero-crossing periods, and an analogous comparison of spectra for a storm recorded in the North Sea in December 1974.

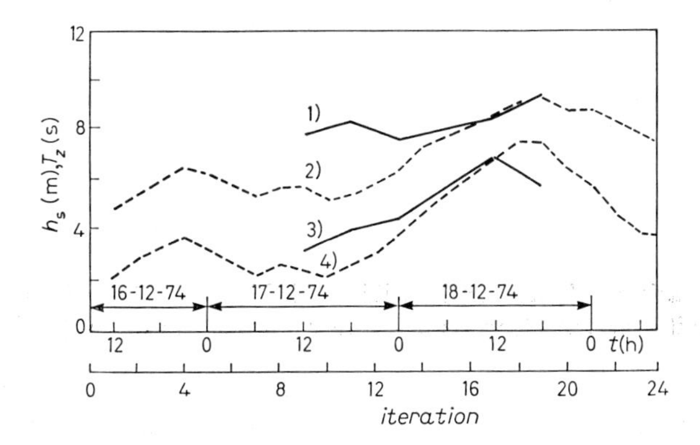

Fig. 4. – Comparison of measured (solid) and hindcast (dashed line) significant wave height and zero-crossing period at station Famita for the storm of December 17-18, 1974 (after [12]): curve 1) measured T_z, curve 2) model T_z, curve 3) measured h_s, curve 4) model h_s.

From the applicative point of view the main limitation of both the purely parametrical and the hybrid models lies in their inability of dealing with the, spatially or temporally, rapidly varying wind fields. If energy generated in a certain zone propagates into another with a completely different wind direction, clearly the hypothesis of the energy going with the wind is unrealistic. Waves need a certain time to adjust to the new situation, during which the estimate is certainly wrong. Thus, in applying this kind of models, we must be cautious about the local conditions, and decide in advance if the conditions are suitable or not for this approach.

I must point out that not all the researchers agree on which is the actual best approach to the problem. CARDONE *et al.* [13] have hindcasted the storms consequent to a number of hurricanes in the Gulf of Mexico, using a source function similar to (2), but without the terms $A + BE$ giving the nonlinear wave-wave interactions. The point is that in a hurricane the wind field, circular and moving with the eye of the hurricane itself, is extremely changeable in space and time, with waves running wildly in all the directions. This makes

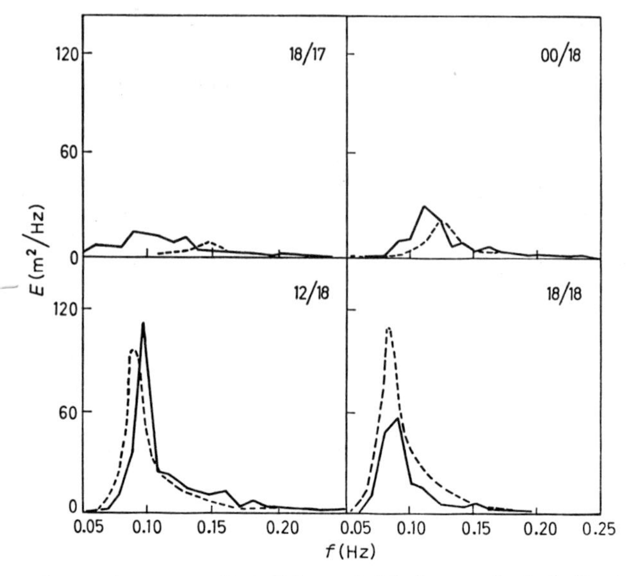

Fig. 5. – Comparison of measured (solid) and hindcast (dashed line) wave spectra at station Famita for the storm of December 17-18, 1974 (after [12]).

the application of a parametric model at least problematic. Figure 6, from [13], shows the predicted and measured wind at a location in the area during the very heavy hurricane Camille. Figure 7, again from [13], shows the hindcasted and measured significant wave height at one oil tower in the Gulf during the same hurricane. I think it worth-while for you to think of which kind of waves people were dealing with there, and to appreciate the capability of some instruments to survive and measure in such conditions.

The variable wave direction is well illustrated by fig. 8, a directional spectrum measured in the Gulf of Mexico during hurricane Delia, and reported by FORRISTALL *et al.* [14]. The parameter s indicates the concentration of energy around the mean direction.

What I have said up to now can give you an idea of what is going on in the field of wave evaluation, but with one restriction. Almost all the models mentioned above are deep-water models. Shallow-water effects are not taken into account. The only exception is the model of Cardone *et al.* [13], applied also in [14]. Working on grid in shallow water, the effects of refraction, shoaling

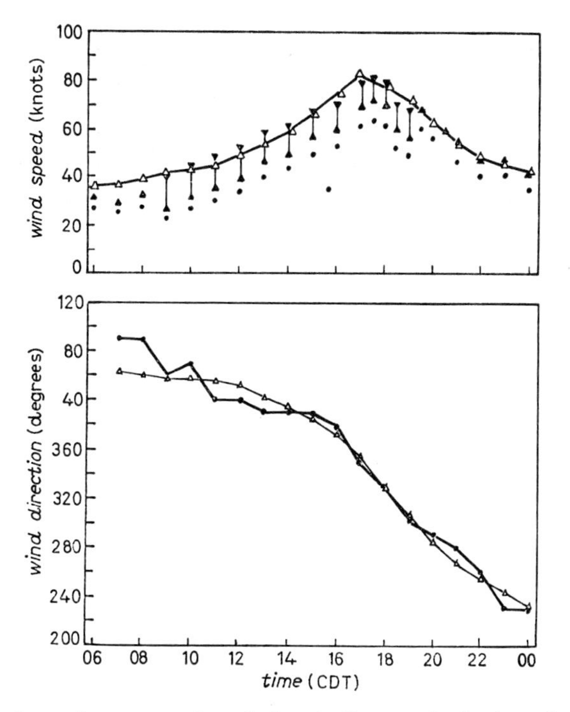

Fig. 6. – Predicted and measured wind at Burwood during hurricane Camille (after [13]): *a*) wind speed: • measured (65 ft), ▾ off land (adjusted), ▴ off water (adjusted). ▵ predicted; *b*) wind direction: • measured, ▵ predicted.

and bottom friction were modelled by the development of a propagation algorithm that represents the effect of 1 h of wave travel over a bottom of variable depth.

A completely different approach for prediction in shallow water was

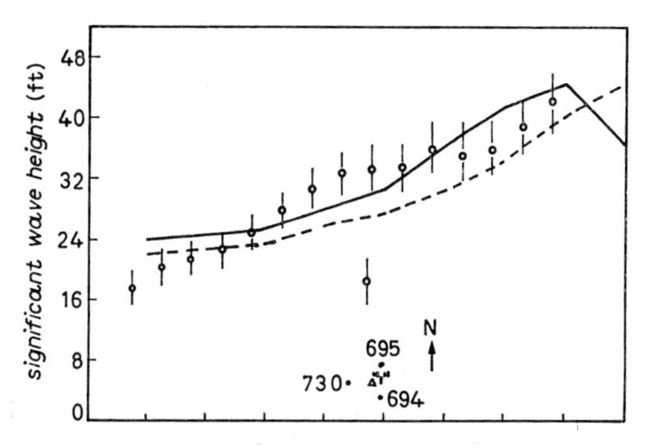

Fig. 7. – Hindicast and significant wave heights at ODGP station 1 during hurricane Camille (after [13]): ○ determined from spectral analysis of data (90% confidence limits shown), ——— hindcast 694, – – – hindcast 695.

proposed by COLLINS [15], and then developed by MALANOTTE RIZZOLI and myself [16]. The idea is the following. Suppose our interest is concentrated on knowing the waves at a single position. We can then consider all the possible wave trains approaching the location from all the directions. In particular, for each frequency we can define a certain number of directions, according to the directional resolution we want. For each direction we can then

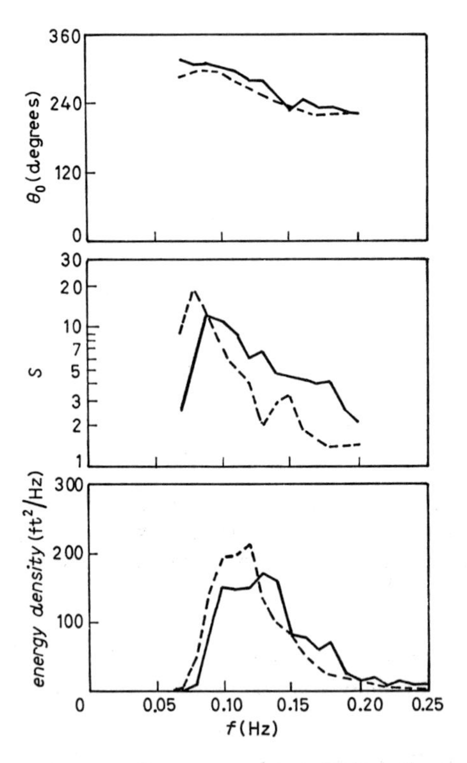

Fig. 8. – Directional spectrum at Buccaneer, 1230 CDT 4 September 1973 (after [14]).

study, by the refraction laws, the path followed by the waves. Under the obvious assumption that the bottom topography does not change, this path is the only possible one for waves of that frequency and approaching the spot of interest from that direction. Figure 9 shows a few rays (the actual net is much denser) evaluated in the Northern Adriatic Sea for $f = 0.14$ Hz, using as focusing point the oceanographic tower of CNR. Consider now one of these rays, or characteristics. When wind blows, waves grow, and, in particular, we can give attention to the wave component associated, for frequency and direction, to that ray. While the wave packet proceeds along the ray, toward the target point, its energy grows, according to the wind input and to eq. (1). But note that the equation is now much simpler, as, just because we are moving with the energy itself, we do not have any input from the boundaries. Thus we are simply left with the equation

$$\frac{\partial E}{\partial t} = S \tag{3}$$

depending only on the source function S. Integration of (3) along each characteristic supplies the final energy at the target point for the corresponding frequency and direction. The overall ensemble of all the results for all the possible characteristics is the directional spectrum at the position and time of interest.

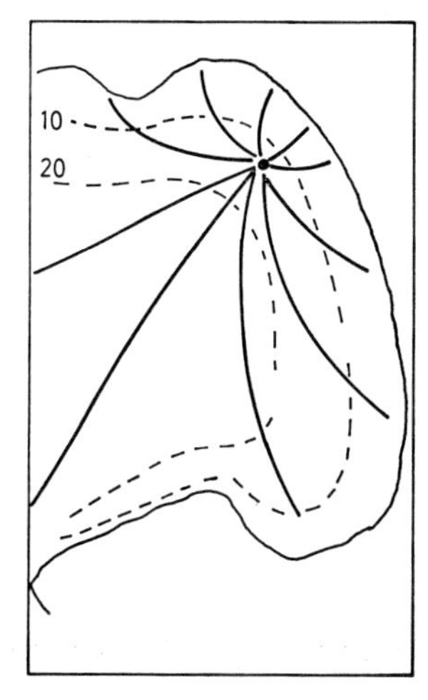

Fig. 9. – Refraction rays in the Northern Adriatic Sea. Frequency is 0.14 Hz. Depth in metres.

This approach to the problem has the advantages of taking fully into account shoaling and refraction, and to require much less computer time with respect to a grid model. On the other hand, the full two-dimensional spectrum is known only as a final result at the target, hence there are no chances of taking into account the nonlinear processes, like the wave-wave interactions

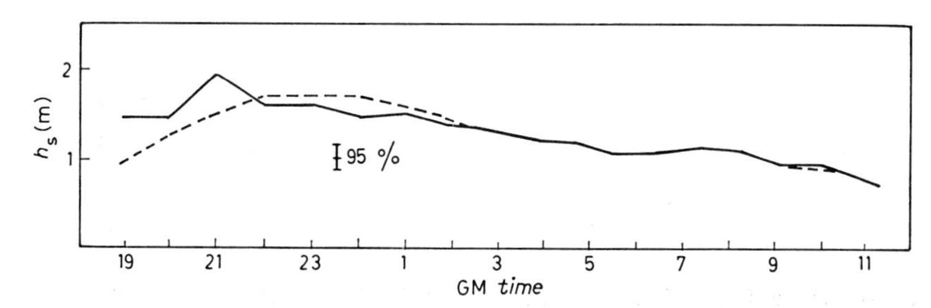

Fig. 10. – Observed (continuous) and computed (dashed line) significant wave heights for records of March 10-11, 1976 (after [16]).

and the bottom friction. Personally we have skipped this point resorting to a first-hand estimate of the wave conditions at each single point by using the results of JONSWAP [10] and evaluating the consequent bottom friction. Nonlinear wave-wave interactions were not taken into account. Figure 10

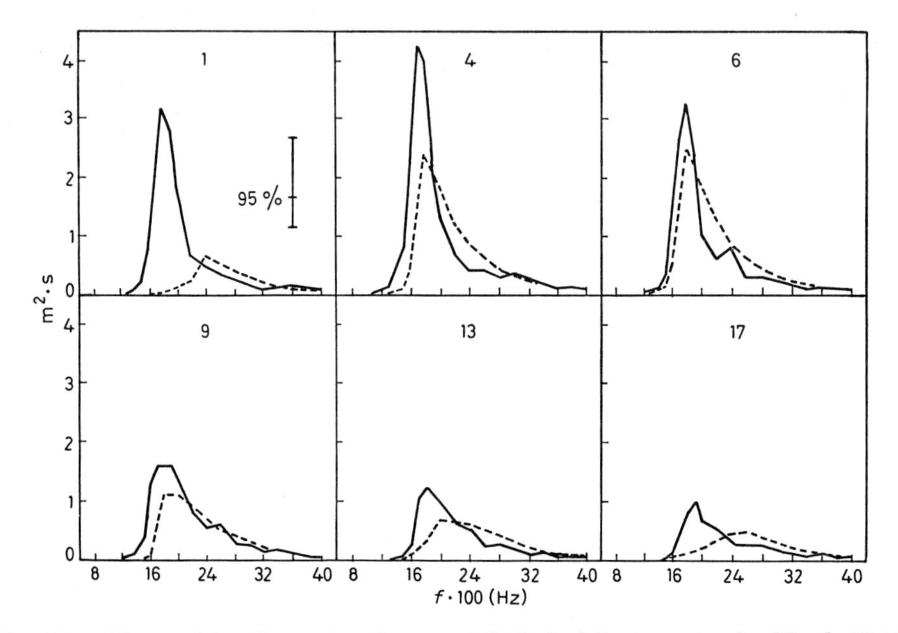

Fig. 11. – Observed (continuous) and computed (dashed line) spectra for March 10-11, 1976 (after [16]).

shows the comparison between the hindcasted and measured significant wave heights at the CNR tower. Figure 11 is an analogous comparison for 6 spectra. Figure 12 compares the mean directions for each frequency for one of the above spectra.

It is now time to summarize the general situation. In deep water we have

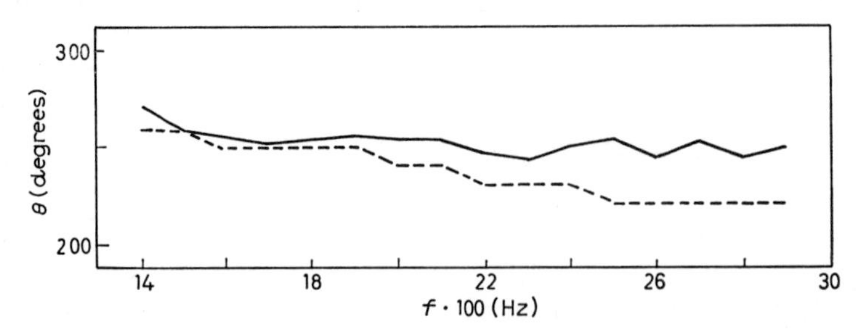

Fig. 12. – Observed (continuous) and computed (dashed line) main directions for each frequency. Vertical scale unit is degrees counterclockwise in respect to grid x-axis (after [16]).

basically two approaches, physical and parametrical. The former has no limitation in use, but uncertainties derive for the practical impossibilities of taking the nonlinear wave-wave interactions correctly into account. On the other hand, the parametric and hybrid models cannot deal well with swell and rapidly varying wind fields. Extension of the models to shallow water is on the way, but it will introduce a remarkable increase in computer time for the estimate of the source function connected to the interactions with the bottom. The ray approach is quite suitable for shallow water, but it cannot handle properly the nonlinear phenomena. It is anyhow probably the best approach when the result has to be provided at a single position.

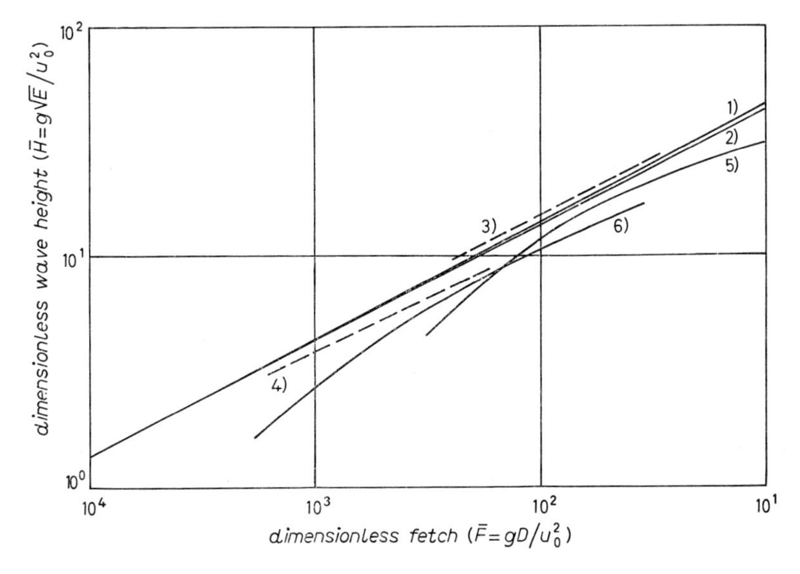

Fig. 13. – Comparison of relationships between nondimensional wave height and nondimensional fetch to those derived empirically by MITSUYSASU and HASSELMANN *et al.* (after [17]): curve 1) Mitsuysasu, $\bar{H} = 1.31 \cdot 10^{-2} \bar{F}^{0.504}$; curve 2) Hasselmann *et al.*, $\bar{H} = 1.27 \cdot 10^{-2} \bar{F}^{0.500}$; curve 3) Barnett, 15 $m \cdot s^{-1}$ test; curve 4) Barnett, 30 $m \cdot s^{-1}$ test; curve 5) 1-*B* model, 15 m·s test; curve 6) 1-*B* model, 25 m·s test.

From a general point of view, most of these models provide quite reasonable results. The difficulty is now to improve them in such a way as to reduce the actual error from, say, 20% to 5%. It is clear that, if we want to have consistently such a figure, all the processes embedded in the model have to be represented with the same accuracy. RESIO and VINCENT [17] have recently pointed out that most of the models behave well in the normal range of application and in the open ocean, conditions usually coincident with those used for calibration. But the models usually diverge when applied to complex shoreline areas and to wind and fetch conditions far from those of calibration. Figure 13, taken from [17], is well representative in this sense. Note that dimensionless fetch and wave height have been used in the figure, D being the actual fetch.

Thus we see we have still a lot to go before we can claim to have solved the problem. In which direction we are going to move it is not easy to say. Especially the parametric approach is too recent to be fully digested and developed, and also for an objective comparison of the different capabilities. As anyhow the economical push on the argument is likely to last for a good while, we will see a corresponding development in this field. But I cannot guess now the final solution.

REFERENCES

[1] H. U. SVERDRUP and W. H. MUNK: *U.S. Navy Hydr. Office Pub.* No. 601 (1947).
[2] R. H. GELCI, H. CAZALE and J. VASSAL: *Extrait du Bull. Inf. Comité Central d'Océanographie*, **9**, 416 (1957).
[3] O. M. PHILLIPS: *J. Fluid Mech.*, **2**, 417 (1957).
[4] J. MILES: *J. Fluid Mech.*, **3**, 185 (1957).
[5] T. P. BARNETT: *J. Geophys. Res.*, **73**, 513 (1968).
[6] K. HASSELMANN: *J. Fluid Mech.*, **12**, 481 (1962).
[7] T. P. BARNETT and J. C. WILKERSON: *J. Marine Res.*, **25**, 3 (1967).
[8] J. A. EWING: *Dtsch. Hydrog. Z.*, **24**, 241 (1971).
[9] O. M. PHILLIPS: *J. Fluid Mech.*, **4**, 426 (1958).
[10] K. HASSELMANN, T. P. BARNETT, E. BOUWS, H. CARLSON, D. E. CARTWRIGHT, K. ENKE, J. A. EWING, H. GIENAPP, D. E. HASSELMANN, P. KRUSEMAN, A. MEERBURG, P. MÜLLER, D. J. OLBERS, K. RITCHER, W. SELL and H. WALDEN: *Ergeb. Dtsch. Hydr. Z., Reihe A* (8°), No. 12 (1973).
[11] K. HASSELMANN, D. B. ROSS, P. MÜLLER and W. SELL: *J. Phys. Ocean.*, **6**, 200 (1976).
[12] H. GÜNTHER, W. ROSENTHAL, T. J. WEARE, B. A. WORTHINGTON, K. HASSELMANN and J. A. EWING: *J. Geophys. Res.*, **84**, 5739 (1979).
[13] V. J. CARDONE, W. J. PIERSON and E. G. WARD: *J. Petrol. Tech.*, **28**, 385 (1976).
[14] G. Z. FORRISTALL, E. G. WARD, V. J. CARDONE and L. E. BORGMANN: *J. Phys. Ocean.*, **8**, 888 (1978).
[15] J. I. COLLINS: *J. Geophys. Res.*, **77**, 2693 (1972).
[16] L. CAVALERI and P. MALANOTTE RIZZOLI: NATO *Symposium on Turbulent Fluxes through the Sea Surface, Wave Dynamics and Prediction*, edited by A. FAVRE and K. HASSELMANN (New York, N. Y., 1978), p. 629.
[17] D. T. RESIO and C. L. VINCENT: *Hydr. Lab. U.S. Army Eng. Water Exp. St.*, Miscellaneous paper H-77-9 (1977).

The Simulation and Measurement of Random Ocean Wave Statistics.

A. R. OSBORNE (*)

Exxon Production Research Company - Houston, Tex. 77001

1. – Introduction.

The development of the mathematical description of random processes has led to the understanding of stochastic problems in many fields of endeavor [1-3]. Included among these fields is oceanographic science, where the theory of random ocean surface waves has developed substantially over recent years [4-6]. It is now understood, for example, that the sea surface elevation may, for certain restricted conditions, be viewed as a linear, stationary, ergodic, Gaussian random process described by a well-behaved power spectrum. While several authors [7-9] have considered cases where some of these conditions are violated, we shall, in this paper, be concerned with the theoretical predictions for random waves whose amplitudes are Gaussian and whose statistical properties do not vary in time. To this end we shall be most concerned with statistical parameters of random waves which are not well understood theoretically, for example the heights and periods of zero-crossing waves; in this particular context we compare our results to the theoretical predictions of Longuet-Higgins [10]. The purpose of this paper is to describe a Monte Carlo simulation technique which allows the rapid and efficient generation of many desired statistical parameters of random waves. Comparison of this model will be made to linear theory and to data.

The Monte Carlo simulation of random ocean surface waves has become an important means of exploring wave properties, especially when theoretical methods fail to provide accurate descriptions of the statistical behavior of several important wave parameters such as zero-crossing crest heights, wave heights and wave periods [11]. Filter simulations have been used extensively by ROBINSON *et al.* [12] and BORGMAN [11], while BORGMAN [11] has also pursued several techniques which employ the method of uniform random phases. However,

---

(*) Present address: Istituto di Cosmogeofisica, 10133 Torino, Italia.

none of the previous simulations have proven entirely adequate for accurately determining the statistical behavior of random ocean wave parameters, particularly in the extreme. The reason that these methods have not been useful for compiling large numbers of wave statistics is due to the limited accuracy of the techniques employed and/or to the large amount of computer time necessary for the accumulation of large numbers of waves. Accuracy problems

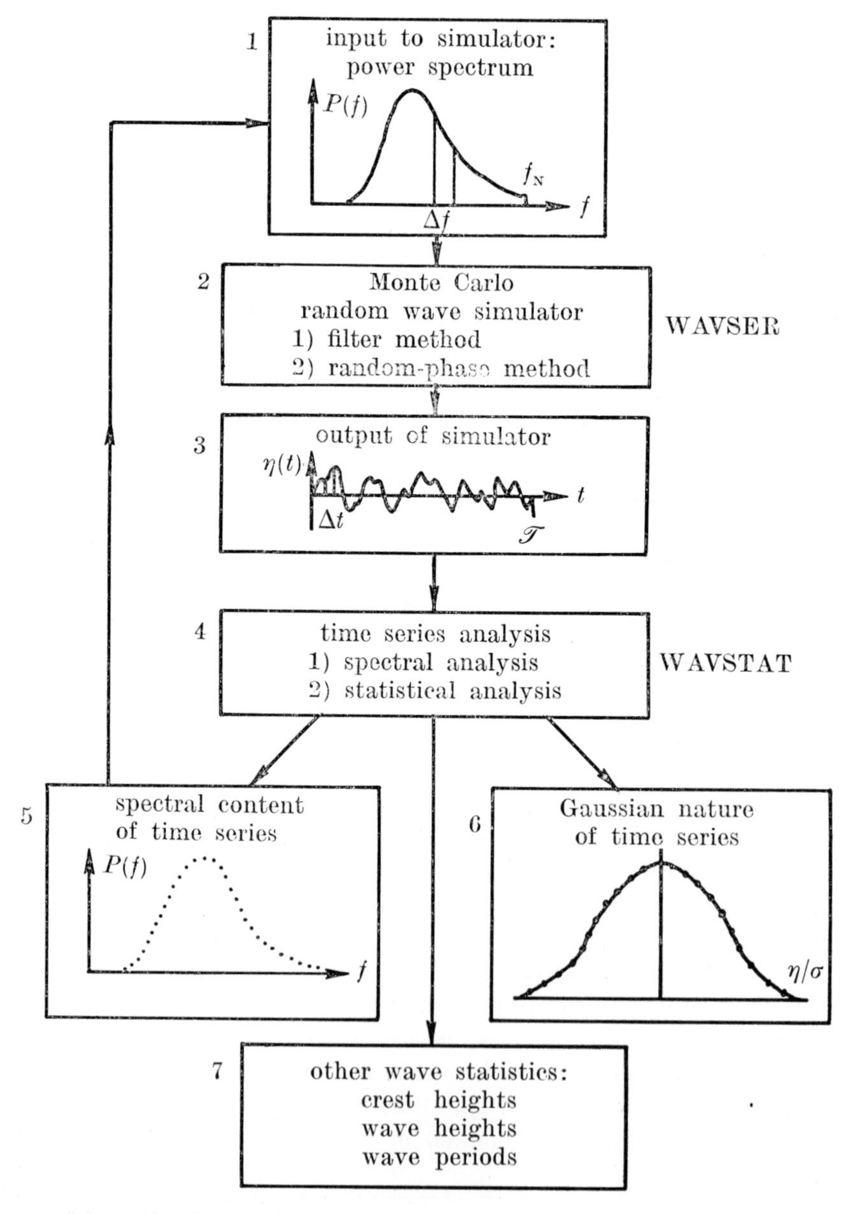

Fig. 1. – Schematic of a random wave simulator.

relate to the fact that some simulations 1) display a degraded behavior in reproducing the Gaussian nature of the simulated sea surface elevation and/or 2) poorly reproduce the energy content of the power spectrum. Large computer time requirements usually result in modifications of the simulation to reduce computer time and result in degraded accuracy. Of all the methods previously developed, the random-phase model of Borgman [11] seems to be the most reliable, as we shall show, and we shall refer to it throughout.

Figure 1 is a schematic which shows a series of steps necessary for the validation of a linear random-wave simulation. First, a power spectrum is selected to characterize the sea state (block 1). The spectrum is supplied as input to the simulator (block 2, representing our computer program WAVSER) which uses a Monte Carlo method (employing random phase or filter techniques) to calculate a realization of the sea surface elevation as a function of time (block 3). This realization is determined by a preselected set of random numbers. Selection of a different set of random numbers will result in an entirely different realization of the wave time series. However, all true realizations of the time series must contain the input power spectrum, which implies that they all have the same significant wave height. Spectral-analysis methods such as the fast Fourier transform (FFT) [13] or the Blackman-Tukey [14] method may be employed to recover the spectral content of the simulated wave record (block 4, our computer program WAVSTAT). The output spectrum (block 5) can then be compared directly to the input spectrum (block 1). If there is good agreement between the two spectra, it can be assumed that the record is described by the desired spectrum. Possible errors can occur not only in the simulator but also in the spectral algorithm; both sources of error contribute to any differences between actual and simulated spectra.

In addition to verification of the spectral content, perhaps the next most important test determining the adequacy of the simulation is to compare a histogram of wave record amplitudes to a Gaussian probability density (block 6). This test is an important one, because, as we shall show, some wave simulators tend to generate fewer large amplitudes (and thus fewer large waves) than predicted by the Gaussian function. Thus the Gaussian test has important implications on the accumulation of the statistics of large waves.

We shall discuss each of the components in fig. 1 in the following sections. First, we briefly describe the two power spectra used in our analysis, namely the Pierson-Moskowitz and JONSWAP spectra. Second, we discuss the random-phase wave simulator developed in this study and compare it to a filter simulation. Third, the spectral content and the amplitudes of a typical wave record are compared to known theoretical results. Fourth, an investigation of the behavior of the statistics of wave crests, wave heights and wave periods is made using the random wave simulation. Finally the results are compared to theory, to data from the work of Haring, Osborne and Spencer [15] and to data from hurricane Eloise.

2. – Sea state spectra.

2·1. *Pierson-Moskowitz spectrum.* – The energy content, as a function of wave frequency, of the sea surface is described by a power spectrum (see, for example, [16]). An analytic form of the spectrum has been developed by PIERSON and MOSKOWITZ [6] for a « fully developed » sea. It is assumed that the Pierson-Moskowitz spectrum is an asymptotic form occurring in the limit of infinite fetch, duration and water depth for a selected constant wind speed. This spectral form is given by

$$P(f) = \frac{\alpha g^2}{(4\pi)^4 f^5} \exp\left[-\frac{5}{4}\left[\frac{f_d}{f}\right]^4\right], \tag{1}$$

where the constants are defined in table I. Also shown in this table are analytic expressions for the significant wave height H_s, the dominant frequency f_d, the

TABLE I. – *Summary of equations describing the Pierson-Moskowitz spectrum.*

the spectrum	$P(f) = \frac{\alpha g^2}{(2\pi)^4 f^5} \exp\left[-\frac{5}{4}\left[\frac{f_d}{f}\right]^4\right] = \frac{a}{f^5} \exp[-b/f^4]$
parameters	$a = \frac{g^2}{(2\pi)^4}\alpha, \qquad b = \beta\left[\frac{g}{2\pi W_s}\right]^4 = \frac{5}{4} f_d^4,$ $g = 9.8\ \text{m/s}^2, \qquad \alpha = 0.0081, \qquad \beta = 0.74$
significant wave height	$H_s = \frac{2}{g}\sqrt{\frac{\alpha}{\beta}}\, W_s^2 = 4\sqrt{m_0}$
dominant frequency	$f_d = \frac{g}{2\pi}\left[\frac{4\beta}{5}\right]^{\frac{1}{4}} \frac{1}{W_s} = \frac{1}{\pi}\left[\frac{g^2\alpha}{5}\right]^{\frac{1}{4}} H_s^{-\frac{1}{2}}$
dominant period	$T_d = \frac{1}{f_d} = \frac{2\pi}{g}\left[\frac{5}{4\beta}\right]^{\frac{1}{4}} W_s = \pi\left[\frac{5}{g^2\alpha}\right]^{\frac{1}{4}} H_s^{\frac{1}{2}}$

dominant period T_d, all as a function of wind speed W_s. The Pierson-Moskowitz spectrum is considered to be « classical » in the sense that (even though it is empirical) it has undergone thorough experimental verification within the assumptions upon which it is founded.

2·2. JONSWAP *spectrum*. – When the sea is not in equilibrium, and thus not fully developed, the spectrum cannot be described by the form of Pierson and Moskowitz. Observations of waves in the North Sea by HASSELMANN *et al.* [17] have shown that an alternate form for the power spectrum of wind waves may be developed to account for nonequilibrium effects. One such form is called the JONSWAP (Joint Oceanographic North Sea Wave Analysis Program) spectrum and is related to the Pierson-Moskowitz form by

$$P(f) = \frac{g^2\alpha^*}{(2\pi)^4 f^5}\exp\left[-\frac{5}{4}\left(\frac{f_d^*}{f}\right)^4\right]\gamma^{\exp\left[-\frac{1}{2}((f-f_d^*)/\sigma_0 f_d^*)^2\right]}. \tag{2}$$

Here α^* and f_d^* are scaling parameters whose properties are discussed in [17]. The constants are normally taken to be

$$\gamma = 3.3\,, \quad \sigma_0 = 0.07 \quad \text{if} \quad f \leqslant f_d^*\,, \quad \sigma_0 = 0.09 \quad \text{if} \quad f > f_d^*\,.$$

The constants γ and σ_0 were obtained by fitting the above function to JONSWAP data. Shown in fig. 2 is a comparison of a Pierson-Moskowitz spectrum (for a significant wave height of 7.13 m and $T_z = 9.5$ s) and a JONSWAP spectrum (here we have set $\gamma = 6.49$, $\alpha = 0.0323$ and $f_d^* = 0.127$, which gives $H_s = 7.25$ m, $T_z = 6.6$ s). Clearly the JONSWAP spectrum is a nonequilibrium

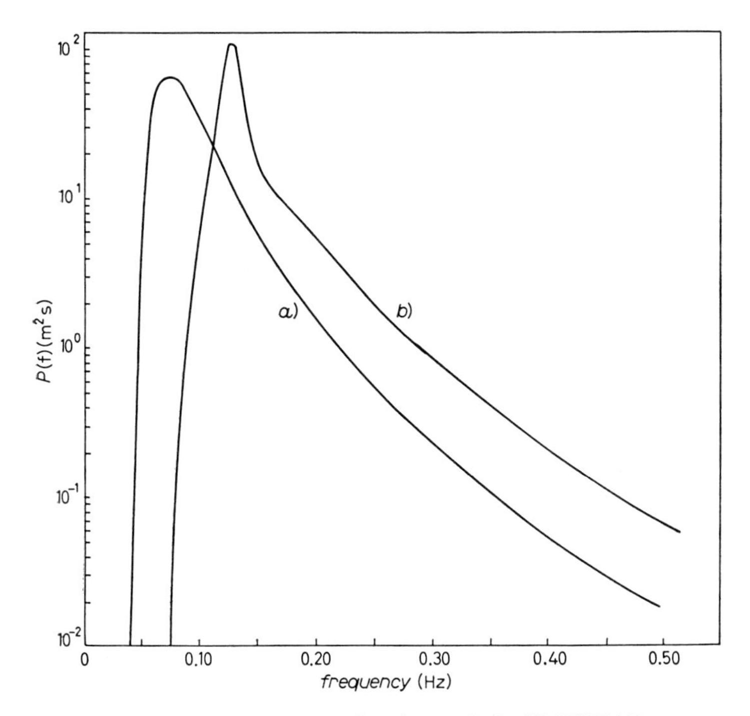

Fig. 2. – Comparison of *a*) Pierson-Moskowitz and *b*) JONSWAP power spectra.

spectrum and is characterized by a narrower peak and a larger dominant frequency than the corresponding equilibrium Pierson-Moskowitz spectrum.

2'3. *Parameters derivable from the spectrum.* – The moments of a power spectrum M_0, M_1, M_2, ... are defined by

$$M_n = (2\pi)^n \int_0^{f_N} f^n P(f)\,df\,, \tag{3}$$

where the upper limit f_N is assumed to be the Nyquist cut-off frequency. The moments may be related to several parameters of interest, a few of which are the zero-crossing period T_z, the significant period $\tilde{T}$, the average wave period $\overline{T}$, the number of zero-crossing waves N_z, the number of waves $\overline{N}$ and the spectral width parameter ε. Table II contains equations relating all of these parameters to the spectral moments. Analytic expressions for the spectral moments have been developed by OSBORNE [18] for the Pierson-Moskowitz spectrum. These equations are provided for convenience in table III; the derivation of the equations is given in appendix A. Briefly we point out that for the Pierson-Moskowitz spectrum all of the spectral moments are a function

TABLE II. – *Parameters derivable from the moments.*

Parameter	Expression
significant wave height	$H_s = 4\sqrt{M_0}$
standard deviation	$\sigma = \sqrt{M_0}$
zero-crossing period	$T_z = 2\pi\sqrt{\dfrac{M_0}{M_2}}$
significant period	$\tilde{T} = 2\pi\dfrac{M_0}{M_1}$
average wave period	$\overline{T} = 2\pi\sqrt{\dfrac{M_2}{M_4}}$
ratio of zero-crossing waves to total waves	$\dfrac{N_z}{\overline{N}} = \dfrac{\overline{T}}{T_z} = \sqrt{\dfrac{M_2^2}{M_0 M_4}}$
spectral-width parameter	$\varepsilon = \sqrt{1-\left(\dfrac{N_z}{\overline{N}}\right)^2} = \sqrt{1-\dfrac{M_2^2}{M_0 M_4}}$

TABLE III. – *Moments of the Pierson-Moskowitz spectrum* (see appendix A).

the spectrum $$P(f) = \frac{\alpha g^2}{(2\pi)^4} f^{-5} \exp\left[-\frac{5}{4}\left[\frac{f_d}{f}\right]^4\right]$$
moments defined $$M_n = \int_0^{\omega_N} \omega^n P(\omega)\,d\omega = (2\pi)^n \int_0^{f_N} f^n P(f)\,df$$
moments for Pierson-Moskowitz spectrum $$M_0 = (H_s/4)^2 \exp[-r_0^4] \simeq 0.0625 H_s^2(1 - r_0^4)$$ $$M_1 = (H_s/4)^{\frac{3}{2}} (g^2\alpha/4)^{\frac{1}{4}} \Gamma(3/4,\ r_0^4) \simeq 0.184\,30 H_s^{\frac{3}{2}}(1 - 1.0881 r_0^3)$$ $$M_2 = (H_s/4)(g^2\alpha/4)^{\frac{1}{2}} \sqrt{\pi}\ \mathrm{erfc}\,(r_0^4) \simeq 0.641\,47 H_s(1 - 1.1284 r_0^2)$$ $$M_3 = (H_s/4)^{\frac{1}{2}} (g^2\alpha/4)^{\frac{3}{4}} \Gamma(1/4, r_0^4) \simeq 3.157\,52 H_s^{\frac{1}{2}}(1 - 1.1033 r_0)$$ $$M_4 = (g^2\alpha/4)\ \mathrm{Ei}\,(r_0^4) \simeq -1.209\,67(1 + 6.9299 \ln r_0 - 1.7325 r_0^4)$$
useful parameters $$r_0 = (5/4)^{\frac{1}{4}} f_d/f_N \simeq 1.0574 f_d/f_N$$ $$f_d = \frac{1}{\pi}\left[\frac{g^2\alpha}{5}\right]^{\frac{1}{4}} H_s^{-\frac{1}{2}} \simeq 0.3622 H_s^{-\frac{1}{2}}$$

of 1) the significant wave height and 2) the Nyquist cut-off frequency of the experimental apparatus used to measure waves. That the values of certain wave parameters are influenced by instrument cut-off has been previously suggested in the literature by WILLIAMS and CARTWRIGHT [19], GODA [20] and RYE and SVEE [21]. The results described in appendix A quantify the effect of significant wave height and instrument cut-off on measured wave parameters. The formulae in table III provide useful checks on the accuracy of the simulation of the Pierson-Moskowitz spectrum and, as we shall later see, also, to a lesser extent, on the JONSWAP spectrum.

3. – The Monte Carlo simulation.

3·1. *The method of random phases.* – Any continuous function of time $\eta(t)$ may be expanded in a Fourier series

$$\eta(t) = \sum_{n=1}^{N} (a_n \cos \omega_n t + b_n \sin \omega_n t)\,. \tag{4}$$

The function $\eta(t)$ is defined on the time interval $(0, \mathscr{T})$, where $\omega_n = 2\pi f_n$, $f_n = n\Delta f$ and $\mathscr{T} = 1/\Delta f$. Thus, given any function of time, suitable coefficients a_n, b_n may be found by the techniques of Fourier analysis, so that eq. (4) describes the function $\eta(t)$ to an arbitrary degree of accuracy. As the number of terms in the series N becomes large, the accuracy improves indefinitely. Another Fourier representation, just as suitable as eq. (4), is given by

$$\eta(t) = \sum_{n=1}^{N} C_n \cos(\omega_n t - \varphi_n)\,, \tag{5}$$

where the constants C_n, φ_n may also be calculated by the methods of Fourier analysis.

If eq. (4) is used to describe a random signal, then the coefficients a_n and b_n, $1 \leqslant n \leqslant N$, may be assumed to be independent random variables distributed about zero by the Gaussian probability density [1]. When this assumption is made, it is found that for a selected set of independent random numbers (a_n, b_n) the function $\eta(t)$ has an appearance very much like a time series trace of random ocean waves, an example of which is shown in fig. 1, block 3. For a particular set of coefficients eq. (4) will describe a particular realization of the ocean surface. A different set of random coefficients will generate a sea surface realization which is separate and distinct from any other; thus there is an infinite number of realizations of a random sea surface. However, we require that every realization must have the same statistical and spectral properties as all the others. This is true provided that the variance σ_n^2 of the Gaussian density is defined to be [1]

$$\sigma_n^2 = P(f_n)\Delta f_n = \overline{a_n^2} = \overline{b_n^2}\,. \tag{6}$$

Equation (5) may also be used to represent a random process if the φ_n are uniformly distributed random numbers on the interval $(0, 2\pi)$, and if $C_n = \sqrt{2P(f_n)\Delta f_n}$ [1]. Of the equally good representations, eqs. (4) and (5), we shall choose eq. (5) for our further consideration; this decision will ultimately save considerable computer time and result in a substantial increase in accuracy.

3'2. *Unequal frequency bins.* – A digital simulation of eq. (5) must compute $\eta(t)$ at discrete time steps Δt, which leads to the following expression for the parameters within eq. (5):

$$\begin{cases} C_n = \sqrt{2P(f_n)\,\Delta f_n}\,, \\ \omega_n = 2\pi f_n, \quad 0 \leqslant f_n \leqslant f_\mathrm{N}, \quad 0 \leqslant n \leqslant N\,, \\ t \;= t_m = m\,\Delta t, \quad 0 \leqslant t_m \leqslant \mathscr{T}, \quad 0 \leqslant m \leqslant M\,, \\ f_\mathrm{N} = 1/2\Delta t = \text{Nyquist frequency}\,, \\ \varphi_n \quad \text{are uniformly distributed random numbers on } (0, 2\pi)\,. \end{cases} \tag{7}$$

We have elected in the above equations to select the frequency f_n at unequal intervals for reasons we shall soon discuss.

Because of the dependence of the coefficients C_n on the power spectrum $P(f_n)$, each realization of the time series $\eta(t)$ will contain the distribution of wave energy as a function of frequency described by the input power spectrum. Figure 3*a*) shows a typical power spectrum describing a random sea with sig-

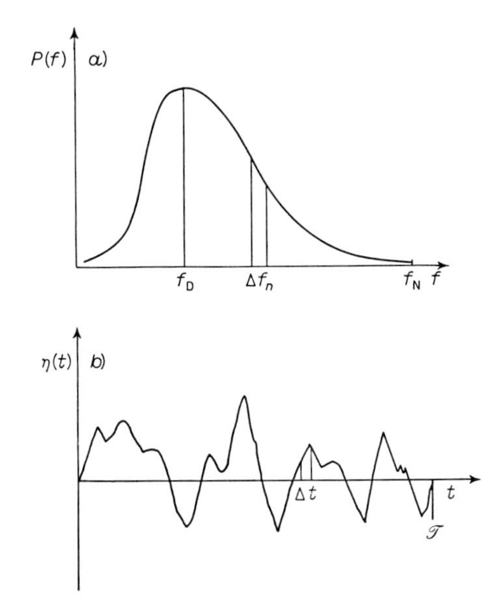

Fig. 3. – *a*) Power spectrum partitioned at frequency intervals Δf_n and characterized by Nyquist frequency f_N. *b*) Time series partitioned at time intervals Δt and characterized by a length $\mathcal{T}$.

nificant wave height $H_s = 4\sqrt{M_0}$. Here M_0 is the area under the power spectrum:

$$M_0 = \int_0^{f_N} P(f)\,\mathrm{d}f\,. \tag{8}$$

The upper limit in the integral is the Nyquist frequency, f_N, which provides an upper cut-off beyond which the wave energy is unknown. In order to evaluate the series representation of $\eta(t)$, we partition the spectrum into N not necessarily equal frequency intervals Δf_n, $1 \leqslant n \leqslant N$. Corresponding to each frequency bin there is an average value of the spectrum $P(f_n)$; thus $P(f_n)\Delta f_n$ is the total wave energy in the frequency bin Δf_n. We then select N uniform random numbers φ_n on the interval $(0, 2\pi)$. Equation (5) may be evaluated for the times $t = t_m$, an example of which is shown in fig. 3*b*). Here $0 \leqslant t \leqslant \mathcal{T}$, where $\mathscr{T}$ is the length of the series in seconds. If we were to choose Δf_n and Δt both constant, then according to Fourier theory

$$f_N = \frac{1}{2\Delta t}, \qquad \mathscr{T} = \frac{1}{\Delta f_n}.$$

The range of m is then given by $0 \leqslant m \leqslant 2N$. Thus, if we let N frequencies (with $\Delta f_n = \Delta f = \text{const}$) describe the power spectrum, we may evaluate the time series out to $2N$ points before it begins to repeat itself (a natural consequence of Fourier analysis theory). Thus, in order to generate a times series of the sea surface elevation with, say, 10 000 points, the sum in eq. (5) must be over $N = 5000$ frequencies. This is an unfortunate circumstance which results in the use of large amounts of computer time.

In order to ensure that the series does not repeat at the fundamental period $\mathscr{T} = 1/\Delta f$ (and to reduce computer time), we elected, with BORGMAN [11], to select unequal frequency intervals corresponding to equal area partitions under the spectrum. To this end we may compute the area of the spectrum by eq. (8) and then divide this area by N to obtain a small subarea A_0:

$$A_0 = M_0/N = \int_{f_n}^{f_{n+1}} P(f)\,\mathrm{d}f \tag{9}$$

for all n, $1 \leqslant n \leqslant N$. Thus for every frequency bin we may define a width Δf_n which ensures that the area under the spectrum, bounded by $(f_n, f_n + \Delta f_n)$, is equal to M_0/N; hence we require

$$\Delta f_n = f_{n+1} - f_n . \tag{10}$$

If we set $f_0 = 0$, then f_1 can be found as the solution to eq. (9). Successive values of f_n may be computed in a like manner. In general we have elected to do the integral numerically (although there is an exact solution for the Pierson-Moskowitz spectrum [11, 18]) and use Newtonian iteration to discover successive values of f_n. Since $A_0 = P(f_n)\,\Delta f_n$, we have $C_n = \sqrt{2A_0} = \text{const} = C_0$, the coefficients may be moved outside the summation of eq. (5) to give

$$\eta(t_m) = C_0 \sum_{n=1}^{N} \cos(\omega_n t_m - \varphi_n) . \tag{11}$$

Hence, not only have we rendered the series nonrepeating, but also we have reduced the computation by $N - 1$ multiplications. Since the series is now nonrepeating, we may ultimately choose the number of frequencies N to be much less than 5000, provided that good frequency resolution is maintained.

3'3. *Recursive computation of trigonometric functions.* – We may rewrite eq. (5) as

$$\eta(t_m) = \sum_{n=1}^{N} C_n \cos(\omega_n t_m) \cos\varphi_n + C_n \sin(\omega_n t_m) \sin\varphi_n . \tag{12}$$

Then with suitable changes of variable we obtain

$$\eta(t_m) = \sum_{n=1}^{N} A_n \cos(m\alpha_n) + B_n \sin(m\alpha_n) , \tag{13}$$

where

$$\left\{ \begin{aligned} \alpha_n &= 2\pi \Delta t f_n , \\ A_n &= C_n \cos\varphi_n , \\ B_n &= C_n \sin\varphi_n . \end{aligned} \right. \tag{14}$$

The motivation for writing $\eta(t)$ in this way is that the trigonometric functions of eq. (13) may be evaluated by recursion relations in the interger m:

$$\left\{ \begin{aligned} \cos(m\alpha_n) &= 2\cos[(m-1)\alpha_n] \cos\alpha_n - \cos[(m-2)\alpha_n] , \\ \sin(m\alpha_n) &= 2\sin[(m-1)\alpha_n] \cos\alpha_n - \sin[(m-2)\alpha_n] . \end{aligned} \right. \tag{15}$$

These equations allow computation of the $2m$ values of $\cos(m\alpha_n)$ and $\sin(m\alpha_n)$ given starter values for $m = 0, 1$. It is clear that computationally the trigonometric functions can then be replaced by two « multiply » instructions and an « add » instruction in the computer, resulting in substantial saving of computer time. In order to use the above equations, it is necessary to determine all of the m values of the sines and cosines for a particular frequency f_n. Thus

$$\eta_{mn} = A_n \cos(m\alpha_n) + B_n \sin(m\alpha_n) \tag{16}$$

for a particular n and all values of m, $0 \leqslant m \leqslant M$. Then the sum over frequencies can be made

$$\eta(t_m) = \sum_{n=1}^{N} \eta_{mn} . \tag{17}$$

Computation of the array η_{mn} can require substantial storage problems (100 frequencies and 10 000 points on the time series would require one million storage locations). This storage problem can be alleviated by computing all points on the series for a single frequency and then summing over successive frequencies.

One additional problem was encountered in the development of the random-phase simulator. Use of « equal area » frequency bins results in an increasing bin width in the right-hand tail of the spectrum. In order to maintain good frequency bin resolution, the area A_0 may be reduced periodically with increasing frequency. The method employed here is to divide A_0 by 2 when the frequency bin width exceeds a preselected value. In this way good spectral resolution can be maintained even in the tail of the spectrum. However, use of this technique means that the Fourier coefficients C_n are no longer constant (they change when A_0 is divided by 2) and thus C_n cannot be brought outside the summation sign as in eq. (11).

A mathematical model of the computer program WAVSER, which uses the methods described above, is provided in appendix B. The program requires about 4 s of computer time (IBM 360/178) to generate a 10 000-point time series. Use of eq. (11), a form proposed by BORGMAN which does not repeat itself at regular intervals, requires about 2.4 min. Direct execution of the Fourier series representation, eq. (5), uses about 45 min of computer time. A summary of the key elements employed in the rapid and efficient random-phase simulation described herein is provided in table IV.

TABLE IV. – *Summary of key elements in the random-phase simulation.*

use unequal frequency bins to prevent series from repeating, equal area bins
use recursion relations for computing sines and cosines
compute all time points in the series for a single frequency and then sum over all frequencies to reduce computer storage
decrease frequency bin size in the spectral tail to maintain good frequency resolution
results: 4 s of IBM 370/168 time needed to generate a 10 000-point time series in double precision for 143 frequencies

3·4. *Verification of spectral content.* – We now make a comparison of the actual and simulated spectra using the above model. Two spectra were simulated; the Pierson-Moskowitz ($H_s = 7.13$ m and $T_z = 9.5$ s) and JONSWAP ($H_s = 7.25$ m, $\gamma = 6.44$, $\alpha = 0.0323$, $f^*_D = 0.127$) analytical forms given by eqs. (1) and (2) were used as inputs to the wave generator. In order to maintain an arbitrary frequency bin resolution of $\Delta f_n \leqslant 0.03$ Hz, the computer program was designed to optimize the number of frequencies within the constraints of the model (see appendix B). For the Pierson-Moskowitz spectrum the number of frequencies selected was 143, while for the JONSWAP spectrum the number was 154. A total of 1.2 million points on each time series ($\Delta t = 0.5$ s, $f_N = 1$ Hz) were generated from each spectrum. The two time series were then analyzed in 10 000-point blocks by a Blackman-Tukey algorithm [14] using 200 lagged products and also by a FFT algorithm [13]; essentially similar results were obtained. Figure 4 shows a 200 s sample of a time series generated

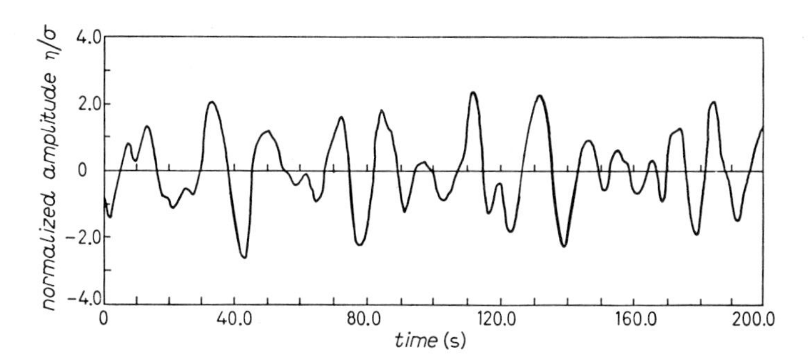

Fig. 4. – Sample of a time series generated by the random-phase method.

by the random-phase method for the Pierson-Moskowitz spectrum. Figure 5 shows a comparison of the actual and simulated Pierson-Moskowitz spectra. It is apparent that the agreement between the two spectra is quite good over the entire frequency range. Note that there is some disagreement at low frequencies, however the effects are slight, since the energy here is only on the order of 0.1 % or less of the peak energy. Table V compares analytical expressions of several parameters for the Pierson-Moskowitz spectrum with their simulated values. Good agreement is found between actual (as computed by

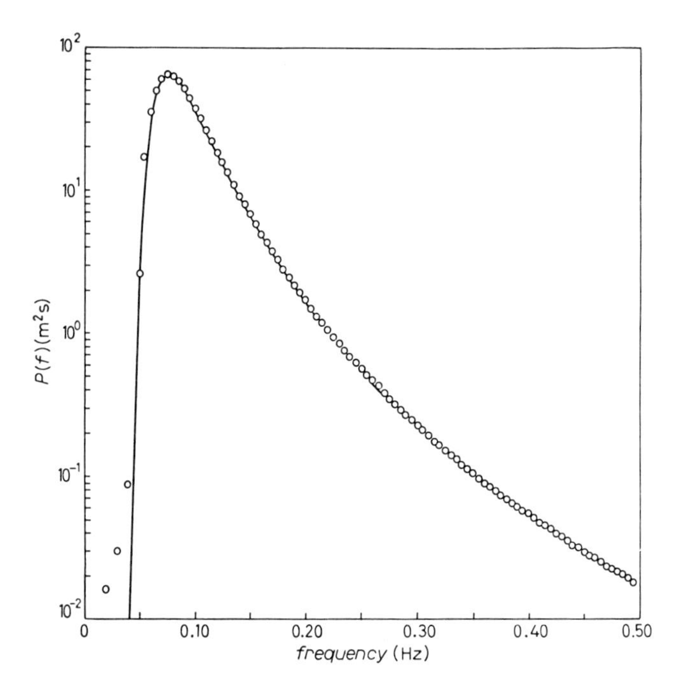

Fig. 5. – Comparison of simulated (continuous curve) and actual (circles) Pierson-Moskowitz spectra for $H_s = 7.13$ m.

TABLE V. – *Actual and simulated values of the parameters of the Pierson-Moskowitz spectrum for a significant wave height of* 7.13 m, $\sigma = 1.701$ m.

Parameter	Actual spectrum	Simulated spectrum	Wave count
H_s	7.135	7.150	6.803
$\tilde{T}$	10.35	10.30	11.74
T_z	9.63	9.57	9.68
$\bar{T}$	6.36	6.32	6.61
M_0	3.18	3.20	—
M_1	1.93	1.95	—
M_2	1.36	1.38	—
M_3	1.17	1.20	—
M_4	1.32	1.36	—
ε	0.7502	0.7506	0.7312
N_z	62 338	62 691	61 846
$\bar{N}$	94 284	90 798	90 797
$N_z/\bar{N}$	0.6612	0.6905	0.6811

the formulae of tables II and III) and simulated parameters (computed by the spectral analysis). Thus the simulated record agrees in detail quite well with the behavior predicted for a random time series governed by the Pierson-Moskowitz spectrum. Also shown in table V are parameters obtained by counting the waves in the simulated record (denoted « wave count » in the table). In this analysis, H_s was obtained by averaging the highest one-third waves ($H_{\frac{1}{3}}$) and $\tilde{T}$ was estimated by averaging the wave periods of the highest one-third waves. Based upon the simulated results it seems that $H_{\frac{1}{3}} = 3.8\sqrt{M_0}$ is a more reliable formula for estimating H_s than the traditional formula $H_{\frac{1}{3}} = 4\sqrt{M_0}$. This is consistent with the non-Rayleigh behavior of wave heights and will be elaborated on in a later section.

Figure 6 shows a comparison of the actual and simulated JONSWAP spectra for a record of 1.2 million points digitized at 0.5 s intervals. The significant wave height of the record is 7.25 m. Comparison of the actual and simulated spectra is quite good, the only substantial difference is near the peak, where the simulated spectrum is about 8 % low. This error is probably bias error in the spectral-analysis algorithm [22]. Table VI compares the simulated parameters of the JONSWAP spectrum with analytic computations based on a Pierson-Moskowitz spectrum (here we used $H_s = 7.25$ m, $f_D = 0.127$ and $f_N = 0.5$ Hz in the formulae of table III).

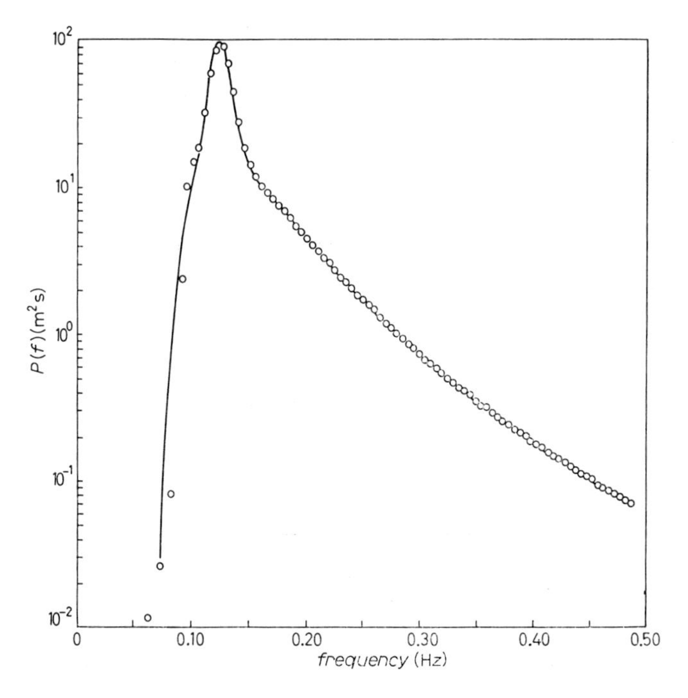

Fig. 6. – Comparison of simulated (circles) and actual (continuous curve) JONSWAP spectra for $H_s = 7.25$ m.

TABLE VI. – *Simulated values of the parameters of the* JONSWAP *spectrum compared to parameters computed from the Pierson-Moskowitz formulae for a significant wave height of* 7.25 m, $\sigma = 1.745$ m.

Parameter	Actual spectrum	Simulated spectrum	Wave count
H_s	7.25	7.24	6.98
$\tilde{T}$	7.04	6.89	7.51
T_z	6.64	6.61	6.70
$\bar{T}$	5.01	5.09	5.27
M_0	3.26	3.28	—
M_1	2.91	2.98	—
M_2	2.92	2.96	—
M_3	3.38	3.35	—
M_4	4.60	4.51	—
ε	0.6567	0.6385	0.6178
N_z	15 060	15 125	14 902
$\bar{N}$	19 971	19 654	18 960
$N_z/\bar{N}$	0.7541	0.7696	0.7860

3·5. *Verification of Gaussian behavior.* – The Gaussian nature of a linear wave record is one of its most fundamental properties (a consequence of the central-limit theorem) and is discussed at length in the literature [1, 4, 16]. We are interested here primarily with a comparison of the Gaussian probability density

$$P(\eta|\sigma) = \frac{1}{\sqrt{2\pi}\,\sigma}\exp\left[-\frac{1}{2}\left(\frac{\eta}{\sigma}\right)^2\right] \tag{18}$$

and a histogram of the normalized sea surface elevation $x = \eta(t_m)/\sigma$. Here σ is the standard deviation of the wave record and is related to the significant wave height H_s and spectral area by $H_s = 4\sqrt{M_0} = 4\sigma$. The record variance σ^2 is

$$\sigma^2 = M_0\,. \tag{19}$$

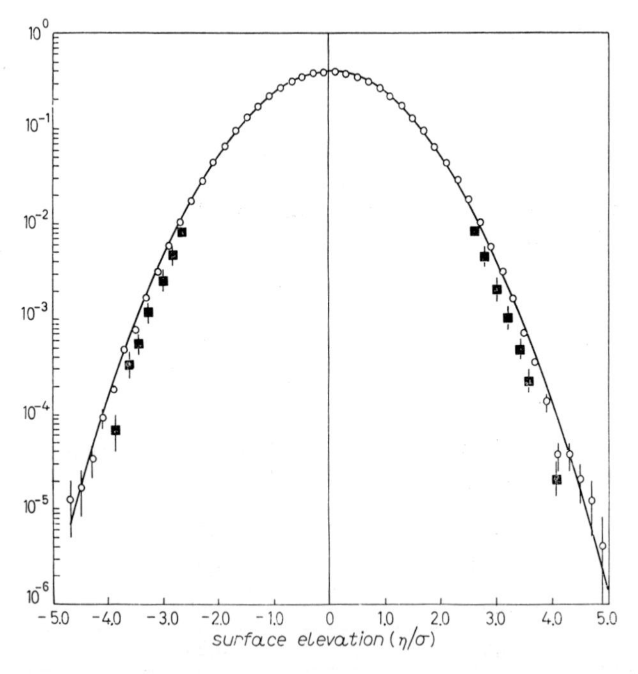

Fig. 7. – Comparison of the Gaussian function (continuous curve) with wave amplitude densities from the random-phase simulation (circles) and from a filter simulation (full squares); $N = 1\,200\,000$, $\varepsilon = 0.7506$, mean $= 0$, standard deviation $= 1.0004$, skewness $= 0.0092$, kurtosis $= 2.9728$.

Figure 7 compares eq. (18) with a histogram (containing $1.2\cdot 10^6$ points) of the normalized sea surface elevation which was simulated using the Pierson-Moskowitz spectrum. The vertical bars through each point on the histogram are « error bars » (see [23]). The length of these bars is based upon the following fact: the probability that a single wave amplitude will lie in the interval $(x, x + dx)$ is small, and hence the number of waves in this range is Poisson distributed. Error bars should be interpreted in the following way: « if the simulation were to be repeated with a different set of uniform random numbers, then each point in the histogram would have about a 30 % probability of moving outside the error bar range ». Thus, from one simulation to another, we should not be surprised to see statistical fluctuations in histogram points which are on the order of the size of the error bars.

Also computed for all histograms in this paper are several statistical parameters of interest (shown, for example, in the caption of fig. 7). The letter N designates the number of points contributing to the histogram $(1.2 \cdot 10^6)$. ε is the value of the spectral-width parameter (0.7506). « Mean » is the mean of the normalized (*i.e.* η/σ) record (0.0). « Standard deviation » is the standard deviation of the normalized record (1.0004). If we define the central moments m_r^c of a histogram (ordinate f_i *vs.* abscissa x_i) about the mean, $\bar{x}$, by

$$m_r^c = \sum_{i=1}^{k} f_i (x_i - \bar{x})^r \Big/ \sum_{i=1}^{k} f_i \,, \qquad r = 0, 1, 2, \ldots,$$

then the following definitions hold:

$$\begin{aligned} &\text{mean} && = m_1^c \equiv x = 0\,, \\ &\text{standard deviation} && = m_2^c \equiv \sigma_0\,, \\ &\text{skewness} && = m_3^c/\sigma^3 \equiv s\,, \\ &\text{kurtosis} && = \mu_4^c/\sigma^4 \equiv k\,. \end{aligned}$$

Here skewness is a measure of right-left asymmetry of the histogram; kurtosis is a measure of peakedness. For a zero-mean Gaussian process η we find that the corresponding process η/σ has the properties $\sigma_0 = 1$, $s = 0$ and $k = 3$.

Figure 7 shows that there is good agreement between the Gaussian density and the simulated points out to five standard deviations (for the Pierson-Moskowitz spectrum). Deviations of the simulated points from the Gaussian function are small and are consistent with the length of the error bars. The statistical parameters of standard deviation, skewness and kurtosis compare well with the values expected for a Gaussian process.

Figure 7 also shows points resulting from a filter simulation due to ROBINSON *et al.* [12]. It can be seen in the figure that the histogram of the sea surface elevation derived from the filter simulation agrees quite well with the Gaussian function out to nearly 2.5 standard deviations. For greater values of x, however, the histogram of the filter simulation is substantially lower than the Gaussian function. Thus there appears to be a tendency for the filter simulation to produce fewer large wave amplitudes than predicted by the

Gaussian density. It is because of this deficiency that the filter method was rejected for use in our analysis of random wave statistics.

Shown in fig. 8 is a comparison of the Gaussian density to the normalized sea surface elevation for the JONSWAP spectrum. Again, there is excellent agreement between desired and simulated results.

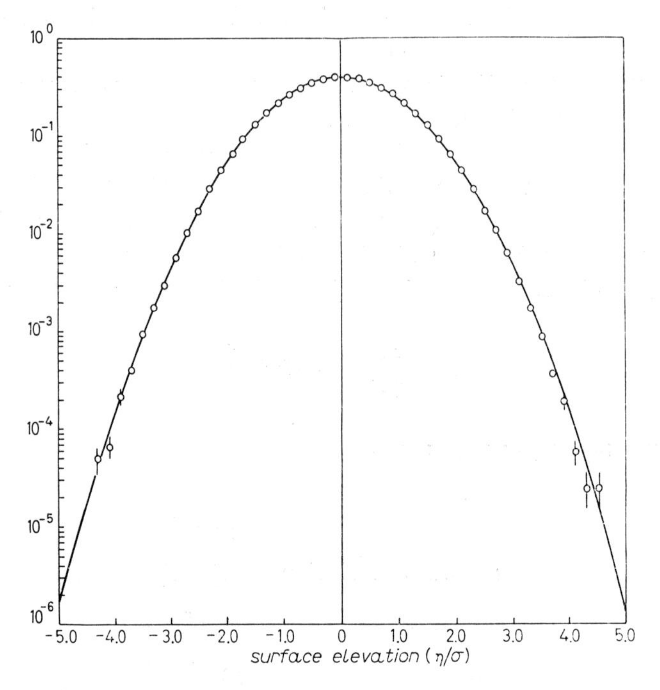

Fig. 8. – Comparison of the Gaussian function to the wave amplitude density from the JONSWAP simulation; $N = 1\,200\,000$, $\varepsilon = 0.7476$, mean $= 0$, standard deviation $= 1.0003$, skewness $= 0.0042$, kurtosis $= 2.9854$.

We have verified the spectral content and the Gaussian nature of the Monte Carlo random-phase simulation of Pierson-Moskowitz and JONSWAP spectra. As a result of this verification there is considerable confidence that the random-phase model is adequate for investigating the extreme statistical properties of random waves. Before proceeding further we shall first define several additional statistical properties important for the understanding of the behavior of linear random waves.

4. – Statistical parameters of a random sea.

4·1. *Statistics of crests and troughs.* – The probability density of crests and troughs in a linear Gaussian wave record has been derived by RICE [1] and by

CARTWRIGHT and LONGUET-HIGGINS [16]:

$$P(\eta_c|\varepsilon) = \frac{1}{\sqrt{2\pi}}\,\varepsilon\exp\left[-\frac{1}{2}\left(\frac{\eta_c}{\varepsilon}\right)^2\right] + \frac{(1-\varepsilon^2)^{\frac{1}{2}}}{\sqrt{2\pi}}\,\eta_c\exp\left[-\frac{1}{2}\eta_c^2\right]\int\limits_{-\infty}^{(\eta_c/\varepsilon)(1-\varepsilon^2)^{\frac{1}{2}}}\exp\left[-\frac{x^2}{2}\right]dx\,. \tag{20}$$

Here the normalized crest height η_c is given by the ratio of crest height amplitude η_{max} (or trough depth η_{min}) divided by the standard deviation of the wave record σ (see fig. 9). Notice that the density function is dependent only upon the normalized crest height η_c and the spectral-width parameter ε. Theoretically ε may vary between 0 and 1. In the limit as ε is allowed to approach zero the density function of eq. (20) approaches Rayleigh. If, however, ε is allowed to approach 1, the density function becomes Gaussian. For values of ε between 0 and 1 the density of crests and troughs lies somewhere between the Rayleigh and Gaussian densities.

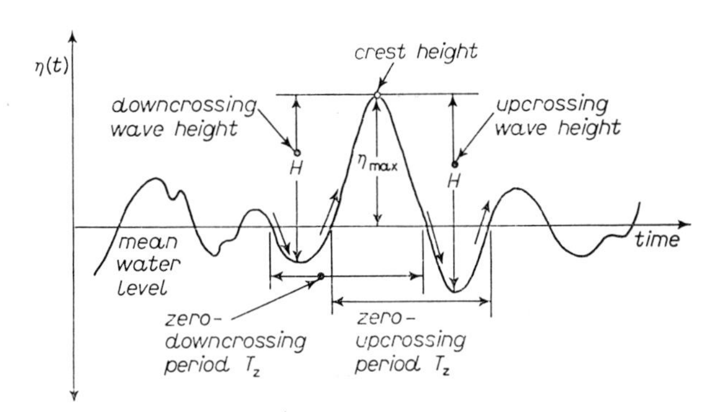

Fig. 9. – Statistical parameters of a random time series.

4·2. *Statistics of wave heights.* – The statistics of wave heights may be defined in terms of upcrossings (H_{up}), downcrossings (H_{down}) and an average between leading and trailing troughs ($(H_{up} + H_{down})/2$). All of these definitions can be seen in fig. 9. The wave height density function is not known in general; however, the function is well known for the case $\varepsilon = 0$:

$$p(h) = h\exp\left[-\frac{h^2}{2}\right]. \tag{21}$$

Here h is the normalized wave height

$$h = H/2\sigma\,, \tag{22}$$

where H is the wave height and σ is the standard deviation of the wave record. The Rayleigh wave height density function has often been assumed to govern the statistics of wave heights, even though it has been derived theoretically only in the narrow spectral sense. Recently HARING *et al.* [15] have shown that eqs. (21), (22) do not agree well with data. In a later section we shall make comparisons of this function with our simulated results and with data.

We also compare to the results of Longuet-Higgins [10], who has shown that, if finite bandwidth effects are included, an alternate form for the wave height density function may be derived which results in a modification of eq. (22) such that σ is replaced by σ_0:

$$h = H/2\sigma_0, \qquad \sigma_0 = \sigma\sqrt{1 - 0.734\nu_0^2}, \qquad \nu_0^2 = M_0 M_2/M_1^2 - 1 . \tag{22'}$$

4.3. *Statistics of zero-crossing period.* – LONGUET-HIGGINS [24] has derived the probability density of zero-crossing periods for a sea state defined by spectral width $\varepsilon = 0$. This expression is approximate and was derived to the same order of approximation as the Rayleigh wave height density function. The period density function is given by

$$p(\tau) = \frac{1}{2(1+\tau^2)^{\frac{3}{2}}}, \tag{23}$$

where

$$\tau = (T - \tilde{T})/\nu\tilde{T} . \tag{24}$$

Here T is the wave period, and $\tilde{T}$ is the « significant period ». Also

$$\nu = \left[\frac{\mu_2}{\mu_0}\right]^{\frac{1}{2}} \frac{\tilde{T}}{2\pi} . \tag{25}$$

The cumulants μ_0 and μ_2 are found from

$$\mu_0 = M_0 , \tag{26}$$

$$\mu_2 = \frac{M_2 M_0 - M_1^2}{M_0}, \tag{27}$$

and finally

$$\tilde{T} = 2\pi M_0/M_1 . \tag{28}$$

Here the M_n are the moments of the power spectrum given by eq. (3).

4.4. *Statistics of crests and troughs of zero-crossing waves.* – At the present time we are unaware of a theory to describe the statistics of crests and troughs of zero-crossing waves. These are maxima or minima for waves which cross zero for successive up or down crossings. Thus small waves superposed on top of large waves tend not to be included in a compilation of zero-crossing statistics. We shall find in a comparison to the random-phase simulation and in a comparison to data that the statistics of crests and troughs of zero-crossing waves may be approximated by the Rayleigh distribution for large amplitudes.

5. – Simulation results.

In this section we examine the statistical parameters of a random ocean surface as predicted by the random-phase simulation. Histograms of these

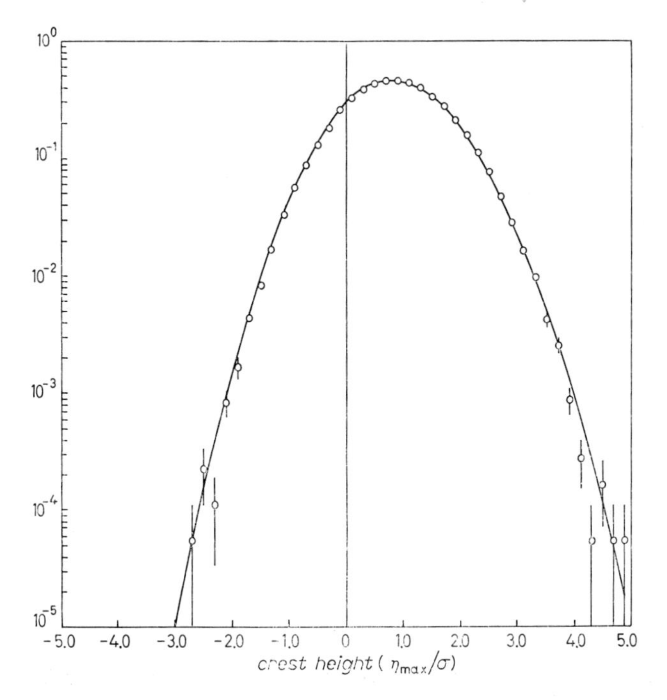

Fig. 10. – Comparison of theory to wave crest probability density from the Pierson-Moskowitz simulation; $N_{max} = 90\,797$, $\varepsilon = 0.7506$, mean $= 0.8467$, standard deviation $= 0.8508$, skewness $= 0.1087$, kurtosis $= 3.0174$.

statistical parameters have been extracted from the simulated records and compared to theoretical expressions for their probability densities. We first consider the density of wave crests. Figure 10 compares the theoretical expression of Cartwright and Longuet-Higgins [16] with the crest height histogram obtained from the random-phase simulation. A Pierson-Moskowitz spectrum was used in the simulation with a significant wave height of 7.13 m. A total of 90 797 wave crests were simulated. The spectral-width parameter ε was 0.75 for the simulation run. The data and the theoretical function have been normalized by the standard deviation of the simulated wave record ($\sigma = 1.7$ m). The results have been graphed on a logarithmic-linear plot, thus allowing a comparison of theory and simulation where the probability density is extremely small in the tail regions. Note the good agreement between theory and simulation even for large maxima. Some statistical fluctuation is present for points above 4 standard deviations and below -2 standard deviations. However, this is consistent with the smaller number of statistics in these regions, as indicated by the length of the error bars.

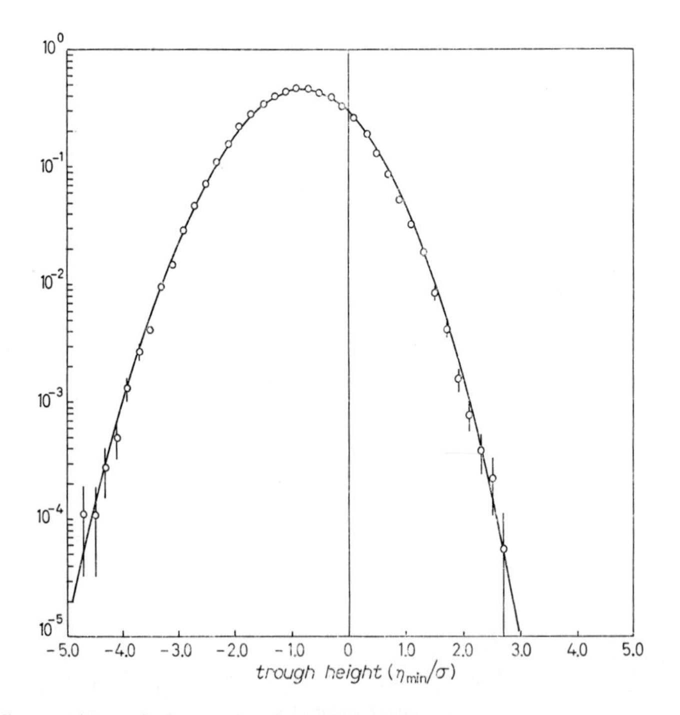

Fig. 11. – Comparison of theory to wave trough probability density from the Pierson-Moskowitz simulation; $N_{\min} = 90\,797$, $\varepsilon = 0.7506$, mean $= -0.8461$, standard deviation $= 0.8489$, skewness $= -0.1042$, kurtosis $= 3.0382$.

The probability density of troughs is investigated in fig. 11, where theoretical and simulated results are compared. Because the simulated record is symmetrical about the mean, the density of troughs has exactly the same form as that of crests except for a co-ordinate inversion of the wave amplitude. Again, there is excellent agreement between theory and simulation. Figure 12 shows a composite of crest and trough statistics obtained by inverting the trough statistics and averaging them with crest statistics. A total of 181 594 wave maxima and minima were thus available for this average histogram. The results are compared to the theory of crests and, again, good agreement is found.

Additional runs have been made using the JONSWAP spectrum as input to the random-phase simulator. Wave crest and trough statistics were found to agree well with theory for this case also. These results will not be presented here, because they are essentially a duplication of fig. 10 through 12.

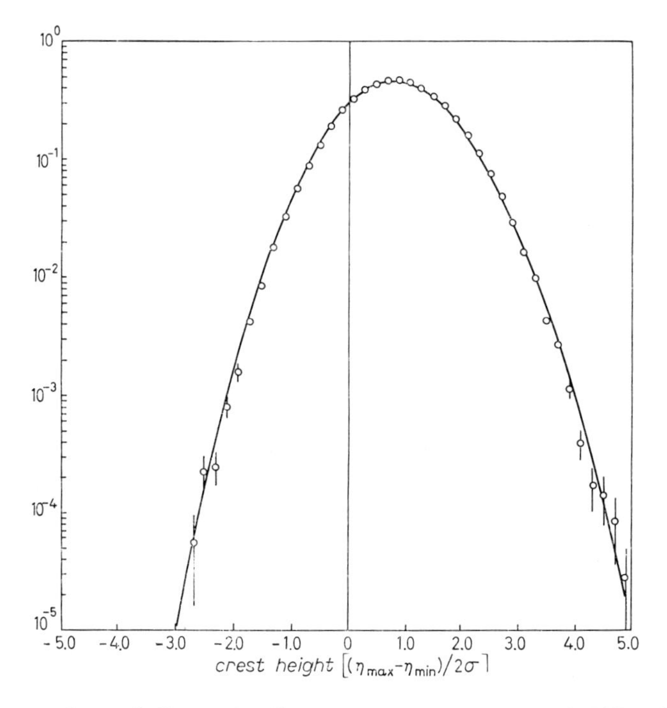

Fig. 12. – Comparison of theory to the average wave crest probability density from the Pierson-Moskowitz simulation; $N_{tot} = 181\,594$, $\varepsilon = 0.7506$, mean $= 0.8464$, standard deviation $= 0.8498$, skewness $= 0.1064$, kurtosis $= 3.0278$.

The statistics of wave heights for simulated Pierson-Moskowitz and JONSWAP computer runs are compared to the Rayleigh distribution function (actually graphed is 1 minus the distribution function) in fig. 13. Both simulated distributions fall well below the Rayleigh function, particularly in the extreme. The results from the Pierson-Moskowitz simulation deviate slightly further from the Rayleigh than did the JONSWAP simulation. For the one-in-a-thousand wave the deviation of wave heights for the Pierson-Moskowitz spectrum from Rayleigh is about 13 %, but only about 11 % for the JONSWAP spectrum. One might expect this deviation to occur based upon the fact that the JONSWAP spectrum is somewhat more narrow-banded than is the Pierson-Moskowitz spectrum. This is due to the fact that extremely narrow-banded processes (*i.e.* those for which the spectral-width parameter is zero) are governed by the Rayleigh distribution exactly. We will see in a following section that these simulated results are consistent with wave height data obtained from several widely separated locations.

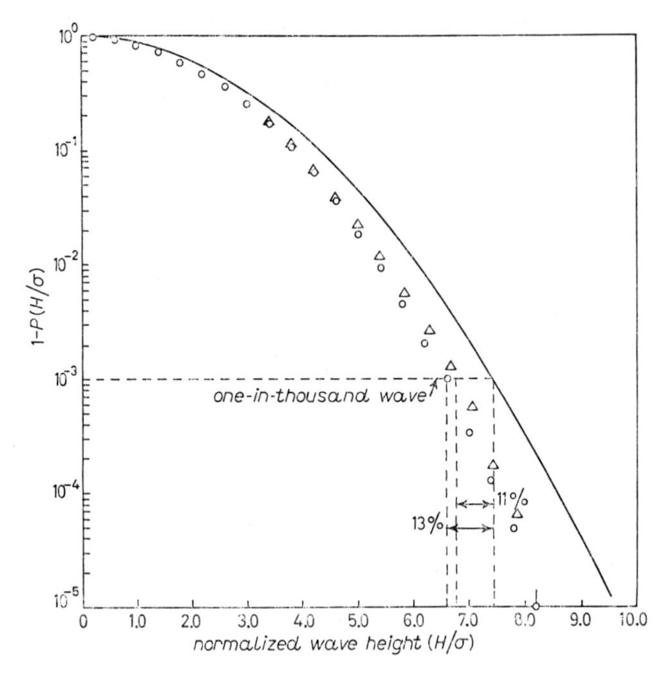

Fig. 13. – Cumulative distribution of zero-crossing wave heights from Pierson-Moskowitz and JONSWAP simulations: ——— Rayleigh distribution, ○ simulated Pierson-Moskowitz, △ simulated JONSWAP.

An improved comparison results when we use eq. (22′) to normalize wave heights as suggested by LONGUET-HIGGINS [10]. Using the spectral moments of tables V and VI in eq. (22′), we find that the wave height distribution for the Pierson-Moskowitz spectrum falls only 7 % below Rayleigh for the one-in-a-thousand wave, while those for the JONSWAP spectrum fall 6 % low. Clearly, while some improvement has occurred, only about 50 % of the original discrepancy has been accounted for. Because we are using a linear model, we feel that our methods may serve as a control for possible future theoretical calculations oriented toward a more complete understanding of the linear theory.

The statistics of zero-crossing wave periods is examined in fig. 14, where the theoretical density function of Longuet-Higgins [24] is compared to the

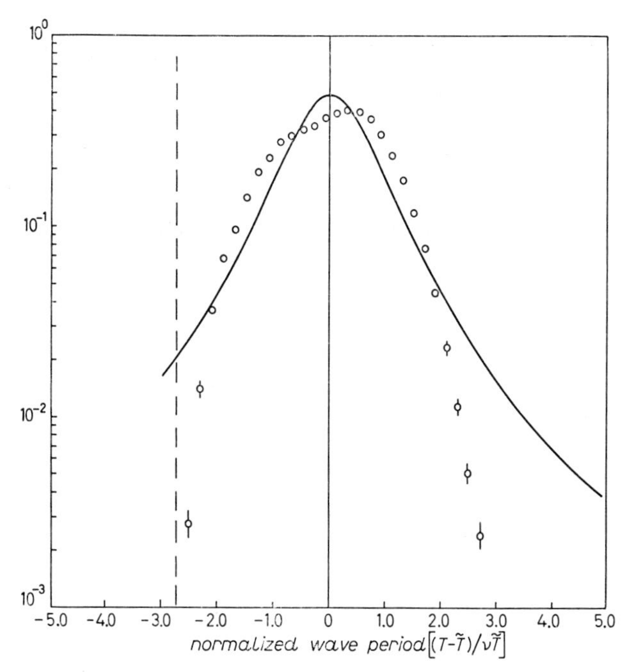

Fig. 14. – Comparison of theory and average wave period probability density from the Pierson-Moskowitz simulation: ——— theory, ∘ simulation; $N_z = 61\,843$, $\varepsilon = 0.7506$, mean $= -0.0052$, standard deviation $= 0.9253$, skewness $= -0.1141$, kurtosis $= 2.5536$.

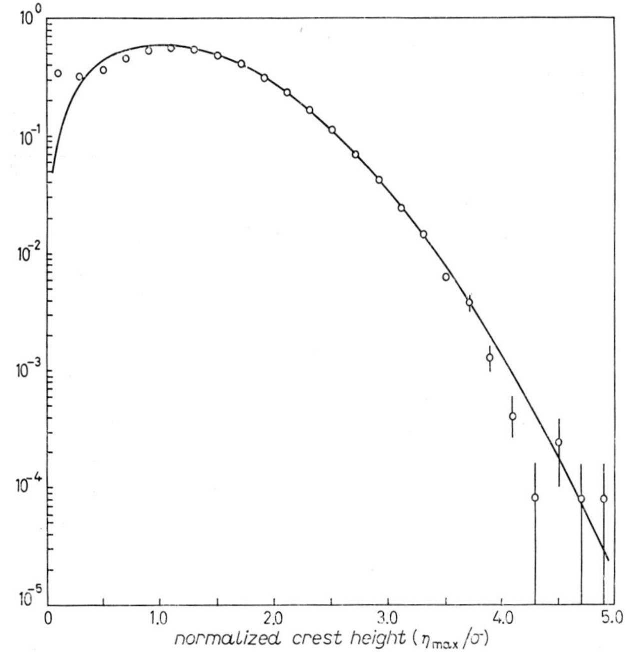

Fig. 15. – Comparison of Rayleigh function and simulated zero-crossing wave crest probability density from Pierson-Moskowitz spectrum; $N = 61\,904$, $\varepsilon = 0.7506$, mean $= 1.2050$, standard deviation $= 0.6934$, skewness $= 0.4930$, kurtosis $= 2.9137$.

random-phase simulation (using the Pierson-Moskowitz spectrum). There is a considerable lack of agreement between the two functions. The simulated histogram falls well below the peak of the theoretical function for normalized wave periods near zero. Substantial disagreement occurs between the two at intermediate and at extreme periods also. This implies that wave periods are considerably shorter in the extreme than those predicted by the theoretical density function. Note that no wave periods appear to the left of the vertical dashed line. This is because wave periods less than zero cannot exist.

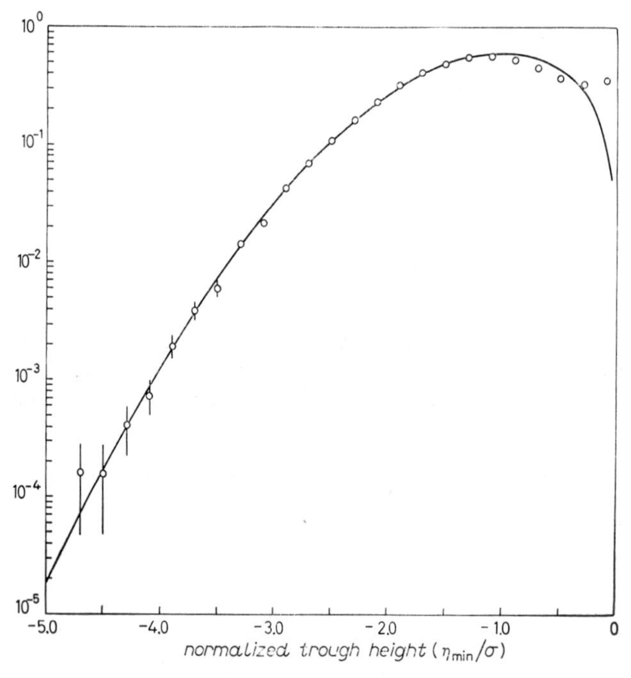

Fig. 16. – Comparison of Rayleigh function and simulated zero-crossing wave trough probability density from Pierson-Moskowitz spectrum; $N_z = 61\,905$, $\varepsilon = 0.7506$, mean $= -1.2031$, standard deviation $= 0.6919$, skewness $= -0.4895$, kurtosis $= 2.9535$.

We now compare crest and trough statistics from the simulation with the Rayleigh distribution. Figure 15 compares the density of crests of zero-crossing waves with the Rayleigh density. While poor agreement is found between simulation and theory for small values of crest height, good agreement is found in the extreme. Some statistical fluctuation is present at large crest amplitude, but these fluctuations are not inconsistent with the uncertainties. Figure 16 compares the statistics of troughs of zero-crossing waves in a like manner. Once again, we find poor agreement for small values and excellent agreement in the extreme. These results confirm the Rayleigh behavior of large crests and troughs for Gaussian random waves.

6. – Comparison to data.

Ocean wave data have been recorded at South Pass in the Gulf of Mexico during hurricane Eloise (as part of the Ocean Current Measurement Program (OCMP), see [25-27]) (*). A three-day record of time series wave data was obtained during hurricane passage in 99 m of water. We have spectrally and statistically analyzed this record using standard procedures. Figure 17 shows

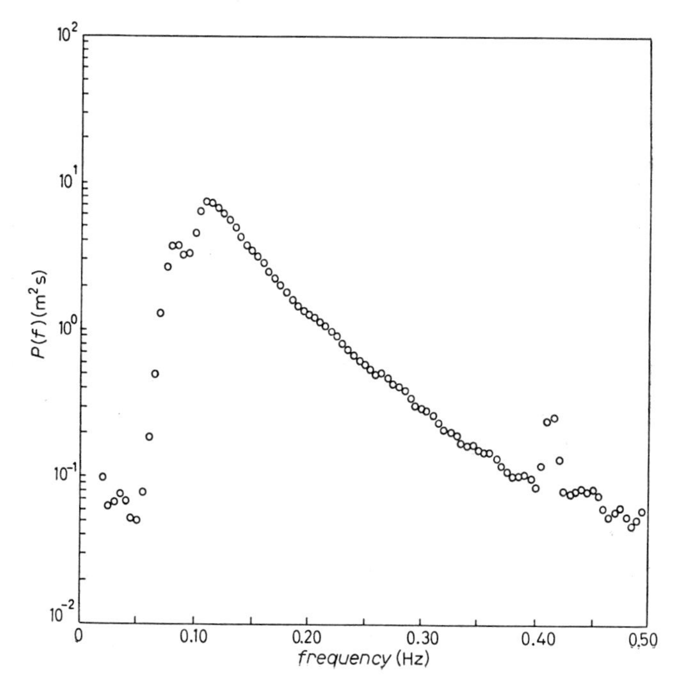

Fig. 17. – Sample power spectrum of hurricane Eloise wave heights recorded at South Pass.

a sample spectrum (from a 40 min record) which occurred during the three-day interval. The spectrum has a dominating frequency near 0.11 Hz. A lobe in the spectrum exists at 0.075 Hz and is probably due to swell. An additional peak of unknown (probably experimental) origin exists at approximately 0.4 Hz. We now investigate the statistical properties of the waves in this

(*) The data collection program was excuted by Shell Oil Company with financial support from Amoco Production Research, Chevron Oil Field Research Company, Exxon Production Research Company, and Mobil Research and Development Company.

record and allude to the importance of nonstationary and nonlinear effects; we further discuss the influence of the spectral shape on the statistical parameters investigated.

Figure 18 compares the probability density of the wave amplitudes with the Gaussian function. Twenty-minute segments of the record were detrended, had their means removed and were divided by the standard deviation of the segment. Figure 18 is a composite of 200 normalized segments of data, about 70 h of real time. Good agreement between the data and the Gaussian function is found for small and intermediate values of wave amplitude. However, for large positive amplitudes, the data are somewhat greater than the Gaus-

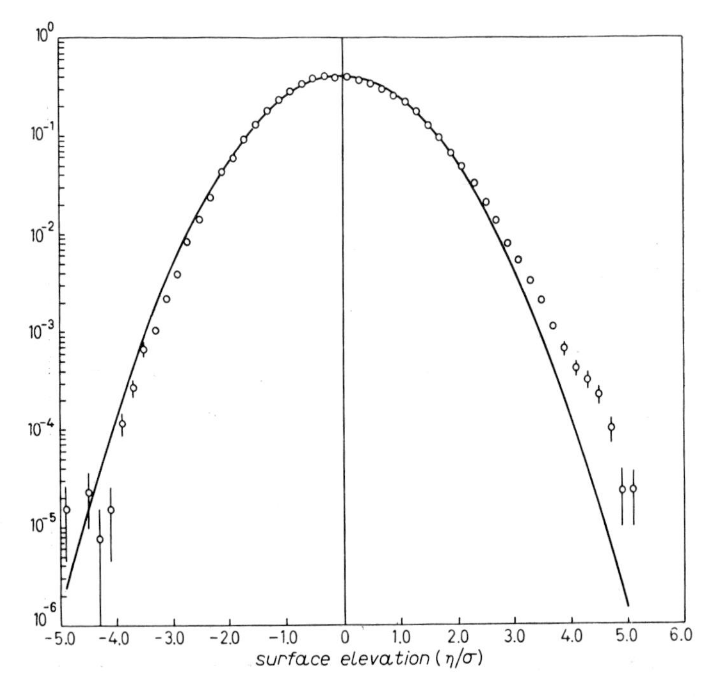

Fig. 18. – Comparison of Gaussian function to wave amplitude density during hurricane Eloise; $N = 645\,000$, $\varepsilon = 0.8362$, mean $= 0$, standard deviation $= 1.0002$, skewness $= 0.2327$, kurtosis $= 3.9625$.

sian function. For large negative amplitudes, the data fall below Gaussian. These results are due to the fact that the sea surface elevation is nonlinear, resulting in more peaked crests and shallower troughs than predicted by linear theory (see [8]). As a result, the probability density of the wave elevation has finite skewness and a value of kurtosis greater than 3.0. The random-phase simulation described in this paper cannot predict this nonlinear behavior. Appropriate modifications will be made and reported at a later time to account for this effect (see [7]).

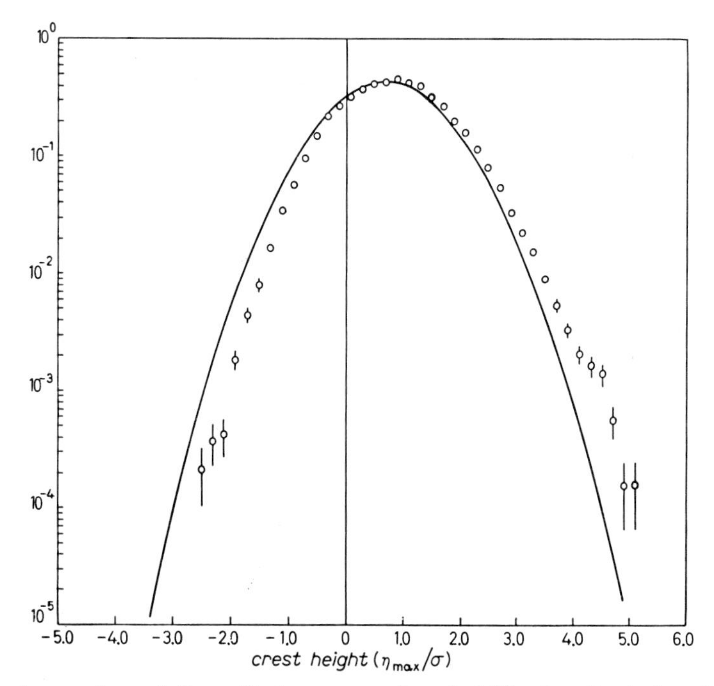

Fig. 19. – Comparison of theoretical wave crest probability density to hurricane Eloise data; $N_{max} = 95\,209$, $\varepsilon = 0.8362$, mean $= 0.8465$, standard deviation $= 0.9038$, skewness $= 0.8987$, kurtosis $= 18.7079$.

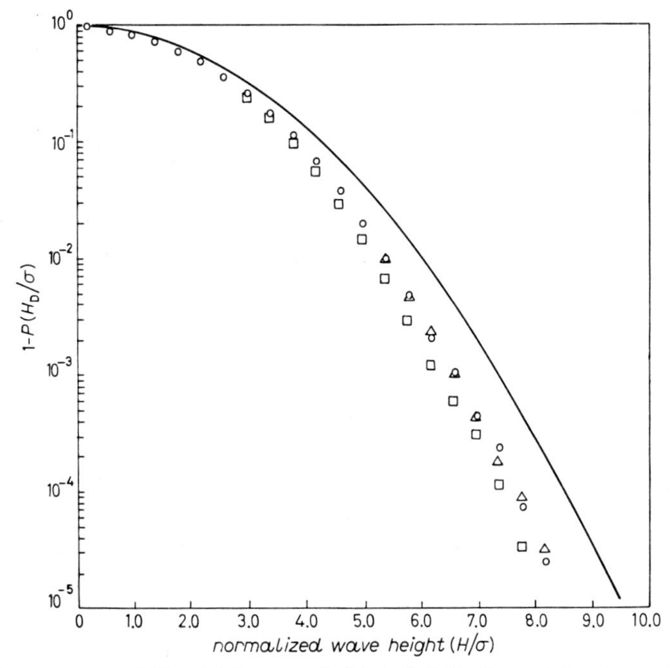

Fig. 20. – Comparison of Rayleigh wave height distribution of hurricane Eloise data: ○ downcrossing waves, △ upcrossing waves, □ average of forward and backward troughs; $N_z = 63\,896$, $\varepsilon = 0.8382$.

Figure 19 compares the wave crest density function of Cartwright and Longuet-Higgins [16] with hurricane Eloise data. It is seen that the data fall somewhat higher than the theory for large crest amplitudes and fall below the theory for negative amplitudes. This result is likely due to the fact that ocean waves behave nonlinearly, with enhanced crests and shallow troughs.

The probability distribution of heights of zero-crossing waves from hurricane Eloise is shown in fig. 20. Three classifications have been investigated:

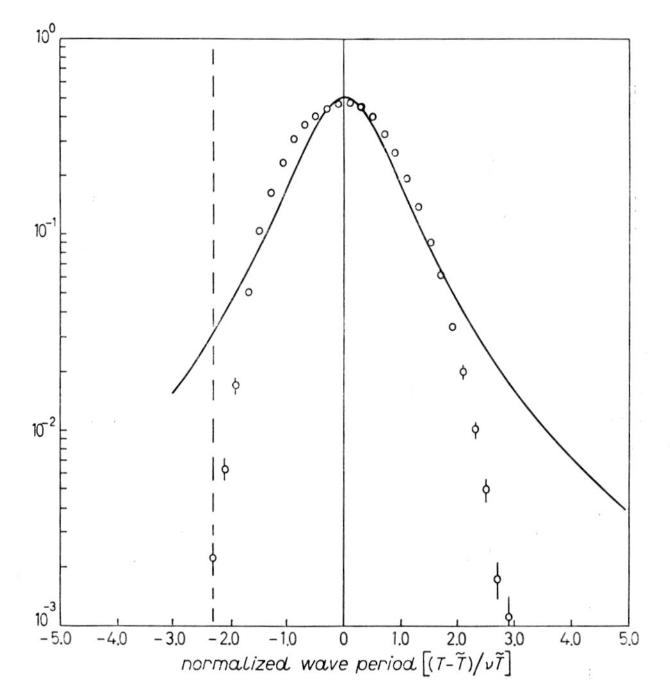

Fig. 21. – Comparison of theoretical wave period probability density to hurricane Eloise data; $N_z = 63\,896$, $\varepsilon = 0.8362$, mean $= -0.0043$, standard deviation $= 0.9720$, skewness $= 24.73$, kurtosis $= 2145.68$.

1) downcrossing waves, 2) upcrossing waves and 3) the wave defined by the average of the forward and backward troughs. For a linear wave record, the statistics of downcrossing and upcrossing waves are exactly the same. However, when nonlinear effects are present, the statistics of downcrossing waves may fall further from the Rayleigh function than do upcrossing waves [15]. Furthermore, the statistics of the average waves fall further from Rayleigh than either of the above two. This last result is predicted by the random-phase simulation; however, the simulation predicts that the statistics of up and downcrossing waves are exactly the same. Little evidence is seen in the Eloise data to support the previously suggested difference between upcrossing

and downcrossing statistics. The main result of the comparison between the data and theory (eqs. (21) and (22)) in fig. 20 is that the data fall about 12% below Rayleigh. This is somewhat greater than the deviation of 8% noted by FORRISTALL [28] in his analysis of hurricane Camille data.

Use of the moments computed from the data gives a range for the non-dimensional parameter ν_0 of $0.31 \leqslant \nu_0 \leqslant 0.47$. Using the results of Longuet-Higgins [10], we can account for about one-half of the 12% difference between Rayleigh and the data shown in fig. 20.

The statistics of periods of zero-downcrossing waves are shown in fig. 21.

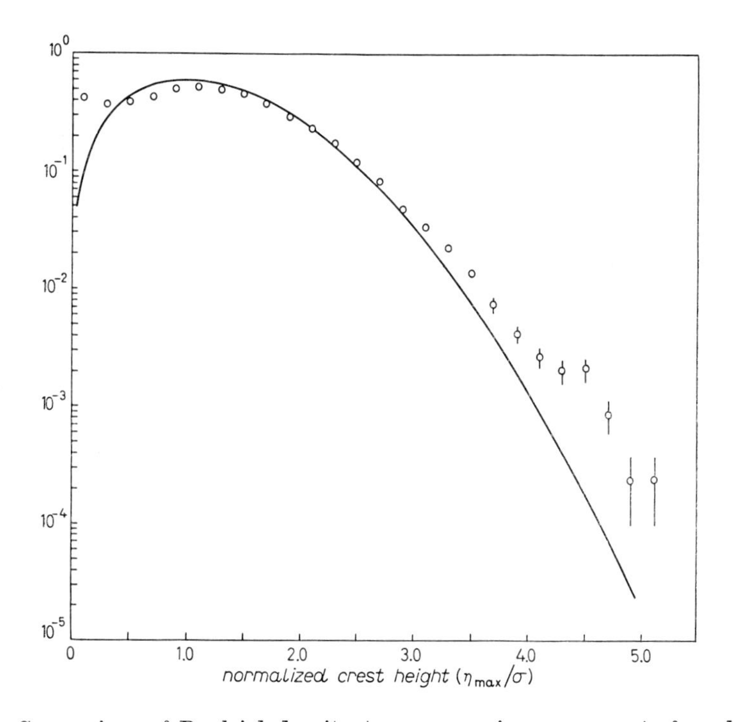

Fig. 22. – Comparison of Rayleigh density to zero-crossing wave crests from hurricane Eloise; $N_z = 63\,965$, $\varepsilon = 0.8362$, mean $= 1.2032$, standard deviation $= 0.7712$, skewness $=$ $= 2.0705$, kurtosis $= 4.4923$.

Of particular interest is the fact that the data fall below the theory for small periods, above the theory for intermediate periods, and substantially below the theory for extreme periods. All of these results are consistent with predictions made by the random-wave simulator. Presumably most of the deviation of the data from simulated results is due to the fact that the wave spectra are different for the two cases; nonlinear effects may also contribute to the discrepancies.

The statistics of the crests of zero-crossing waves obtained from the hurricane Eloise record are shown in fig. 22. The curve for the Rayleigh function is also shown. The results are very similar to those predicted by the random-phase simulation; however, a large deviation exists in the extreme tail, where nonlinear effects are greatest.

We last investigate the statistics of wave heights for a wide variety of locations and conditions (see fig. 23). The details of this investigation can be found in [15]. We compare the data to results from the random-phase simulation using an average Pierson-Moskowitz spectrum as input (we used the average significant wave height of all the data); note that good agreement is found in this comparison. There is little influence of nonlinear effects

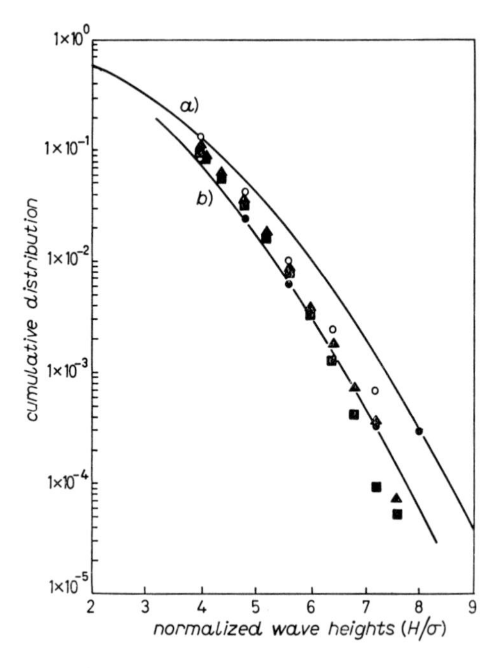

Fig. 23. – Comparison of Rayleigh wave height distribution to data and to the random-phase simulation: *a*) Rayleigh distribution, *b*) simulated distribution ($N_z = 61\,691$), data from ref. [23].

on wave height statistics as evidenced by the above comparison. This result is probably due to the fact that nonlinear wave crests are more peaked and their corresponding troughs are more shallow than for linear waves. Thus nonlinear wave height statistics deviate little from the linear wave height statistics predicted by the simulator. These results are further consistent with the observation that the wave height probability distribution in the normalized form of fig. 23 is likely not very sensitive to the exact form of the power spectrum.

The present results are consistent with the work of Longuet-Higgins [10], although there is some evidence (based primarily on simulation results) that his calculations may not completely account for the effects of finite bandwidth.

7. – Conclusions.

A fast, accurate Monte Carlo random-phase simulation of ocean waves has been used to study the statistical properties of the ocean surface. In particular, the distributions of crest heights, wave heights and wave periods have been investigated. Comparisons to data verify the validity of the model to within the linearity assumptions upon which it is founded.

Appendix A

Moments of the Pierson-Moskowitz spectrum.

We have elected to modify the Pierson-Moskowitz spectrum for non-equilibrium effects [18] by writing

$$P(f) = A_0 B f^{\Delta l - 5} \exp\left[B f^{-4}\right], \tag{A.1}$$

where Δl is an empirical constant which may be set to zero to obtain the usual Pierson-Moskowitz form; $B = 5f_d^4/4$. The moments of the above spectrum are defined by

$$M_{n\Delta l} = (2\pi)^n \int_0^{f_N} f^n P(f)\, \mathrm{d}f. \tag{A.2}$$

Substitution of (A.1) into (A.2) gives

$$M_{n\Delta l} = (2\pi)^n A_0 B \int_0^{f_N} f^{-5+n+\Delta l} \exp\left[-B f^{-4}\right]. \tag{A.3}$$

If we use the substitution $x = Bf^{-4}$ in (A.3), we find

$$M_{n\Delta l} = (4g^2\alpha)^{n/4} \frac{H_s^{2-n/2}}{16} R[n, \Delta l, r_0], \tag{A.4}$$

where

$$r_0 = [(5-\Delta l)/4]^{\frac{1}{4}} f_d/f_N \,, \tag{A.5}$$

$$R[n, \Delta l, r_0] = \Gamma[1-(n+\Delta l)/4, r_0^4]/\Gamma[1-\Delta l/4]\,. \tag{A.6}$$

Here $\Gamma(a, x)$ is the incomplete gamma-function and $\Gamma(x)$ is the gamma-function [29]. We have used the fact that

$$A_0 = A/B^{\Delta l/4}\Gamma(1-\Delta l/4)\,, \qquad A = H_s^2/4\,, \tag{A.7}$$

$$B = [(5-\Delta l)/4] f_d^4\,. \tag{A.8}$$

By employing (A.7) and (A.8), we ensure, for $f_d/f_N \to 0$, that

$$M_{0\Delta l} = H_s^2/16\,. \tag{A.9}$$

Equation (A.9) is, of course, required by assumption, *i.e.* that the significant wave height be simply related to the area under the power spectrum.

It is instructive to examine the function $R[n, \Delta l, r_0]$, in particular when $r_0 \ll 1$. This is true for nominal wave heights, $0.5\text{ m} \leqslant H_s < \infty$, and Nyquist frequencies, $2\text{ Hz} \leqslant f_N \leqslant 5\text{ Hz}$. According to [29] the incomplete gamma-function can be written

$$\Gamma(a, x) = \Gamma(a) - \gamma(a, x)\,, \tag{A.10}$$

where

$$\gamma(a, x) = \int_0^x \exp[-z] z^{a-1} dz\,. \tag{A.11}$$

Furthermore, $\gamma(a, x)$ can be expressed as a confluent hypergeometric function:

$$\gamma(a, x) = x^a \exp[-x] M(1, 1+a, x)/a\,. \tag{A.12}$$

For small arguments

$$M(\delta, \varrho, x) = 1 + (\delta/\varrho)x + [\delta(\delta+1)/\varrho(\varrho+1)]x^2 + \dots\,. \tag{A.13}$$

Finally we obtain an approximate form for the incomplete gamma-function for $x \ll 1$

$$\Gamma(a, x) = \Gamma(a) - \frac{x^a \exp[-x]}{a}\left\{1 + \frac{x}{1+a} + \frac{x^2}{(1+a)(2+a)} + \dots\right\}. \tag{A.14}$$

Using the above expression when $\Delta l = 0$ (see (A.6)), we find the approximate forms for the moments of the Pierson-Moskowitz spectrum as given in table III.

Appendix B - Flowchart of the computer algorithm.

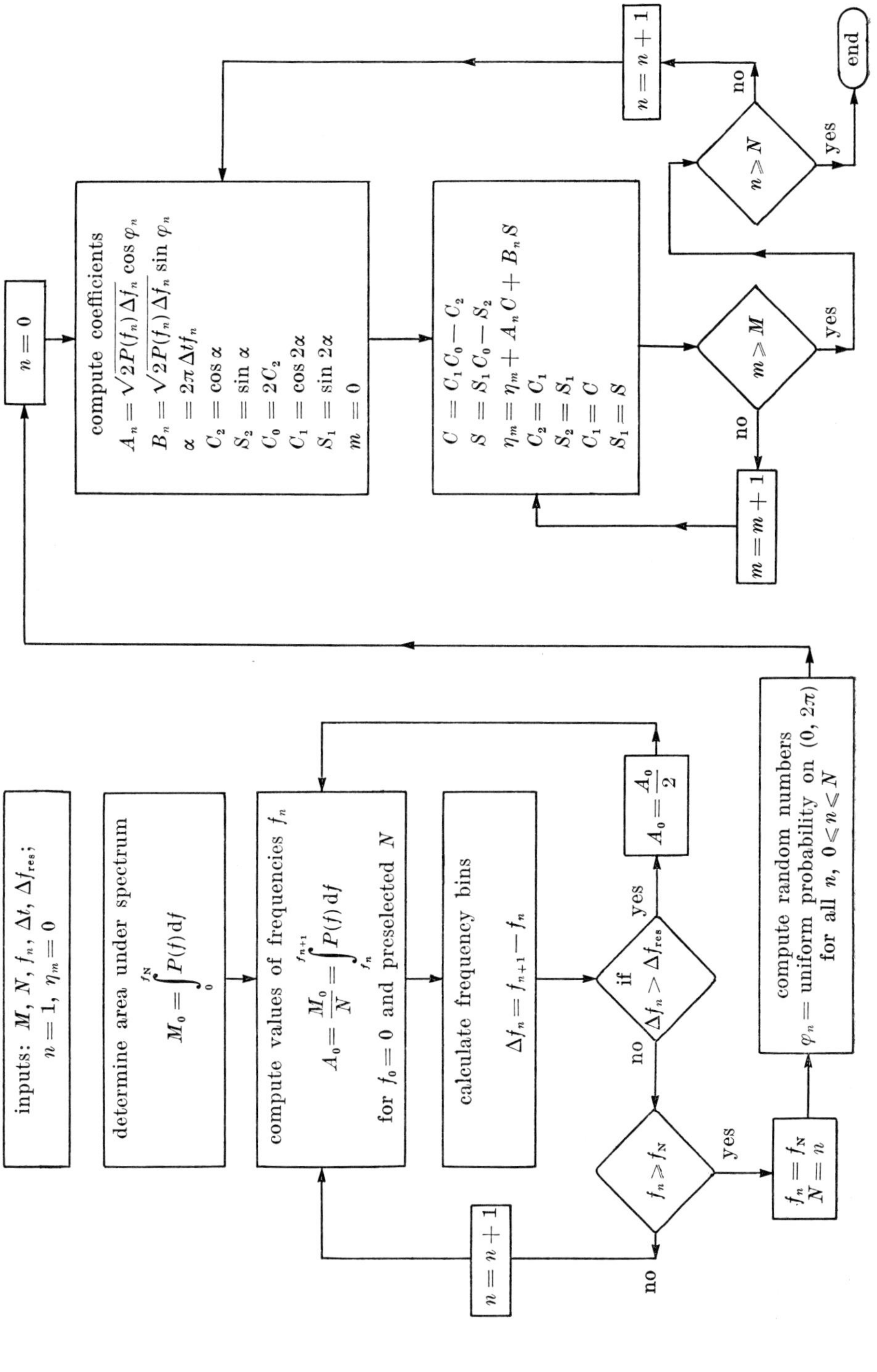

REFERENCES

[1] S. O. Rice: *Bell Syst. Tech. J.*, **23**, 282 (1944); **24**, 46 (1945).
[2] S. Chandrasekhar: *Rev. Mod. Phys.*, **15**, 1 (1943).
[3] J. L. Doob: *Ann. Math.*, **43**, 351 (1942).
[4] M. S. Longuet-Higgins: *J. Mar. Res.*, **11**, 245 (1952).
[5] W. J. Pierson: *Adv. Geophys.*, **2**, 93 (1955).
[6] W. J. Pierson and L. Moskowitz: *J. Geophys. Res.*, **69**, 5181 (1964).
[7] L. J. Tick: *J. Math. Mech.*, **8**, 643 (1959).
[8] M. S. Longuet-Higgins: *J. Fluid Mech.*, **17**, 359 (1963).
[9] O. M. Phillips: *The dynamics of random finite amplitude gravity waves*, in *Ocean Wave Spectra* (Englewood Cliffs, N. J., 1963), p. 171.
[10] M. S. Longuet-Higgins: *J. Geophys. Res.*, **85**, 1519 (1980).
[11] L. E. Borgman: *Ocean wave simulation for engineering design*, University of California Techanical Report, HEL-9-13 (1967).
[12] R. J. Robinson, H. R. Brannon and G. W. Kattawar: *Soc. Petr. Eng. J.*, **7**, 87 (1967).
[13] J. W. Cooley and J. W. Tukey: *Math. Comp.*, **19**, 1 (1965).
[14] R. B. Blackman and J. W. Tukey: *The Measurement of Power Spectra* (New York, N. Y., 1958).
[15] R. E. Haring, A. R. Osborne and L. P. Spencer: *Extreme wave parameters based on continental shelf storm wave records*, in ASCE *XV Coastal Engineering Conference* (Hawaii, 1976), p. 151.
[16] D. E. Cartwright and M. S. Longuet-Higgins: *Proc. R. Soc. London Ser. A*, **237**, 212 (1959).
[17] K. Hasselmann: *Dtsche Hydrogr. Z.*, **8**, 10 (1973).
[18] A. R. Osborne: unpublished (1980).
[19] A. J. Williams and D. E. Cartwright: *Trans. Am. Geophys. Union*, **38**, 864 (1957).
[20] Y. Goda: *Estimation of wave statistics from spectral information*, in *Proceedings of the Conference on Ocean Wave Measurement and Analysis* (New Orleans, La., 1974), p. 320.
[21] H. Rye and R. Svee: *On the parametric representation of a wind wave field*, in ASCE *XV Costal Engineering Conference* (Hawaii, 1976), p. 183.
[22] J. S. Bendat and A. G. Piersol: *Random Data: Analysis and Measurement Procedures* (New York, N. Y., 1971), p. 100.
[23] P. R. Bevington: *Data Reduction and Error Analysis for the Physical Sciences* (New York, N. Y., 1969).
[24] M. S. Longuet-Higgins: *J. Geophys. Res.*, **80**, 2688 (1975).
[25] G. Z. Forristall: *A two-layer model for hurricane driven currents on an irregular grid*, submitted to *J. Phys. Oceanogr.* (1980).
[26] G. Z. Forristall and R. C. Hamilton: *Current measurements in support of fixed platform design and construction*, in *Working Conference on Current Measurements*, sponsored by the NOAA Office of Ocean Engineering and the Delaware Sea Grant College Program (1978), p. 199.
[27] G. Z. Forristall, E. G. Ward and V. J. Cardone: *Directional wave spectra and wave kinematics in hurricanes Carmen and Eloise*, to be published in *Proceedings of the VII International Conference on Coastal Engineering* (Sydney, 1980).
[28] G. Z. Forristall: *J. Geophys. Res.*, **83**, 2353 (1978).
[29] M. Abramowitz and I. A. Stegun, editors: *Handbook of Mathematical Functions*, National Bureau of Standards Applied Mathematics Series 55 (Washington, D. C., 1964).

PROCEEDINGS OF THE INTERNATIONAL SCHOOL OF PHYSICS
« ENRICO FERMI »

Course I
Questioni relative alla rivelazione delle particelle elementari, con particolare riguardo alla radiazione cosmica
edited by G. PUPPI

Course II
Questioni relative alla rivelazione delle particelle elementari, e alle loro interazioni con particolare riguardo alle particelle artificialmente prodotte ed accelerate
edited by G. PUPPI

Course III
Questioni di struttura nucleare e dei processi nucleari alle basse energie
edited by G. SALVETTI

Course IV
Proprietà magnetiche della materia
edited by L. GIULOTTO

Course V
Fisica dello stato solido
edited by F. FUMI

Course VI
Fisica del plasma e applicazioni astrofisiche
edited by G. RIGHINI

Course VII
Teoria della informazione
edited by E. R. CAIANIELLO

Course VIII
Problemi matematici della teoria quantistica delle particelle e dei campi
edited by A. BORSELLINO

Course IX
Fisica dei pioni
edited by B. TOUSCHEK

Course X
Thermodynamics of Irreversible Processes
edited by S. R. DE GROOT

Course XI
Weak Interactions
edited by L. A. RADICATI

Course XII
Solar Radioastronomy
edited by G. RIGHINI

Course XIII
Physics of Plasma: Experiments and Techniques
edited by H. ALFVÉN

Course XIV
Ergodic Theories
edited by P. CALDIROLA

Course XV
Nuclear Spectroscopy
edited by G. RACAH

Course XVI
Physicomathematical Aspects of Biology
edited by N. RASHEVSKY

Course XVII
Topics of Radiofrequency Spectroscopy
edited by A. GOZZINI

Course XVIII
Physics of Solids (Radiation Damage in Solids)
edited by D. S. BILLINGTON

Course XIX
Cosmic Rays, Solar Particles and Space Research
edited by B. PETERS

Course XX
Evidence for Gravitational Theories
edited by C. MØLLER

Course XXI
Liquid Helium
edited by G. CARERI

Course XXII
Semiconductors
edited by R. A. SMITH

Course XXIII
Nuclear Physics
edited by V. F. WEISSKOPF

Course XXIV
Space Exploration and the Solar System
edited by B. Rossi

Course XXV
Advanced Plasma Theory
edited by M. N. Rosenbluth

Course XXVI
Selected Topics on Elementary Particle Physics
edited by M. Conversi

Course XXVII
Dispersion and Absorption of Sound by Molecular Processes
edited by D. Sette

Course XXVIII
Star Evolution
edited by L. Gratton

Course XXIX
Dispersion Relations and Their Connection with Causality
edited by E. P. Wigner

Course XXX
Radiation Dosimetry
edited by F. W. Spiers and G. W. Reed

Course XXXI
Quantum Electronics and Coherent Light
edited by C. H. Townes and P. A. Miles

Course XXXII
Weak Interactions and High-Energy Neutrino Physics
edited by T. D. Lee

Course XXXIII
Strong Interactions
edited by L. W. Alvarez

Course XXXIV
The Optical Properties of Solids
edited by J. Tauc

Course XXXV
High-Energy Astrophysics
edited by L. Gratton

Course XXXVI
Many-Body Description of Nuclear Structure and Reactions
edited by C. Bloch

Course XXXVII
Theory of Magnetism in Transition Metals
edited by W. Marshall

Course XXXVIII
Interaction of High-Energy Particles with Nuclei
edited by T. E. O. Ericson

Course XXXIX
Plasma Astrophysics
edited by P. A. Sturrock

Course XL
Nuclear Structure and Nuclear Reactions
edited by M. Jean

Course XLI
Selected Topics in Particle Physics
edited by J. Steinberger

Course XLII
Quantum Optics
edited by R. J. Glauber

Course XLIII
Processing of Optical Data by Organisms and by Machines
edited by W. Reichardt

Course XLIV
Molecular Beams and Reaction Kinetics
edited by Ch. Schlier

Course XLV
Local Quantum Theory
edited by R. Jost

Course XLVI
Physics with Storage Rings
edited by B. Touschek

Course XLVII
General Relativity and Cosmology
edited by R. K. Sachs

Course XLVIII
Physics of High Energy Density
edited by P. Caldirola and H. Knoepfel

Course IL
Foundations of Quantum Mechanics
edited by B. d'Espagnat

Course L
Mantle and Core in Planetary Physics
edited by J. Coulomb and M. Caputo

Course LI
Critical Phenomena
edited by M. S. Green

Course LII
Atomic Structure and Properties of Solids
edited by E. Burstein

Course LIII
Developments and Borderlines of Nuclear Physics
edited by H. MORINAGA

Course LIV
Developments in High-Energy Physics
edited by R. R. GATTO

Course LV
Lattice Dynamics and Intermolecular Forces
edited by S. CALIFANO

Course LVI
Experimental Gravitation
edited by B. BERTOTTI

Course LVII
Topics in the History of 20th Century Physics
edited by C. WEINER

Course LVIII
Dynamic Aspects of Surface Physics
edited by F. O. GOODMAN

Course LIX
Local Properties at Phase Transitions
edited by K. A. MÜLLER

Course LX
C*-Algebras and their Applications to Statistical Mechanics and Quantum Field Theory
edited by D. KASTLER

Course LXI
Atomic Structure and Mechanical Properties of Metals
edited by G. CAGLIOTI

Course LXII
Nuclear Spectroscopy and Nuclear Reactions with Heavy Ions
edited by H. FARAGGI and R. A. RICCI

Course LXIII
New Directions in Physical Acoustics
edited by D. SETTE

Course LXIV
Nonlinear Spectroscopy
edited by N. BLOEMBERGEN

Course LXV
Physics and Astrophysics of Neutron Stars and Black Holes
edited by R. GIACCONI and R. RUFFINI

Course LXVI
Health and Medical Physics
edited by J. BAARLI

Course LXVII
Isolated Gravitating Systems in General Relativity
edited by J. EHLERS

Course LXVIII
Metrology and Fundamental Constants
edited by A. FERRO MILONE, P. GIACOMO and S. LESCHIUTTA

Course LXIX
Elementary Modes of Excitation in Nuclei
edited by A. BOHR and R. A. BROGLIA

Course LXX
Physics of Magnetic Garnets
edited by A. PAOLETTI

Course LXXI
Weak Interactions
edited by M. BALDO CEOLIN

Course LXXII
Problems in the Foundations of Physics
edited by G. TORALDO DI FRANCIA

Course LXXIII
Early Solar System Processes and the Present Solar System
edited by D. LAL

Course LXXIV
Development of High-Power Lasers and their Applications
edited by C. PELLEGRINI

Course LXXV
Intermolecular Spectroscopy and Dynamical Properties of Dense Systems
edited by J. VAN KRANENDONK

Course LXXVI
Medical Physics
edited by J. R. GREENING

Course LXXVII
Nuclear Structure and Heavy-Ion Collisions
edited by R. A. BROGLIA, R. A. RICCI and C. H. DASSO

Course LXXVIII
Physics of the Earth's Interior
edited by A. M. DZIEWONSKI and E. BOSCHI

Course LXXIX
From Nuclei to Particles
edited by A. MOLINARI

TIPOGRAFIA COMPOSITORI - BOLOGNA